Engineering Mechanics
STATICS

Engineering Mechanics
STATICS

FIFTH EDITION

Anthony Bedford • Wallace Fowler

University of Texas at Austin

PEARSON

Prentice
Hall

Upper Saddle River, New Jersey 07458

Library of Congress Cataloging-in-Publication Data

Bedford, A.

  Engineering mechanics: statics / Anthony Bedford, Wallace Fowler.—5th ed.
     p.   cm.
  Includes index.
  ISBN 978-0-13-612915-8
  1. Mechanics, Applied.   2. Statics.   I. Fowler, Wallace L.   II. Title.   III. Title: Statics.
  TA351.B43 2006
  620.1'03--dc22

                                    2007009935

Vice President and Editorial Director, ECS: *Marcia J. Horton*
Acquisitions Editor: *Tacy Quinn*
Associate Editor: *Dee Bernhard*
Managing Editor: *Scott Disanno*
Media Editor: *David Alick*
Marketing Manager: *Tim Galligan*
Production Editor: *Craig Little*
Director of Creative Services: *Paul Belfanti*
Creative Director: *Juan Lopez*
Art Director: *Jonathan Boylan*
Interior Designer: *Kenny Beck*
Cover Designer: *Jonathan Boylan*
Art Editor: *Xiaohong Zhu*
Manufacturing Manager: *Alexis Heydt-Long*
Manufacturing Buyer: *Lisa McDowell*

About the Cover: Courtesy Getty Images/Photographer's Choice. Photographer: Lester Lefkowitz. Used by permission.

© 2008 by Pearson Education, Inc.
Pearson Prentice Hall
Pearson Education, Inc.
Upper Saddle River, NJ 07458

The author and publisher of this book have used their best efforts in preparing this book. These efforts include the development, research, and testing of the theories and programs to determine their effectiveness. The author and publisher make no warranty of any kind, expressed or implied, with regard to these programs or the documentation contained in this book. The author and publisher shall not be liable in any event for incidental or consequential damages in connection with, or arising out of, the furnishing, performance, or use of these programs.

MATLAB is a registered trademark of The MathWorks, Inc., 3 Apple Hill Drive, Natick, MA 01760-2098. Mathcad is a registered trademark of MathSoft Engineering and Education, 101 Main St., Cambridge, MA 02142-1521. All other product names, brand names, and company names are trademarks or registered trademarks of their respective owners.

Printed in the United States of America

10 9 8 7 6 5 4 3 2

ISBN 0-13-612915-3
      978-0-13-612915-8

Pearson Education Ltd., *London*
Pearson Education Australia Pty., Limited, *Sydney*
Pearson Education Singapore, Pte. Ltd
Pearson Education North Asia Ltd., *Hong Kong*
Pearson Education Canada, Ltd., *Toronto*
Pearson Educación de Mexico, S.A. de C.V.
Pearson Education—Japan, *Tokyo*
Pearson Education Malaysia, Pte. Ltd.
Pearson Education, *Upper Saddle River, New Jersey*

Contents

APPENDICES

Preface

We began the development of the Fifth Editions of *Engineering Mechanics: Statics* and *Dynamics* by asking ourselves how our textbooks could be restructured to help students learn mechanics more *effectively* and *efficiently*.

From the first editions, our objective has been to present material in a way that emulates the teacher's development of concepts in the classroom and emphasizes visual analysis to enhance student understanding.

Now, based on our classroom experiences and insights provided by colleagues and students over many years, we have designed the fifth editions to conform more closely to the way today's students actually use textbooks in learning mechanics. In developing the new elements described below, we have continued to adhere to our original goals of teaching effective problem-solving procedures and the central importance of free-body diagrams.

New to this Edition

Active Examples

A new example format designed to help students learn concepts and methods and test their understanding. Discussions are visually related to figures and equations in a new *integrated text/art* format for efficient reading. A "Practice Problem" is provided at the end of the Active Example so that students will be motivated to spend more time working with the example and checking whether they understood the material. They can easily assess their understanding by referring to the answer to the Practice Problem that is provided on the page, or by studying the *complete solution* that is presented in an appendix in the same text/art integrated format as the Active Example.

Example-Focused Problems

New homework problems designed to encourage students to study given examples and expand their understanding of concepts. The numbers of these problems are cited at the beginning of each example so that teachers can easily use them to encourage study of selected topics.

Results

Most sections of the text now conclude with a new Results subsection, a self-contained and complete description of the results required to understand the following examples and problems. They are presented in the same integrated text/art format used in the Active Examples for easier comprehension. Students can efficiently refer to these subsections while studying examples and working problems.

Problem Sets

Thirty percent of the problems are new in the statics text. Problems that are relatively lengthier or more difficult have been marked with an asterisk. Additional problems can be generated using the online homework system with its algorithmic capabilities.

Hallmark Elements of the Text

Examples

In addition to the new Active Examples, we maintain our examples that follow a three-part framework—**Strategy/ Solution/Critical Thinking**—designed to help students develop engineering problem skills. In the Strategy sections, we demonstrate how to plan the solution to a problem. The Solution presents the detailed steps needed to arrive at the required results.

Some of the examples have a focus on design and provide detailed discussions of applications in statics in engineering design.

Computational Mechanics

Some instructors prefer to teach statics without emphasizing the use of the computer. Others use statics as an opportunity to introduce students to the use of computers in engineering, having them either write their own programs in a lower level language of use higher level problem-solving software. Our book is suitable for both approaches. Optional, self-contained Computational Mechanics material is available for this text on the Companion Website, including tutorials using Mathcad and MATLAB. See the supplements section for further information.

Art Program

We recognize the importance of helping students visualize problems in mechanics. Students prefer, and are more motivated by, realistic situations. Our texts include many photographs as well as figures with "photo-realistic" rendering to help students visualize applications and provide a stronger connection to actual engineering practice.

Consistent Use of Color

To help students recognize and interpret elements of figures, we use consistent identifying colors:

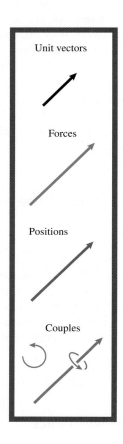

Unit vectors

Forces

Positions

Couples

Triple Accuracy Checking—Commitment to Students and Instructors

It is our commitment to students and instructors to take precautions to ensure the accuracy of the text and its solutions manual to the best of our ability. We use a system of triple accuracy checking in which three parties, in addition to the authors, solve the problems in an effort to be sure that their answers are correct and that they are of an appropriate level of difficulty. Our accuracy team consists of:

- Scott Hendricks of Virginia Polytechnic University
- Karim Nohra of the University of South Florida
- Kurt Norlin of Laurel Technical Services

The parties further verified the text, examples, problems, and solutions manuals to help ensure accuracy. Any errors remain the responsibility of the authors. We welcome communication from students and instructors concerning errors or areas for improvement. Our mailing address is Department of Aerospace Engineering and Engineering Mechanics, University of Texas at Austin, Austin, Texas 78712. Our email address is abedford@mail.utexas.edu

Instructor & Student Resources

Student Resources

The *Statics* Study Pack is designed to give students the tools to improve their skills drawing free-body diagrams, and to help them review for tests. It contains a tutorial on free-body diagrams with fifty practice problems of increasing difficulty with complete solutions. Further strategies and tips help students understand how to use the diagrams in solving the accompanying problems. This supplement and accompanying chapter by chapter review material was prepared by Peter Schiavone of the University of Alberta. An access code for the Companion Website is included inside the Study Pack.

The Statics Study Pack is also available as a stand-alone item. Order stand-alone Study Packs with the ISBN 0-13-614002-5.

Web Assessment and Tutorial Resources—Students can access tutorial resources such as supplemental practice problems on the Companion Website for this text.

www.prenhall.com/bedford

Additionally, instructors can assign online homework for students using PH GradeAssist. Answers are graded and results are recorded electronically.

Each tutorial discusses a basic mechanics concept, and then shows how to solve a specific problem related to this concept using MATLAB and Mathcad. There are twenty tutorials each for MATLAB and Mathcad, and are available to instructors in PDF format for distribution to students. Worksheets were developed by Ronald Larsen and Stephen Hunt of Montana State University–Bozeman.

Instructor Resources

Instructor's Solutions Manual—This supplement, available to instructors, contains completely worked out solutions. Each solution comes with the problem statement as well as associated artwork. The ISBN for the printed manual is 0-13-614003-3. Solutions are also available electronically for instructors at www.prenhall.com

Instructor's Resource Center on CD—This CD contains PowerPoint slides and JPEG files of all art from the text. It also contains sets of PowerPoint slides showing each example.

Web Assessment and Tutorial Resources—Through Prentice Hall Grade Assist, instructors can create online assignments for students using problems from the text. PH GradeAssist also offers every problem in an algorithmic format so that each student can work with slightly different numbers. Students also benefit from an integrated e-book. Answers to problems are recorded in an online grade book that can be downloaded into Excel. For additional tutorial resources, students should access the Companion Website where they can find supplemental problem sets and information. Contact your Prentice Hall representative for details or a demonstration.

Ordering Options

Engineering Mechanics: Statics with Study Pack (0-13-600042-8). Engineering Mechanics: Statics with Study Pack and PHGA (0-13-135456-6).

Acknowledgments

The following colleagues provided reviews based on their knowledge and teaching experience that greatly helped us in preparing the current and previous editions.

Shaaban Abdallah
 University of Cincinnati

Edward E. Adams
 Michigan Technological University

George G. Adams
 Northeastern University

Raid S. Al-Akkad
 University of Dayton

Jerry L. Anderson
 Memphis State University

James G. Andrews
 University of Iowa

Robert J. Asaro
 University of California, San Diego

Leonard B. Baldwin
 University of Wyoming

Haim Baruh
 Rutgers University

Gautam Batra
 University of Nebraska

David M. Bayer
 University of North Carolina

Glenn Beltz
 University of California–Santa Barbara

Mary Bergs
 Marquette University

Don L. Boyer
 Arizona State University

Spencer Brinkerhoff
 Northern Arizona University

L. M. Brock
 University of Kentucky

William (Randy) Burkett
 Texas Tech University

Donald Carlson
 University of Illinois

Major Robert M. Carpenter
 U.S. Military Academy

Douglas Carroll
 University of Missouri, Rolla

Paul C. Chan
 New Jersey Institute of Technology

Namas Chandra
 Florida State University

James Cheney
 University of California, Davis

Ravinder Chona
 Texas A & M University

Daniel C. Deckler
 The University of Akron Wayne College

Anthony DeLuzio
 Merrimack College

Mitsunori Denda
 Rutgers University

James F. Devine
 University of South Florida

Craig Douglas
 University of Massachusetts, Lowell

Marijan Dravinski
 University of Southern California

S. Olani Durrant
 Brigham Young University

Estelle Eke
 California State University, Sacramento

Bogdan I. Epureanu
 University of Michigan

William Ferrante
 University of Rhode Island

Robert W. Fitzgerald
Worcester Polytechnic Institute

George T. Flowers
Auburn University

Mark Frisina
Wentworth Institute

Robert W. Fuessle
Bradley University

Walter Gerstle
University of New Mexico

William Gurley
University of Tennessee, Chattanooga

John Hansberry
University of Massachusetts, Dartmouth

Mark J. Harper
United States Naval Academy

W. C. Hauser
California Polytechnic University, Pomona

Linda Hayes
University of Texas–Austin

R. Craig Henderson
Tennessee Technological University

Paul R. Heyliger
Colorado State University

James Hill
University of Alabama

Robert W. Hinks
Arizona State University

Allen Hoffman
Worcester Polytechnic Institute

Edward E. Hornsey
University of Missouri, Rolla

Robert A. Howland
University of Notre Dame

Joe Ianelli
University of Tennessee, Knoxville

Ali Iranmanesh
Gadsden State Community College

David B. Johnson
Southern Methodist University

E. O. Jones, Jr.
Auburn University

Serope Kalpakjian
Illinois Institute of Technology

Kathleen A. Keil
California Polytechnic University, San Luis Obispo

Yohannes Ketema
University of Minnesota

Seyyed M. H. Khandani
Diablo Valley College

Charles M. Krousgrill
Purdue University

B. Kent Lall
Portland State University

Chad M. Landis
Rice Unversity

Kenneth W. Lau
University of Massachusetts, Lowell

Norman Laws
University of Pittsburgh

William M. Lee
U.S. Naval Academy

Donald G. Lemke
University of Illinois, Chicago

Richard J. Leuba
North Carolina State University

Richard Lewis
Louisiana Technological University

John B. Ligon
Michigan Tech University

Bertram Long
Northeastern University

V. J. Lopardo
U.S. Naval Academy

Frank K. Lu
University of Texas, Arlington

Mark T. Lusk
Colorado School of Mines

K. Madhaven
Christian Brothers College

Nels Madsen
Auburn University

James R. Matthews
University of New Mexico

Gary H. McDonald
University of Tennessee

James McDonald
Texas Technical University

Jim Meagher
California Polytechnic State University, San Luis Obispo

Lee Minardi
Tufts University

Norman Munroe
Florida International University

Shanti Nair
University of Massachusetts, Amherst

Saeed Niku
*California Polytechnic State University,
San Luis Obispo*

Mohammad Noori
North Carolina State University

Harinder Singh Oberoi
Western Washington University

James O'Connor
University of Texas, Austin

Samuel P. Owusu-Ofori
North Carolina A & T State University

Venkata Panchakarla
Florida State University

Assimina A. Pelegri
Rutgers University

Noel C. Perkins
University of Michigan

Corrado Poli
University of Massachusetts–Amherst

David J. Purdy
Rose-Hulman Institute of Technology

Yitshak Ram
Louisiana State University

Colin E. Ratcliffe
U.S. Naval Academy

Daniel Riahi
University of Illinois

Charles Ritz
California Polytechnic State University, Pomona

George Rosborough
University of Colorado, Boulder

Edwin C. Rossow
Northwestern University

Kenneth Sawyers
Lehigh University

Robert Schmidt
University of Detroit

Robert J. Schultz
Oregon State University

Richard A. Scott
University of Michigan

Brian Self
U.S. Air Force Academy

William Semke
University of North Dakota

Patricia M. Shamamy
Lawrence Technological University

Sorin Siegler
Drexel University

Peng Song
Rutgers State University

Candace S. Sulzbach
Colorado School of Mines

L. N. Tao
Illinois Institute of Technology

Craig Thompson
Western Wyoming Community College

John Tomko
Cleveland State University

Kevin Z. Truman
Washington University

John Valasek
Texas A & M University

Christine Valle
Georgia Institute of Technology

Dennis VandenBrink
Western Michigan University

Thomas J. Vasko
University of Hartford

Mark R. Virkler
University of Missouri, Columbia

William H. Walston, Jr.
University of Maryland

Andrew J. Walters
Mississippi State University

Reynolds Watkins
Utah State University

Charles White
Northeastern University

Norman Wittels
Worcester Polytechnic Institute

Julius P. Wong
University of Louisville

T. W. Wu
University of Kentucky

Constance Ziemian
Bucknell University

The new elements that make this edition such a departure from our previous ones, particularly the integration of text and art, were developed with the help of students, colleagues and our publisher. Our reviewers of the early samples offered encouragement and suggested helpful refinements. Once the new format had been established, the support we received from Prentice Hall in the development of the books was typically superb. Our editor Tacy Quinn smoothly organized the large team effort that books of this kind require and gave us enthusiastic help and insightful advice. Marcia Horton and Tim Galligan were dedicated to this major revision from our initial conversations about our ideas through the text's publication. Craig Little continued to teach us about the details of book production and was instrumental in keeping the project on schedule. Xiaohong Zhu again provided consummate support on art and photography issues. Dee Bernhard and Mack Patterson managed our communications with reviewers and users of the books. Jennifer Lonschein provided editorial and production support. David Alick, Ben Paris, and Kristin Mayo coordinated the development of the online resources that have become such essential tools for users. Jonathan Boylan designed the covers. We thank Peter Schiavone for developing the study packs that accompany the books, and Stephen Hunt and Ronald Larsen for writing the MATLAB/Mathcad tutorials. Scott Hendricks, Karim Nohra, and Kurt Norlin, valued colleagues from previous campaigns, advised us on style and clarity, corrected many of our errors, and revised the solution manuals. We are responsible for errors that remain. Nancy Bedford gave us editorial advice and help with checking. Many other talented and professional people at Prentice Hall and elsewhere contributed, and we thank them. And once again we thank our families, especially Nancy and Marsha, for their forbearance and understanding of the realities of new editions.

Anthony Bedford and Wallace Fowler
Austin, Texas

About the Authors

Anthony Bedford (*l*) and Wallace T. Fowler

Anthony Bedford is Professor Emeritus of Aerospace Engineering and Engineering Mechanics at the University of Texas at Austin, and has been on the faculty since 1968. A member of the University of Texas Academy of Distinguished Teachers, his main professional activity has been education and research in engineering mechanics. He has written papers on mixture theory, wave propagation, and the mechanics of high velocity impacts and is the author of the books *Hamilton's Principle in Continuum Mechanics, Introduction to Elastic Wave Propagation* (with D. S. Drumheller), and *Mechanics of Materials* (with K. M. Liechti). He has industrial experience at Douglas Aircraft Company, TRW, and Sandia National Laboratories.

Wallace T. Fowler is the Paul D. & Betty Robertson Meek Centennial Professor of Engineering at the University of Texas and is Director of the Texas Space Grant Consortium. He is a fellow of both the American Institute of Aeronautics and Astronautics (AIAA) and the American Society for Engineering Education (ASEE). Dr. Fowler was the recipient of the 1976 General Dynamics Teaching Excellence Award, the 1985 AIAA–ASEE John Leland Atwood Award (for National Outstanding Aerospace Engineering Educator), the 1990–91 University of Texas System Chancellor's Council Teaching Award, and the 1994 ASEE Fred Merryfield Design Teaching Award. In 1997, he was elected to membership in the University of Texas Academy of Distinguished Teachers. Dr. Fowler also served as 2000–2001 national president of the American Society for Engineering Education. Dr. Fowler's research and teaching interests at UT Austin focus on space systems engineering and design.

Photo Credits

Engineering Mechanics
STATICS

CHAPTER

1

Introduction

How do engineers design and construct the devices we use, from simple objects such as chairs and pencil sharpeners to complicated ones such as dams, cars, airplanes, and spacecraft? They must have a deep understanding of the physics underlying the design of such devices and must be able to use mathematical models to predict their behavior. Students of engineering begin to learn how to analyze and predict the behaviors of physical systems by studying mechanics.

◀ Engineers are guided by the principles of statics during each step of the design and assembly of a structure. Statics is one of the sciences underlying the art of structural design.

1.1 Engineering and Mechanics

BACKGROUND

How can engineers design complex systems and predict their characteristics before they are constructed? Engineers have always relied on their knowledge of previous designs, experiments, ingenuity, and creativity to develop new designs. Modern engineers add a powerful technique: They develop mathematical equations based on the physical characteristics of the devices they design. With these mathematical models, engineers predict the behavior of their designs, modify them, and test them prior to their actual construction. Aerospace engineers use mathematical models to predict the paths the space shuttle will follow in flight. Civil engineers use mathematical models to analyze the effects of loads on buildings and foundations.

At its most basic level, mechanics is the study of forces and their effects. Elementary mechanics is divided into *statics*, the study of objects in equilibrium, and *dynamics*, the study of objects in motion. The results obtained in elementary mechanics apply directly to many fields of engineering. Mechanical and civil engineers designing structures use the equilibrium equations derived in statics. Civil engineers analyzing the responses of buildings to earthquakes and aerospace engineers determining the trajectories of satellites use the equations of motion derived in dynamics.

Mechanics was the first analytical science. As a result, fundamental concepts, analytical methods, and analogies from mechanics are found in virtually every field of engineering. Students of chemical and electrical engineering gain a deeper appreciation for basic concepts in their fields, such as equilibrium, energy, and stability, by learning them in their original mechanical contexts. By studying mechanics, they retrace the historical development of these ideas.

Mechanics consists of broad principles that govern the behavior of objects. In this book we describe these principles and provide examples that demonstrate some of their applications. Although it is essential that you practice working problems similar to these examples, and we include many problems of this kind, our objective is to help you understand the principles well enough to apply them to situations that are new to you. Each generation of engineers confronts new problems.

Problem Solving

In the study of mechanics you learn problem-solving procedures that you will use in succeeding courses and throughout your career. Although different types of problems require different approaches, the following steps apply to many of them:

- Identify the information that is given and the information, or answer, you must determine. It's often helpful to restate the problem in your own words. When appropriate, make sure you understand the physical system or model involved.
- Develop a *strategy* for the problem. This means identifying the principles and equations that apply and deciding how you will use them to solve the problem. Whenever possible, draw diagrams to help visualize and solve the problem.
- Whenever you can, try to predict the answer. This will develop your intuition and will often help you recognize an incorrect answer.
- Solve the equations and, whenever possible, interpret your results and compare them with your prediction. This last step is a *reality check*. Is your answer reasonable?

Numbers

Engineering measurements, calculations, and results are expressed in numbers. You need to know how we express numbers in the examples and problems and how to express the results of your own calculations.

Significant Digits   This term refers to the number of meaningful (that is, accurate) digits in a number, counting to the right starting with the first nonzero digit. The two numbers 7.630 and 0.007630 are each stated to four significant digits. If only the first four digits in the number 7,630,000 are known to be accurate, this can be indicated by writing the number in scientific notation as 7.630×10^6.

If a number is the result of a measurement, the significant digits it contains are limited by the accuracy of the measurement. If the result of a measurement is stated to be 2.43, this means that the actual value is believed to be closer to 2.43 than to 2.42 or 2.44.

Numbers may be rounded off to a certain number of significant digits. For example, we can express the value of π to three significant digits, 3.14, or we can express it to six significant digits, 3.14159. When you use a calculator or computer, the number of significant digits is limited by the number of digits the machine is designed to carry.

Use of Numbers in This Book   You should treat numbers given in problems as exact values and not be concerned about how many significant digits they contain. If a problem states that a quantity equals 32.2, you can assume its value is 32.200. . . . We generally express intermediate results and answers in the examples and the answers to the problems to at least three significant digits. If you use a calculator, your results should be that accurate. Be sure to avoid round-off errors that occur if you round off intermediate results when making a series of calculations. Instead, carry through your calculations with as much accuracy as you can by retaining values in your calculator.

Space and Time

Space simply refers to the three-dimensional universe in which we live. Our daily experiences give us an intuitive notion of space and the locations, or positions, of points in space. The distance between two points in space is the length of the straight line joining them.

Measuring the distance between points in space requires a unit of length. We use both the International System of units, or SI units, and U.S. Customary units. In SI units, the unit of length is the meter (m). In U.S. Customary units, the unit of length is the foot (ft).

Time is, of course, familiar—our lives are measured by it. The daily cycles of light and darkness and the hours, minutes, and seconds measured by our clocks and watches give us an intuitive notion of time. Time is measured by the intervals between repeatable events, such as the swings of a clock pendulum or the vibrations of a quartz crystal in a watch. In both SI units and U.S. Customary units, the unit of time is the second (s). The minute (min), hour (h), and day are also frequently used.

If the position of a point in space relative to some reference point changes with time, the rate of change of its position is called its *velocity*, and the rate of change of its velocity is called its *acceleration*. In SI units, the velocity is expressed in meters per second (m/s) and the acceleration is expressed in meters per second per second, or meters per second squared (m/s^2). In U.S.

Customary units, the velocity is expressed in feet per second (ft/s) and the acceleration is expressed in feet per second squared (ft/s²).

Newton's Laws

Elementary mechanics was established on a firm basis with the publication in 1687 of *Philosophiae Naturalis Principia Mathematica*, by Isaac Newton. Although highly original, it built on fundamental concepts developed by many others during a long and difficult struggle toward understanding (Fig. 1.1).

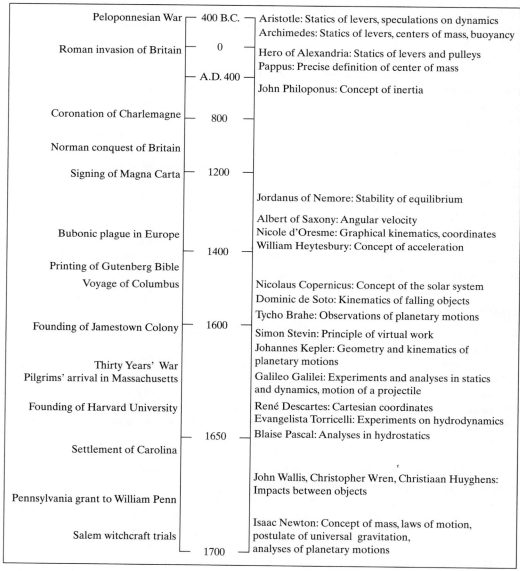

Figure 1.1

Chronology of developments in mechanics up to the publication of Newton's *Principia* in relation to other events in history.

Newton stated three "laws" of motion, which we express in modern terms:

1. *When the sum of the forces acting on a particle is zero, its velocity is constant. In particular, if the particle is initially stationary, it will remain stationary.*

2. *When the sum of the forces acting on a particle is not zero, the sum of the forces is equal to the rate of change of the linear momentum of the particle. If the mass is constant, the sum of the forces is equal to the product of the mass of the particle and its acceleration.*

3. *The forces exerted by two particles on each other are equal in magnitude and opposite in direction.*

Notice that we did not define force and mass before stating Newton's laws. The modern view is that these terms are defined by the second law. To demonstrate, suppose that we choose an arbitrary object and define it to have unit mass. Then we define a unit of force to be the force that gives our unit mass an acceleration of unit magnitude. In principle, we can then determine the mass of any object: We apply a unit force to it, measure the resulting acceleration, and use the second law to determine the mass. We can also determine the magnitude of any force: We apply it to our unit mass, measure the resulting acceleration, and use the second law to determine the force.

Thus Newton's second law gives precise meanings to the terms *mass* and *force*. In SI units, the unit of mass is the kilogram (kg). The unit of force is the newton (N), which is the force required to give a mass of one kilogram an acceleration of one meter per second squared. In U.S. Customary units, the unit of force is the pound (lb). The unit of mass is the slug, which is the amount of mass accelerated at one foot per second squared by a force of one pound.

Although the results we discuss in this book are applicable to many of the problems met in engineering practice, there are limits to the validity of Newton's laws. For example, they don't give accurate results if a problem involves velocities that are not small compared to the velocity of light $(3 \times 10^8$ m/s$)$. Einstein's special theory of relativity applies to such problems. Elementary mechanics also fails in problems involving dimensions that are not large compared to atomic dimensions. Quantum mechanics must be used to describe phenomena on the atomic scale.

International System of Units

In SI units, length is measured in meters (m) and mass in kilograms (kg). Time is measured in seconds (s), although other familiar measures such as minutes (min), hours (h), and days are also used when convenient. Meters, kilograms, and seconds are called the *base units* of the SI system. Force is measured in newtons (N). Recall that these units are related by Newton's second law: One newton is the force required to give an object of one kilogram mass an acceleration of one meter per second squared:

$$1 \text{ N} = (1 \text{ kg})(1 \text{ m/s}^2) = 1 \text{ kg-m/s}^2.$$

Because the newton can be expressed in terms of the base units, it is called a *derived unit*.

To express quantities by numbers of convenient size, multiples of units are indicated by prefixes. The most common prefixes, their abbreviations, and the multiples they represent are shown in Table 1.1. For example, 1 km is 1 kilometer, which is 1000 m, and 1 Mg is 1 megagram, which is 10^6 g, or 1000 kg. We frequently use kilonewtons (kN).

Table 1.1 The common prefixes used in SI units and the multiples they represent.

Prefix	Abbreviation	Multiple
nano-	n	10^{-9}
micro-	μ	10^{-6}
milli-	m	10^{-3}
kilo-	k	10^{3}
mega-	M	10^{6}
giga-	G	10^{9}

U.S. Customary Units

In U.S. Customary units, length is measured in feet (ft) and force is measured in pounds (lb). Time is measured in seconds (s). These are the base units of the U.S. Customary system. In this system of units, mass is a derived unit. The unit of mass is the slug, which is the mass of material accelerated at one foot per second squared by a force of one pound. Newton's second law states that

$$1 \text{ lb} = (1 \text{ slug})(1 \text{ ft/s}^2).$$

From this expression we obtain

$$1 \text{ slug} = 1 \text{ lb-s}^2/\text{ft}.$$

We use other U.S. Customary units such as the mile (1 mi = 5280 ft) and the inch (1 ft = 12 in). We also use the kilopound (kip), which is 1000 lb.

Angular Units

In both SI and U.S. Customary units, angles are normally expressed in radians (rad). We show the value of an angle θ in radians in Fig. 1.2. It is defined to be the ratio of the part of the circumference subtended by θ to the radius of the circle. Angles are also expressed in degrees. Since there are 360 degrees (360°) in a complete circle, and the complete circumference of the circle is $2\pi R$, 360° equals 2π rad.

Equations containing angles are nearly always derived under the assumption that angles are expressed in radians. Therefore, when you want to substitute the value of an angle expressed in degrees into an equation, you should first convert it into radians. A notable exception to this rule is that many calculators are designed to accept angles expressed in either degrees or radians when you use them to evaluate functions such as $\sin \theta$.

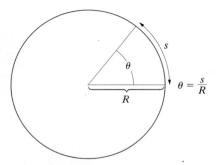

Figure 1.2
Definition of an angle in radians.

Conversion of Units

Many situations arise in engineering practice that require values expressed in one kind of unit to be converted into values in other units. For example, if some of the data to be used in an equation are given in SI units and some are given in U.S. Customary units, they must all be expressed in terms of one system of units before they are substituted into the equation. Converting units is straightforward, although it must be done with care.

Suppose that we want to express 1 mile per hour (mi/h) in terms of feet per second (ft/s). Because 1 mile equals 5280 feet and 1 hour equals 3600 seconds, we can treat the expressions

$$\left(\frac{5280 \text{ ft}}{1 \text{ mi}}\right) \quad \text{and} \quad \left(\frac{1 \text{ h}}{3600 \text{ s}}\right)$$

as ratios whose values are 1. In this way, we obtain

$$1 \text{ mi/h} = (1 \text{ mi/h})\left(\frac{5280 \text{ ft}}{1 \text{ mi}}\right)\left(\frac{1 \text{ h}}{3600 \text{ s}}\right) = 1.47 \text{ ft/s}.$$

Some useful unit conversions are given in Table 1.2.

Table 1.2 Unit conversions.

Time	1 minute	=	60 seconds
	1 hour	=	60 minutes
	1 day	=	24 hours
Length	1 foot	=	12 inches
	1 mile	=	5280 feet
	1 inch	=	25.4 millimeters
	1 foot	=	0.3048 meters
Angle	2π radians	=	360 degrees
Mass	1 slug	=	14.59 kilograms
Force	1 pound	=	4.448 newtons

RESULTS

- Identify the given information and the answer that must be determined.
- Develop a strategy; identify principles and equations that apply and how they will be used.
- Try to predict the answer whenever possible.
- Obtain the answer and, whenever possible, interpret it and compare it with the prediction.

> Problem Solving: These steps apply to many types of problems.

SI Units—The *base units* are time in seconds (s), length in meters (m), and mass in kilograms (kg). The unit of force is the newton (N), which is the force required to accelerate a mass of one kilogram at one meter per second squared.

> Systems of units.

U.S. Customary Units—The base units are time in seconds (s), length in feet (ft), and force in pounds (lb). The unit of mass is the slug, which is the mass accelerated at one foot per second squared by a force of one pound.

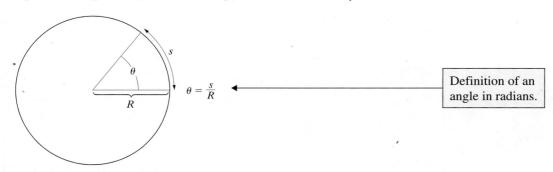

$$\theta = \frac{s}{R}$$

> Definition of an angle in radians.

Equivalent quantities, such as 1 hour = 60 minutes, can be written as ratios whose values are 1:

$$\left(\frac{1\ h}{60\ min}\right) = 1,$$

and used to convert units. For example,

$$15\ min = 15\ min\left(\frac{1\ h}{60\ min}\right) = 0.25\ h.$$

> Conversion of units.

A comprehensive resource on units has been compiled by Russ Rowlett of the University of North Carolina at Chapel Hill and made available online at www.unc.edu/~rowlett/units.

Active Example 1.1	Converting Units (▶ *Related Problem 1.11*)

A man is riding a bicycle at a speed of 6 meters per second (m/s). How fast is he going in kilometers per hour (km/h)?

Strategy
One kilometer is 1000 meters and one hour is 60 minutes × 60 seconds = 3600 seconds. We can use these unit conversions to determine his speed in km/h.

Solution

Convert meters to kilometers.

Convert seconds to hours.

$$6 \text{ m/s} = 6 \text{ m/s} \left(\frac{1 \text{ km}}{1000 \text{ m}}\right)\left(\frac{3600 \text{ s}}{1 \text{ h}}\right)$$

$$= 21.6 \text{ km/h}.$$

Practice Problem A man is riding a bicycle at a speed of 10 feet per second (ft/s). How fast is he going in miles per hour (mi/h)?

Answer: 6.82 mi/h.

Example 1.2	Converting Units of Pressure (▶ *Related Problem 1.16*)

Deep Submersible Vehicle.

The pressure exerted at a point of the hull of the deep submersible vehicle is 3.00×10^6 Pa (pascals). A pascal is 1 newton per square meter. Determine the pressure in pounds per square foot.

Strategy
From Table 1.2, 1 pound = 4.448 newtons and 1 foot = 0.3048 meters. With these unit conversions we can calculate the pressure in pounds per square foot.

Solution
The pressure (to three significant digits) is

$$3.00 \times 10^6 \text{ N/m}^2 = (3.00 \times 10^6 \text{ N/m}^2)\left(\frac{1 \text{ lb}}{4.448 \text{ N}}\right)\left(\frac{0.3048 \text{ m}}{1 \text{ ft}}\right)^2$$

$$= 62{,}700 \text{ lb/ft}^2.$$

Critical Thinking
How could we have obtained this result in a more direct way? Notice from the table of unit conversions in the inside front cover that 1 Pa = 0.0209 lb/ft². Therefore,

$$3.00 \times 10^6 \text{ N/m}^2 = (3.00 \times 10^6 \text{ N/m}^2)\left(\frac{0.0209 \text{ lb/ft}^2}{1 \text{ N/m}^2}\right)$$

$$= 62{,}700 \text{ lb/ft}^2.$$

| **Example 1.3** | **Determining Units from an Equation** (▶ *Related Problem 1.20*) |

Suppose that in Einstein's equation

$$E = \dot{m}c^2,$$

the mass m is in kilograms and the velocity of light c is in meters per second.

(a) What are the SI units of E?

(b) If the value of E in SI units is 20, what is its value in U.S. Customary base units?

Strategy

(a) Since we know the units of the terms m and c, we can deduce the units of E from the given equation.

(b) We can use the unit conversions for mass and length from Table 1.2 to convert E from SI units to U.S. Customary units.

Solution

(a) From the equation for E,

$$E = (m \text{ kg})(c \text{ m/s})^2,$$

the SI units of E are kg-m^2/s^2.

(b) From Table 1.2, 1 slug = 14.59 kg and 1 ft = 0.3048 m. Therefore,

$$1 \text{ kg-m}^2/\text{s}^2 = (1 \text{ kg-m}^2/\text{s}^2)\left(\frac{1 \text{ slug}}{14.59 \text{ kg}}\right)\left(\frac{1 \text{ ft}}{0.3048 \text{ m}}\right)^2$$

$$= 0.738 \text{ slug-ft}^2/\text{s}^2.$$

The value of E in U.S. Customary units is

$$E = (20)(0.738) = 14.8 \text{ slug-ft}^2/\text{s}^2.$$

Critical Thinking

In part (a), how did we know that we could determine the units of E by determining the units of mc^2? The dimensions, or units, of each term in an equation must be the same. For example, in the equation $a + b = c$, the dimensions of each of the terms a, b, and c must be the same. The equation is said to be *dimensionally homogeneous*. This requirement is expressed by the colloquial phrase "Don't compare apples and oranges."

Problems

1.1 The value of π is 3.14159265.... If C is the circumference of a circle and r is its radius, determine the value of r/C to four significant digits.

Problem 1.1

1.2 The base of natural logarithms is $e = 2.718281828\ldots$.

(a) Express e to five significant digits.

(b) Determine the value of e^2 to five significant digits.

(c) Use the value of e you obtained in part (a) to determine the value of e^2 to five significant digits.

[Part (c) demonstrates the hazard of using rounded-off values in calculations.]

1.3 A machinist drills a circular hole in a panel with a nominal radius $r = 5$ mm. The actual radius of the hole is in the range $r = 5 \pm 0.01$ mm.

(a) To what number of significant digits can you express the radius?

(b) To what number of significant digits can you express the area of the hole?

Problem 1.3

1.4 The opening in the soccer goal is 24 ft wide and 8 ft high, so its area is 24 ft $\times$ 8 ft = 192 ft^2. What is its area in m^2 to three significant digits?

Problem 1.4

1.5 The Burj Dubai, scheduled for completion in 2008, will be the world's tallest building with a height of 705 m. The area of its ground footprint will be 8000 m^2. Convert its height and footprint area to U.S. Customary units to three significant digits.

Problem 1.5

1.6 Suppose that you have just purchased a Ferrari F355 coupe and you want to know whether you can use your set of SAE (U.S. Customary unit) wrenches to work on it. You have wrenches with widths w = 1/4 in, 1/2 in, 3/4 in, and 1 in, and the car has nuts with dimensions n = 5 mm, 10 mm, 15 mm, 20 mm, and 25 mm. Defining a wrench to fit if w is no more than 2% larger than n, which of your wrenches can you use?

Problem 1.6

1.7 Suppose that the height of Mt. Everest is known to be between 29,032 ft and 29,034 ft. Based on this information, to how many significant digits can you express the height (a) in feet? (b) in meters?

1.8 The maglev (magnetic levitation) train from Shanghai to the airport at Pudong reaches a speed of 430 km/h. Determine its speed (a) in mi/h; (b) in ft/s.

Problem 1.8

1.9 In the 2006 Winter Olympics, the men's 15-km cross-country skiing race was won by Andrus Veerpalu of Estonia in a time of 38 minutes, 1.3 seconds. Determine his average speed (the distance traveled divided by the time required) to three significant digits (a) in km/h; (b) in mi/h.

1.10 The Porsche's engine exerts 229 ft-lb (foot-pounds) of torque at 4600 rpm. Determine the value of the torque in N-m (newton-meters).

Problem 1.10

▶ **1.11** The *kinetic energy* of the man in Active Example 1.1 is defined by $\frac{1}{2}mv^2$, where m is his mass and v is his velocity. The man's mass is 68 kg and he is moving at 6 m/s, so his kinetic energy is $\frac{1}{2}(68 \text{ kg})(6 \text{ m/s})^2 = 1224 \text{ kg-m}^2/\text{s}^2$. What is his kinetic energy in U.S. Customary units?

1.12 The acceleration due to gravity at sea level in SI units is g = 9.81 m/s^2. By converting units, use this value to determine the acceleration due to gravity at sea level in U.S. Customary units.

1.13 A *furlong per fortnight* is a facetious unit of velocity, perhaps made up by a student as a satirical comment on the bewildering variety of units engineers must deal with. A furlong is 660 ft (1/8 mile). A fortnight is 2 weeks (14 nights). If you walk to class at 2 m/s, what is your speed in furlongs per fortnight to three significant digits?

1.14 Determine the cross-sectional area of the beam (a) in m^2; (b) in in^2.

Problem 1.14

1.15 The cross-sectional area of the C12×30 American Standard Channel steel beam is $A = 8.81$ in^2. What is its cross-sectional area in mm^2?

Problem 1.15

▶ **1.16** A pressure transducer measures a value of 300 lb/in^2. Determine the value of the pressure in pascals. A pascal (Pa) is one newton per square meter.

1.17 A horsepower is 550 ft-lb/s. A watt is 1 N-m/s. Determine how many watts are generated by the engines of the passenger jet if they are producing 7000 horsepower.

Problem 1.17

1.18 Chapter 7 discusses distributed loads that are expressed in units of force per unit length. If the value of a distributed load is 400 N/m, what is its value in lb/ft?

1.19 The moment of inertia of the rectangular area about the x axis is given by the equation

$$I = \tfrac{1}{3}bh^3.$$

The dimensions of the area are $b = 200$ mm and $h = 100$ mm. Determine the value of I to four significant digits in terms of (a) mm^4, (b) m^4, and (c) in^4.

Problem 1.19

▶ **1.20** In Example 1.3, instead of Einstein's equation consider the equation $L = mc$, where the mass m is in kilograms and the velocity of light c is in meters per second. (a) What are the SI units of L? (b) If the value of L in SI units is 12, what is its value in U.S. Customary base units?

1.21 The equation

$$\sigma = \frac{My}{I}$$

is used in the mechanics of materials to determine normal stresses in beams.

(a) When this equation is expressed in terms of SI base units, M is in newton-meters (N-m), y is in meters (m), and I is in meters to the fourth power (m^4). What are the SI units of σ?
(b) If $M = 2000$ N-m, $y = 0.1$ m, and $I = 7 \times 10^{-5}$ m^4, what is the value of σ in U.S. Customary base units?

1.2 Newtonian Gravitation

BACKGROUND

Newton postulated that the gravitational force between two particles of mass m_1 and m_2 that are separated by a distance r (Fig. 1.3) is

$$F = \frac{Gm_1m_2}{r^2},\qquad(1.1)$$

where G is called the universal gravitational constant. The value of G in SI units is 6.67×10^{-11} N-m^2/kg^2. Based on this postulate, he calculated the gravitational force between a particle of mass m_1 and a homogeneous sphere of mass m_2 and found that it is also given by Eq. (1.1), with r denoting the distance from the particle to the center of the sphere. Although the earth is not a homogeneous sphere, we can use this result to approximate the weight of an object of mass m due to the gravitational attraction of the earth. We have

$$W = \frac{Gmm_E}{r^2},\qquad(1.2)$$

where m_E is the mass of the earth and r is the distance from the center of the earth to the object. Notice that the weight of an object depends on its location relative to the center of the earth, whereas the mass of the object is a measure of the amount of matter it contains and doesn't depend on its position.

When an object's weight is the only force acting on it, the resulting acceleration is called the acceleration due to gravity. In this case, Newton's second law states that $W = ma$, and from Eq. (1.2) we see that the acceleration due to gravity is

$$a = \frac{Gm_E}{r^2}.\qquad(1.3)$$

The *acceleration due to gravity at sea level* is denoted by g. Denoting the radius of the earth by R_E, we see from Eq. (1.3) that $Gm_E = gR_E^2$. Substituting this result into Eq. (1.3), we obtain an expression for the acceleration due to gravity at a distance r from the center of the earth in terms of the acceleration due to gravity at sea level:

$$a = g\frac{R_E^2}{r^2}.\qquad(1.4)$$

Since the weight of the object $W = ma$, the weight of an object at a distance r from the center of the earth is

$$W = mg\frac{R_E^2}{r^2}.\qquad(1.5)$$

At sea level $(r = R_E)$, the weight of an object is given in terms of its mass by the simple relation

$$W = mg.\qquad(1.6)$$

The value of g varies from location to location on the surface of the earth. The values we use in examples and problems are $g = 9.81$ m/s^2 in SI units and $g = 32.2$ ft/s^2 in U.S. Customary units.

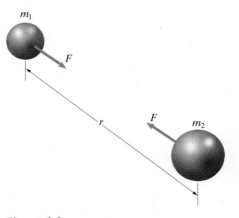

Figure 1.3
The gravitational forces between two particles are equal in magnitude and directed along the line between them.

RESULTS

The gravitational force between two particles of mass m_1 and m_2 that are separated by a distance r is

$$F = \frac{Gm_1m_2}{r^2},\qquad(1.1)$$

where G is the universal gravitational constant. The value of G in SI units is 6.67×10^{-11} N-m^2/kg^2.

> Newtonian gravitation.

When the earth is modeled as a homogeneous sphere of radius R_E, the acceleration due to gravity at a distance r from the center is

$$a = g\frac{R_E^2}{r^2},\qquad(1.4)$$

where g is the acceleration due to gravity at sea level.

> Acceleration due to gravity of the earth.

$$W = mg,\qquad(1.6)$$

where m is the mass of the object and g is the acceleration due to gravity at sea level.

> Weight of an object at sea level.

Active Example 1.4   Weight and Mass (▶ *Related Problem 1.22*)

The C-clamp weighs 14 oz at sea level. [16 oz (ounces) = 1 lb.] The acceleration due to gravity at sea level is $g = 32.2$ ft/s^2. What is the mass of the C-clamp in slugs?

Strategy
We must first determine the weight of the C-clamp in pounds. Then we can use Eq. (1.6) to determine the mass in slugs.

Solution

$$14\text{ oz} = 14\text{ oz}\left(\frac{1\text{ lb}}{16\text{ oz}}\right) = 0.875\text{ lb.}$$

> Convert the weight from ounces to pounds.

$$m = \frac{W}{g} = \frac{0.875\text{ lb}}{32.2\text{ ft/s}^2} = 0.0272\text{ slug.}$$

> Use Eq. (1.6) to calculate the mass in slugs.

Practice Problem The mass of the C-clamp is 0.397 kg. The acceleration due to gravity at sea level is $g = 9.81$ m/s^2. What is the weight of the C-clamp at sea level in newtons?

Answer: 3.89 N.

| **Example 1.5** | **Determining an Object's Weight** (▶ *Related Problem 1.27*) |

When the Mars Exploration Rover was fully assembled, its mass was 180 kg. The acceleration due to gravity at the surface of Mars is 3.68 m/s^2 and the radius of Mars is 3390 km.

(a) What was the rover's weight when it was at sea level on Earth?

(b) What is the rover's weight on the surface of Mars?

(c) The entry phase began when the spacecraft reached the Mars atmospheric entry interface point at 3522 km from the center of Mars. What was the rover's weight at that point?

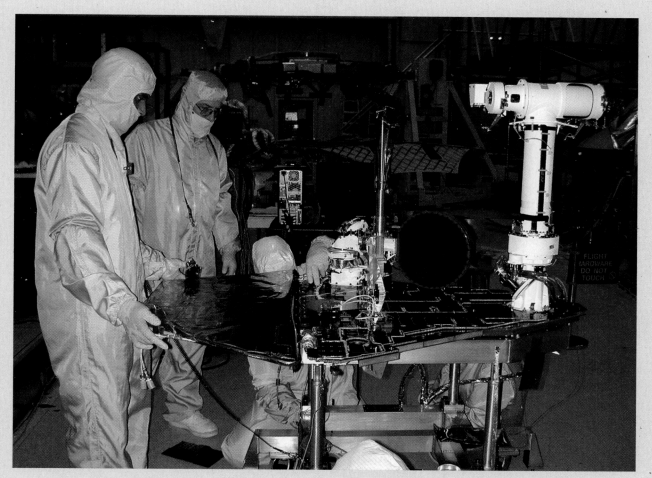

Mars Exploration Rover being assembled.

Strategy

The rover's weight at sea level on Earth is given by Eq. (1.6) with $g = 9.81$ m/s^2.

We can determine the weight on the surface of Mars by using Eq. (1.6) with the acceleration due to gravity equal to 3.68 m/s^2.

To determine the rover's weight as it began the entry phase, we can write an equation for Mars equivalent to Eq. (1.5).

Solution

(a) The weight at sea level on Earth is

$$W = mg$$
$$= (180 \text{ kg})(9.81 \text{ m/s}^2)$$
$$= 1770 \text{ N } (397 \text{ lb}).$$

(b) Let $g_M = 3.68$ m/s^2 be the acceleration due to gravity at the surface of Mars. Then the weight of the rover on the surface of Mars is

$$W = mg_M$$
$$= (180 \text{ kg})(3.68 \text{ m/s}^2)$$
$$= 662 \text{ N } (149 \text{ lb}).$$

(c) Let $R_M = 3390$ km be the radius of Mars. From Eq. (1.5), the rover's weight when it is 3522 km above the center of Mars is

$$W = mg_M \frac{R_M^2}{r^2}$$
$$= (180 \text{ kg})(3.68 \text{ m/s}^2)\frac{(3{,}390{,}000 \text{ m})^2}{(3{,}522{,}000 \text{ m})^2}$$
$$= 614 \text{ N } (138 \text{ lb}).$$

Critical Thinking

In part (c), how did we know that we could apply Eq. (1.5) to Mars? Equation (1.5) is applied to Earth based on modeling it as a homogeneous sphere. It can be applied to other celestial objects under the same assumption. The accuracy of the results depends on how aspherical and inhomogeneous the object is.

Problems

▶ **1.22** The acceleration due to gravity on the surface of the moon is 1.62 m/s². (a) What would the mass of the C-clamp in Active Example 1.4 be on the surface of the moon? (b) What would the weight of the C-clamp in newtons be on the surface of the moon?

1.23 The 1 ft × 1 ft × 1 ft cube of iron weighs 490 lb at sea level. Determine the weight in newtons of a 1 m × 1 m × 1 m cube of the same material at sea level.

1 ft

1 ft   **1 ft**

Problem 1.23

1.24 The area of the Pacific Ocean is 64,186,000 square miles and its average depth is 12,925 ft. Assume that the weight per unit volume of ocean water is 64 lb/ft³. Determine the mass of the Pacific Ocean (a) in slugs; (b) in kilograms.

1.25 The acceleration due to gravity at sea level is $g = 9.81$ m/s². The radius of the earth is 6370 km. The universal gravitational constant $G = 6.67 \times 10^{-11}$ N-m²/kg². Use this information to determine the mass of the earth.

1.26 A person weighs 180 lb at sea level. The radius of the earth is 3960 mi. What force is exerted on the person by the gravitational attraction of the earth if he is in a space station in orbit 200 mi above the surface of the earth?

▶ **1.27** The acceleration due to gravity on the surface of the moon is 1.62 m/s². The moon's radius is $R_M = 1738$ km. (See Example 1.5.)

(a) What is the weight in newtons on the surface of the moon of an object that has a mass of 10 kg?

(b) Using the approach described in Example 1.5, determine the force exerted on the object by the gravity of the moon if the object is located 1738 km above the moon's surface.

1.28 If an object is near the surface of the earth, the variation of its weight with distance from the center of the earth can often be neglected. The acceleration due to gravity at sea level is $g = 9.81$ m/s². The radius of the earth is 6370 km. The weight of an object at sea level is mg, where m is its mass. At what height above the surface of the earth does the weight of the object decrease to $0.99mg$?

1.29 The planet Neptune has an equatorial diameter of 49,532 km and its mass is 1.0247×10^{26} kg. If the planet is modeled as a homogeneous sphere, what is the acceleration due to gravity at its surface? (The universal gravitational constant is $G = 6.67 \times 10^{-11}$ N-m²/kg².)

Problem 1.29

1.30 At a point between the earth and the moon, the magnitude of the force exerted on an object by the earth's gravity equals the magnitude of the force exerted on the object by the moon's gravity. What is the distance from the center of the earth to that point to three significant digits? The distance from the center of the earth to the center of the moon is 383,000 km, and the radius of the earth is 6370 km. The radius of the moon is 1738 km, and the acceleration due to gravity at its surface is 1.62 m/s².

Vectors

If an object is subjected to several forces that have different magnitudes and act in different directions, how can the magnitude and direction of the resulting total force on the object be determined? Forces are vectors and must be added according to the definition of vector addition. In engineering we deal with many quantities that have both magnitude and direction and can be expressed and analyzed as vectors. In this chapter we review vector operations, express vectors in terms of components, and present examples of engineering applications of vectors.

◀ Fields of vectors show the velocities and directions of a gas flow at three vertical positions. Vectors are used to describe and analyze quantities that have magnitude and direction, including positions, forces, moments, velocities, and accelerations.

Figure 2.1
(**a**) Two points A and B of a mechanism.
(**b**) The vector $\mathbf{r}_{AB}$ from A to B.

Figure 2.2
Representing the force cable AB exerts on
the tower by a vector $\mathbf{F}$.

2.1 Scalars and Vectors

BACKGROUND

A physical quantity that is completely described by a real number is called a *scalar*. Time is a scalar quantity. Mass is also a scalar quantity. For example, you completely describe the mass of a car by saying that its value is 1200 kg.

In contrast, you have to specify both a nonnegative real number, or *magnitude,* and a direction to describe a vector quantity. Two vector quantities are equal only if both their magnitudes and their directions are equal.

The position of a point in space relative to another point is a vector quantity. To describe the location of a city relative to your home, it is not enough to say that it is 100 miles away. You must say that it is 100 miles west of your home. Force is also a vector quantity. When you push a piece of furniture across the floor, you apply a force of magnitude sufficient to move the furniture and you apply it in the direction you want the furniture to move.

We will represent vectors by boldfaced letters, $\mathbf{U}$, $\mathbf{V}$, $\mathbf{W}$, ..., and will denote the magnitude of a vector $\mathbf{U}$ by $|\mathbf{U}|$. A vector is represented graphically by an arrow. The direction of the arrow indicates the direction of the vector, and the length of the arrow is defined to be proportional to the magnitude. For example, consider the points A and B of the mechanism in Fig. 2.1a. We can specify the position of point B relative to point A by the vector $\mathbf{r}_{AB}$ in Fig. 2.1b. The direction of $\mathbf{r}_{AB}$ indicates the direction from point A to point B. If the distance between the two points is 200 mm, the magnitude $|\mathbf{r}_{AB}| = 200$ mm.

The cable AB in Fig. 2.2 helps support the television transmission tower. We can represent the force the cable exerts on the tower by a vector $\mathbf{F}$ as shown. If the cable exerts an 800-N force on the tower, $|\mathbf{F}| = 800$ N. (A cable suspended in this way will exhibit some sag, or curvature, and the tension will vary along its length. For now, we assume that the curvature in suspended cables and ropes and the variations in their tensions can be neglected. This assumption is approximately valid if the weight of the rope or cable is small in comparison to the tension. We discuss and analyze suspended cables and ropes in more detail in Chapter 10.)

Vectors are a convenient means for representing physical quantities that have magnitude and direction, but that is only the beginning of their usefulness. Just as real numbers are manipulated with the familiar rules for addition, subtraction, multiplication, and so forth, there are rules for manipulating vectors. These rules provide powerful tools for engineering analysis.

Vector Addition

When an object moves from one location in space to another, we say it undergoes a *displacement*. If we move a book (or, speaking more precisely, some point of a book) from one location on a table to another, as shown in Fig. 2.3a, we can represent the displacement by the vector $\mathbf{U}$. The direction of $\mathbf{U}$ indicates the direction of the displacement, and $|\mathbf{U}|$ is the distance the book moves.

Suppose that we give the book a second displacement $\mathbf{V}$, as shown in Fig. 2.3b. The two displacements $\mathbf{U}$ and $\mathbf{V}$ are equivalent to a single displacement of the book from its initial position to its final position, which we represent by the vector $\mathbf{W}$ in Fig. 2.3c. Notice that the final position of the book is the same whether we first give it the displacement $\mathbf{U}$ and then the displacement $\mathbf{V}$ or we first give it the displacement $\mathbf{V}$ and then the displacement $\mathbf{U}$ (Fig. 2.3d). The displacement $\mathbf{W}$ is defined to be the sum of the displacements $\mathbf{U}$ and $\mathbf{V}$:

$$\mathbf{U} + \mathbf{V} = \mathbf{W}.$$

Figure 2.3
(**a**) A displacement represented by the vector **U**.
(**b**) The displacement **U** followed by the displacement **V**.
(**c**) The displacements **U** and **V** are equivalent to the displacement **W**.
(**d**) The final position of the book doesn't depend on the order of the displacements.

Figure 2.4
(**a**) Two vectors **U** and **V**.
(**b**) The head of **U** placed at the tail of **V**.
(**c**) The triangle rule for obtaining the sum of **U** and **V**.
(**d**) The sum is independent of the order in which the vectors are added.
(**e**) The parallelogram rule for obtaining the sum of **U** and **V**.

The definition of vector addition is motivated by the addition of displacements. Consider the two vectors **U** and **V** shown in Fig. 2.4a. If we place them head to tail (Fig. 2.4b), their sum is defined to be the vector from the tail of **U** to the head of **V** (Fig. 2.4c). This is called the *triangle rule* for vector addition. Figure 2.4d demonstrates that the sum is independent of the order in which the vectors are placed head to tail. From this figure we obtain the *parallelogram rule* for vector addition (Fig. 2.4e).

The definition of vector addition implies that

$$\mathbf{U} + \mathbf{V} = \mathbf{V} + \mathbf{U} \quad \text{Vector addition is commutative.} \tag{2.1}$$

and

$$(\mathbf{U} + \mathbf{V}) + \mathbf{W} = \mathbf{U} + (\mathbf{V} + \mathbf{W}) \quad \text{Vector addition is associative.} \tag{2.2}$$

for any vectors **U**, **V**, and **W**: These results mean that when two or more vectors are added, the order in which they are added doesn't matter. The sum can be obtained by placing the vectors head to tail in any order, and the vector from the tail of the first vector to the head of the last one is the sum (Fig. 2.5). If the sum of two or more vectors is zero, they form a closed polygon when they are placed head to tail (Fig. 2.6).

Figure 2.5
Sum of the three vectors **U**, **V**, and **W**.

Figure 2.6
Three vectors **U**, **V**, and **W** whose sum is zero.

Figure 2.7
Arrows denoting the relative positions of
points are vectors.

$\mathbf{U}$ $2\mathbf{U}$ $-\mathbf{U} = (-1)\mathbf{U}$ $\dfrac{\mathbf{U}}{2} = \dfrac{1}{2}\mathbf{U}$

Figure 2.8
A vector $\mathbf{U}$ and some of its scalar multiples.

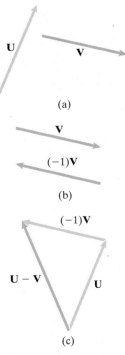

(a)

(b)

(c)

Figure 2.9
(a) Two vectors $\mathbf{U}$ and $\mathbf{V}$.
(b) The vectors $\mathbf{V}$ and $(-1)\mathbf{V}$.
(c) The sum of $\mathbf{U}$ and $(-1)\mathbf{V}$ is the vector
    difference $\mathbf{U} - \mathbf{V}$.

A physical quantity is called a vector if it has magnitude and direction and obeys the definition of vector addition. We have seen that displacement is a vector. The position of a point in space relative to another point is also a vector quantity. In Fig. 2.7, the vector $\mathbf{r}_{AC}$ from A to C is the sum of $\mathbf{r}_{AB}$ and $\mathbf{r}_{BC}$. A force has direction and magnitude, but do forces obey the definition of vector addition? For now we will assume that they do. When we discuss dynamics, we will show that Newton's second law implies that force is a vector.

Product of a Scalar and a Vector

The product of a scalar (real number) a and a vector $\mathbf{U}$ is a vector written as $a\mathbf{U}$. Its magnitude is $|a||\mathbf{U}|$, where $|a|$ is the absolute value of the scalar a. The direction of $a\mathbf{U}$ is the same as the direction of $\mathbf{U}$ when a is positive and is opposite to the direction of $\mathbf{U}$ when a is negative.

The product $(-1)\mathbf{U}$ is written as $-\mathbf{U}$ and is called "the negative of the vector $\mathbf{U}$." It has the same magnitude as $\mathbf{U}$ but the opposite direction. The division of a vector $\mathbf{U}$ by a scalar a is defined to be the product

$$\frac{\mathbf{U}}{a} = \left(\frac{1}{a}\right)\mathbf{U}.$$

Figure 2.8 shows a vector $\mathbf{U}$ and the products of $\mathbf{U}$ with the scalars 2, -1, and 1/2.
The definitions of vector addition and the product of a scalar and a vector imply that

$$a(b\mathbf{U}) = (ab)\mathbf{U}, \quad \text{The product is associative with} \atop \text{respect to scalar multiplication.} \tag{2.3}$$

$$(a + b)\mathbf{U} = a\mathbf{U} + b\mathbf{U}, \quad \text{The products are distributive} \atop \text{with respect to scalar addition.} \tag{2.4}$$

and

$$a(\mathbf{U} + \mathbf{V}) = a\mathbf{U} + a\mathbf{V}, \quad \text{The products are distributive} \atop \text{with respect to vector addition.} \tag{2.5}$$

for any scalars a and b and vectors $\mathbf{U}$ and $\mathbf{V}$. We will need these results when we discuss components of vectors.

Vector Subtraction

The difference of two vectors $\mathbf{U}$ and $\mathbf{V}$ is obtained by adding $\mathbf{U}$ to the vector $(-1)\mathbf{V}$:

$$\mathbf{U} - \mathbf{V} = \mathbf{U} + (-1)\mathbf{V}. \tag{2.6}$$

Consider the two vectors $\mathbf{U}$ and $\mathbf{V}$ shown in Fig. 2.9a. The vector $(-1)\mathbf{V}$ has the same magnitude as the vector $\mathbf{V}$ but is in the opposite direction (Fig. 2.9b). In Fig. 2.9c, we add the vector $\mathbf{U}$ to the vector $(-1)\mathbf{V}$ to obtain $\mathbf{U} - \mathbf{V}$.

Unit Vectors

A *unit vector* is simply a vector whose magnitude is 1. A unit vector specifies a direction and also provides a convenient way to express a vector that has a particular direction. If a unit vector $\mathbf{e}$ and a vector $\mathbf{U}$ have the same direction, we can write $\mathbf{U}$ as the product of its magnitude $|\mathbf{U}|$ and the unit vector $\mathbf{e}$ (Fig. 2.10),

$$\mathbf{U} = |\mathbf{U}|\mathbf{e}.$$

Any vector **U** *can be regarded as the product of its magnitude and a unit vector that has the same direction as* **U**. Dividing both sides of this equation by $|\mathbf{U}|$ yields

$$\frac{\mathbf{U}}{|\mathbf{U}|} = \mathbf{e},$$

so *dividing any vector by its magnitude yields a unit vector that has the same direction.*

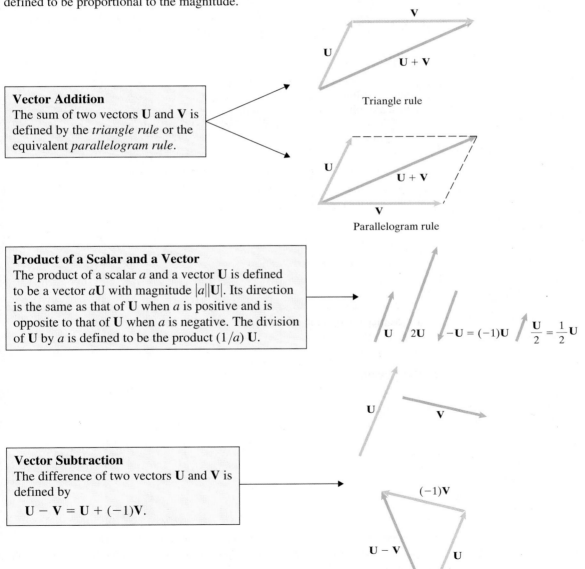

Figure 2.10
Since **U** and **e** have the same direction, the vector **U** equals the product of its magnitude with **e**.

RESULTS

A physical quantity that is completely described by a real number is called a *scalar*. A *vector* has both magnitude and direction and satisfies a defined rule of addition. A vector is represented graphically by an arrow whose length is defined to be proportional to the magnitude.

Vector Addition
The sum of two vectors **U** and **V** is defined by the *triangle rule* or the equivalent *parallelogram rule*.

Triangle rule

Parallelogram rule

Product of a Scalar and a Vector
The product of a scalar *a* and a vector **U** is defined to be a vector $a\mathbf{U}$ with magnitude $|a||\mathbf{U}|$. Its direction is the same as that of **U** when *a* is positive and is opposite to that of **U** when *a* is negative. The division of **U** by *a* is defined to be the product $(1/a)\,\mathbf{U}$.

$\mathbf{U}$ $2\mathbf{U}$ $-\mathbf{U} = (-1)\mathbf{U}$ $\dfrac{\mathbf{U}}{2} = \dfrac{1}{2}\mathbf{U}$

Vector Subtraction
The difference of two vectors **U** and **V** is defined by
$$\mathbf{U} - \mathbf{V} = \mathbf{U} + (-1)\mathbf{V}.$$

$(-1)\mathbf{V}$

$\mathbf{U} - \mathbf{V}$ $\mathbf{U}$

Unit Vectors

A unit vector is a vector whose magnitude is 1.
Any vector **U** can be expressed as |**U**|**e**, where
e is a unit vector with the same direction as **U**.
Dividing a vector **U** by its magnitude yields a
unit vector with the same direction as **U**.

Active Example 2.1	**Vector Operations** (▶ *Related Problem 2.1*)

The magnitudes of the vectors shown are |**U**| = 8 and |**V**| = 3. The vector **V** is
vertical. Graphically determine the magnitude of the vector **U** + 2**V**.

Strategy

By drawing the vectors to scale and applying the triangle rule for addition, we
can measure the magnitude of the vector **U** + 2**V**.

Solution

Drawing the vectors **U** and 2**V**
to scale, place them head to tail.

The measured value of
|**U** + 2**V**| is 13.0.

Practice Problem The magnitudes of the vectors shown are |**U**| = 8 and |**V**| = 3.
The vector **V** is vertical. Graphically determine the magnitude of the vector **U** − 2**V**.

Answer: |**U** − 2**V**| = 5.7.

| **Example 2.2** | **Adding Vectors** (▶ *Related Problem 2.2*) |

Part of the roof of a sports stadium is to be supported by the cables *AB* and *AC*. The forces the cables exert on the pylon to which they are attached are represented by the vectors $\mathbf{F}_{AB}$ and $\mathbf{F}_{AC}$. The magnitudes of the forces are $|\mathbf{F}_{AB}| = 100$ kN and $|\mathbf{F}_{AC}| = 60$ kN. Determine the magnitude and direction of the sum of the forces exerted on the pylon by the cables.

Strategy
By drawing the parallelogram rule for adding the two forces *with the vectors drawn to scale*, we can measure the magnitude and direction of their sum.

Solution
We graphically construct the parallelogram rule for obtaining the sum of the two forces with the lengths of $\mathbf{F}_{AB}$ and $\mathbf{F}_{AC}$ proportional to their magnitudes (Fig. a). By measuring the figure, we estimate the magnitude of the vector $\mathbf{F}_{AB} + \mathbf{F}_{AC}$ to be 155 kN and its direction to be 19° above the horizontal.

(a) Graphical solution.

Critical Thinking
In engineering applications, vector operations are nearly always done analytically. So why is it worthwhile to gain experience with graphical methods? Doing so enhances your intuition about vectors and helps you understand vector operations. Also, sketching out a graphical solution can often help you formulate an analytical solution.

Problems

▶ **2.1** In Active Example 2.1, suppose that the vectors **U** and **V** are reoriented as shown. The vector **V** is vertical. The magnitudes are $|\mathbf{U}| = 8$ and $|\mathbf{V}| = 3$. Graphically determine the magnitude of the vector **U** + 2**V**.

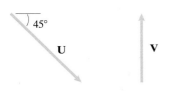

Problem 2.1

▶ **2.2** Suppose that the pylon in Example 2.2 is moved closer to the stadium so that the angle between the forces $\mathbf{F}_{AB}$ and $\mathbf{F}_{AC}$ is 50°. Draw a sketch of the new situation. The magnitudes of the forces are $|\mathbf{F}_{AB}| = 100$ kN and $|\mathbf{F}_{AC}| = 60$ kN. Graphically determine the magnitude and direction of the sum of the forces exerted on the pylon by the cables.

Refer to the following diagram when solving Problems 2.3 through 2.5. The force vectors F$_A$, F$_B$, and F$_C$ lie in the same plane.

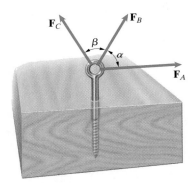

Problems 2.3–2.5

2.3 The magnitude $|\mathbf{F}_A| = 80$ lb and the angle $\alpha = 65°$. The magnitude $|\mathbf{F}_A + \mathbf{F}_B| = 120$ lb. Graphically determine the magnitude of $\mathbf{F}_B$.

2.4 The magnitudes $|\mathbf{F}_A| = 40$ N, $|\mathbf{F}_B| = 50$ N, and $|\mathbf{F}_C| = 40$ N. The angles $\alpha = 50°$ and $\beta = 80°$. Graphically determine the magnitude of $\mathbf{F}_A + \mathbf{F}_B + \mathbf{F}_C$.

2.5 The magnitudes $|\mathbf{F}_A| = |\mathbf{F}_B| = |\mathbf{F}_C| = 100$ lb, and the angle $\alpha = 30°$. Graphically determine the value of the angle β for which the magnitude $|\mathbf{F}_A + \mathbf{F}_B + \mathbf{F}_C|$ is a minimum and the minimum value of $|\mathbf{F}_A + \mathbf{F}_B + \mathbf{F}_C|$.

2.6 The angle $\theta = 50°$. Graphically determine the magnitude of the vector $\mathbf{r}_{AC}$.

Problem 2.6

2.7 The vectors $\mathbf{F}_A$ and $\mathbf{F}_B$ represent the forces exerted on the pulley by the belt. Their magnitudes are $|\mathbf{F}_A| = 80$ N and $|\mathbf{F}_B| = 60$ N. Graphically determine the magnitude of the total force the belt exerts on the pulley.

Problem 2.7

2.8 The sum of the forces $\mathbf{F}_A + \mathbf{F}_B + \mathbf{F}_C = \mathbf{0}$. The magnitude $|\mathbf{F}_A| = 100$ N and the angle $\alpha = 60°$. Graphically determine the magnitudes $|\mathbf{F}_B|$ and $|\mathbf{F}_C|$.

2.9 The sum of the forces $\mathbf{F}_A + \mathbf{F}_B + \mathbf{F}_C = \mathbf{0}$. The magnitudes $|\mathbf{F}_A| = 100$ N and $|\mathbf{F}_B| = 80$ N. Graphically determine the magnitude $|\mathbf{F}_C|$ and the angle α.

Problems 2.8/2.9

2.10 The forces acting on the sailplane are represented by three vectors. The lift **L** and drag **D** are perpendicular. The magnitude of the weight **W** is 500 lb. The sum of the forces **W** + **L** + **D** = **0**. Graphically determine the magnitudes of the lift and drag.

Problem 2.10

2.11 A spherical storage tank is suspended from cables. The tank is subjected to three forces, the forces $\mathbf{F}_A$ and $\mathbf{F}_B$ exerted by the cables and its weight **W**. The weight of the tank is $|\mathbf{W}| = 600$ lb. The vector sum of the forces acting on the tank equals zero. Graphically determine the magnitudes of $\mathbf{F}_A$ and $\mathbf{F}_B$.

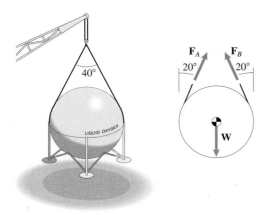

Problem 2.11

2.12 The rope ABC exerts forces $\mathbf{F}_{BA}$ and $\mathbf{F}_{BC}$ of equal magnitude on the block at B. The magnitude of the total force exerted on the block by the two forces is 200 lb. Graphically determine $|\mathbf{F}_{BA}|$.

Problem 2.12

2.13 Two snowcats tow an emergency shelter to a new location near McMurdo Station, Antarctica. (The top view is shown. The cables are horizontal.) The total force $\mathbf{F}_A + \mathbf{F}_B$ exerted on the shelter is in the direction parallel to the line L and its magnitude is 400 lb. Graphically determine the magnitudes of $\mathbf{F}_A$ and $\mathbf{F}_B$.

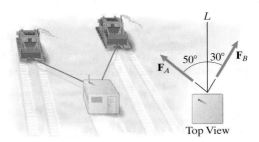

Problem 2.13

2.14 A surveyor determines that the horizontal distance from A to B is 400 m and the horizontal distance from A to C is 600 m. Graphically determine the magnitude of the vector $\mathbf{r}_{BC}$ and the angle α.

Problem 2.14

2.15 The vector **r** extends from point A to the midpoint between points B and C. Prove that

$$\mathbf{r} = \tfrac{1}{2}(\mathbf{r}_{AB} + \mathbf{r}_{AC}).$$

Problem 2.15

2.16 By drawing sketches of the vectors, explain why

$$\mathbf{U} + (\mathbf{V} + \mathbf{W}) = (\mathbf{U} + \mathbf{V}) + \mathbf{W}.$$

2.2 Components in Two Dimensions

BACKGROUND

Vectors are much easier to work with when they are expressed in terms of mutually perpendicular vector components. Here we explain how to express vectors in cartesian components and give examples of vector manipulations using components.

Consider the vector **U** in Fig. 2.11a. By placing a cartesian coordinate system so that the vector **U** is parallel to the x–y plane, we can write it as the sum of perpendicular *vector components* $\mathbf{U}_x$ and $\mathbf{U}_y$ that are parallel to the x and y axes (Fig. 2.11b):

$$\mathbf{U} = \mathbf{U}_x + \mathbf{U}_y.$$

Then by introducing a unit vector **i** defined to point in the direction of the positive x axis and a unit vector **j** defined to point in the direction of the positive y axis (Fig. 2.11c), we can express the vector **U** in the form

$$\mathbf{U} = U_x \mathbf{i} + U_y \mathbf{j}. \tag{2.7}$$

The scalars U_x and U_y are called *scalar components of* **U**. *When we refer simply to the components of a vector, we will mean its scalar components.* We will refer to U_x and U_y as the x and y components of **U**.

The components of a vector specify both its direction relative to the cartesian coordinate system and its magnitude. From the right triangle formed by the vector **U** and its vector components (Fig. 2.11c), we see that the magnitude of **U** is given in terms of its components by the Pythagorean theorem:

$$|\mathbf{U}| = \sqrt{U_x^2 + U_y^2}. \tag{2.8}$$

With this equation the magnitude of a vector can be determined when its components are known.

Manipulating Vectors in Terms of Components

The sum of two vectors **U** and **V** in terms of their components is

$$\begin{aligned}\mathbf{U} + \mathbf{V} &= (U_x \mathbf{i} + U_y \mathbf{j}) + (V_x \mathbf{i} + V_y \mathbf{j}) \\ &= (U_x + V_x)\mathbf{i} + (U_y + V_y)\mathbf{j}.\end{aligned} \tag{2.9}$$

(a)

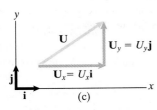

Figure 2.11
(**a**) A vector **U**.
(**b**) The vector components $\mathbf{U}_x$ and $\mathbf{U}_y$.
(**c**) The vector components can be expressed in terms of **i** and **j**.

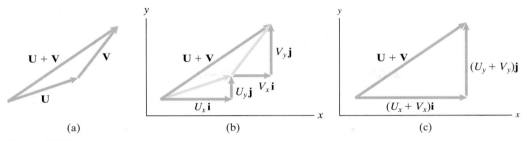

Figure 2.12
(**a**) The sum of **U** and **V**. (**b**) The vector components of **U** and **V**. (**c**) The sum of the components in each coordinate direction equals the component of **U** + **V** in that direction.

The components of **U** + **V** are the sums of the components of the vectors **U** and **V**. Notice that in obtaining this result we used Eqs. (2.2), (2.4), and (2.5).

It is instructive to derive Eq. (2.9) graphically. The summation of **U** and **V** is shown in Fig. 2.12a. In Fig. 2.12b we introduce a coordinate system and show the components **U** and **V**. In Fig. 2.12c we add the x and y components, obtaining Eq. (2.9).

The product of a number a and a vector **U** in terms of the components of **U** is

$$a\mathbf{U} = a(U_x\mathbf{i} + U_y\mathbf{j}) = aU_x\mathbf{i} + aU_y\mathbf{j}.$$

The component of $a\mathbf{U}$ in each coordinate direction equals the product of a and the component of **U** in that direction. We used Eqs. (2.3) and (2.5) to obtain this result.

Position Vectors in Terms of Components

We can express the position vector of a point relative to another point in terms of the cartesian coordinates of the points. Consider point A with coordinates (x_A, y_A) and point B with coordinates (x_B, y_B). Let $\mathbf{r}_{AB}$ be the vector that specifies the position of B relative to A (Fig. 2.13a). That is, we denote the vector *from* a point A *to* a point B by $\mathbf{r}_{AB}$. We see from Fig. 2.13b that $\mathbf{r}_{AB}$ is given in terms of the coordinates of points A and B by

$$\mathbf{r}_{AB} = (x_B - x_A)\mathbf{i} + (y_B - y_A)\mathbf{j}. \tag{2.10}$$

Notice that the x component of the position vector from a point A to a point B is obtained by subtracting the x coordinate of A from the x coordinate of B, and the y component is obtained by subtracting the y coordinate of A from the y coordinate of B.

(a)

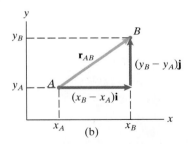

Figure 2.13
(**a**) Two points A and B and the position vector $\mathbf{r}_{AB}$ from A to B.
(**b**) The components of $\mathbf{r}_{AB}$ can be determined from the coordinates of points A and B.

RESULTS

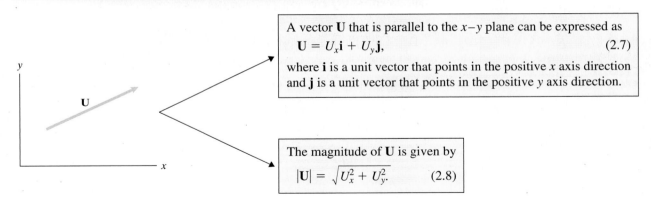

A vector **U** that is parallel to the x–y plane can be expressed as
$$\mathbf{U} = U_x\mathbf{i} + U_y\mathbf{j}, \tag{2.7}$$
where **i** is a unit vector that points in the positive x axis direction and **j** is a unit vector that points in the positive y axis direction.

The magnitude of **U** is given by
$$|\mathbf{U}| = \sqrt{U_x^2 + U_y^2}. \tag{2.8}$$

Manipulating Vectors in Terms of Components

Vector addition (or subtraction) and multiplication of a vector by a number can be carried out in terms of components.

$$
\begin{aligned}
\mathbf{U} + \mathbf{V} &= (U_x\mathbf{i} + U_y\mathbf{j}) + (V_x\mathbf{i} + V_y\mathbf{j}) \\
&= (U_x + V_x)\mathbf{i} + (U_y + V_y)\mathbf{j}, \quad (2.9) \\
a\mathbf{U} &= a(U_x\mathbf{i} + U_y\mathbf{j}) \\
&= aU_x\mathbf{i} + aU_y\mathbf{j}.
\end{aligned}
$$

Position Vectors in Terms of Components

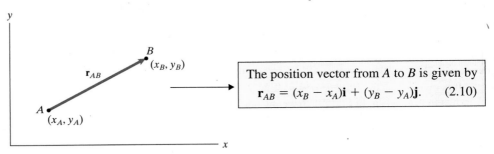

The position vector from A to B is given by

$$\mathbf{r}_{AB} = (x_B - x_A)\mathbf{i} + (y_B - y_A)\mathbf{j}. \quad (2.10)$$

Active Example 2.3 **Determining Components** (▶ *Related Problem 2.31*)

The cable from point A to point B exerts a 900-N force on the top of the television transmission tower that is represented by the vector $\mathbf{F}$. Express $\mathbf{F}$ in terms of components using the coordinate system shown.

Strategy

We will determine the components of the vector $\mathbf{F}$ in two ways. In the first method, we will determine the angle between $\mathbf{F}$ and the y axis and use trigonometry to determine the components. In the second method, we will use the given slope of the cable AB and apply similar triangles to determine the components of $\mathbf{F}$.

Solution

First Method

Determine the angle between $\mathbf{F}$ and the y axis:

$$\alpha = \arctan\left(\frac{40}{80}\right) = 26.6°.$$

Use trigonometry to determine **F** in terms of its components:

$$\mathbf{F} = |\mathbf{F}|\sin\alpha\,\mathbf{i} - |\mathbf{F}|\cos\alpha\,\mathbf{j}$$
$$= 900\sin 26.6°\,\mathbf{i} - 900\cos 26.6°\,\mathbf{j}\ (\text{N})$$
$$= 402\mathbf{i} - 805\mathbf{j}\ (\text{N}).$$

Second Method

Using the given dimensions, calculate the distance from A to B:

$$\sqrt{(40\ \text{m})^2 + (80\ \text{m})^2} = 89.4\ \text{m}.$$

Use similar triangles to determine the components of **F**:

$$\frac{|F_x|}{|\mathbf{F}|} = \frac{40\ \text{m}}{89.4\ \text{m}} \quad \text{and} \quad \frac{|F_y|}{|\mathbf{F}|} = \frac{80\ \text{m}}{89.4\ \text{m}},$$

so

$$\mathbf{F} = \frac{40}{89.4}(900\ \text{N})\mathbf{i} - \frac{80}{89.4}(900\ \text{N})\mathbf{j}$$
$$= 402\mathbf{i} - 805\mathbf{j}\ (\text{N}).$$

Practice Problem The cable from point A to point B exerts a 900-N force on the top of the television transmission tower that is represented by the vector **F**. Suppose that you change the placement of point B so that the magnitude of the y component of **F** is three times the magnitude of the x component of **F**. Express **F** in terms of its components. How far along the x axis from the origin of the coordinate system should B be placed?

Answer: $\mathbf{F} = 285\mathbf{i} - 854\mathbf{j}$ (N). Place point B at 26.7 m from the origin.

| Example 2.4 | Determining Components in Terms of an Angle (▶ *Related Problem 2.33*) |

Hydraulic cylinders are used to exert forces in many mechanical devices. The force is exerted by pressurized liquid (hydraulic fluid) pushing against a piston within the cylinder. The hydraulic cylinder AB exerts a 4000-lb force $\mathbf{F}$ on the bed of the dump truck at B. Express $\mathbf{F}$ in terms of components using the coordinate system shown.

Strategy
When the direction of a vector is specified by an angle, as in this example, we can determine the values of the components from the right triangle formed by the vector and its components.

Solution
We draw the vector $\mathbf{F}$ and its vector components in Fig. a. From the resulting right triangle, we see that the magnitude of $\mathbf{F}_x$ is

$$|\mathbf{F}_x| = |\mathbf{F}| \cos 30° = (4000 \text{ lb}) \cos 30° = 3460 \text{ lb}.$$

$\mathbf{F}_x$ points in the negative x direction, so

$$\mathbf{F}_x = -3460\mathbf{i} \text{ (lb)}.$$

The magnitude of $\mathbf{F}_y$ is

$$|\mathbf{F}_y| = |\mathbf{F}| \sin 30° = (4000 \text{ lb}) \sin 30° = 2000 \text{ lb}.$$

The vector component $\mathbf{F}_y$ points in the positive y direction, so

$$\mathbf{F}_y = 2000\mathbf{j} \text{ (lb)}.$$

The vector $\mathbf{F}$, in terms of its components, is

$$\mathbf{F} = \mathbf{F}_x + \mathbf{F}_y = -3460\mathbf{i} + 2000\mathbf{j} \text{ (lb)}.$$

The x component of $\mathbf{F}$ is -3460 lb, and the y component is 2000 lb.

(a) The force $\mathbf{F}$ and its components form a right triangle.

Critical Thinking
When you have determined the components of a given vector, you should make sure they appear reasonable. In this example you can see from the vector's direction that the x component should be negative and the y component positive. You can also make sure that the components yield the correct magnitude. In this example,

$$|\mathbf{F}| = \sqrt{(-3460 \text{ lb})^2 + (2000 \text{ lb})^2} = 4000 \text{ lb}.$$

| **Example 2.5** | **Determining an Unknown Vector Magnitude** (▶ *Related Problem 2.47*) |

The cables A and B exert forces $\mathbf{F}_A$ and $\mathbf{F}_B$ on the hook. The magnitude of $\mathbf{F}_A$ is 100 lb. The tension in cable B has been adjusted so that the total force $\mathbf{F}_A + \mathbf{F}_B$ is perpendicular to the wall to which the hook is attached.

(a) What is the magnitude of $\mathbf{F}_B$?

(b) What is the magnitude of the total force exerted on the hook by the two cables?

Strategy
The vector sum of the two forces is perpendicular to the wall, so the sum of the components parallel to the wall equals zero. From this condition we can obtain an equation for the magnitude of $\mathbf{F}_B$.

Solution
(a) In terms of the coordinate system shown in Fig. a, the components of $\mathbf{F}_A$ and $\mathbf{F}_B$ are

$$\mathbf{F}_A = |\mathbf{F}_A| \sin 40°\mathbf{i} + |\mathbf{F}_A| \cos 40°\mathbf{j},$$

$$\mathbf{F}_B = |\mathbf{F}_B| \sin 20°\mathbf{i} - |\mathbf{F}_B| \cos 20°\mathbf{j}.$$

The total force is

$$\mathbf{F}_A + \mathbf{F}_B = (|\mathbf{F}_A| \sin 40° + |\mathbf{F}_B| \sin 20°)\mathbf{i}$$
$$+ (|\mathbf{F}_A| \cos 40° - |\mathbf{F}_B| \cos 20°)\mathbf{j}.$$

Now we set the component of the total force parallel to the wall (the y component) equal to zero:

$$|\mathbf{F}_A| \cos 40° - |\mathbf{F}_B| \cos 20° = 0,$$

We thus obtain an equation for the magnitude of $\mathbf{F}_B$:

$$|\mathbf{F}_B| = \frac{|\mathbf{F}_A| \cos 40°}{\cos 20°} = \frac{(100 \text{ lb})\cos 40°}{\cos 20°} = 81.5 \text{ lb.}$$

(b) Since we now know the magnitude of $\mathbf{F}_B$, we can determine the total force acting on the hook:

$$\mathbf{F}_A + \mathbf{F}_B = (|\mathbf{F}_A| \sin 40° + |\mathbf{F}_B| \sin 20°)\mathbf{i}$$
$$= [(100 \text{ lb})\sin 40° + (81.5 \text{ lb})\sin 20°]\mathbf{i} = 92.2\mathbf{i} \text{ (lb).}$$

The magnitude of the total force is 92.2 lb.

Critical Thinking
We can obtain the solution to (a) in a less formal way. If the component of the total force parallel to the wall is zero, we see in Fig. a that the magnitude of the vertical component of $\mathbf{F}_A$ must equal the magnitude of the vertical component of $\mathbf{F}_B$:

$$|\mathbf{F}_A| \cos 40° = |\mathbf{F}_B| \cos 20°.$$

Therefore the magnitude of $\mathbf{F}_B$ is

$$|\mathbf{F}_B| = \frac{|\mathbf{F}_A| \cos 40°}{\cos 20°} = \frac{(100 \text{ lb}) \cos 40°}{\cos 20°} = 81.5 \text{ lb.}$$

(a) Resolving $\mathbf{F}_A$ and $\mathbf{F}_B$ into components parallel and perpendicular to the wall.

Problems

2.17 A force $\mathbf{F} = 40\mathbf{i} - 20\mathbf{j}$ (N). What is its magnitude $|\mathbf{F}|$?

Strategy: The magnitude of a vector in terms of its components is given by Eq. (2.8).

2.18 An engineer estimating the components of a force $\mathbf{F} = F_x\mathbf{i} + F_y\mathbf{j}$ acting on a bridge abutment has determined that $F_x = 130$ MN, $|\mathbf{F}| = 165$ MN, and F_y is negative. What is F_y?

2.19 A support is subjected to a force $\mathbf{F} = F_x\mathbf{i} + 80\mathbf{j}$ (N). If the support will safely support a force of magnitude 100 N, what is the allowable range of values of the component F_x?

2.20 If $\mathbf{F}_A = 600\mathbf{i} - 800\mathbf{j}$ (kip) and $\mathbf{F}_B = 200\mathbf{i} - 200\mathbf{j}$ (kip), what is the magnitude of the force $\mathbf{F} = \mathbf{F}_A - 2\mathbf{F}_B$?

2.21 The forces acting on the sailplane are its weight $\mathbf{W} = -500\mathbf{j}$ (lb), the drag $\mathbf{D} = -200\mathbf{i} + 100\mathbf{j}$ (lb), and the lift $\mathbf{L}$. The sum of the forces $\mathbf{W} + \mathbf{L} + \mathbf{D} = \mathbf{0}$. Determine the components and the magnitude of $\mathbf{L}$.

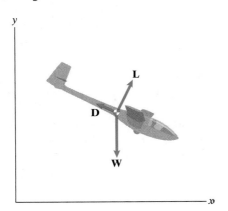

Problem 2.21

2.22 Two perpendicular vectors $\mathbf{U}$ and $\mathbf{V}$ lie in the x–y plane. The vector $\mathbf{U} = 6\mathbf{i} - 8\mathbf{j}$ and $|\mathbf{V}| = 20$. What are the components of $\mathbf{V}$?

2.23 A fish exerts a 10-lb force on the line that is represented by the vector $\mathbf{F}$. Express $\mathbf{F}$ in terms of components using the coordinate system shown.

Problem 2.23

2.24 A man exerts a 60-lb force $\mathbf{F}$ to push a crate onto a truck. (a) Express $\mathbf{F}$ in terms of components using the coordinate system shown. (b) The weight of the crate is 100 lb. Determine the magnitude of the sum of the forces exerted by the man and the crate's weight.

Problem 2.24

2.25 The missile's engine exerts a 260-kN force $\mathbf{F}$. (a) Express $\mathbf{F}$ in terms of components using the coordinate system shown. (b) The mass of the missile is 8800 kg. Determine the magnitude of the sum of the forces exerted by the engine and the missile's weight.

Problem 2.25

2.26 For the truss shown, express the position vector $\mathbf{r}_{AD}$ from point A to point D in terms of components. Use your result to determine the distance from point A to point D.

Problem 2.26

2.27 The points $A, B, \ldots$ are the joints of the hexagonal structural element. Let $\mathbf{r}_{AB}$ be the position vector from joint A to joint B, $\mathbf{r}_{AC}$ the position vector from joint A to joint C, and so forth. Determine the components of the vectors $\mathbf{r}_{AC}$ and $\mathbf{r}_{AF}$.

2.28 Determine the components of the vector $\mathbf{r}_{AB} - \mathbf{r}_{BC}$.

Problems 2.27/2.28

2.29 The coordinates of point A are $(1.8, 3.0)$ ft. The y coordinate of point B is 0.6 ft. The vector $\mathbf{r}_{AB}$ has the same direction as the unit vector $\mathbf{e}_{AB} = 0.616\mathbf{i} - 0.788\mathbf{j}$. What are the components of $\mathbf{r}_{AB}$?

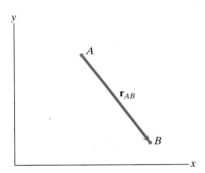

Problem 2.29

2.30 (a) Express the position vector from point A of the front-end loader to point B in terms of components.

(b) Express the position vector from point B to point C in terms of components.

(c) Use the results of (a) and (b) to determine the distance from point A to point C.

Problem 2.30

▶ **2.31** In Active Example 2.3, the cable AB exerts a 900-N force on the top of the tower. Suppose that the attachment point B is moved in the horizontal direction farther from the tower, and assume that the magnitude of the force $\mathbf{F}$ the cable exerts on the top of the tower is proportional to the length of the cable. (a) What is the distance from the tower to point B if the magnitude of the force is 1000 N? (b) Express the 1000-N force $\mathbf{F}$ in terms of components using the coordinate system shown.

2.32 Determine the position vector $\mathbf{r}_{AB}$ in terms of its components if (a) $\theta = 30°$; (b) $\theta = 225°$.

Problem 2.32

▶ **2.33** In Example 2.4, the coordinates of the fixed point A are $(17, 1)$ ft. The driver lowers the bed of the truck into a new position in which the coordinates of point B are $(9, 3)$ ft. The magnitude of the force $\mathbf{F}$ exerted on the bed by the hydraulic cylinder when the bed is in the new position is 4800 lb. Draw a sketch of the new situation. Express $\mathbf{F}$ in terms of components.

2.34 A surveyor measures the location of point A and determines that $\mathbf{r}_{OA} = 400\mathbf{i} + 800\mathbf{j}$ (m). He wants to determine the location of a point B so that $|\mathbf{r}_{AB}| = 400$ m and $|\mathbf{r}_{OA} + \mathbf{r}_{AB}| = 1200$ m. What are the cartesian coordinates of point B?

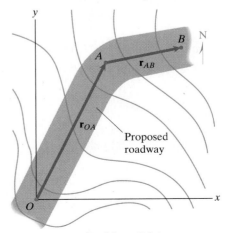

Problem 2.34

2.35 The magnitude of the position vector $\mathbf{r}_{BA}$ from point B to point A is 6 m and the magnitude of the position vector $\mathbf{r}_{CA}$ from point C to point A is 4 m. What are the components of $\mathbf{r}_{BA}$?

2.36 In Problem 2.35, determine the components of a unit vector $\mathbf{e}_{CA}$ that points from point C toward point A.

Strategy: Determine the components of $\mathbf{r}_{CA}$ and then divide the vector $\mathbf{r}_{CA}$ by its magnitude.

Problems 2.35/2.36

2.37 The x and y coordinates of points A, B, and C of the sailboat are shown.

(a) Determine the components of a unit vector that is parallel to the forestay AB and points from A toward B.

(b) Determine the components of a unit vector that is parallel to the backstay BC and points from C toward B.

Problem 2.37

2.38 The length of the bar AB is 0.6 m. Determine the components of a unit vector $\mathbf{e}_{AB}$ that points from point A toward point B.

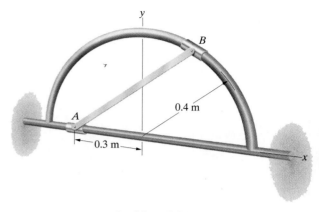

Problem 2.38

2.39 Determine the components of a unit vector that is parallel to the hydraulic actuator BC and points from B toward C.

2.40 The hydraulic actuator BC exerts a 1.2-kN force $\mathbf{F}$ on the joint at C that is parallel to the actuator and points from B toward C. Determine the components of $\mathbf{F}$.

Problems 2.39/2.40

2.41 A surveyor finds that the length of the line *OA* is 1500 m and the length of the line *OB* is 2000 m.

(a) Determine the components of the position vector from point *A* to point *B*.

(b) Determine the components of a unit vector that points from point *A* toward point *B*.

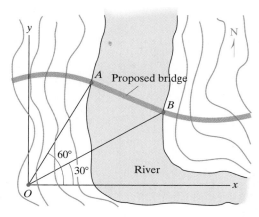

Problem 2.41

2.42 The magnitudes of the forces exerted by the cables are $|\mathbf{T}_1| = 2800$ lb, $|\mathbf{T}_2| = 3200$ lb, $|\mathbf{T}_3| = 4000$ lb, and $|\mathbf{T}_4| = 5000$ lb. What is the magnitude of the total force exerted by the four cables?

2.43 The tensions in the four cables are equal: $|\mathbf{T}_1| = |\mathbf{T}_2| = |\mathbf{T}_3| = |\mathbf{T}_4| = T$. Determine the value of T so that the four cables exert a total force of 12,500-lb magnitude on the support.

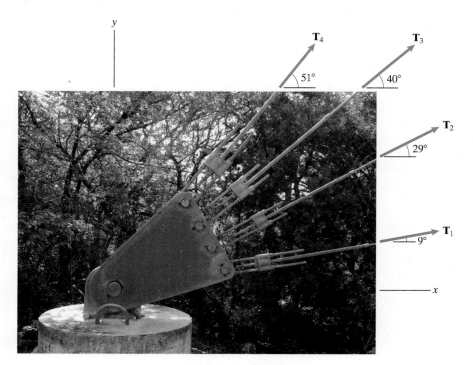

Problems 2.42/2.43

2.44 The rope ABC exerts forces $\mathbf{F}_{BA}$ and $\mathbf{F}_{BC}$ on the block at B. Their magnitudes are equal: $|\mathbf{F}_{BA}| = |\mathbf{F}_{BC}|$. The magnitude of the total force exerted on the block at B by the rope is $|\mathbf{F}_{BA} + \mathbf{F}_{BC}| = 920$ N. Determine $|\mathbf{F}_{BA}|$ by expressing the forces $\mathbf{F}_{BA}$ and $\mathbf{F}_{BC}$ in terms of components.

Problem 2.44

2.45 The magnitude of the horizontal force $\mathbf{F}_1$ is 5 kN and $\mathbf{F}_1 + \mathbf{F}_2 + \mathbf{F}_3 = \mathbf{0}$. What are the magnitudes of $\mathbf{F}_2$ and $\mathbf{F}_3$?

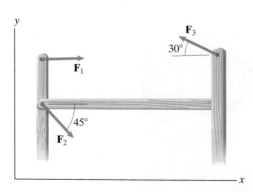

Problem 2.45

2.46 Four groups engage in a tug-of-war. The magnitudes of the forces exerted by groups B, C, and D are $|\mathbf{F}_B| = 800$ lb, $|\mathbf{F}_C| = 1000$ lb, and $|\mathbf{F}_D| = 900$ lb. If the vector sum of the four forces equals zero, what is the magnitude of $\mathbf{F}_A$ and the angle α?

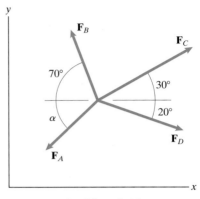

Problem 2.46

▶ **2.47** In Example 2.5, suppose that the attachment point of cable A is moved so that the angle between the cable and the wall increases from 40° to 55°. Draw a sketch showing the forces exerted on the hook by the two cables. If you want the total force $\mathbf{F}_A + \mathbf{F}_B$ to have a magnitude of 200 lb and be in the direction perpendicular to the wall, what are the necessary magnitudes of $\mathbf{F}_A$ and $\mathbf{F}_B$?

2.48 The bracket must support the two forces shown, where $|\mathbf{F}_1| = |\mathbf{F}_2| = 2$ kN. An engineer determines that the bracket will safely support a total force of magnitude 3.5 kN in any direction. Assume that $0 \leq \alpha \leq 90°$. What is the safe range of the angle α?

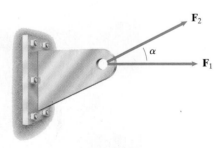

Problem 2.48

2.49 The figure shows three forces acting on a joint of a structure. The magnitude of $\mathbf{F}_C$ is 60 kN, and $\mathbf{F}_A + \mathbf{F}_B + \mathbf{F}_C = \mathbf{0}$. What are the magnitudes of $\mathbf{F}_A$ and $\mathbf{F}_B$?

Problem 2.49

2.50 Four coplanar forces act on a beam. The forces $\mathbf{F}_B$ and $\mathbf{F}_C$ are vertical. The vector sum of the forces is zero. The magnitudes $|\mathbf{F}_B| = 10$ kN and $|\mathbf{F}_C| = 5$ kN. Determine the magnitudes of $\mathbf{F}_A$ and $\mathbf{F}_D$.

Problem 2.50

2.51 Six forces act on a beam that forms part of a building's frame. The vector sum of the forces is zero. The magnitudes $|\mathbf{F}_B| = |\mathbf{F}_E| = 20$ kN, $|\mathbf{F}_C| = 16$ kN, and $|\mathbf{F}_D| = 9$ kN. Determine the magnitudes of $\mathbf{F}_A$ and $\mathbf{F}_G$.

Problem 2.51

2.52 The total weight of the man and parasail is $|\mathbf{W}| = 230$ lb. The drag force $\mathbf{D}$ is perpendicular to the lift force $\mathbf{L}$. If the vector sum of the three forces is zero, what are the magnitudes of $\mathbf{L}$ and $\mathbf{D}$?

Problem 2.52

2.53 The three forces acting on the car are shown. The force $\mathbf{T}$ is parallel to the x axis and the magnitude of the force $\mathbf{W}$ is 14 kN. If $\mathbf{T} + \mathbf{W} + \mathbf{N} = \mathbf{0}$, what are the magnitudes of the forces $\mathbf{T}$ and $\mathbf{N}$?

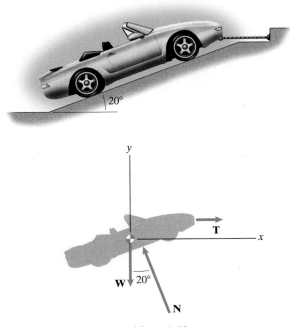

Problem 2.53

2.54 The cables A, B, and C help support a pillar that forms part of the supports of a structure. The magnitudes of the forces exerted by the cables are equal: $|\mathbf{F}_A| = |\mathbf{F}_B| = |\mathbf{F}_C|$. The magnitude of the vector sum of the three forces is 200 kN. What is $|\mathbf{F}_A|$?

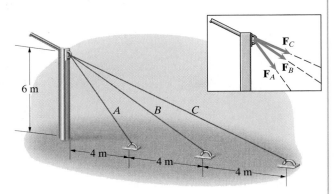

Problem 2.54

2.55 The total force exerted on the top of the mast B by the sailboat's forestay AB and backstay BC is $180\mathbf{i} - 820\mathbf{j}$ (N). What are the magnitudes of the forces exerted at B by the cables AB and BC?

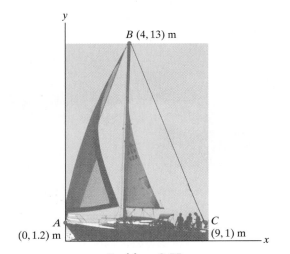

Problem 2.55

2.56 The structure shown forms part of a truss designed by an architectural engineer to support the roof of an orchestra shell. The members AB, AC, and AD exert forces $\mathbf{F}_{AB}$, $\mathbf{F}_{AC}$, and $\mathbf{F}_{AD}$ on the joint A. The magnitude $|\mathbf{F}_{AB}| = 4$ kN. If the vector sum of the three forces equals zero, what are the magnitudes of $\mathbf{F}_{AC}$ and $\mathbf{F}_{AD}$?

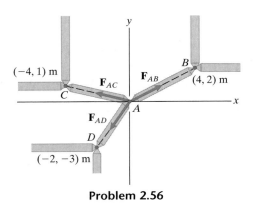

Problem 2.56

2.57 The distance $s = 45$ in.

(a) Determine the unit vector $\mathbf{e}_{BA}$ that points from B toward A.

(b) Use the unit vector you obtained in (a) to determine the coordinates of the collar C.

2.58 Determine the x and y coordinates of the collar C as functions of the distance s.

Problems 2.57/2.58

2.59 The position vector **r** goes from point A to a point on the straight line between B and C. Its magnitude is $|\mathbf{r}| = 6$ ft. Express **r** in terms of components.

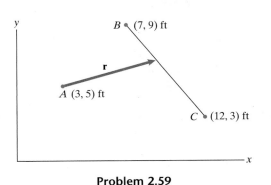

Problem 2.59

2.60 Let **r** be the position vector from point C to the point that is a distance s meters from point A along the straight line between A and B. Express **r** in terms of components. (Your answer will be in terms of s.)

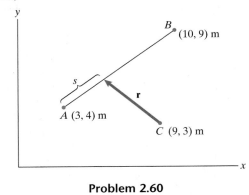

Problem 2.60

2.3 Components in Three Dimensions

BACKGROUND

Many engineering applications require vectors to be expressed in terms of components in a three-dimensional coordinate system. In this section we explain this technique and demonstrate vector operations in three dimensions.

We first review how to draw objects in three dimensions. Consider a three-dimensional object such as a cube. If we draw the cube as it appears when the point of view is perpendicular to one of its faces, we obtain Fig. 2.14a. In this view, the cube appears two dimensional. The dimension perpendicular to the page cannot be seen. To remedy this, we move the point of view upward and to the right, obtaining Fig. 2.14b. In this *oblique* view, the third dimension is visible. The hidden edges of the cube are shown as dashed lines.

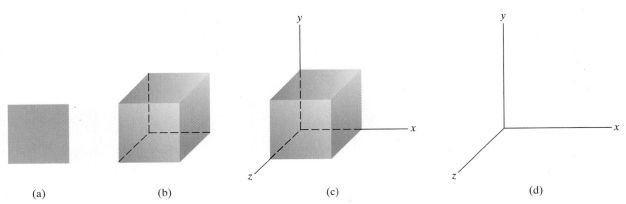

(a)   (b)   (c)   (d)

Figure 2.14
(**a**) A cube viewed with the line of sight perpendicular to a face.
(**b**) An oblique view of the cube.
(**c**) A cartesian coordinate system aligned with the edges of the cube.
(**d**) Three-dimensional representation of the coordinate system.

Figure 2.15
Recognizing a right-handed coordinate system.

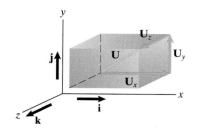

Figure 2.16
A vector **U** and its vector components.

We can use this approach to draw three-dimensional coordinate systems. In Fig. 2.14c we align the x, y, and z axes of a three-dimensional cartesian coordinate system with the edges of the cube. The three-dimensional representation of the coordinate system alone is shown in Fig. 2.14d. The coordinate system shown is said to be *right handed*. If the fingers of the right hand are pointed in the direction of the positive x axis and then bent (as in preparing to make a fist) toward the positive y axis, the thumb points in the direction of the positive z axis (Fig. 2.15). Otherwise, the coordinate system is left handed. Because some equations used in mathematics and engineering do not yield correct results when they are applied using a left-handed coordinate system, we use only right-handed coordinate systems.

We can express a vector **U** in terms of vector components $\mathbf{U}_x$, $\mathbf{U}_y$, and $\mathbf{U}_z$ parallel to the x, y, and z axes, respectively (Fig. 2.16), as

$$\mathbf{U} = \mathbf{U}_x + \mathbf{U}_y + \mathbf{U}_z. \tag{2.11}$$

(We have drawn a box around the vector to help in visualizing the directions of the vector components.) By introducing unit vectors **i**, **j**, and **k** that point in the positive x, y, and z directions, we can express **U** in terms of scalar components as

$$\mathbf{U} = U_x\mathbf{i} + U_y\mathbf{j} + U_z\mathbf{k}. \tag{2.12}$$

We will refer to the scalars U_x, U_y, and U_z as the x, y, and z components of **U**.

Magnitude of a Vector in Terms of Components

Consider a vector **U** and its vector components (Fig. 2.17a). From the right triangle formed by the vectors $\mathbf{U}_y$, $\mathbf{U}_z$, and their sum $\mathbf{U}_y + \mathbf{U}_z$ (Fig. 2.17b), we can see that

$$|\mathbf{U}_y + \mathbf{U}_z|^2 = |\mathbf{U}_y|^2 + |\mathbf{U}_z|^2. \tag{2.13}$$

The vector **U** is the sum of the vectors $\mathbf{U}_x$ and $\mathbf{U}_y + \mathbf{U}_z$. These three vectors form a right triangle (Fig. 2.17c), from which we obtain

$$|\mathbf{U}|^2 = |\mathbf{U}_x|^2 + |\mathbf{U}_y + \mathbf{U}_z|^2.$$

Substituting Eq. (2.13) into this result yields the equation

$$|\mathbf{U}|^2 = |\mathbf{U}_x|^2 + |\mathbf{U}_y|^2 + |\mathbf{U}_z|^2 = U_x^2 + U_y^2 + U_z^2.$$

(a) (b) (c)

Figure 2.17
(**a**) A vector **U** and its vector components.
(**b**) The right triangle formed by the vectors $\mathbf{U}_y$, $\mathbf{U}_z$, and $\mathbf{U}_y + \mathbf{U}_z$.
(**c**) The right triangle formed by the vectors **U**, $\mathbf{U}_x$, and $\mathbf{U}_y + \mathbf{U}_z$.

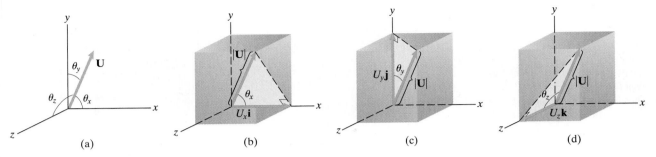

Figure 2.18
(a) A vector $\mathbf{U}$ and the angles θ_x, θ_y, and θ_z.
(b)–(d) The angles θ_x, θ_y, and θ_z and the vector components of $\mathbf{U}$.

Thus, the magnitude of a vector $\mathbf{U}$ is given in terms of its components in three dimensions by

$$|\mathbf{U}| = \sqrt{U_x^2 + U_y^2 + U_z^2}. \qquad (2.14)$$

Direction Cosines

We described the direction of a vector relative to a two-dimensional cartesian coordinate system by specifying the angle between the vector and one of the coordinate axes. One of the ways we can describe the direction of a vector in three dimensions is by specifying the angles θ_x, θ_y, and θ_z between the vector and the positive coordinate axes (Fig. 2.18a).

In Figs. 2.18b–d, we demonstrate that the components of the vector $\mathbf{U}$ are respectively given in terms of the angles θ_x, θ_y, and θ_z, by

$$U_x = |\mathbf{U}| \cos \theta_x, \quad U_y = |\mathbf{U}| \cos \theta_y, \quad U_z = |\mathbf{U}| \cos \theta_z. \qquad (2.15)$$

The quantities $\cos \theta_x$, $\cos \theta_y$, and $\cos \theta_z$ are called the *direction cosines* of $\mathbf{U}$. The direction cosines of a vector are not independent. If we substitute Eqs. (2.15) into Eq. (2.14), we find that the direction cosines satisfy the relation

$$\cos^2 \theta_x + \cos^2 \theta_y + \cos^2 \theta_z = 1. \qquad (2.16)$$

Suppose that $\mathbf{e}$ is a unit vector with the same direction as $\mathbf{U}$, so that

$$\mathbf{U} = |\mathbf{U}|\mathbf{e}.$$

In terms of components, this equation is

$$U_x\mathbf{i} + U_y\mathbf{j} + U_z\mathbf{k} = |\mathbf{U}|(e_x\mathbf{i} + e_y\mathbf{j} + e_z\mathbf{k}).$$

Thus the relations between the components of $\mathbf{U}$ and $\mathbf{e}$ are

$$U_x = |\mathbf{U}|e_x, \quad U_y = |\mathbf{U}|e_y, \quad U_z = |\mathbf{U}|e_z.$$

By comparing these equations to Eqs. (2.15), we see that

$$\cos \theta_x = e_x, \quad \cos \theta_y = e_y, \quad \cos \theta_z = e_z.$$

The direction cosines of a vector $\mathbf{U}$ are the components of a unit vector with the same direction as $\mathbf{U}$.

Figure 2.19
(**a**) The position vector from point A to point B.
(**b**) The components of $\mathbf{r}_{AB}$ can be determined from the coordinates of points A and B.

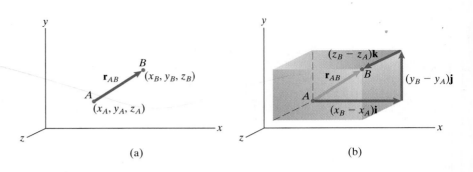

(a)

(b)

Position Vectors in Terms of Components

Generalizing the two-dimensional case, we consider a point A with coordinates (x_A, y_A, z_A) and a point B with coordinates (x_B, y_B, z_B). The position vector $\mathbf{r}_{AB}$ from A to B, shown in Fig. 2.19a, is given in terms of the coordinates of A and B by

$$\mathbf{r}_{AB} = (x_B - x_A)\mathbf{i} + (y_B - y_A)\mathbf{j} + (z_B - z_A)\mathbf{k}. \tag{2.17}$$

The components are obtained by subtracting the coordinates of point A from the coordinates of point B (Fig. 2.19b).

Components of a Vector Parallel to a Given Line

In three-dimensional applications, the direction of a vector is often defined by specifying the coordinates of two points on a line that is parallel to the vector. This information can be used to determine the components of the vector.

Suppose that we know the coordinates of two points A and B on a line parallel to a vector $\mathbf{U}$ (Fig. 2.20a). We can use Eq. (2.17) to determine the

(a)

(b)

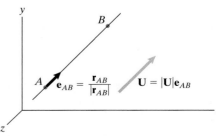

(c)

Figure 2.20
(**a**) Two points A and B on a line parallel to $\mathbf{U}$.
(**b**) The position vector from A to B.
(**c**) The unit vector $\mathbf{e}_{AB}$ that points from A toward B.

position vector $\mathbf{r}_{AB}$ from A to B (Fig. 2.20b). We can divide $\mathbf{r}_{AB}$ by its magnitude to obtain a unit vector $\mathbf{e}_{AB}$ that points from A toward B (Fig. 2.20c). Since $\mathbf{e}_{AB}$ has the same direction as $\mathbf{U}$, we can determine $\mathbf{U}$ in terms of its scalar components by expressing it as the product of its magnitude and $\mathbf{e}_{AB}$.

More generally, suppose that we know the magnitude of a vector $\mathbf{U}$ and the components of any vector $\mathbf{V}$ that has the same direction as $\mathbf{U}$. Then $\mathbf{V}/|\mathbf{V}|$ is a unit vector with the same direction as $\mathbf{U}$, and we can determine the components of $\mathbf{U}$ by expressing it as $\mathbf{U} = |\mathbf{U}|(\mathbf{V}/|\mathbf{V}|)$.

RESULTS

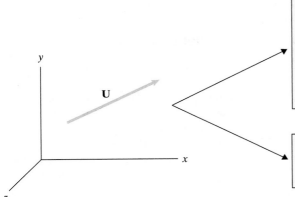

Any vector $\mathbf{U}$ can be expressed as

$$\mathbf{U} = U_x\mathbf{i} + U_y\mathbf{j} + U_z\mathbf{k}, \qquad (2.12)$$

where $\mathbf{i}$ is a unit vector that points in the positive x axis direction, $\mathbf{j}$ is a unit vector that points in the positive y axis direction, and $\mathbf{k}$ is a unit vector that points in the positive z axis direction.

The magnitude of $\mathbf{U}$ is given by

$$|\mathbf{U}| = \sqrt{U_x^2 + U_y^2 + U_z^2}. \qquad (2.14)$$

Direction Cosines

The direction of a vector $\mathbf{U}$ relative to a given coordinate system can be specified by the angles $\theta_x, \theta_y,$ and θ_z between the vector and the positive coordinate axes.

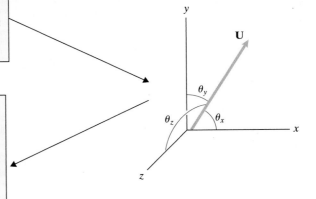

The components of $\mathbf{U}$ are given by

$$U_x = |\mathbf{U}|\cos\theta_x,$$
$$U_y = |\mathbf{U}|\cos\theta_y, \qquad (2.15)$$
$$U_z = |\mathbf{U}|\cos\theta_z.$$

The terms $\cos\theta_x$, $\cos\theta_y$, and $\cos\theta_z$ are called the *direction cosines* of $\mathbf{U}$. The direction cosines are the components of a unit vector with the same direction as $\mathbf{U}$.

Position Vectors in Terms of Components

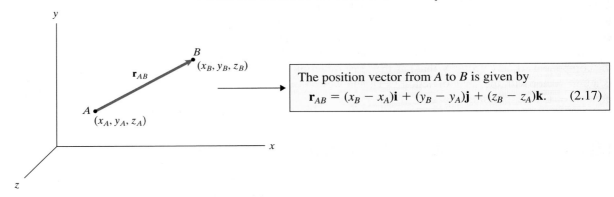

The position vector from A to B is given by

$$\mathbf{r}_{AB} = (x_B - x_A)\mathbf{i} + (y_B - y_A)\mathbf{j} + (z_B - z_A)\mathbf{k}. \qquad (2.17)$$

Components of a Vector Parallel to a Given Line

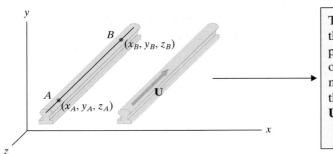

The vector $\mathbf{U}$ is parallel to the line through points A and B. Obtain the position vector $\mathbf{r}_{AB}$ from A to B in terms of its components. Divide $\mathbf{r}_{AB}$ by its magnitude to obtain a unit vector $\mathbf{e}_{AB}$ that is parallel to the line. Then the vector $\mathbf{U}$ in terms of its components is given by

$$\mathbf{U} = |\mathbf{U}|\mathbf{e}_{AB}.$$

Active Example 2.6 **Direction Cosines** (▶ *Related Problem 2.67*)

The coordinates of point C of the truss are $x_C = 4$ m, $y_C = 0, z_C = 0$, and the coordinates of point D are $x_D = 2$ m, $y_D = 3$ m, $z_D = 1$ m. What are the direction cosines of the position vector $\mathbf{r}_{CD}$ from point C to point D?

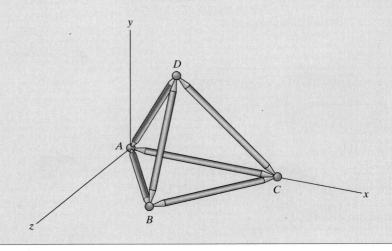

Strategy

Knowing the coordinates of points C and D, we can determine $\mathbf{r}_{CD}$ in terms of its components. Then we can calculate the magnitude of $\mathbf{r}_{CD}$ (the distance from C to D) and use Eqs. (2.15) to obtain the direction cosines.

Solution

D (2, 3, 1) m

$\mathbf{r}_{CD}$

(4, 0, 0) m

C

$$\mathbf{r}_{CD} = (x_D - x_C)\mathbf{i} + (y_D - y_C)\mathbf{j} + (z_D - z_C)\mathbf{k}.$$
$$= (2 - 4)\mathbf{i} + (3 - 0)\mathbf{j} + (1 - 0)\mathbf{k} \text{ (m)}$$
$$= -2\mathbf{i} + 3\mathbf{j} + \mathbf{k} \text{ (m)}.$$

Determine the position vector $\mathbf{r}_{CD}$ in terms of its components.

$$|\mathbf{r}_{CD}| = \sqrt{r_{CDx}^2 + r_{CDy}^2 + r_{CDz}^2}$$
$$= \sqrt{(-2 \text{ m})^2 + (3 \text{ m})^2 + (1 \text{ m})^2}$$
$$= 3.74 \text{ m}.$$

Calculate the magnitude of $\mathbf{r}_{CD}$.

$$\cos\theta_x = \frac{r_{CDx}}{|\mathbf{r}_{CD}|} = \frac{-2 \text{ m}}{3.74 \text{ m}} = -0.535,$$

$$\cos\theta_y = \frac{r_{CDy}}{|\mathbf{r}_{CD}|} = \frac{3 \text{ m}}{3.74 \text{ m}} = 0.802,$$

Determine the direction cosines.

$$\cos\theta_z = \frac{r_{CDz}}{|\mathbf{r}_{CD}|} = \frac{1 \text{ m}}{3.74 \text{ m}} = 0.267,$$

Practice Problem The coordinates of point B of the truss are $x_B = 2.4$ m, $y_B = 0$, $z_B = 3$ m. Determine the components of a unit vector $\mathbf{e}_{BD}$ that points from point B toward point D.

Answer: $\mathbf{e}_{BD} = -0.110\mathbf{i} + 0.827\mathbf{j} - 0.551\mathbf{k}$.

| Example 2.7 | **Determining Components in Three Dimensions** (▶ *Related Problem 2.76*) |

The crane exerts a 600-lb force **F** on the caisson. The angle between **F** and the *x* axis is 54°, and the angle between **F** and the *y* axis is 40°. The *z* component of **F** is positive. Express **F** in terms of components.

Strategy

Only two of the angles between the vector and the positive coordinate axes are given, but we can use Eq. (2.16) to determine the third angle. Then we can determine the components of **F** by using Eqs. (2.15).

Solution

The angles between **F** and the positive coordinate axes are related by

$$\cos^2 \theta_x + \cos^2 \theta_y + \cos^2 \theta_z = (\cos 54°)^2 + (\cos 40°)^2 + \cos^2 \theta_z = 1.$$

Solving this equation for $\cos \theta_z$, we obtain the two solutions $\cos \theta_z = 0.260$ and $\cos \theta_z = -0.260$, which tells us that $\theta_z = 74.9°$ or $\theta_z = 105.1°$. The *z* component of the vector **F** is positive, so the angle between **F** and the positive *z* axis is less than 90°. Therefore $\theta_z = 74.9°$.

The components of **F** are

$$F_x = |\mathbf{F}| \cos \theta_x = 600 \cos 54° = 353 \text{ lb,}$$

$$F_y = |\mathbf{F}| \cos \theta_y = 600 \cos 40° = 460 \text{ lb,}$$

$$F_z = |\mathbf{F}| \cos \theta_z = 600 \cos 74.9° = 156 \text{ lb.}$$

Critical Thinking

You are aware that knowing the square of a number does not tell you the value of the number uniquely. If $a^2 = 4$, the number a can be either 2 or −2. In this example, knowledge of the angles θ_x and θ_y allowed us to solve Eq. (2.16) for the value of $\cos^2 \theta_z$, which resulted in two possible values of the angle θ_z. There is a simple geometrical explanation for why this happened. The two angles θ_x and θ_y are sufficient to define a line parallel to the vector **F**, *but not the direction of* **F** *along that line.* The two values of θ_z we obtained correspond to the two possible directions of **F** along the line. Additional information is needed to indicate the direction. In this example, the additional information was supplied by stating that the *z* component of **F** is positive.

Example 2.8   Determining Components in Three Dimensions (▶ *Related Problem 2.86*)

The tether of the balloon exerts an 800-N force **F** on the hook at *O*. The vertical line *AB* intersects the *x–z* plane at point *A*. The angle between the *z* axis and the line *OA* is 60°, and the angle between the line *OA* and **F** is 45°. Express **F** in terms of components.

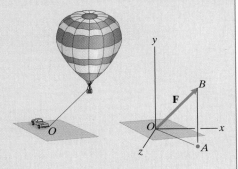

Strategy
We can determine the components of **F** from the given geometric information in two steps. First, we express **F** as the sum of two vector components parallel to the lines *OA* and *AB*. The component parallel to *AB* is the vector component **F**$_y$. Then we can use the component parallel to *OA* to determine the vector components **F**$_x$ and **F**$_z$.

Solution
In Fig. a, we express **F** as the sum of its *y* component **F**$_y$ and the component **F**$_h$ parallel to *OA*. The magnitude of **F**$_y$ is

$$|\mathbf{F}_y| = |\mathbf{F}| \sin 45° = (800 \text{ N}) \sin 45° = 566 \text{ N},$$

and the magnitude of **F**$_h$ is

$$|\mathbf{F}_h| = |\mathbf{F}| \cos 45° = (800 \text{ N}) \cos 45° = 566 \text{ N}.$$

In Fig. b, we express **F**$_h$ in terms of the vector components **F**$_x$ and **F**$_z$. The magnitude of **F**$_x$ is

$$|\mathbf{F}_x| = |\mathbf{F}_h| \sin 60° = (566 \text{ N}) \sin 60° = 490 \text{ N},$$

and the magnitude of **F**$_z$ is

$$|\mathbf{F}_z| = |\mathbf{F}_h| \cos 60° = (566 \text{ N}) \cos 60° = 283 \text{ N}.$$

The vector components **F**$_x$, **F**$_y$, and **F**$_z$ all point in the positive axis directions, so the scalar components of **F** are positive:

$$\mathbf{F} = 490\mathbf{i} + 566\mathbf{j} + 283\mathbf{k} \text{ (N)}.$$

(a) Resolving **F** into vector components parallel to *OA* and *OB*.

(b) Resolving **F**$_h$ into vector components parallel to the *x* and *z* axes.

Critical Thinking
As this example demonstrates, two angles are required to specify a vector's direction relative to a three-dimensional coordinate system. The two angles used may not be defined in the same way as in the example, but however they are defined, you can determine the components of the vector in terms of the magnitude and the two specified angles by a procedure similar to the one we used here.

Example 2.9	**Determining Components in Three Dimensions** (▶ *Related Problem 2.90*)

The rope extends from point B through a metal loop attached to the wall at A to point C. The rope exerts forces $\mathbf{F}_{AB}$ and $\mathbf{F}_{AC}$ on the loop at A with magnitudes $|\mathbf{F}_{AB}| = |\mathbf{F}_{AC}| = 200$ lb. What is the magnitude of the total force $\mathbf{F} = \mathbf{F}_{AB} + \mathbf{F}_{AC}$ exerted on the loop by the rope?

Strategy

The force $\mathbf{F}_{AB}$ is parallel to the line from A to B, and the force $\mathbf{F}_{AC}$ is parallel to the line from A to C. Since we can determine the coordinates of points A, B, and C from the given dimensions, we can determine the components of unit vectors that have the same directions as the two forces and use them to express the forces in terms of scalar components.

Solution

Let $\mathbf{r}_{AB}$ be the position vector from point A to point B and let $\mathbf{r}_{AC}$ be the position vector from point A to point C (Fig. a). From the given dimensions, the coordinates of points A, B, and C are

$$A: (6, 7, 0) \text{ ft}, \qquad B: (2, 0, 4) \text{ ft}, \qquad C: (12, 0, 6) \text{ ft}.$$

(a) The position vectors $\mathbf{r}_{AB}$ and $\mathbf{r}_{AC}$.

Therefore, the components of $\mathbf{r}_{AB}$ and $\mathbf{r}_{AC}$, with the coordinates in ft, are given by

$$
\begin{aligned}
\mathbf{r}_{AB} &= (x_B - x_A)\mathbf{i} + (y_B - y_A)\mathbf{j} + (z_B - z_A)\mathbf{k} \\
&= (2 - 6)\mathbf{i} + (0 - 7)\mathbf{j} + (4 - 0)\mathbf{k} \\
&= -4\mathbf{i} - 7\mathbf{j} + 4\mathbf{k} \text{ (ft)}
\end{aligned}
$$

and

$$\mathbf{r}_{AC} = (x_C - x_A)\mathbf{i} + (y_C - y_A)\mathbf{j} + (z_C - z_A)\mathbf{k}$$
$$= (12 - 6)\mathbf{i} + (0 - 7)\mathbf{j} + (6 - 0)\mathbf{k}$$
$$= 6\mathbf{i} - 7\mathbf{j} + 6\mathbf{k} \text{ (ft)}.$$

Their magnitudes are $|\mathbf{r}_{AB}| = 9$ ft and $|\mathbf{r}_{AC}| = 11$ ft. By dividing $\mathbf{r}_{AB}$ and $\mathbf{r}_{AC}$ by their magnitudes, we obtain unit vectors $\mathbf{e}_{AB}$ and $\mathbf{e}_{AC}$ that point in the directions of $\mathbf{F}_{AB}$ and $\mathbf{F}_{AC}$ (Fig. b):

$$\mathbf{e}_{AB} = \frac{\mathbf{r}_{AB}}{|\mathbf{r}_{AB}|} = -0.444\mathbf{i} - 0.778\mathbf{j} + 0.444\mathbf{k},$$

$$\mathbf{e}_{AC} = \frac{\mathbf{r}_{AC}}{|\mathbf{r}_{AC}|} = 0.545\mathbf{i} - 0.636\mathbf{j} + 0.545\mathbf{k}.$$

(b) The unit vectors $\mathbf{e}_{AB}$ and $\mathbf{e}_{AC}$.

The forces $\mathbf{F}_{AB}$ and $\mathbf{F}_{AC}$ are

$$\mathbf{F}_{AB} = (200 \text{ lb})\mathbf{e}_{AB} = -88.9\mathbf{i} - 155.6\mathbf{j} + 88.9\mathbf{k} \text{ (lb)},$$
$$\mathbf{F}_{AC} = (200 \text{ lb})\mathbf{e}_{AC} = 109.1\mathbf{i} - 127.3\mathbf{j} + 109.1\mathbf{k} \text{ (lb)}.$$

The total force exerted on the loop by the rope is

$$\mathbf{F} = \mathbf{F}_{AB} + \mathbf{F}_{AC} = 20.2\mathbf{i} - 282.8\mathbf{j} + 198.0\mathbf{k} \text{ (lb)},$$

and its magnitude is

$$|\mathbf{F}| = \sqrt{(20.2)^2 + (-282.8)^2 + (198.0)^2} = 346 \text{ lb}.$$

Critical Thinking

How do you know that the magnitude and direction of the total force exerted on the metal loop at A by the rope is given by the magnitude and direction of the vector $\mathbf{F} = \mathbf{F}_{AB} + \mathbf{F}_{AC}$? At this point in our development of mechanics, we assume that force is a vector, but have provided no proof. In the study of dynamics it is shown that Newton's second law implies that force is a vector.

Example 2.10	**Determining Components of a Force** (▶ *Related Problem 2.95*)

The cable AB exerts a 50-N force $\mathbf{T}$ on the collar at A. Express $\mathbf{T}$ in terms of components.

Strategy

Let $\mathbf{r}_{AB}$ be the position vector from A to B. We will divide $\mathbf{r}_{AB}$ by its magnitude to obtain a unit vector $\mathbf{e}_{AB}$ having the same direction as the force $\mathbf{T}$. Then we can obtain $\mathbf{T}$ in terms of scalar components by expressing it as the product of its magnitude and $\mathbf{e}_{AB}$. To begin this procedure, we must first determine the coordinates of the collar A. We will do so by obtaining a unit vector $\mathbf{e}_{CD}$ pointing from C toward D and multiplying it by 0.2 m to determine the position of the collar A relative to C.

Solution

Determining the Coordinates of Point A   The position vector from C to D, with the coordinates in meters, is

$$\mathbf{r}_{CD} = (0.2 - 0.4)\mathbf{i} + (0 - 0.3)\mathbf{j} + (0.25 - 0)\mathbf{k}$$
$$= -0.2\mathbf{i} - 0.3\mathbf{j} + 0.25\mathbf{k} \ (\text{m}).$$

Dividing this vector by its magnitude, we obtain the unit vector $\mathbf{e}_{CD}$ (Fig. a):

$$\mathbf{e}_{CD} = \frac{\mathbf{r}_{CD}}{|\mathbf{r}_{CD}|} = \frac{-0.2\mathbf{i} - 0.3\mathbf{j} + 0.25\mathbf{k}}{\sqrt{(-0.2)^2 + (-0.3)^2 + (0.25)^2}}$$
$$= -0.456\mathbf{i} - 0.684\mathbf{j} + 0.570\mathbf{k}.$$

Using this vector, we obtain the position vector from C to A:

$$\mathbf{r}_{CA} = (0.2 \text{ m})\mathbf{e}_{CD} = -0.091\mathbf{i} - 0.137\mathbf{j} + 0.114\mathbf{k} \ (\text{m}).$$

The position vector from the origin of the coordinate system to C is $\mathbf{r}_{OC} = 0.4\mathbf{i} + 0.3\mathbf{j}$ (m), so the position vector from the origin to A is

$$\mathbf{r}_{OA} = \mathbf{r}_{OC} + \mathbf{r}_{CA} = (0.4\mathbf{i} + 0.3\mathbf{j}) + (-0.091\mathbf{i} - 0.137\mathbf{j} + 0.114\mathbf{k})$$
$$= 0.309\mathbf{i} + 0.163\mathbf{j} + 0.114\mathbf{k} \ (\text{m}).$$

The coordinates of A are $(0.309, 0.163, 0.114)$ m.

(a) The unit vectors $\mathbf{e}_{AB}$ and $\mathbf{e}_{CD}$.

Determining the Components of T   Using the coordinates of point A, we find that the position vector from A to B is

$$\mathbf{r}_{AB} = (0 - 0.309)\mathbf{i} + (0.5 - 0.163)\mathbf{j} + (0.15 - 0.114)\mathbf{k}$$
$$= -0.309\mathbf{i} + 0.337\mathbf{j} + 0.036\mathbf{k} \text{ (m)}.$$

Dividing this vector by its magnitude, we obtain the unit vector $\mathbf{e}_{AB}$ (Fig. a).

$$\mathbf{e}_{AB} = \frac{\mathbf{r}_{AB}}{|\mathbf{r}_{AB}|} = \frac{-0.309\mathbf{i} + 0.337\mathbf{j} + 0.036\mathbf{k} \text{ (m)}}{\sqrt{(-0.309 \text{ m})^2 + (0.337 \text{ m})^2 + (0.036 \text{ m})^2}}$$
$$= -0.674\mathbf{i} + 0.735\mathbf{j} + 0.079\mathbf{k}.$$

The force $\mathbf{T}$ is

$$\mathbf{T} = |\mathbf{T}|\mathbf{e}_{AB} = (50 \text{ N})(-0.674\mathbf{i} + 0.735\mathbf{j} + 0.079\mathbf{k})$$
$$= -33.7\mathbf{i} + 36.7\mathbf{j} + 3.9\mathbf{k} \text{ (N)}.$$

Critical Thinking
Look at the two ways unit vectors were used in this example. The unit vector $\mathbf{e}_{CD}$ was used to obtain the components of the position vector $\mathbf{r}_{CA}$, which made it possible to determine the coordinates of point A. The coordinates of point A were then used to determine the unit vector $\mathbf{e}_{AB}$, which was used to express the force $\mathbf{T}$ in terms of its components.

Problems

2.61 A vector $\mathbf{U} = 3\mathbf{i} - 4\mathbf{j} - 12\mathbf{k}$. What is its magnitude?

Strategy: The magnitude of a vector is given in terms of its components by Eq. (2.14).

2.62 The vector $\mathbf{e} = \frac{1}{3}\mathbf{i} + \frac{2}{3}\mathbf{j} + e_z\mathbf{k}$ is a unit vector. Determine the component e_z.

2.63 An engineer determines that the attachment point will be subjected to a force $\mathbf{F} = 20\mathbf{i} + F_y\mathbf{j} - 45\mathbf{k}$ (kN). If the attachment point will safely support a force of 80-kN magnitude in any direction, what is the acceptable range of values of F_y?

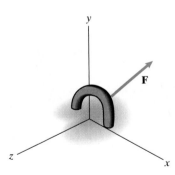

Problem 2.63

2.64 A vector $\mathbf{U} = U_x\mathbf{i} + U_y\mathbf{j} + U_z\mathbf{k}$. Its magnitude $|\mathbf{U}| = 30$. Its components are related by the equations $U_y = -2U_x$ and $U_z = 4U_y$. Determine the components.

2.65 An object is acted upon by two forces $\mathbf{F}_1 = 20\mathbf{i} + 30\mathbf{j} - 24\mathbf{k}$ (kN) and $\mathbf{F}_2 = -60\mathbf{i} + 20\mathbf{j} + 40\mathbf{k}$ (kN). What is the magnitude of the total force acting on the object?

2.66 Two vectors $\mathbf{U} = 3\mathbf{i} - 2\mathbf{j} + 6\mathbf{k}$ and $\mathbf{V} = 4\mathbf{i} + 12\mathbf{j} - 3\mathbf{k}$.
(a) Determine the magnitudes of $\mathbf{U}$ and $\mathbf{V}$.
(b) Determine the magnitude of the vector $3\mathbf{U} + 2\mathbf{V}$.

▶ **2.67** In Active Example 2.6, suppose that you want to redesign the truss, changing the position of point D so that the magnitude of the vector $\mathbf{r}_{CD}$ from point C to point D is 3 m. To accomplish this, let the coordinates of point D be $(2, y_D, 1)$ m, and determine the value of y_D so that $|\mathbf{r}_{CD}| = 3$ m. Draw a sketch of the truss with point D in its new position. What are the new direction cosines of $\mathbf{r}_{CD}$?

2.68 A force vector is given in terms of its components by $\mathbf{F} = 10\mathbf{i} - 20\mathbf{j} - 20\mathbf{k}$ (N).
(a) What are the direction cosines of $\mathbf{F}$?
(b) Determine the components of a unit vector $\mathbf{e}$ that has the same direction as $\mathbf{F}$.

2.69 The cable exerts a force **F** on the hook at O whose magnitude is 200 N. The angle between the vector **F** and the x axis is 40°, and the angle between the vector **F** and the y axis is 70°.

(a) What is the angle between the vector **F** and the z axis?

(b) Express **F** in terms of components.

    Strategy: (a) Because you know the angles between the vector **F** and the x and y axes, you can use Eq. (2.16) to determine the angle between **F** and the z axis. (Observe from the figure that the angle between **F** and the z axis is clearly within the range $0 < \theta_z < 180°$.) (b) The components of **F** can be obtained with Eqs. (2.15).

Problem 2.69

2.70 A unit vector has direction cosines $\cos \theta_x = -0.5$ and $\cos \theta_y = 0.2$. Its z component is positive. Express it in terms of components.

2.71 The airplane's engines exert a total thrust force **T** of 200-kN magnitude. The angle between **T** and the x axis is 120°, and the angle between **T** and the y axis is 130°. The z component of **T** is positive.

(a) What is the angle between **T** and the z axis?

(b) Express **T** in terms of components.

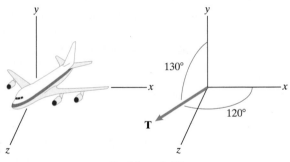

Problem 2.71

Refer to the following diagram when solving Problems 2.72 through 2.75.

Problems 2.72–2.75

2.72 Determine the components of the position vector $\mathbf{r}_{BD}$ from point B to point D. Use your result to determine the distance from B to D.

2.73 What are the direction cosines of the position vector $\mathbf{r}_{BD}$ from point B to point D?

2.74 Determine the components of the unit vector $\mathbf{e}_{CD}$ that points from point C toward point D.

2.75 What are the direction cosines of the unit vector $\mathbf{e}_{CD}$ that points from point C toward point D?

▶ **2.76** In Example 2.7, suppose that the caisson shifts on the ground to a new position. The magnitude of the force **F** remains 600 lb. In the new position, the angle between the force **F** and the x axis is 60° and the angle between **F** and the z axis is 70°. Express **F** in terms of components.

2.77 Astronauts on the space shuttle use radar to determine the magnitudes and direction cosines of the position vectors of two satellites A and B. The vector $\mathbf{r}_A$ from the shuttle to satellite A has magnitude 2 km and direction cosines $\cos \theta_x = 0.768$, $\cos \theta_y = 0.384$, $\cos \theta_z = 0.512$. The vector $\mathbf{r}_B$ from the shuttle to satellite B has magnitude 4 km and direction cosines $\cos \theta_x = 0.743$, $\cos \theta_y = 0.557$, $\cos \theta_z = -0.371$. What is the distance between the satellites?

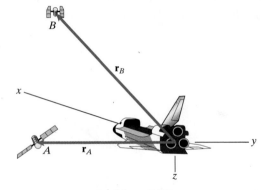

Problem 2.77

2.78 Archaeologists measure a pre-Columbian ceremonial structure and obtain the dimensions shown. Determine (a) the magnitude and (b) the direction cosines of the position vector from point A to point B.

Problem 2.78

2.79 Consider the structure described in Problem 2.78. After returning to the United States, an archaeologist discovers that a graduate student has erased the only data file containing the dimension b. But from recorded GPS data he is able to calculate that the distance from point B to point C is 16.61 m.

(a) What is the distance b?

(b) Determine the direction cosines of the position vector from B to C.

2.80 Observers at A and B use theodolites to measure the direction from their positions to a rocket in flight. If the coordinates of the rocket's position at a given instant are (4, 4, 2) km, determine the direction cosines of the vectors $\mathbf{r}_{AR}$ and $\mathbf{r}_{BR}$ that the observers would measure at that instant.

2.81* Suppose that the coordinates of the rocket's position are unknown. At a given instant, the person at A determines that the direction cosines of $\mathbf{r}_{AR}$ are $\cos \theta_x = 0.535$, $\cos \theta_y = 0.802$, and $\cos \theta_z = 0.267$, and the person at B determines that the direction cosines of $\mathbf{r}_{BR}$ are $\cos \theta_x = -0.576$, $\cos \theta_y = 0.798$, and $\cos \theta_z = -0.177$. What are the coordinates of the rocket's position at that instant?

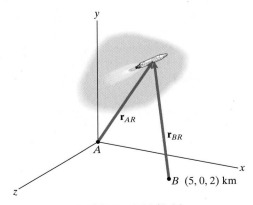

Problems 2.80/2.81

2.82* The height of Mount Everest was originally measured by a surveyor in the following way. He first measured the altitudes of two points and the horizontal distance between them. For example, suppose that the points A and B are 3000 m above sea level and are 10,000 m apart. He then used a theodolite to measure the direction cosines of the vector $\mathbf{r}_{AP}$ from point A to the top of the mountain P and the vector $\mathbf{r}_{BP}$ from point B to P. Suppose that the direction cosines of $\mathbf{r}_{AP}$ are $\cos \theta_x = 0.5179$, $\cos \theta_y = 0.6906$, and $\cos \theta_z = 0.5048$, and the direction cosines of $\mathbf{r}_{BP}$ are $\cos \theta_x = -0.3743$, $\cos \theta_y = 0.7486$, and $\cos \theta_z = 0.5472$. Using this data, determine the height of Mount Everest above sea level.

Problem 2.82

2.83 The distance from point O to point A is 20 ft. The straight line AB is parallel to the y axis, and point B is in the x–z plane. Express the vector $\mathbf{r}_{OA}$ in terms of components.

Strategy: You can express $\mathbf{r}_{OA}$ as the sum of a vector from O to B and a vector from B to A. You can then express the vector from O to B as the sum of vector components parallel to the x and z axes. See Example 2.8.

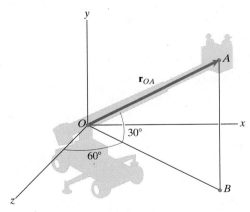

Problem 2.83

2.84 The magnitudes of the two force vectors are $|\mathbf{F}_A| = 140$ lb and $|\mathbf{F}_B| = 100$ lb. Determine the magnitude of the sum of the forces $\mathbf{F}_A + \mathbf{F}_B$.

2.85 Determine the direction cosines of the vectors $\mathbf{F}_A$ and $\mathbf{F}_B$.

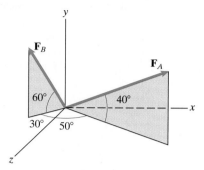

Problems 2.84/2.85

▶**2.86** In Example 2.8, suppose that a change in the wind causes a change in the position of the balloon and increases the magnitude of the force $\mathbf{F}$ exerted on the hook at O to 900 N. In the new position, the angle between the vector component $\mathbf{F}_h$ and $\mathbf{F}$ is 35°, and the angle between the vector components $\mathbf{F}_h$ and $\mathbf{F}_z$ is 40°. Draw a sketch showing the relationship of these angles to the components of $\mathbf{F}$. Express $\mathbf{F}$ in terms of its components.

2.87 An engineer calculates that the magnitude of the axial force in one of the beams of a geodesic dome is $|\mathbf{P}| = 7.65$ kN. The cartesian coordinates of the endpoints A and B of the straight beam are $(-12.4, 22.0, -18.4)$ m and $(-9.2, 24.4, -15.6)$ m, respectively. Express the force $\mathbf{P}$ in terms of components.

Problem 2.87

2.88 The cable BC exerts an 8-kN force $\mathbf{F}$ on the bar AB at B.
(a) Determine the components of a unit vector that points from point B toward point C.
(b) Express $\mathbf{F}$ in terms of components.

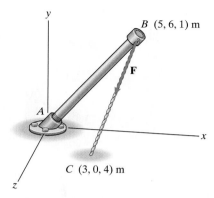

Problem 2.88

2.89 A cable extends from point C to point E. It exerts a 50-lb force $\mathbf{T}$ on the plate at C that is directed along the line from C to E. Express $\mathbf{T}$ in terms of components.

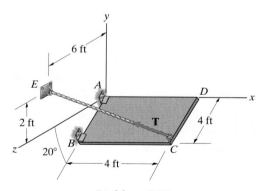

Problem 2.89

▶ **2.90** In Example 2.9, suppose that the metal loop at A is moved upward so that the vertical distance to A increases from 7 ft to 8 ft. As a result, the magnitudes of the forces $\mathbf{F}_{AB}$ and $\mathbf{F}_{AC}$ increase to $|\mathbf{F}_{AB}| = |\mathbf{F}_{AC}| = 240$ lb. What is the magnitude of the total force $\mathbf{F} = \mathbf{F}_{AB} + \mathbf{F}_{AC}$ exerted on the loop by the rope?

2.91 The cable AB exerts a 200-lb force $\mathbf{F}_{AB}$ at point A that is directed along the line from A to B. Express $\mathbf{F}_{AB}$ in terms of components.

2.92 Cable AB exerts a 200-lb force $\mathbf{F}_{AB}$ at point A that is directed along the line from A to B. The cable AC exerts a 100-lb force $\mathbf{F}_{AC}$ at point A that is directed along the line from A to C. Determine the magnitude of the total force exerted at point A by the two cables.

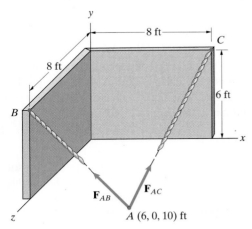

Problems 2.91/2.92

2.93 The 70-m-tall tower is supported by three cables that exert forces $\mathbf{F}_{AB}$, $\mathbf{F}_{AC}$, and $\mathbf{F}_{AD}$ on it. The magnitude of each force is 2 kN. Express the total force exerted on the tower by the three cables in terms of components.

2.94 The magnitude of the force $\mathbf{F}_{AB}$ is 2 kN. The x and z components of the vector sum of the forces exerted on the tower by the three cables are zero. What are the magnitudes of $\mathbf{F}_{AC}$ and $\mathbf{F}_{AD}$?

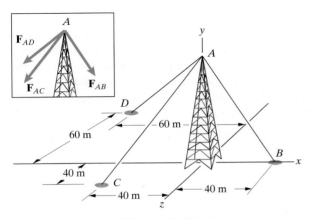

Problems 2.93/2.94

▶ **2.95** In Example 2.10, suppose that the distance from point C to the collar A is increased from 0.2 m to 0.3 m, and the magnitude of the force $\mathbf{T}$ increases to 60 N. Express $\mathbf{T}$ in terms of its components.

2.96 The cable AB exerts a 32-lb force $\mathbf{T}$ on the collar at A. Express $\mathbf{T}$ in terms of components.

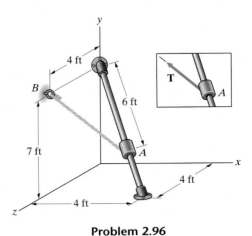

Problem 2.96

2.97 The circular bar has a 4-m radius and lies in the x–y plane. Express the position vector from point B to the collar at A in terms of components.

2.98 The cable AB exerts a 60-N force $\mathbf{T}$ on the collar at A that is directed along the line from A toward B. Express $\mathbf{T}$ in terms of components.

Problems 2.97/2.98

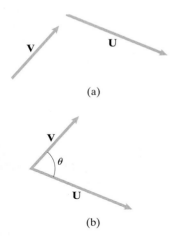

(a)

(b)

Figure 2.21
(a) The vectors **U** and **V**.
(b) The angle θ between **U** and **V** when the two vectors are placed tail to tail.

2.4 Dot Products

BACKGROUND

Two kinds of products of vectors, the dot and cross products, have been found to have applications in science and engineering, especially in mechanics and electromagnetic field theory. We use both of these products in Chapter 4 to evaluate moments of forces about points and lines.

The dot product of two vectors has many uses, including determining the components of a vector parallel and perpendicular to a given line and determining the angle between two lines in space.

Definition

Consider two vectors **U** and **V** (Fig. 2.21a). The *dot product* of **U** and **V**, denoted by **U** · **V** (hence the name "dot product"), is defined to be the product of the magnitude of **U**, the magnitude of **V**, and the cosine of the angle θ between **U** and **V** when they are placed tail to tail (Fig. 2.21b):

$$\mathbf{U} \cdot \mathbf{V} = |\mathbf{U}||\mathbf{V}| \cos \theta. \tag{2.18}$$

Because the result of the dot product is a scalar, the dot product is sometimes called the scalar product. The units of the dot product are the product of the units of the two vectors. *Notice that the dot product of two nonzero vectors is equal to zero if and only if the vectors are perpendicular.*

The dot product has the properties

$$\mathbf{U} \cdot \mathbf{V} = \mathbf{V} \cdot \mathbf{U}, \quad \text{The dot product is commutative.} \tag{2.19}$$

$$a\,(\mathbf{U} \cdot \mathbf{V}) = (a\mathbf{U}) \cdot \mathbf{V} = \mathbf{U} \cdot (a\mathbf{V}), \quad \begin{array}{l}\text{The dot product is associative} \\ \text{with respect to scalar} \\ \text{multiplication.}\end{array} \tag{2.20}$$

and

$$\mathbf{U} \cdot (\mathbf{V} + \mathbf{W}) = \mathbf{U} \cdot \mathbf{V} + \mathbf{U} \cdot \mathbf{W}, \quad \begin{array}{l}\text{The dot product is associative} \\ \text{with respect to vector addition.}\end{array} \tag{2.21}$$

for any scalar a and vectors **U**, **V**, and **W**.

Dot Products in Terms of Components

In this section we derive an equation that allows you to determine the dot product of two vectors if you know their scalar components. The derivation also results in an equation for the angle between the vectors. The first step is to determine the dot products formed from the unit vectors **i**, **j**, and **k**. Let us evaluate the dot product **i** · **i**. The magnitude $|\mathbf{i}| = 1$, and the angle between two identical vectors placed tail to tail is zero, so we obtain

$$\mathbf{i} \cdot \mathbf{i} = |\mathbf{i}||\mathbf{i}| \cos (0) = (1)(1)(1) = 1.$$

The dot product of **i** and **j** is

$$\mathbf{i} \cdot \mathbf{j} = |\mathbf{i}||\mathbf{j}| \cdot \cos (90°) = (1)(1)(0) = 0.$$

Continuing in this way, we obtain

$$\begin{array}{lll} \mathbf{i} \cdot \mathbf{i} = 1, & \mathbf{i} \cdot \mathbf{j} = 0, & \mathbf{i} \cdot \mathbf{k} = 0, \\ \mathbf{j} \cdot \mathbf{i} = 0, & \mathbf{j} \cdot \mathbf{j} = 1, & \mathbf{j} \cdot \mathbf{k} = 0, \\ \mathbf{k} \cdot \mathbf{i} = 0, & \mathbf{k} \cdot \mathbf{j} = 0, & \mathbf{k} \cdot \mathbf{k} = 1. \end{array} \tag{2.22}$$

The dot product of two vectors **U** and **V**, expressed in terms of their components, is

$$
\begin{aligned}
\mathbf{U} \cdot \mathbf{V} &= (U_x\mathbf{i} + U_y\mathbf{j} + U_z\mathbf{k}) \cdot (V_x\mathbf{i} + V_y\mathbf{j} + V_z\mathbf{k}) \\
&= U_xV_x(\mathbf{i} \cdot \mathbf{i}) + U_xV_y(\mathbf{i} \cdot \mathbf{j}) + U_xV_z(\mathbf{i} \cdot \mathbf{k}) \\
&\quad + U_yV_x(\mathbf{j} \cdot \mathbf{i}) + U_yV_y(\mathbf{j} \cdot \mathbf{j}) + U_yV_z(\mathbf{j} \cdot \mathbf{k}) \\
&\quad + U_zV_x(\mathbf{k} \cdot \mathbf{i}) + U_zV_y(\mathbf{k} \cdot \mathbf{j}) + U_zV_z(\mathbf{k} \cdot \mathbf{k}).
\end{aligned}
$$

In obtaining this result, we used Eqs. (2.20) and (2.21). Substituting Eqs. (2.22) into this expression, we obtain an equation for the dot product in terms of the scalar components of the two vectors:

$$
\mathbf{U} \cdot \mathbf{V} = U_xV_x + U_yV_y + U_zV_z. \tag{2.23}
$$

To obtain an equation for the angle θ in terms of the components of the vectors, we equate the expression for the dot product given by Eq. (2.23) to the definition of the dot product, Eq. (2.18), and solve for $\cos\theta$:

$$
\cos\theta = \frac{\mathbf{U} \cdot \mathbf{V}}{|\mathbf{U}||\mathbf{V}|} = \frac{U_xV_x + U_yV_y + U_zV_z}{|\mathbf{U}||\mathbf{V}|}. \tag{2.24}
$$

Vector Components Parallel and Normal to a Line

In some engineering applications a vector must be expressed in terms of vector components that are parallel and normal (perpendicular) to a given line. The component of a vector parallel to a line is called the *projection* of the vector onto the line. For example, when the vector represents a force, the projection of the force onto a line is the component of the force in the direction of the line.

We can determine the components of a vector parallel and normal to a line by using the dot product. Consider a vector **U** and a straight line L (Fig. 2.22a). We can express **U** as the sum of vector components $\mathbf{U}_p$ and $\mathbf{U}_n$ that are parallel and normal to L (Fig. 2.22b).

The Parallel Component   In terms of the angle θ between **U** and the vector component $\mathbf{U}_p$, the magnitude of $\mathbf{U}_p$ is

$$
|\mathbf{U}_p| = |\mathbf{U}| \cos\theta. \tag{2.25}
$$

Let **e** be a unit vector parallel to L (Fig. 2.23). The dot product of **e** and **U** is

$$
\mathbf{e} \cdot \mathbf{U} = |\mathbf{e}||\mathbf{U}| \cos\theta = |\mathbf{U}| \cos\theta.
$$

Comparing this result with Eq. (2.25), we see that the magnitude of $\mathbf{U}_p$ is

$$
|\mathbf{U}_p| = \mathbf{e} \cdot \mathbf{U}.
$$

Therefore the parallel vector component, or projection of **U** onto L, is

$$
\mathbf{U}_p = (\mathbf{e} \cdot \mathbf{U})\mathbf{e}. \tag{2.26}
$$

(This equation holds even if **e** doesn't point in the direction of $\mathbf{U}_p$. In that case, the angle $\theta > 90°$ and $\mathbf{e} \cdot \mathbf{U}$ is negative.) When the components of a vector and the components of a unit vector **e** parallel to a line L are known, we can use Eq. (2.26) to determine the component of the vector parallel to L.

The Normal Component   Once the parallel vector component has been determined, we can obtain the normal vector component from the relation $\mathbf{U} = \mathbf{U}_p + \mathbf{U}_n$:

$$
\mathbf{U}_n = \mathbf{U} - \mathbf{U}_p. \tag{2.27}
$$

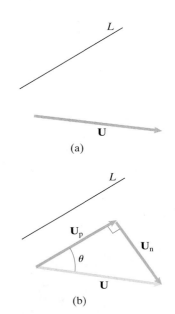

Figure 2.22
(**a**) A vector **U** and line L.
(**b**) Resolving **U** into components parallel and normal to L.

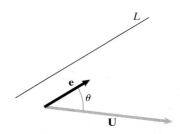

Figure 2.23
The unit vector **e** is parallel to L.

RESULTS

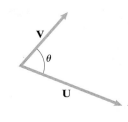

Dot Product

The dot product of two vectors **U** and **V** is defined by

$$\mathbf{U} \cdot \mathbf{V} = |\mathbf{U}||\mathbf{V}|\cos\theta, \qquad (2.18)$$

where θ is the angle between the vectors when they are placed tail to tail. Notice that $\mathbf{U} \cdot \mathbf{U} = |\mathbf{U}|^2$. If $|\mathbf{U}| \neq 0$ and $|\mathbf{V}| \neq 0$, $\mathbf{U} \cdot \mathbf{V} = 0$ if and only if **U** and **V** are perpendicular.

Dot Product in Terms of Components

The dot product of **U** and **V** is given in terms of the components of the vectors by

$$\mathbf{U} \cdot \mathbf{V} = U_x V_x + U_y V_y + U_z V_z. \qquad (2.23)$$

Vector Components Parallel and Normal to a Line

A vector **U** can be resolved into a vector component $\mathbf{U}_p$ that is parallel to a given line L and a vector component $\mathbf{U}_n$ that is normal to L. If **e** is a unit vector that is parallel to L, the parallel component of **U** is given by

$$\mathbf{U}_p = (\mathbf{e} \cdot \mathbf{U})\,\mathbf{e}. \qquad (2.26)$$

The normal component can be obtained from the relation

$$\mathbf{U}_n = \mathbf{U} - \mathbf{U}_p. \qquad (2.27)$$

Active Example 2.11   **Dot Products** (▶ *Related Problem 2.99*)

The components of two vectors **U** and **V** are $\mathbf{U} = 6\mathbf{i} - 5\mathbf{j} - 3\mathbf{k}$ and $\mathbf{V} = 4\mathbf{i} + 2\mathbf{j} + 2\mathbf{k}$. (a) What is the value of $\mathbf{U} \cdot \mathbf{V}$? (b) What is the angle between **U** and **V** when they are placed tail to tail?

Strategy

Knowing the components of **U** and **V**, we can use Eq. (2.23) to determine the value of $\mathbf{U} \cdot \mathbf{V}$. Then we can use the definition of the dot product, Eq. (2.18), to calculate the angle between the vectors.

Solution

$$\mathbf{U} \cdot \mathbf{V} = U_x V_x + U_y V_y + U_z V_z$$
$$= (6)(4) + (-5)(2) + (-3)(2)$$
$$= 8.$$

> Use the components of the vectors to determine the value of $\mathbf{U} \cdot \mathbf{V}$.

$$\mathbf{U} \cdot \mathbf{V} = |\mathbf{U}||\mathbf{V}|\cos \theta,$$

so

$$\cos \theta = \frac{\mathbf{U} \cdot \mathbf{V}}{|\mathbf{U}||\mathbf{V}|}$$

> Use the definition of $\mathbf{U} \cdot \mathbf{V}$ to determine θ.

$$= \frac{8}{\sqrt{(6)^2 + (-5)^2 + (-3)^2} \ \sqrt{(4)^2 + (2)^2 + (2)^2}}$$

$$= 0.195.$$

Therefore $\theta = 78.7°$.

Practice Problem The components of two vectors $\mathbf{U}$ and $\mathbf{V}$ are $\mathbf{U} = 6\mathbf{i} - 5\mathbf{j} - 3\mathbf{k}$ and $\mathbf{V} = V_x\mathbf{i} + 2\mathbf{j} + 2\mathbf{k}$. Determine the value of the component V_x so that the vectors $\mathbf{U}$ and $\mathbf{V}$ are perpendicular.

Answer: $V_x = 2.67$.

Example 2.12   Using the Dot Product to Determine an Angle (▶ *Related Problem 2.100*)

What is the angle θ between the lines AB and AC?

Strategy
We know the coordinates of the points A, B, and C, so we can determine the components of the vector $\mathbf{r}_{AB}$ from A to B and the vector $\mathbf{r}_{AC}$ from A to C (Fig. a). Then we can use Eq. (2.24) to determine θ.

Solution
The vectors $\mathbf{r}_{AB}$ and $\mathbf{r}_{AC}$, with the coordinates in meters, are

$$\mathbf{r}_{AB} = (6 - 4)\mathbf{i} + (1 - 3)\mathbf{j} + (-2 - 2)\mathbf{k} = 2\mathbf{i} - 2\mathbf{j} - 4\mathbf{k} \ (\text{m}),$$
$$\mathbf{r}_{AC} = (8 - 4)\mathbf{i} + (8 - 3)\mathbf{j} + (4 - 2)\mathbf{k} = 4\mathbf{i} + 5\mathbf{j} + 2\mathbf{k} \ (\text{m}).$$

Their magnitudes are

$$|\mathbf{r}_{AB}| = \sqrt{(2 \ \text{m})^2 + (-2 \ \text{m})^2 + (-4 \ \text{m})^2} = 4.90 \ \text{m},$$
$$|\mathbf{r}_{AC}| = \sqrt{(4 \ \text{m})^2 + (5 \ \text{m})^2 + (2 \ \text{m})^2} = 6.71 \ \text{m}.$$

The dot product of $\mathbf{r}_{AB}$ and $\mathbf{r}_{AC}$ is

$$\mathbf{r}_{AB} \cdot \mathbf{r}_{AC} = (2 \ \text{m})(4 \ \text{m}) + (-2 \ \text{m})(5 \ \text{m}) + (-4 \ \text{m})(2 \ \text{m}) = -10 \ \text{m}^2.$$

Therefore,

$$\cos \theta = \frac{\mathbf{r}_{AB} \cdot \mathbf{r}_{AC}}{|\mathbf{r}_{AB}||\mathbf{r}_{AC}|} = \frac{-10 \ \text{m}^2}{(4.90 \ \text{m})(6.71 \ \text{m})} = -0.304.$$

The angle $\theta = \arccos(-0.304) = 107.7°$.

Critical Thinking
What does it mean if the dot product of two vectors is negative? From Eq. (2.18) and the graph of the cosine (Fig. b), you can see that the dot product is negative, as it is in this example, only if the enclosed angle between the two vectors is greater than 90°.

(a) The position vectors $\mathbf{r}_{AB}$ and $\mathbf{r}_{AC}$.

(b) Graph of $\cos \theta$.

Example 2.13   Vector Components Parallel and Normal to a Line (▶ *Related Problem 2.111*)

Suppose that you pull on the cable *OA*, exerting a 50-N force **F** at *O*. What are the vector components of **F** parallel and normal to the cable *OB*?

Strategy

Expressing **F** as the sum of vector components parallel and normal to *OB* (Fig. a), we can determine the vector components by using Eqs. (2.26) and (2.27). But to apply them, we must first express **F** in terms of scalar components and determine the scalar components of a unit vector parallel to *OB*. We can obtain the components of **F** by determining the components of the unit vector pointing from *O* toward *A* and multiplying them by $|\mathbf{F}|$.

Solution

The position vectors from *O* to *A* and from *O* to *B* are (Fig. b)

$$\mathbf{r}_{OA} = 6\mathbf{i} + 6\mathbf{j} - 3\mathbf{k} \ (\text{m}),$$

$$\mathbf{r}_{OB} = 10\mathbf{i} - 2\mathbf{j} + 3\mathbf{k} \ (\text{m}).$$

Their magnitudes are $|\mathbf{r}_{OA}| = 9$ m and $|\mathbf{r}_{OB}| = 10.6$ m. Dividing these vectors by their magnitudes, we obtain unit vectors that point from the origin toward *A* and *B* (Fig. c):

(a) The components of **F** parallel and normal to *OB*.

$$\mathbf{e}_{OA} = \frac{\mathbf{r}_{OA}}{|\mathbf{r}_{OA}|} = \frac{6\mathbf{i} + 6\mathbf{j} - 3\mathbf{k} \ (\text{m})}{9 \ \text{m}} = 0.667\mathbf{i} + 0.667\mathbf{j} - 0.333\mathbf{k},$$

$$\mathbf{e}_{OB} = \frac{\mathbf{r}_{OB}}{|\mathbf{r}_{OB}|} = \frac{10\mathbf{i} - 2\mathbf{j} + 3\mathbf{k} \ (\text{m})}{10.6 \ \text{m}} = 0.941\mathbf{i} - 0.188\mathbf{j} + 0.282\mathbf{k}.$$

The force **F** in terms of scalar components is

$$\mathbf{F} = |\mathbf{F}|\mathbf{e}_{OA} = (50 \ \text{N})(0.667\mathbf{i} + 0.667\mathbf{j} - 0.333\mathbf{k})$$
$$= 33.3\mathbf{i} + 33.3\mathbf{j} - 16.7\mathbf{k} \ (\text{N}).$$

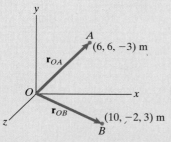

(b) The position vectors $\mathbf{r}_{OA}$ and $\mathbf{r}_{OB}$.

Taking the dot product of $\mathbf{e}_{OB}$ and **F**, we obtain

$$\mathbf{e}_{OB} \cdot \mathbf{F} = (0.941)(33.3 \ \text{N}) + (-0.188)(33.3 \ \text{N}) + (0.282)(-16.7 \ \text{N})$$
$$= 20.4 \ \text{N}.$$

The parallel vector component of **F** is

$$\mathbf{F}_p = (\mathbf{e}_{OB} \cdot \mathbf{F})\mathbf{e}_{OB} = (20.4 \ \text{N})(0.941\mathbf{i} - 0.188\mathbf{j} + 0.282\mathbf{k})$$
$$= 19.2\mathbf{i} - 3.83\mathbf{j} + 5.75\mathbf{k} \ (\text{N}),$$

and the normal vector component is

$$\mathbf{F}_n = \mathbf{F} - \mathbf{F}_p = 14.2\mathbf{i} + 37.2\mathbf{j} - 22.4\mathbf{k} \ (\text{N}).$$

(c) The unit vectors $\mathbf{e}_{OA}$ and $\mathbf{e}_{OB}$.

Critical Thinking

How can you confirm that two vectors are perpendicular? It is clear from Eq. (2.18) that the dot product of two nonzero vectors is zero if and only if the enclosed angle between them is 90°. We can use this diagnostic test to confirm that the components of **F** determined in this example are perpendicular. Evaluating the dot product of $\mathbf{F}_p$ and $\mathbf{F}_n$ in terms of their components in newtons, we obtain

$$\mathbf{F}_p \cdot \mathbf{F}_n = (19.2)(14.2) + (-3.83)(37.2) + (5.75)(-22.4) = 0.$$

Problems

▶ **2.99** In Active Example 2.11, suppose that the vector **V** is changed to **V** = 4**i** − 6**j** − 10**k**.
(a) What is the value of **U** · **V**?
(b) What is the angle between **U** and **V** when they are placed tail to tail?

▶ **2.100** In Example 2.12, suppose that the coordinates of point B are changed to (6, 4, 4) m. What is the angle θ between the lines AB and AC?

2.101 What is the dot product of the position vector **r** = −10**i** + 25**j** (m) and the force vector **F** = 300**i** + 250**j** + 300**k** (N)?

2.102 Suppose that the dot product of two vectors **U** and **V** is **U** · **V** = 0. If |**U**| ≠ 0, what do you know about the vector **V**?

2.103 Two *perpendicular* vectors are given in terms of their components by **U** = U_x**i** − 4**j** + 6**k** and **V** = 3**i** + 2**j** − 3**k**. Use the dot product to determine the component U_x.

2.104 The three vectors

$$\mathbf{U} = U_x\mathbf{i} + 3\mathbf{j} + 2\mathbf{k},$$

$$\mathbf{V} = -3\mathbf{i} + V_y\mathbf{j} + 3\mathbf{k},$$

$$\mathbf{W} = -2\mathbf{i} + 4\mathbf{j} + W_z\mathbf{k}$$

are mutually perpendicular. Use the dot product to determine the components U_x, V_y, and W_z.

2.105 The magnitudes |**U**| = 10 and |**V**| = 20.
(a) Use Eq. (2.18) to determine **U** · **V**.
(b) Use Eq. (2.23) to determine **U** · **V**.

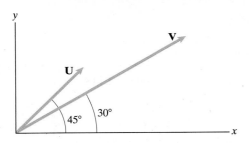

Problem 2.105

2.106 By evaluating the dot product **U** · **V**, prove the identity cos(θ₁ − θ₂) = cos θ₁ cos θ₂ + sin θ₁ sin θ₂.

Strategy: Evaluate the dot product both by using Eq. (2.18) and by using Eq. (2.23).

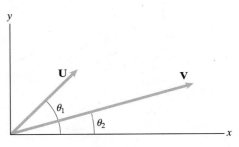

Problem 2.106

2.107 Use the dot product to determine the angle between the forestay (cable AB) and the backstay (cable BC) of the sailboat.

Problem 2.107

2.108 Determine the angle θ between the lines AB and AC
(a) by using the law of cosines (see Appendix A);
(b) by using the dot product.

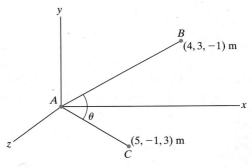

Problem 2.108

2.109 The ship O measures the positions of the ship A and the airplane B and obtains the coordinates shown. What is the angle θ between the lines of sight OA and OB?

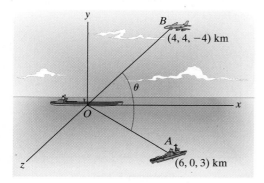

Problem 2.109

2.110 Astronauts on the space shuttle use radar to determine the magnitudes and direction cosines of the position vectors of two satellites A and B. The vector $\mathbf{r}_A$ from the shuttle to satellite A has magnitude 2 km and direction cosines $\cos\theta_x = 0.768$, $\cos\theta_y = 0.384$, $\cos\theta_z = 0.512$. The vector $\mathbf{r}_B$ from the shuttle to satellite B has magnitude 4 km and direction cosines $\cos\theta_x = 0.743$, $\cos\theta_y = 0.557$, $\cos\theta_z = -0.371$. What is the angle θ between the vectors $\mathbf{r}_A$ and $\mathbf{r}_B$?

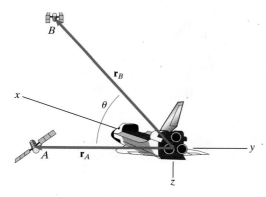

Problem 2.110

▶ **2.111** In Example 2.13, if you shift your position and the coordinates of point A where you apply the 50-N force become $(8, 3, -3)$ m, what is the vector component of $\mathbf{F}$ parallel to the cable OB?

2.112 The person exerts a force $\mathbf{F} = 60\mathbf{i} - 40\mathbf{j}$ (N) on the handle of the exercise machine. Use Eq. (2.26) to determine the vector component of $\mathbf{F}$ that is parallel to the line from the origin O to where the person grips the handle.

Problem 2.112

2.113 At the instant shown, the Harrier's thrust vector is $\mathbf{T} = 17{,}000\mathbf{i} + 68{,}000\mathbf{j} - 8{,}000\mathbf{k}$ (N) and its velocity vector is $\mathbf{v} = 7.3\mathbf{i} + 1.8\mathbf{j} - 0.6\mathbf{k}$ (m/s). The quantity $P = |\mathbf{T}_p||\mathbf{v}|$, where $\mathbf{T}_p$ is the vector component of $\mathbf{T}$ parallel to $\mathbf{v}$, is the power currently being transferred to the airplane by its engine. Determine the value of P.

Problem 2.113

2.114 Cables extend from A to B and from A to C. The cable AC exerts a 1000-lb force **F** at A.

(a) What is the angle between the cables AB and AC?

(b) Determine the vector component of **F** parallel to the cable AB.

2.115 Let $\mathbf{r}_{AB}$ be the position vector from point A to point B. Determine the vector component of $\mathbf{r}_{AB}$ parallel to the cable AC.

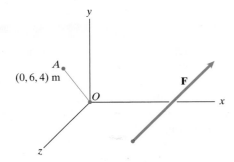

Problems 2.114/2.115

2.116 The force $\mathbf{F} = 10\mathbf{i} + 12\mathbf{j} - 6\mathbf{k}$ (N). Determine the vector components of **F** parallel and normal to the line OA.

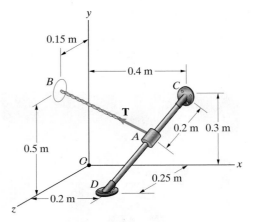

Problem 2.116

2.117 The rope AB exerts a 50-N force **T** on collar A. Determine the vector component of **T** parallel to the bar CD.

2.118 In Problem 2.117, determine the vector component of **T** normal to the bar CD.

Problems 2.117/2.118

2.119 The disk A is at the midpoint of the sloped surface. The string from A to B exerts a 0.2-lb force **F** on the disk. If you express **F** in terms of vector components parallel and normal to the sloped surface, what is the component normal to the surface?

2.120 In Problem 2.119, what is the vector component of **F** parallel to the surface?

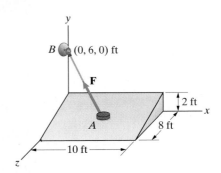

Problems 2.119/2.120

2.121 An astronaut in a maneuvering unit approaches a space station. At the present instant, the station informs him that his position relative to the origin of the station's coordinate system is $\mathbf{r}_G = 50\mathbf{i} + 80\mathbf{j} + 180\mathbf{k}$ (m) and his velocity is $\mathbf{v} = -2.2\mathbf{j} - 3.6\mathbf{k}$ (m/s). The position of an airlock is $\mathbf{r}_A = -12\mathbf{i} + 20\mathbf{k}$ (m). Determine the angle between his velocity vector and the line from his position to the airlock's position.

2.122 In Problem 2.121, determine the vector component of the astronaut's velocity parallel to the line from his position to the airlock's position.

2.123 Point P is at longitude 30°W and latitude 45°N on the Atlantic Ocean between Nova Scotia and France. Point Q is at longitude 60°E and latitude 20°N in the Arabian Sea. Use the dot product to determine the shortest distance along the surface of the earth from P to Q in terms of the radius of the earth R_E.

Strategy: Use the dot product to determine the angle between the lines OP and OQ; then use the definition of an angle in radians to determine the distance along the surface of the earth from P to Q.

Problems 2.121/2.122

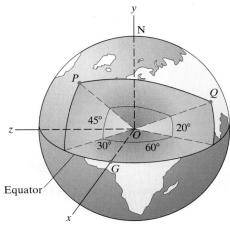

Problem 2.123

2.5 Cross Products

BACKGROUND

Like the dot product, the cross product of two vectors has many applications, including determining the rate of rotation of a fluid particle and calculating the force exerted on a charged particle by a magnetic field. Because of its usefulness for determining moments of forces, the cross product is an indispensable tool in mechanics. In this section we show you how to evaluate cross products and give examples of simple applications.

Definition

Consider two vectors $\mathbf{U}$ and $\mathbf{V}$ (Fig. 2.24a). The *cross product* of $\mathbf{U}$ and $\mathbf{V}$, denoted $\mathbf{U} \times \mathbf{V}$, is defined by

$$\mathbf{U} \times \mathbf{V} = |\mathbf{U}||\mathbf{V}| \sin \theta \, \mathbf{e}. \qquad (2.28)$$

The angle θ is the angle between $\mathbf{U}$ and $\mathbf{V}$ when they are placed tail to tail (Fig. 2.24b). The vector $\mathbf{e}$ is a unit vector defined to be perpendicular to both $\mathbf{U}$ and $\mathbf{V}$. Since this leaves two possibilities for the direction of $\mathbf{e}$, the vectors $\mathbf{U}$, $\mathbf{V}$, and $\mathbf{e}$ are defined to be a right-handed system. The *right-hand rule* for determining the direction of $\mathbf{e}$ is shown in Fig. 2.24c. If the fingers of the right hand are pointed in the direction of the vector $\mathbf{U}$ (the first vector in the cross product) and then bent toward the vector $\mathbf{V}$ (the second vector in the cross product), the thumb points in the direction of $\mathbf{e}$.

Because the result of the cross product is a vector, it is sometimes called the vector product. The units of the cross product are the product of the units of the two vectors. Notice that the cross product of two nonzero vectors is equal to zero if and only if the two vectors are parallel.

An interesting property of the cross product is that it is *not* commutative. Eq. (2.28) implies that the magnitude of the vector $\mathbf{U} \times \mathbf{V}$ is equal to the magnitude of the vector $\mathbf{V} \times \mathbf{U}$, but the right-hand rule indicates that they are opposite in direction (Fig. 2.25). That is,

$$\mathbf{U} \times \mathbf{V} = -\mathbf{V} \times \mathbf{U}. \quad \text{The cross product is \textit{not} commutative.} \qquad (2.29)$$

The cross product also satisfies the relations

$$a\,(\mathbf{U} \times \mathbf{V}) = (a\mathbf{U}) \times \mathbf{V} = \mathbf{U} \times (a\mathbf{V}) \quad \begin{array}{l}\text{The cross product is}\\ \text{associative with}\\ \text{respect to scalar}\\ \text{multiplication.}\end{array} \qquad (2.30)$$

and

$$\mathbf{U} \times (\mathbf{V} + \mathbf{W}) = (\mathbf{U} \times \mathbf{V}) + (\mathbf{U} \times \mathbf{W}) \quad \begin{array}{l}\text{The cross product is}\\ \text{distributive with}\\ \text{respect to vector}\\ \text{addition.}\end{array} \qquad (2.31)$$

for any scalar a and vectors $\mathbf{U}$, $\mathbf{V}$, and $\mathbf{W}$.

Cross Products in Terms of Components

To obtain an equation for the cross product of two vectors in terms of their components, we must determine the cross products formed from the unit vectors $\mathbf{i}$, $\mathbf{j}$, and $\mathbf{k}$. Since the angle between two identical vectors placed tail to tail is zero, it follows that

$$\mathbf{i} \times \mathbf{i} = |\mathbf{i}||\mathbf{i}| \sin(0)\mathbf{e} = \mathbf{0}.$$

The cross product $\mathbf{i} \times \mathbf{j}$ is

$$\mathbf{i} \times \mathbf{j} = |\mathbf{i}||\mathbf{j}| \sin 90°\mathbf{e} = \mathbf{e},$$

where $\mathbf{e}$ is a unit vector perpendicular to $\mathbf{i}$ and $\mathbf{j}$. Either $\mathbf{e} = \mathbf{k}$ or $\mathbf{e} = -\mathbf{k}$. Applying the right-hand rule, we find that $\mathbf{e} = \mathbf{k}$ (Fig. 2.26). Therefore,

$$\mathbf{i} \times \mathbf{j} = \mathbf{k}.$$

Continuing in this way, we obtain

$$\begin{array}{lll}\mathbf{i} \times \mathbf{i} = \mathbf{0}, & \mathbf{i} \times \mathbf{j} = \mathbf{k}, & \mathbf{i} \times \mathbf{k} = -\mathbf{j},\\ \mathbf{j} \times \mathbf{i} = -\mathbf{k}, & \mathbf{j} \times \mathbf{j} = \mathbf{0}, & \mathbf{j} \times \mathbf{k} = \mathbf{i},\\ \mathbf{k} \times \mathbf{i} = \mathbf{j}, & \mathbf{k} \times \mathbf{j} = -\mathbf{i}, & \mathbf{k} \times \mathbf{k} = \mathbf{0}.\end{array} \qquad (2.32)$$

These results can be remembered easily by arranging the unit vectors in a circle, as shown in Fig. 2.27a. The cross product of adjacent vectors is equal to the third vector with a positive sign if the order of the vectors in the cross product is the order indicated by the arrows and a negative sign otherwise. For example, in Fig. 2.27b we see that $\mathbf{i} \times \mathbf{j} = \mathbf{k}$, but $\mathbf{i} \times \mathbf{k} = -\mathbf{j}$.

The cross product of two vectors $\mathbf{U}$ and $\mathbf{V}$, expressed in terms of their components, is

$$\begin{aligned}\mathbf{U} \times \mathbf{V} = {}& (U_x\mathbf{i} + U_y\mathbf{j} + U_z\mathbf{k}) \times (V_x\mathbf{i} + V_y\mathbf{j} + V_z\mathbf{k})\\ = {}& U_xV_x(\mathbf{i} \times \mathbf{i}) + U_xV_y(\mathbf{i} \times \mathbf{j}) + U_xV_z(\mathbf{i} \times \mathbf{k})\\ & + U_yV_x(\mathbf{j} \times \mathbf{i}) + U_yV_y(\mathbf{j} \times \mathbf{j}) + U_yV_z(\mathbf{j} \times \mathbf{k})\\ & + U_zV_x(\mathbf{k} \times \mathbf{i}) + U_zV_y(\mathbf{k} \times \mathbf{j}) + U_zV_z(\mathbf{k} \times \mathbf{k}).\end{aligned}$$

(a)

(b)

(c)

Figure 2.24
(a) The vectors $\mathbf{U}$ and $\mathbf{V}$.
(b) The angle θ between the vectors when they are placed tail to tail.
(c) Determining the direction of $\mathbf{e}$ by the right-hand rule.

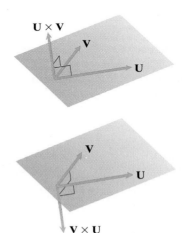

Figure 2.25
Directions of $\mathbf{U} \times \mathbf{V}$ and $\mathbf{V} \times \mathbf{U}$.

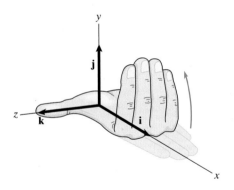

Figure 2.26
The right-hand rule indicates that $\mathbf{i} \times \mathbf{j} = \mathbf{k}$.

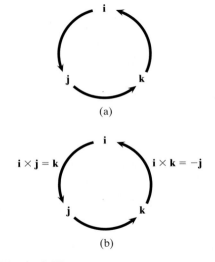

Figure 2.27
(a) Arrange the unit vectors in a circle with arrows to indicate their order.
(b) You can use the circle to determine their cross products.

By substituting Eqs. (2.32) into this expression, we obtain the equation

$$\mathbf{U} \times \mathbf{V} = (U_y V_z - U_z V_y)\mathbf{i} - (U_x V_z - U_z V_x)\mathbf{j}$$
$$+ (U_x V_y - U_y V_x)\mathbf{k}. \tag{2.33}$$

This result can be compactly written as the determinant

$$\mathbf{U} \times \mathbf{V} = \begin{vmatrix} \mathbf{i} & \mathbf{j} & \mathbf{k} \\ U_x & U_y & U_z \\ V_x & V_y & V_z \end{vmatrix}. \tag{2.34}$$

This equation is based on Eqs. (2.32), which we obtained using a right-handed coordinate system. It gives the correct result for the cross product only if a right-handed coordinate system is used to determine the components of $\mathbf{U}$ and $\mathbf{V}$.

Evaluating a 3 × 3 Determinant

A 3 × 3 determinant can be evaluated by repeating its first two columns and evaluating the products of the terms along the six diagonal lines:

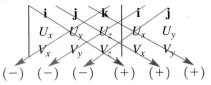

Adding the terms obtained from the diagonals that run downward to the right (blue arrows) and subtracting the terms obtained from the diagonals that run downward to the left (red arrows) gives the value of the determinant:

$$\begin{vmatrix} \mathbf{i} & \mathbf{j} & \mathbf{k} \\ U_x & U_y & U_z \\ V_x & V_y & V_z \end{vmatrix} = \begin{matrix} U_y V_z \mathbf{i} + U_z V_x \mathbf{j} + U_x V_y \mathbf{k} \\ -U_y V_x \mathbf{k} - U_z V_y \mathbf{i} - U_x V_z \mathbf{j}. \end{matrix}$$

A 3 × 3 determinant can also be evaluated by expressing it as

$$\begin{vmatrix} \mathbf{i} & \mathbf{j} & \mathbf{k} \\ U_x & U_y & U_z \\ V_x & V_y & V_z \end{vmatrix} = \mathbf{i} \begin{vmatrix} U_y & U_z \\ V_y & V_z \end{vmatrix} - \mathbf{j} \begin{vmatrix} U_x & U_z \\ V_x & V_z \end{vmatrix} + \mathbf{k} \begin{vmatrix} U_x & U_y \\ V_x & V_y \end{vmatrix}.$$

The terms on the right are obtained by multiplying each element of the first row of the 3 × 3 determinant by the 2 × 2 determinant obtained by crossing out that element's row and column. For example, the first element of the first row, $\mathbf{i}$, is multiplied by the 2 × 2 determinant

$$\begin{vmatrix} \mathbf{i} & \mathbf{j} & \mathbf{k} \\ U_x & U_y & U_z \\ V_x & V_y & V_z \end{vmatrix}.$$

Be sure to remember that the second term is subtracted. Expanding the 2 × 2 determinants, we obtain the value of the determinant:

$$\begin{vmatrix} \mathbf{i} & \mathbf{j} & \mathbf{k} \\ U_x & U_y & U_z \\ V_x & V_y & V_z \end{vmatrix} = (U_y V_z - U_z V_y)\mathbf{i} - (U_x V_z - U_z V_x)\mathbf{j} + (U_x V_y - U_y V_x)\mathbf{k}.$$

Mixed Triple Products

In Chapter 4, when we discuss the moment of a force about a line, we will use an operation called the *mixed triple product*, defined by

$$\mathbf{U} \cdot (\mathbf{V} \times \mathbf{W}). \tag{2.35}$$

In terms of the scalar components of the vectors,

$$\mathbf{U} \cdot (\mathbf{V} \times \mathbf{W}) = (U_x \mathbf{i} + U_y \mathbf{j} + U_z \mathbf{k}) \cdot \begin{vmatrix} \mathbf{i} & \mathbf{j} & \mathbf{k} \\ V_x & V_y & V_z \\ W_x & W_y & W_z \end{vmatrix}$$

$$= (U_x \mathbf{i} + U_y \mathbf{j} + U_z \mathbf{k}) \cdot [(V_y W_z - V_z W_y)\mathbf{i}$$
$$- (V_x W_z - V_z W_x)\mathbf{j} + (V_x W_y - V_y W_x)\mathbf{k}]$$
$$= U_x(V_y W_z - V_z W_y) - U_y(V_x W_z - V_z W_x)$$
$$+ U_z(V_x W_y - V_y W_x).$$

This result can be expressed as the determinant

$$\mathbf{U} \cdot (\mathbf{V} \times \mathbf{W}) = \begin{vmatrix} U_x & U_y & U_z \\ V_x & V_y & V_z \\ W_x & W_y & W_z \end{vmatrix}. \tag{2.36}$$

Interchanging any two of the vectors in the mixed triple product changes the sign but not the absolute value of the result. For example,

$$\mathbf{U} \cdot (\mathbf{V} \times \mathbf{W}) = -\mathbf{W} \cdot (\mathbf{V} \times \mathbf{U}).$$

If the vectors $\mathbf{U}$, $\mathbf{V}$, and $\mathbf{W}$ in Fig. 2.28 form a right-handed system, it can be shown that the volume of the parallelepiped equals $\mathbf{U} \cdot (\mathbf{V} \times \mathbf{W})$.

Figure 2.28
Parallelepiped defined by the vectors $\mathbf{U}$, $\mathbf{V}$, and $\mathbf{W}$.

RESULTS

Cross Product
The cross product of two vectors $\mathbf{U}$ and $\mathbf{V}$ is defined by

$$\mathbf{U} \times \mathbf{V} = |\mathbf{U}||\mathbf{V}|\sin \theta \, \mathbf{e}. \tag{2.28}$$

As in the dot product, θ is the angle between the vectors when they are placed tail to tail. The unit vector $\mathbf{e}$ is defined to be perpendicular to $\mathbf{U}$, perpendicular to $\mathbf{V}$, and directed so that $\mathbf{U}$, $\mathbf{V}$, $\mathbf{e}$ form a right-handed system. If $|\mathbf{U}| \neq 0$ and $|\mathbf{V}| \neq 0$, $\mathbf{U} \times \mathbf{V} = \mathbf{0}$ if and only if $\mathbf{U}$ and $\mathbf{V}$ are parallel.

Cross Product in Terms of Components
The cross product of $\mathbf{U}$ and $\mathbf{V}$ is given in terms of the components of the vectors by

$$\mathbf{U} \times \mathbf{V} = (U_y V_z - U_z V_y)\mathbf{i} - (U_x V_z - U_z V_x)\mathbf{j}$$
$$+ (U_x V_y - U_y V_x)\mathbf{k} \tag{2.33}$$

$$= \begin{vmatrix} \mathbf{i} & \mathbf{j} & \mathbf{k} \\ U_x & U_y & U_z \\ V_x & V_y & V_z \end{vmatrix} \tag{2.34}$$

Mixed Triple Product
The operation $\mathbf{U} \cdot (\mathbf{V} \times \mathbf{W})$ is called the mixed triple product of the vectors $\mathbf{U}$, $\mathbf{V}$, and $\mathbf{W}$. It can be expressed in terms of the components of the vectors by the determinant

$$\mathbf{U} \cdot (\mathbf{V} \times \mathbf{W}) = \begin{vmatrix} U_x & U_y & U_z \\ V_x & V_y & V_z \\ W_x & W_y & W_z \end{vmatrix}. \qquad (2.36)$$

When $\mathbf{U}$, $\mathbf{V}$, $\mathbf{W}$ form a right-handed system, the volume of the parallelepiped shown equals $\mathbf{U} \cdot (\mathbf{V} \times \mathbf{W})$.

Active Example 2.14   Cross Products (▶ *Related Problem 2.124*)

The components of two vectors $\mathbf{U}$ and $\mathbf{V}$ are $\mathbf{U} = 6\mathbf{i} - 5\mathbf{j} - \mathbf{k}$ and $\mathbf{V} = 4\mathbf{i} + 2\mathbf{j} + 2\mathbf{k}$. (a) Determine the cross product $\mathbf{U} \times \mathbf{V}$. (b) Use the dot product to prove that $\mathbf{U} \times \mathbf{V}$ is perpendicular to $\mathbf{U}$.

Strategy
(a) Knowing the components of $\mathbf{U}$ and $\mathbf{V}$, we can use Eq. (2.33) to determine $\mathbf{U} \times \mathbf{V}$. (b) Once we have determined the components of the vector $\mathbf{U} \times \mathbf{V}$, we can prove that it is perpendicular to $\mathbf{U}$ by showing that $(\mathbf{U} \times \mathbf{V}) \cdot \mathbf{U} = 0$.

Solution

$$\begin{aligned} \mathbf{U} \times \mathbf{V} &= (U_y V_z - U_z V_y)\mathbf{i} - (U_x V_z - U_z V_x)\mathbf{j} \\ &\quad + (U_x V_y - U_y V_x)\mathbf{k} \\ &= [(-5)(2) - (-1)(2)]\mathbf{i} - [(6)(2) - (-1)(4)]\mathbf{j} \\ &\quad + [(6)(2) - (-5)(4)]\mathbf{k} \\ &= -8\mathbf{i} - 16\mathbf{j} + 32\mathbf{k}. \end{aligned}$$

(a) Use the components of the vectors to determine $\mathbf{U} \times \mathbf{V}$.

$$\begin{aligned} (\mathbf{U} \times \mathbf{V}) \cdot \mathbf{U} &= (\mathbf{U} \times \mathbf{V})_x U_x + (\mathbf{U} \times \mathbf{V})_y U_y + (\mathbf{U} \times \mathbf{V})_z U_z \\ &= (-8)(6) + (-16)(-5) + (32)(-1) \\ &= 0. \end{aligned}$$

(b) Show that $(\mathbf{U} \times \mathbf{V}) \cdot \mathbf{U} = 0$.

Practice Problem The components of two vectors $\mathbf{U}$ and $\mathbf{V}$ are $\mathbf{U} = 3\mathbf{i} + 2\mathbf{j} - \mathbf{k}$ and $\mathbf{V} = 5\mathbf{i} - 3\mathbf{j} - 4\mathbf{k}$. Determine the components of a unit vector that is perpendicular to $\mathbf{U}$ and perpendicular to $\mathbf{V}$.

Answer: $\mathbf{e} = -0.477\mathbf{i} + 0.304\mathbf{j} - 0.825\mathbf{k}$ or $\mathbf{e} = 0.477\mathbf{i} - 0.304\mathbf{j} + 0.825\mathbf{k}$.

Example 2.15   Minimum Distance from a Point to a Line (▶ Related Problem 2.133)

Consider the straight lines OA and OB.
(a) Determine the components of a unit vector that is perpendicular to both OA and OB.
(b) What is the minimum distance from point A to the line OB?

Strategy
(a) Let $\mathbf{r}_{OA}$ and $\mathbf{r}_{OB}$ be the position vectors from O to A and from O to B (Fig. a). Since the cross product $\mathbf{r}_{OA} \times \mathbf{r}_{OB}$ is perpendicular to $\mathbf{r}_{OA}$ and $\mathbf{r}_{OB}$, we will determine it and divide it by its magnitude to obtain a unit vector perpendicular to the lines OA and OB.
(b) The minimum distance from A to the line OB is the length d of the straight line from A to OB that is perpendicular to OB (Fig. b). We can see that $d = |\mathbf{r}_{OA}| \sin \theta$, where θ is the angle between $\mathbf{r}_{OA}$ and $\mathbf{r}_{OB}$. From the definition of the cross product, the magnitude of $\mathbf{r}_{OA} \times \mathbf{r}_{OB}$ is $|\mathbf{r}_{OA}||\mathbf{r}_{OB}| \sin \theta$, so we can determine d by dividing the magnitude of $\mathbf{r}_{OA} \times \mathbf{r}_{OB}$ by the magnitude of $\mathbf{r}_{OB}$.

Solution
(a) The components of $\mathbf{r}_{OA}$ and $\mathbf{r}_{OB}$ are

$$\mathbf{r}_{OA} = 10\mathbf{i} - 2\mathbf{j} + 3\mathbf{k} \text{ (m)},$$
$$\mathbf{r}_{OB} = 6\mathbf{i} + 6\mathbf{j} - 3\mathbf{k} \text{ (m)}.$$

By using Eq. (2.34), we obtain $\mathbf{r}_{OA} \times \mathbf{r}_{OB}$:

$$\mathbf{r}_{OA} \times \mathbf{r}_{OB} = \begin{vmatrix} \mathbf{i} & \mathbf{j} & \mathbf{k} \\ 10 & -2 & 3 \\ 6 & 6 & -3 \end{vmatrix} = -12\mathbf{i} + 48\mathbf{j} + 72\mathbf{k} \text{ (m}^2\text{)}.$$

This vector is perpendicular to $\mathbf{r}_{OA}$ and $\mathbf{r}_{OB}$. Dividing it by its magnitude, we obtain a unit vector $\mathbf{e}$ that is perpendicular to the lines OA and OB:

$$\mathbf{e} = \frac{\mathbf{r}_{OA} \times \mathbf{r}_{OB}}{|\mathbf{r}_{OA} \times \mathbf{r}_{OB}|} = \frac{-12\mathbf{i} + 48\mathbf{j} + 72\mathbf{k} \text{ (m}^2\text{)}}{\sqrt{(-12 \text{ m}^2)^2 + (48 \text{ m}^2)^2 + (72 \text{ m}^2)^2}}$$

$$= -0.137\mathbf{i} + 0.549\mathbf{j} + 0.824\mathbf{k}.$$

(b) From Fig. b, the minimum distance d is

$$d = |\mathbf{r}_{OA}| \sin \theta.$$

The magnitude of $\mathbf{r}_{OA} \times \mathbf{r}_{OB}$ is

$$|\mathbf{r}_{OA} \times \mathbf{r}_{OB}| = |\mathbf{r}_{OA}||\mathbf{r}_{OB}| \sin \theta.$$

Solving this equation for $\sin \theta$, we find that the distance d is

$$d = |\mathbf{r}_{OA}|\left(\frac{|\mathbf{r}_{OA} \times \mathbf{r}_{OB}|}{|\mathbf{r}_{OA}||\mathbf{r}_{OB}|}\right) = \frac{|\mathbf{r}_{OA} \times \mathbf{r}_{OB}|}{|\mathbf{r}_{OB}|}$$

$$= \frac{\sqrt{(-12 \text{ m}^2)^2 + (48 \text{ m}^2)^2 + (72 \text{ m}^2)^2}}{\sqrt{(6 \text{ m})^2 + (6 \text{ m})^2 + (-3 \text{ m})^2}} = 9.71 \text{ m}.$$

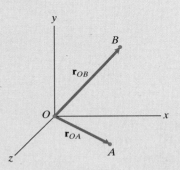

(a) The vectors $\mathbf{r}_{OA}$ and $\mathbf{r}_{OB}$.

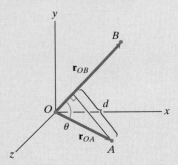

(b) The minimum distance d from A to the line OB.

Critical Thinking
This example is an illustration of the power of vector methods. Determining the minimum distance from point A to the line OB can be formulated as a minimization problem in differential calculus, but the vector solution we present is far simpler.

Example 2.16   Component of a Vector Perpendicular to a Plane (▶ *Related Problem 2.139*)

(a) Determining a unit vector perpendicular to the door.

The rope *CE* exerts a 500-N force **T** on the door *ABCD*. What is the magnitude of the component of **T** perpendicular to the door?

Strategy

We are given the coordinates of the corners *A*, *B*, and *C* of the door. By taking the cross product of the position vector $\mathbf{r}_{CB}$ from *C* to *B* and the position vector $\mathbf{r}_{CA}$ from *C* to *A*, we will obtain a vector that is perpendicular to the door. We can divide the resulting vector by its magnitude to obtain a unit vector perpendicular to the door and then apply Eq. (2.26) to determine the component of **T** perpendicular to the door.

Solution

The components of $\mathbf{r}_{CB}$ and $\mathbf{r}_{CA}$ are

$$\mathbf{r}_{CB} = 0.35\mathbf{i} - 0.2\mathbf{j} + 0.2\mathbf{k} \text{ (m)},$$
$$\mathbf{r}_{CA} = 0.5\mathbf{i} - 0.2\mathbf{j} \text{ (m)}.$$

Their cross product is

$$\mathbf{r}_{CB} \times \mathbf{r}_{CA} = \begin{vmatrix} \mathbf{i} & \mathbf{j} & \mathbf{k} \\ 0.35 & -0.2 & 0.2 \\ 0.5 & -0.2 & 0 \end{vmatrix} = 0.04\mathbf{i} + 0.1\mathbf{j} + 0.03\mathbf{k} \text{ (m}^2\text{)}.$$

Dividing this vector by its magnitude, we obtain a unit vector **e** that is perpendicular to the door (Fig. a):

$$\mathbf{e} = \frac{\mathbf{r}_{CB} \times \mathbf{r}_{CA}}{|\mathbf{r}_{CB} \times \mathbf{r}_{CA}|} = \frac{0.04\mathbf{i} + 0.1\mathbf{j} + 0.03\mathbf{k} \text{ (m}^2\text{)}}{\sqrt{(0.04 \text{ m}^2)^2 + (0.1 \text{ m}^2)^2 + (0.03 \text{ m}^2)^2}}$$
$$= 0.358\mathbf{i} + 0.894\mathbf{j} + 0.268\mathbf{k}.$$

To use Eq. (2.26), we must express **T** in terms of its scalar components. The position vector from *C* to *E* is

$$\mathbf{r}_{CE} = 0.2\mathbf{i} + 0.2\mathbf{j} - 0.1\mathbf{k} \text{ (m)},$$

so we can express the force **T** as

$$\mathbf{T} = |\mathbf{T}|\frac{\mathbf{r}_{CE}}{|\mathbf{r}_{CE}|} = (500 \text{ N})\frac{0.2\mathbf{i} + 0.2\mathbf{j} - 0.1\mathbf{k} \text{ (m)}}{\sqrt{(0.2 \text{ m})^2 + (0.2 \text{ m})^2 + (-0.1 \text{ m})^2}}$$
$$= 333\mathbf{i} + 333\mathbf{j} - 167\mathbf{k} \text{ (N)}.$$

The component of **T** parallel to the unit vector **e**, which is the component of **T** perpendicular to the door, is

$$(\mathbf{e} \cdot \mathbf{T})\mathbf{e} = [(0.358)(333 \text{ N}) + (0.894)(333 \text{ N}) + (0.268)(-167 \text{ N})]\mathbf{e}$$
$$= 373\mathbf{e} \text{ (N)}.$$

The magnitude of the component of **T** perpendicular to the door is 373 N.

Critical Thinking

Why is it useful to determine the component of the force **T** perpendicular to the door? If the *y* axis is vertical and the rope *CE* is the only thing preventing the hinged door from falling, you can see intuitively that it is the component of the force perpendicular to the door that holds it in place. We analyze problems of this kind in Chapter 5.

Problems

▶ **2.124** In Active Example 2.14, suppose that the vector **V** is changed to **V** = 4**i** − 6**j** − 10**k**. (a) Determine the cross product **U** × **V**. (b) Use the dot product to prove that **U** × **V** is perpendicular to **V**.

2.125 Two vectors **U** = 3**i** + 2**j** and **V** = 2**i** + 4**j**.
(a) What is the cross product **U** × **V**?
(b) What is the cross product **V** × **U**?

2.126 The two segments of the L-shaped bar are parallel to the x and z axes. The rope AB exerts a force of magnitude $|\mathbf{F}| = 500$ lb on the bar at A. Determine the cross product $\mathbf{r}_{CA} \times \mathbf{F}$, where $\mathbf{r}_{CA}$ is the position vector from point C to point A.

2.127 The two segments of the L-shaped bar are parallel to the x and z axes. The rope AB exerts a force of magnitude $|\mathbf{F}| = 500$ lb on the bar at A. Determine the cross product $\mathbf{r}_{CB} \times \mathbf{F}$, where $\mathbf{r}_{CB}$ is the position vector from point C to point B. Compare your answer to the answer to Problem 2.126.

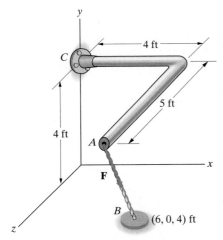

Problems 2.126/2.127

2.128 Suppose that the cross product of two vectors **U** and **V** is **U** × **V** = **0**. If $|\mathbf{U}| \neq 0$, what do you know about the vector **V**?

2.129 The cross product of two vectors **U** and **V** is **U** × **V** = −30**i** + 40**k**. The vector **V** = 4**i** − 2**j** + 3**k**. The vector **U** = 4**i** + U_y**j** + U_z**k**. Determine U_y and U_z.

2.130 The magnitudes $|\mathbf{U}| = 10$ and $|\mathbf{V}| = 20$.
(a) Use the definition of the cross product to determine **U** × **V**.
(b) Use the definition of the cross product to determine **V** × **U**.
(c) Use Eq. (2.34) to determine **U** × **V**.
(d) Use Eq. (2.34) to determine **V** × **U**.

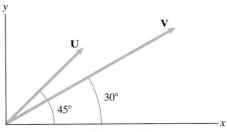

Problem 2.130

2.131 The force **F** = 10**i** − 4**j** (N). Determine the cross product $\mathbf{r}_{AB} \times \mathbf{F}$.

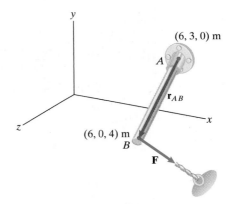

Problem 2.131

2.132 By evaluating the cross product **U** × **V**, prove the identity $\sin(\theta_1 - \theta_2) = \sin\theta_1 \cos\theta_2 - \cos\theta_1 \sin\theta_2$.

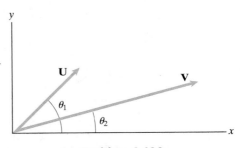

Problem 2.132

▶ **2.133** In Example 2.15, what is the minimum distance from point B to the line OA?

2.134 (a) What is the cross product $\mathbf{r}_{OA} \times \mathbf{r}_{OB}$? (b) Determine a unit vector $\mathbf{e}$ that is perpendicular to $\mathbf{r}_{OA}$ and $\mathbf{r}_{OB}$.

2.135 Use the cross product to determine the length of the shortest straight line from point B to the straight line that passes through points O and A.

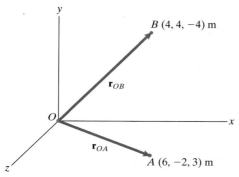

Problems 2.134/2.135

2.136 The cable BC exerts a 1000-lb force $\mathbf{F}$ on the hook at B. Determine $\mathbf{r}_{AB} \times \mathbf{F}$.

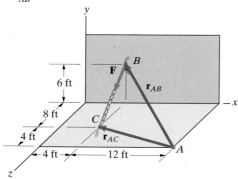

Problem 2.136

2.137 The force vector $\mathbf{F}$ points along the straight line from point A to point B. Its magnitude is $|\mathbf{F}| = 20$ N. The coordinates of points A and B are $x_A = 6$ m, $y_A = 8$ m, $z_A = 4$ m and $x_B = 8$ m, $y_B = 1$ m, $z_B = -2$ m.

(a) Express the vector $\mathbf{F}$ in terms of its components.

(b) Use Eq. (2.34) to determine the cross products $\mathbf{r}_A \times \mathbf{F}$ and $\mathbf{r}_B \times \mathbf{F}$.

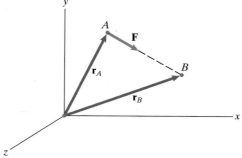

Problem 2.137

2.138 The rope AB exerts a 50-N force $\mathbf{T}$ on the collar at A. Let $\mathbf{r}_{CA}$ be the position vector from point C to point A. Determine the cross product $\mathbf{r}_{CA} \times \mathbf{T}$.

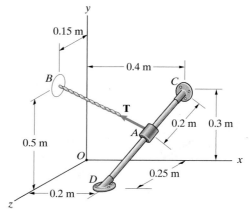

Problem 2.138

▶ **2.139** In Example 2.16, suppose that the attachment point E is moved to the location $(0.3, 0.3, 0)$ m and the magnitude of $\mathbf{T}$ increases to 600 N. What is the magnitude of the component of $\mathbf{T}$ perpendicular to the door?

2.140 The bar AB is 6 m long and is perpendicular to the bars AC and AD. Use the cross product to determine the coordinates x_B, y_B, z_B of point B.

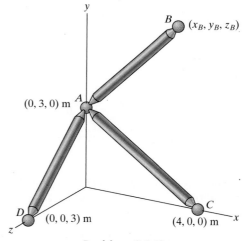

Problem 2.140

2.141* Determine the minimum distance from point P to the plane defined by the three points A, B, and C.

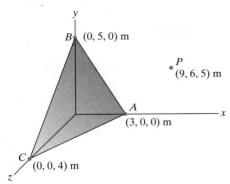

Problem 2.141

2.142* The force vector $\mathbf{F}$ points along the straight line from point A to point B. Use Eqs. (2.28)–(2.31) to prove that

$$\mathbf{r}_B \times \mathbf{F} = \mathbf{r}_A \times \mathbf{F}.$$

Strategy: Let $\mathbf{r}_{AB}$ be the position vector from point A to point B. Express $\mathbf{r}_B$ in terms of $\mathbf{r}_A$ and $\mathbf{r}_{AB}$. Notice that the vectors $\mathbf{r}_{AB}$ and $\mathbf{F}$ are parallel.

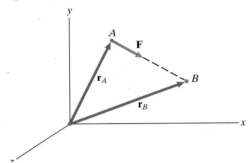

Problem 2.142

2.143 For the vectors $\mathbf{U} = 6\mathbf{i} + 2\mathbf{j} - 4\mathbf{k}$, $\mathbf{V} = 2\mathbf{i} + 7\mathbf{j}$, and $\mathbf{W} = 3\mathbf{i} + 2\mathbf{k}$, evaluate the following mixed triple products:
(a) $\mathbf{U} \cdot (\mathbf{V} \times \mathbf{W})$;
(b) $\mathbf{W} \cdot (\mathbf{V} \times \mathbf{U})$;
(c) $\mathbf{V} \cdot (\mathbf{W} \times \mathbf{U})$.

2.144 Use the mixed triple product to calculate the volume of the parallelepiped.

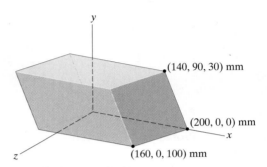

Problem 2.144

2.145 By using Eqs. (2.23) and (2.34), show that

$$\mathbf{U} \cdot (\mathbf{V} \times \mathbf{W}) = \begin{vmatrix} U_x & U_y & U_z \\ V_x & V_y & V_z \\ W_x & W_y & W_z \end{vmatrix}.$$

2.146 The vectors $\mathbf{U} = \mathbf{i} + U_y\mathbf{j} + 4\mathbf{k}$, $\mathbf{V} = 2\mathbf{i} + \mathbf{j} - 2\mathbf{k}$, and $\mathbf{W} = -3\mathbf{i} + \mathbf{j} - 2\mathbf{k}$ are coplanar (they lie in the same plane). What is the component U_y?

Review Problems

2.147 The magnitude of $\mathbf{F}$ is 8 kN. Express $\mathbf{F}$ in terms of scalar components.

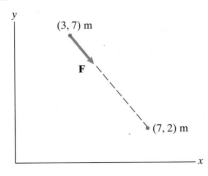

Problem 2.147

2.148 The magnitude of the vertical force $\mathbf{W}$ is 600 lb, and the magnitude of the force $\mathbf{B}$ is 1500 lb. Given that $\mathbf{A} + \mathbf{B} + \mathbf{W} = 0$, determine the magnitude of the force $\mathbf{A}$ and the angle α.

Problem 2.148

2.149 The magnitude of the vertical force vector **A** is 200 lb. If $\mathbf{A} + \mathbf{B} + \mathbf{C} = \mathbf{0}$, what are the magnitudes of the force vectors **B** and **C**?

2.150 The magnitude of the horizontal force vector **D** is 280 lb. If $\mathbf{D} + \mathbf{E} + \mathbf{F} = \mathbf{0}$, what are the magnitudes of the force vectors **E** and **F**?

Problems 2.149/2.150

Refer to the following diagram when solving Problems 2.151 through 2.157.

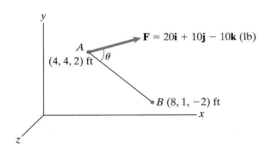

Problems 2.151–2.157

2.151 What are the direction cosines of **F**?

2.152 Determine the components of a unit vector parallel to line AB that points from A toward B.

2.153 What is the angle θ between the line AB and the force **F**?

2.154 Determine the vector component of **F** that is parallel to the line AB.

2.155 Determine the vector component of **F** that is normal to the line AB.

2.156 Determine the vector $\mathbf{r}_{BA} \times \mathbf{F}$, where $\mathbf{r}_{BA}$ is the position vector from B to A.

2.157 (a) Write the position vector $\mathbf{r}_{AB}$ from point A to point B in terms of components.
(b) A vector **R** has magnitude $|\mathbf{R}| = 200$ lb and is parallel to the line from A to B. Write **R** in terms of components.

2.158 The rope exerts a force of magnitude $|\mathbf{F}| = 200$ lb on the top of the pole at B.
(a) Determine the vector $\mathbf{r}_{AB} \times \mathbf{F}$, where $\mathbf{r}_{AB}$ is the position vector from A to B.
(b) Determine the vector $\mathbf{r}_{AC} \times \mathbf{F}$, where $\mathbf{r}_{AC}$ is the position vector from A to C.

Problem 2.158

2.159 The pole supporting the sign is parallel to the x axis and is 6 ft long. Point A is contained in the y–z plane. (a) Express the vector **r** in terms of components. (b) What are the direction cosines of **r**?

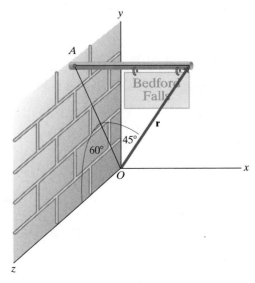

Problem 2.159

2.160   The z component of the force **F** is 80 lb. (a) Express **F** in terms of components. (b) What are the angles θ_x, θ_y, and θ_z between **F** and the positive coordinate axes?

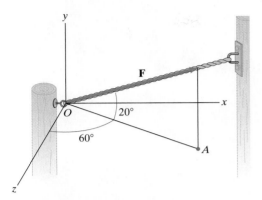

Problem 2.160

2.161   The magnitude of the force vector $\mathbf{F}_B$ is 2 kN. Express it in terms of components.

2.162   The magnitude of the vertical force vector **F** is 6 kN. Determine the vector components of **F** parallel and normal to the line from B to D.

2.163   The magnitude of the vertical force vector **F** is 6 kN. Given that $\mathbf{F} + \mathbf{F}_A + \mathbf{F}_B + \mathbf{F}_C = 0$, what are the magnitudes of $\mathbf{F}_A$, $\mathbf{F}_B$, and $\mathbf{F}_C$?

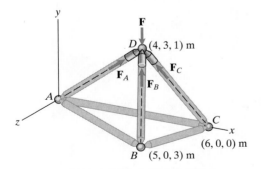

Problems 2.161–2.163

2.164   The magnitude of the vertical force **W** is 160 N. The direction cosines of the position vector from A to B are $\cos \theta_x = 0.500$, $\cos \theta_y = 0.866$, and $\cos \theta_z = 0$, and the direction cosines of the position vector from B to C are $\cos \theta_x = 0.707$, $\cos \theta_y = 0.619$, and $\cos \theta_z = -0.342$. Point G is the midpoint of the line from B to C. Determine the vector $\mathbf{r}_{AG} \times \mathbf{W}$, where $\mathbf{r}_{AG}$ is the position vector from A to G.

Problem 2.164

2.165   The rope CE exerts a 500-N force **T** on the hinged door.
(a) Express **T** in terms of components.
(b) Determine the vector component of **T** parallel to the line from point A to point B.

2.166   In Problem 2.165, let $\mathbf{r}_{BC}$ be the position vector from point B to point C. Determine the cross product $\mathbf{r}_{BC} \times \mathbf{T}$.

Problems 2.165/2.166

Forces

In Chapter 2 we represented forces by vectors and used vector addition to sum forces. In this chapter we discuss forces in more detail and introduce two of the most important concepts in mechanics, equilibrium and the free-body diagram. We will use free-body diagrams to identify the forces on objects and use equilibrium to determine unknown forces.

◄ The forces due to the weight of the bridge are transferred to the vertical support towers by cables. In this chapter we use free-body diagrams to analyze the forces acting on objects in equilibrium.

Figure 3.1
A force **F** and its line of action.

3.1 Forces, Equilibrium, and Free-Body Diagrams

BACKGROUND

Force is a familiar concept, as is evident from the words push, pull, and lift used in everyday conversation. In engineering we deal with different types of forces having a large range of magnitudes. In this section we define some terms used to describe forces, discuss particular forces that occur frequently in engineering applications, and introduce the concepts of equilibrium and free-body diagrams.

Terminology

Line of Action   When a force is represented by a vector, the straight line collinear with the vector is called the *line of action* of the force (Fig. 3.1).

Systems of Forces   A *system of forces* is simply a particular set of forces. A system of forces is *coplanar*, or *two dimensional*, if the lines of action of the forces lie in a plane. Otherwise it is *three dimensional*. A system of forces is *concurrent* if the lines of action of the forces intersect at a point (Fig. 3.2a) and *parallel* if the lines of action are parallel (Fig. 3.2b).

External and Internal Forces   We say that a given object is subjected to an *external force* if the force is exerted by a different object. When one part of a given object is subjected to a force by another part of the same object, we say it is subjected to an *internal force*. These definitions require that you clearly define the object you are considering. For example, suppose that you are the object. When you are standing, the floor—a different object—exerts an external force on your feet. If you press your hands together, your left hand exerts an internal force on your right hand. However, if your right hand is the object you are considering, the force exerted by your left hand is an external force.

Body and Surface Forces   A force acting on an object is called a *body force* if it acts on the volume of the object and a *surface force* if it acts on its surface. The gravitational force on an object is a body force. A surface force can be exerted on an object by contact with another object. Both body and surface forces can result from electromagnetic effects.

Gravitational Forces

You are aware of the force exerted on an object by the earth's gravity whenever you pick up something heavy. We can represent the gravitational force, or weight, of an object by a vector (Fig. 3.3).

Figure 3.2
(a) Concurrent forces.
(b) Parallel forces.

(a)

(b)

Figure 3.3
Representing an object's weight by a vector.

(a)   (b)   (c)

Figure 3.4
(a) Exerting a contact force on a wall by pushing on it.
(b) The vector **F** represents the force you exert on the wall.
(c) The wall exerts a force $-\mathbf{F}$ on your hand.

The magnitude of an object's weight is related to its mass m by

$$|\mathbf{W}| = mg, \qquad (3.1)$$

where g is the acceleration due to gravity at sea level. We will use the values $g = 9.81 \text{ m/s}^2$ in SI units and $g = 32.2 \text{ ft/s}^2$ in U.S. Customary units.

Gravitational forces, and also electromagnetic forces, act at a distance. The objects they act on are not necessarily in contact with the objects exerting the forces. In the next section we discuss forces resulting from contacts between objects.

Contact Forces

Contact forces are the forces that result from contacts between objects. For example, you exert a contact force when you push on a wall (Fig. 3.4a). The surface of your hand exerts a force on the surface of the wall that can be represented by a vector **F** (Fig. 3.4b). The wall exerts an equal and opposite force $-\mathbf{F}$ on your hand (Fig. 3.4c). (Recall Newton's third law: The forces exerted on each other by any two particles are equal in magnitude and opposite in direction. If you have any doubt that the wall exerts a force on your hand, try pushing on the wall while standing on roller skates.)

We will be concerned with contact forces exerted on objects by contact with the surfaces of other objects and by ropes, cables, and springs.

Surfaces Consider two plane surfaces in contact (Fig. 3.5a). We represent the force exerted on the right surface by the left surface by the vector **F** in Fig. 3.5b. We can resolve **F** into a component **N** that is normal to the surface and a component **f** that is parallel to the surface (Fig. 3.5c). The component **N** is called the *normal force*, and the component **f** is called the *friction force*. We sometimes assume that the friction force between two surfaces is negligible in comparison to the normal force, a condition we describe by saying that the surfaces are *smooth*. In this case we show only the normal force (Fig. 3.5d). When the friction force cannot be neglected, we say the surfaces are *rough*.

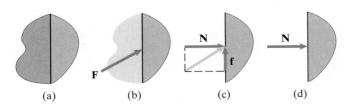

(a)   (b)   (c)   (d)

Figure 3.5
(a) Two plane surfaces in contact.
(b) The force **F** exerted on the right surface.
(c) The force **F** resolved into components normal and parallel to the surface.
(d) Only the normal force is shown when friction is neglected.

Figure 3.6
(a) Curved contacting surfaces. The dashed line indicates the plane tangent to the surfaces at their point of contact.
(b) The normal force and friction force on the right surface.

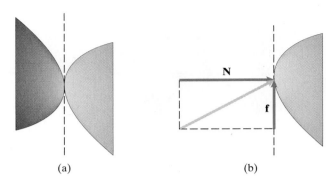

(a)   (b)

If the contacting surfaces are curved (Fig. 3.6a), the normal force and the friction force are perpendicular and parallel to the plane tangent to the surfaces at their point of contact (Fig. 3.6b).

Ropes and Cables   A contact force can be exerted on an object by attaching a rope or cable to the object and pulling on it. In Fig. 3.7a, the crane's cable is attached to a container of building materials. We can represent the force the cable exerts on the container by a vector **T** (Fig. 3.7b). The magnitude of **T** is called the *tension* in the cable, and the line of action of **T** is collinear with the cable. The cable exerts an equal and opposite force −**T** on the crane (Fig. 3.7c).

Notice that we have assumed that the cable is straight and that the tension where the cable is connected to the container equals the tension near the crane. This is approximately true if the weight of the cable is small compared to the tension. Otherwise, the cable will sag significantly and the tension will vary along its length. In Chapter 9 we will discuss ropes and cables whose weights are not small in comparison to their tensions. For now, we assume that ropes and cables are straight and that their tensions are constant along their lengths.

A *pulley* is a wheel with a grooved rim that can be used to change the direction of a rope or cable (Fig. 3.8a). For now, we assume that the tension is

(a)

(b)

(c)

Figure 3.7
(a) A crane with its cable attached to a container.
(b) The force **T** exerted on the container by the cable.
(c) The force −**T** exerted on the crane by the cable.

Figure 3.8
(a) A pulley changes the direction of a rope or cable.
(b) For now, you should assume that the tensions on each side of the pulley are equal.

$|\mathbf{T}_1| = |\mathbf{T}_2|$

Figure 3.9
Coil springs in car suspensions. The arrangement on the right is called a MacPherson strut.

the same on both sides of a pulley (Fig. 3.8b). This is true, or at least approximately true, when the pulley can turn freely and the rope or cable either is stationary or turns the pulley at a constant rate.

Springs Springs are used to exert contact forces in mechanical devices, for example, in the suspensions of cars (Fig. 3.9). Let's consider a coil spring whose unstretched length, the length of the spring when its ends are free, is L_0 (Fig. 3.10a). When the spring is stretched to a length L greater than L_0 (Fig. 3.10b), it pulls on the object to which it is attached with a force $\mathbf{F}$ (Fig. 3.10c). The object exerts an equal and opposite force $-\mathbf{F}$ on the spring (Fig. 3.10d). When the spring is compressed to a length L less than L_0 (Figs. 3.11a, b), the spring pushes on the object with a force $\mathbf{F}$ and the object exerts an equal and opposite force $-\mathbf{F}$ on the spring (Figs. 3.11c, d). If a spring is compressed too much, it may buckle (Fig. 3.11e). A spring designed to exert a force by being compressed is often provided with lateral support to prevent buckling, for example, by enclosing it in a cylindrical sleeve. In the car suspensions shown in Fig. 3.9, the shock absorbers within the coils prevent the springs from buckling.

The magnitude of the force exerted by a spring depends on the material it is made of, its design, and how much it is stretched or compressed relative to its unstretched length. When the change in length is not too large compared to the unstretched length, the coil springs commonly used in mechanical devices exert a force approximately proportional to the change in length:

$$|\mathbf{F}| = k|L - L_0|. \tag{3.2}$$

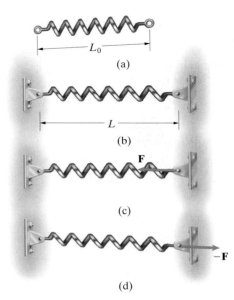

Figure 3.10
(a) A spring of unstretched length L_0.
(b) The spring stretched to a length $L > L_0$.
(c, d) The force $\mathbf{F}$ exerted by the spring and the force $-\mathbf{F}$ on the spring.

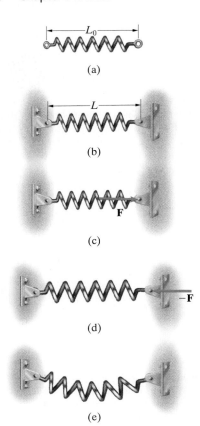

Figure 3.11
(a) A spring of length L_0.
(b) The spring compressed to a length $L < L_0$.
(c, d) The spring pushes on an object with a force **F**, and the object exerts a force $-\mathbf{F}$ on the spring.
(e) A coil spring will buckle if it is compressed too much.

Because the force is a linear function of the change in length (Fig. 3.12), a spring that satisfies this relation is called a *linear spring*. The value of the *spring constant k* depends on the material and design of the spring. Its dimensions are (force)/(length). Notice from Eq. (3.2) that k equals the magnitude of the force required to stretch or compress the spring a unit of length.

Suppose that the unstretched length of a spring is $L_0 = 1$ m and $k = 3000$ N/m. If the spring is stretched to a length $L = 1.2$ m, the magnitude of the pull it exerts is

$$k|L - L_0| = 3000(1.2 - 1) = 600 \text{ N}.$$

Although coil springs are commonly used in mechanical devices, we are also interested in them for a different reason. Springs can be used to *model* situations in which forces depend on displacements. For example, the force necessary to bend the steel beam in Fig. 3.13a is a linear function of the displacement δ, or

$$|\mathbf{F}| = k\delta,$$

if δ is not too large. Therefore we can model the force-deflection behavior of the beam with a linear spring (Fig. 3.13b).

Equilibrium

In everyday conversation, equilibrium means an unchanging state—a state of balance. Before we state precisely what this term means in mechanics, let us consider some familiar examples. If you are in a building as you read this, objects you observe around you that are *at rest (stationary) relative to the building*, such as pieces of furniture, are in equilibrium. A person sitting or standing at rest relative to the building is also in equilibrium. If a train travels at constant speed on a straight track, objects within the train that are at rest relative to the train, such as the passenger seats or a passenger standing in the aisle (Fig. 3.14a), are in equilibrium. *The person at rest relative to the building and also the passenger at rest relative to the train are not accelerating.*

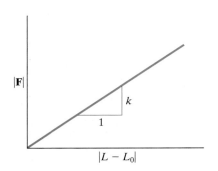

Figure 3.12
The graph of the force exerted by a linear spring as a function of its stretch or compression is a straight line with slope k.

Figure 3.13
(a) A steel beam deflected by a force.
(b) Modeling the beam's behavior with a linear spring.

However, if the train should begin increasing or decreasing its speed, the person standing in the aisle of the train would no longer be in equilibrium and might lose his balance (Fig. 3.14b).

We define an object to be in *equilibrium* only if each point of the object has the same constant velocity, which is referred to as *steady translation*. The velocity must be measured relative to a frame of reference in which Newton's laws are valid. Such a frame is called a *Newtonian* or *inertial reference frame*. In many engineering applications, a frame of reference that is fixed with respect to the earth can be regarded as inertial. Therefore, objects in steady translation relative to the earth can be assumed to be in equilibrium. We make this assumption throughout this book. In the examples cited in the previous paragraph, the furniture and person at rest in a building and also the passenger seats and passenger at rest within the train moving at constant speed are in steady translation relative to the earth and so are in equilibrium.

The vector sum of the external forces acting on an object in equilibrium is zero. We will use the symbol $\Sigma\mathbf{F}$ to denote the sum of the external forces. Thus, when an object is in equilibrium,

$$\Sigma\mathbf{F} = \mathbf{0}. \tag{3.3}$$

In some situations we can use this *equilibrium equation* to determine unknown forces acting on an object in equilibrium. The first step will be to draw a *free-body diagram* of the object to identify the external forces acting on it.

Free-Body Diagrams

A free-body diagram serves to focus attention on the object of interest and helps identify the external forces acting on it. Although in statics we are concerned only with objects in equilibrium, free-body diagrams are also used in dynamics to study the motions of objects.

Although it is one of the most important tools in mechanics, a free-body diagram is a simple concept. It is a drawing of an object and the external forces acting on it. Otherwise, nothing other than the object of interest is included. The drawing shows the object *isolated*, or *freed*, from its surroundings.

Drawing a free-body diagram involves three steps:

1. *Identify the object you want to isolate*—As the following examples show, your choice is often dictated by particular forces you want to determine.
2. *Draw a sketch of the object isolated from its surroundings, and show relevant dimensions and angles*—Your drawing should be reasonably accurate, but it can omit irrelevant details.
3. *Draw vectors representing all of the external forces acting on the isolated object, and label them*—Don't forget to include the gravitational force if you are not intentionally neglecting it.

A coordinate system is necessary to express the forces on the isolated object in terms of components. Often it is convenient to choose the coordinate system before drawing the free-body diagram, but in some situations the best choice of a coordinate system will not be apparent until after it has been drawn.

A simple example demonstrates how you can choose free-body diagrams to determine particular forces and also that you must distinguish carefully between external and internal forces. Two stationary blocks of equal weight W are suspended by cables in Fig. 3.15. The system is in equilibrium. Suppose that we want to determine the tensions in the two cables.

(a)

(b)

Figure 3.14
(a) While the train moves at a constant speed, a person standing in the aisle is in equilibrium.
(b) If the train starts to speed up, the person is no longer in equilibrium.

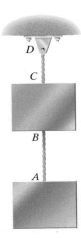

Figure 3.15
Stationary blocks suspended by cables.

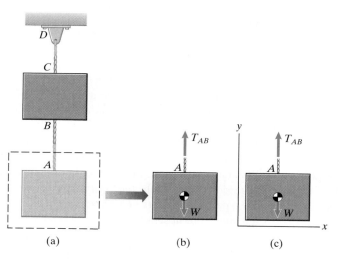

Figure 3.16
(a) Isolating the lower block and part of cable AB.
(b) Indicating the external forces completes the free-body diagram.
(c) Introducing a coordinate system.

To determine the tension in cable AB, we first isolate an "object" consisting of the lower block and part of cable AB (Fig. 3.16a). We then ask ourselves what forces can be exerted on our isolated object by objects not included in the diagram. The earth exerts a gravitational force of magnitude W on the block. Also, where we "cut" cable AB, the cable is subjected to a contact force equal to the tension in the cable (Fig. 3.16b). The arrows in this figure indicate the directions of the forces. The scalar W is the weight of the block and T_{AB} is the tension in cable AB. We assume that the weight of the part of cable AB included in the free-body diagram can be neglected in comparison to the weight of the block.

Since the free-body diagram is in equilibrium, the sum of the external forces equals zero. In terms of a coordinate system with the y axis upward (Fig. 3.16c), we obtain the equilibrium equation

$$\Sigma\mathbf{F} = T_{AB}\mathbf{j} - W\mathbf{j} = (T_{AB} - W)\mathbf{j} = \mathbf{0}.$$

Thus, the tension in cable AB is $T_{AB} = W$.

We can determine the tension in cable CD by isolating the upper block (Fig. 3.17a). The external forces are the weight of the upper block and the tensions in the two cables (Fig. 3.17b). In this case we obtain the equilibrium equation

$$\Sigma\mathbf{F} = T_{CD}\mathbf{j} - T_{AB}\mathbf{j} - W\mathbf{j} = (T_{CD} - T_{AB} - W)\mathbf{j} = \mathbf{0}.$$

Since $T_{AB} = W$, we find that $T_{CD} = 2W$.

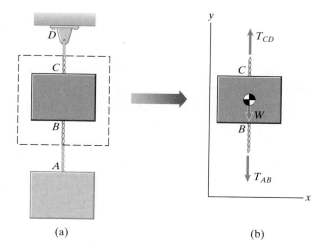

Figure 3.17
(a) Isolating the upper block to determine the tension in cable CD.
(b) Free-body diagram of the upper block.

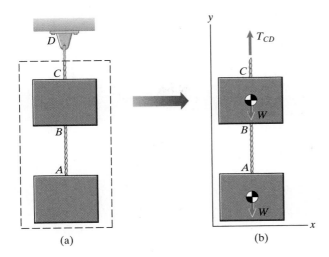

(a)   (b)

Figure 3.18
(a) An alternative choice for determining the tension in cable CD.
(b) Free-body diagram including both blocks and cable AB.

We could also have determined the tension in cable CD by treating the two blocks and the cable AB as a single object (Figs. 3.18a, b). The equilibrium equation is

$$\Sigma\mathbf{F} = T_{CD}\mathbf{j} - W\mathbf{j} - W\mathbf{j} = (T_{CD} - 2W)\mathbf{j} = \mathbf{0},$$

and we again obtain $T_{CD} = 2W$.

Why doesn't the tension in cable AB appear on the free-body diagram in Fig. 3.18b? Remember that only external forces are shown on free-body diagrams. Since cable AB is part of the free-body diagram in this case, the forces it exerts on the upper and lower blocks are internal forces.

RESULTS

Line of Action
The straight line collinear with a vector representing a force is the *line of action* of the force.

Systems of Forces
A system of forces is *two dimensional* if the lines of action of the forces lie in a plane. Otherwise it is *three dimensional*. A system of forces is *concurrent* if the lines of action intersect at a point and is *parallel* if the lines of action are parallel.

External and Internal Forces
An object is subjected to an *external force* if the force is exerted by a different object. A force exerted on part of an object by a different part of the same object is an *internal force*.

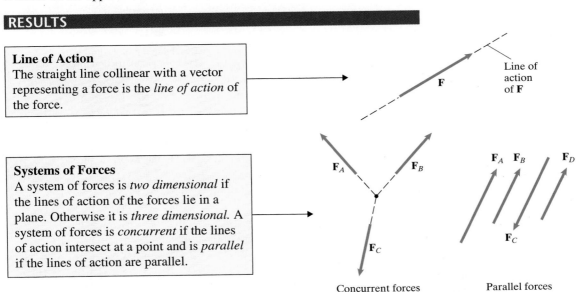

Gravitational Forces
The weight of an object can be represented by a vector. Its magnitude at sea level is related to the mass m of the object by

$$|\mathbf{W}| = mg, \qquad (3.1)$$

where g is the acceleration due to gravity at sea level.

Contact Forces
Contacting objects exert equal and opposite forces on each other.

Objects A and B with plane surfaces in contact.

The contact forces A and B exert on each other.

Resolving the force on B into the normal and friction forces.

When friction is neglected there is only a normal force.

Ropes and Cables
If the weight of a rope or cable connecting two objects is negligible in comparison to its tension, it exerts equal and opposite forces on the objects that are parallel to the rope or cable.

Objects A and B connected by a cable.

The forces exerted on A and B.

Objects A and B connected by a spring.

The forces exerted on A and B.

Linear Springs
The magnitude of the equal and opposite forces exerted on two objects connected by a *linear springs* is

$$|\mathbf{F}| = k|L - L_0|, \qquad (3.2)$$

where k is the *spring constant*, L is the length of the spring, and L_0 is its unstretched length.

Equilibrium
An object is in equilibrium if it is in *steady translation* (each point of the object has the same constant velocity) relative to an inertial reference frame. The sum of the external forces acting on an object in equilibrium is zero:

$$\Sigma \mathbf{F} = \mathbf{0}. \qquad (3.3)$$

Free-Body Diagrams
A *free-body diagram* is a drawing of an object, isolated from its surroundings, that shows the external forces acting on it. Drawing a free-body diagram involves three steps.

1. Identify the object you want to isolate.
2. Draw a sketch of the object isolated from its surroundings.
3. Draw vectors representing the external forces acting on the object.

3.2 Two-Dimensional Force Systems

Suppose that the system of external forces acting on an object in equilibrium is two dimensional (coplanar). By orienting a coordinate system so that the forces lie in the x–y plane, we can express the sum of the external forces as

$$\Sigma \mathbf{F} = (\Sigma F_x)\mathbf{i} + (\Sigma F_y)\mathbf{j} = \mathbf{0},$$

where ΣF_x and ΣF_y are the sums of the x and y components of the forces. Since a vector is zero only if each of its components is zero, we obtain two scalar equilibrium equations:

$$\Sigma F_x = 0, \qquad \Sigma F_y = 0. \qquad (3.4)$$

The sums of the x and y components of the external forces acting on an object in equilibrium must each equal zero.

Active Example 3.1   **Using Equilibrium to Determine Forces** (▶ *Related Problem 3.1*)

The 1440-kg car is held in place on the inclined ramp by the horizontal cable from *A* to *B*. The car's brakes are not engaged, so the tires exert only normal forces on the ramp. Determine the magnitude of the force exerted on the car by the cable.

Strategy
Because the car is in equilibrium, we can draw its free-body diagram and use Eqs. (3.4) to determine the force exerted by the cable.

Solution
Draw the Free-Body Diagram of the Car

Draw a sketch of the isolated car.	⟶

Complete the free-body diagram by showing the forces exerted on the car by its weight, the cable, and the ramp.	⟶

Apply the Equilibrium Equations

$$\Sigma F_x = T - N \sin 20° = 0,$$

$$\Sigma F_y = N \cos 20° - mg = 0.$$

Eliminating N yields

$$T = \frac{mg \sin 20°}{\cos 20°}$$

$$= \frac{(1440 \text{ kg})(9.81 \text{ m/s}^2)\sin 20°}{\cos 20°}$$

$$= 5140 \text{ N}.$$

Practice Problem Suppose that the cable attachment point *B* is moved upward so that the cable is parallel to the ramp. Determine the magnitude of the force exerted on the car by the cable.

Answer: 4830 N.

| **Example 3.2** | **Choosing a Free-Body Diagram** (▶ *Related Problem 3.3*) |

The automobile engine block is suspended by a system of cables. The mass of the block is 200 kg. The system is stationary. What are the tensions in cables AB and AC?

Strategy
We need a free-body diagram that is subjected to the forces we want to determine. By isolating part of the cable system near point A where the cables are joined, we can obtain a free-body diagram that is subjected to the weight of the block and the unknown tensions in cables AB and AC.

Solution
Draw the Free-Body Diagram   Isolating part of the cable system near point A (Fig. a), we obtain a free-body diagram subjected to the weight of the block $W = mg = (200 \text{ kg}) (9.81 \text{ m/s}^2) = 1962$ N and the tensions in cables AB and AC (Fig. b).

(a)   (b)

(a) Isolating part of the cable system.
(b) The completed free-body diagram.

(c) Selecting a coordinate system and resolving the forces into components.

Apply the Equilibrium Equations   We select the coordinate system shown in Fig. c and resolve the cable tensions into x and y components. The resulting equilibrium equations are

$$\Sigma F_x = T_{AC} \cos 45° - T_{AB} \cos 60° = 0,$$
$$\Sigma F_y = T_{AC} \sin 45° + T_{AB} \sin 60° - 1962 \text{ N} = 0.$$

Solving these equations, we find that the tensions in the cables are $T_{AB} = 1436$ N and $T_{AC} = 1016$ N.

Critical Thinking
How can you choose a free-body diagram that permits you to determine particular unknown forces? There are no definite rules for choosing free-body diagrams. You will learn what to do in many cases from the examples we present, but you will also encounter new situations. It may be necessary to try several free-body diagrams before finding one that provides the information you need. Remember that forces you want to determine should appear as external forces on your free-body diagram, and your objective is to obtain a number of equilibrium equations equal to the number of unknown forces.

| Example 3.3 | Applying Equilibrium to a System of Pulleys (▶ *Related Problem 3.54*) |

The mass of each pulley of the system is m, and the mass of the suspended object A is m_A. Determine the force T necessary for the system to be in equilibrium.

Strategy

By drawing free-body diagrams of the individual pulleys and applying equilibrium, we can relate the force T to the weights of the pulleys and the object A.

Solution

We first draw a free-body diagram of the pulley C to which the force T is applied (Fig. a). Notice that we assume the tension in the cable supported by the pulley to equal T on both sides (see Fig. 3.8). From the equilibrium equation

$$T_D - T - T - mg = 0,$$

we determine that the tension in the cable supported by pulley D is

$$T_D = 2T + mg.$$

We now know the tensions in the cables extending from pulleys C and D to pulley B in terms of T. Drawing the free-body diagram of pulley B (Fig. b), we obtain the equilibrium equation

$$T + T + 2T + mg - mg - m_A g = 0.$$

Solving, we obtain $T = m_A g/4$.

Critical Thinking

Notice that the objects we isolate in Figs. a and b include parts of the cables. The weights of those parts of cable are external forces acting on the free-body diagrams. Why didn't we include them? We tacitly assumed that the weights of those parts of cable could be neglected in comparison to the weights of the pulleys and the suspended object A. You will notice throughout the book that weights of objects are often neglected in analyzing the forces acting on them. This is a valid approximation for a given object if its weight is small compared to the other forces acting on it. But in any real engineering application, this assumption must be carefully evaluated. We discuss the weights of objects in more detail in Chapter 7.

(a) Free-body diagram of pulley C.
(b) Free-body diagram of pulley B.

| Example 3.4 | **Forces on an Airplane in Equilibrium** (▶ *Related Problems 3.60–3.62*) |

The figure shows an airplane flying in the vertical plane and its free-body diagram. The forces acting on the airplane are its weight W, the thrust T exerted by its engines, and aerodynamic forces resulting from the pressure distribution on the airplane's surface. The dashed line indicates the path along which the airplane is moving. The aerodynamic forces are resolved into a component perpendicular to the path, the lift L, and a component parallel to the path, the drag D. The angle γ between the horizontal and the path is called the flight path angle, and α is the angle of attack. If the airplane remains in equilibrium for an interval of time, it is said to be in steady flight. If $\gamma = 6°$, $D = 125$ kN, $L = 680$ kN, and the mass of the airplane is 72,000 kg, what values of T and α are necessary to maintain steady flight?

Strategy

The airplane is assumed to be in equilibrium. By applying Eqs. (3.4) to the given free-body diagram, we will obtain two equations with which to determine T and α.

Solution

In terms of the coordinate system in the figure, the equilibrium equations are

$$\Sigma F_x = T \cos \alpha - D - W \sin \gamma = 0, \tag{1}$$
$$\Sigma F_y = T \sin \alpha + L - W \cos \gamma = 0, \tag{2}$$

where the airplane's weight is $W = (72{,}000 \text{ kg})(9.81 \text{ m/s}^2) = 706{,}000$ N. We solve Eq. (2) for $\sin \alpha$, solve Eq. (1) for $\cos \alpha$, and divide to obtain an equation for $\tan \alpha$:

$$\tan \alpha = \frac{W \cos \gamma - L}{W \sin \gamma + D}$$

$$= \frac{(706{,}000 \text{ N}) \cos 6° - 680{,}000 \text{ N}}{(706{,}000 \text{ N}) \sin 6° + 125{,}000 \text{ N}} = 0.113.$$

The angle of attack $\alpha = \arctan(0.113) = 6.44°$. Now we use Eq. (1) to determine the thrust:

$$T = \frac{W \sin \gamma + D}{\cos \alpha}$$

$$= \frac{(706{,}000 \text{ N}) \sin 6° + 125{,}000 \text{ N}}{\cos 6.44°} = 200{,}000 \text{ N}.$$

Notice that the thrust necessary for steady flight is 28% of the airplane's weight.

Problems

▶ **3.1** In Active Example 3.1, suppose that the angle between the ramp supporting the car is increased from 20° to 30°. Draw the free-body diagram of the car showing the new geometry. Suppose that the cable from A to B must exert a 1900-lb horizontal force on the car to hold it in place. Determine the car's weight in pounds.

3.2 The ring weighs 5 lb and is in equilibrium. The force $F_1 = 4.5$ lb. Determine the force F_2 and the angle α.

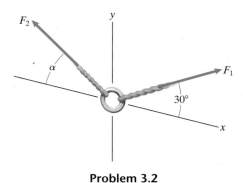

Problem 3.2

▶ **3.3** In Example 3.2, suppose that the attachment point C is moved to the right and cable AC is extended so that the angle between cable AC and the ceiling decreases from 45° to 35°. The angle between cable AB and the ceiling remains 60°. What are the tensions in cables AB and AC?

3.4 The 200-kg engine block is suspended by the cables AB and AC. The angle $\alpha = 40°$. The free-body diagram obtained by isolating the part of the system within the dashed line is shown. Determine the forces T_{AB} and T_{AC}.

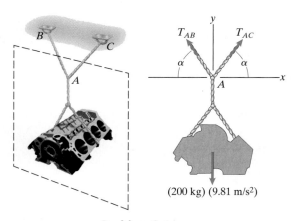

Problem 3.4

3.5 A heavy rope used as a mooring line for a cruise ship sags as shown. If the mass of the rope is 90 kg, what are the tensions in the rope at A and B?

Problem 3.5

3.6 A physiologist estimates that the masseter muscle of a predator, *Martes*, is capable of exerting a force M as large as 900 N. Assume that the jaw is in equilibrium and determine the necessary force T that the temporalis muscle exerts and the force P exerted on the object being bitten.

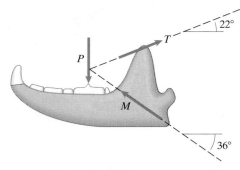

Problem 3.6

3.7 The two springs are identical, with unstretched lengths 250 mm and spring constants $k = 1200$ N/m.

(a) Draw the free-body diagram of block A.

(b) Draw the free-body diagram of block B.

(c) What are the masses of the two blocks?

3.8 The two springs are identical, with unstretched lengths of 250 mm. Suppose that their spring constant k is unknown and the sum of the masses of blocks A and B is 10 kg. Determine the value of k and the masses of the two blocks.

Problems 3.7/3.8

3.9 The inclined surface is smooth. (Remember that "smooth" means that friction is negligible.) The two springs are identical, with unstretched lengths of 250 mm and spring constants $k = 1200$ N/m. What are the masses of blocks A and B?

Problem 3.9

3.10 The mass of the crane is 20,000 kg. The crane's cable is attached to a caisson whose mass is 400 kg. The tension in the cable is 1 kN.

(a) Determine the magnitudes of the normal and friction forces exerted on the crane by the level ground.

(b) Determine the magnitudes of the normal and friction forces exerted on the caisson by the level ground.

Strategy: To do part (a), draw the free-body diagram of the crane and the part of its cable within the dashed line.

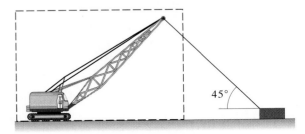

Problem 3.10

3.11 The inclined surface is smooth. The 100-kg crate is held stationary by a force T applied to the cable.

(a) Draw the free-body diagram of the crate.

(b) Determine the force T.

Problem 3.11

3.12 The 1200-kg car is stationary on the sloping road.

(a) If $\alpha = 20°$, what are the magnitudes of the total normal and friction forces exerted on the car's tires by the road?

(b) The car can remain stationary only if the total friction force necessary for equilibrium is not greater than 0.6 times the total normal force. What is the largest angle α for which the car can remain stationary?

Problem 3.12

3.13 The 100-lb crate is in equilibrium on the smooth surface. The spring constant is $k = 400$ lb/ft. Let S be the stretch of the spring. Obtain an equation for S (in feet) as a function of the angle α.

Problem 3.13

3.14 The 600-lb box is held in place on the smooth bed of the dump truck by the rope AB.

(a) If $\alpha = 25°$, what is the tension in the rope?

(b) If the rope will safely support a tension of 400 lb, what is the maximum allowable value of α?

Problem 3.14

3.15 The 80-lb box is held in place on the smooth inclined surface by the rope AB. Determine the tension in the rope and the normal force exerted on the box by the inclined surface.

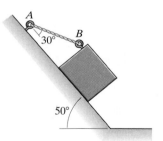

Problem 3.15

3.16 The 1360-kg car and the 2100-kg tow truck are stationary. The muddy surface on which the car's tires rest exerts negligible friction forces on them. What is the tension in the tow cable?

Problem 3.16

3.17 Each box weighs 40 lb. The angles are measured relative to the horizontal. The surfaces are smooth. Determine the tension in the rope A and the normal force exerted on box B by the inclined surface.

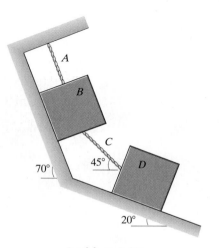

Problem 3.17

3.18 A 10-kg painting is hung with a wire supported by a nail. The length of the wire is 1.3 m.

(a) What is the tension in the wire?

(b) What is the magnitude of the force exerted on the nail by the wire?

Problem 3.18

3.19 A 10-kg painting is hung with a wire supported by two nails. The length of the wire is 1.3 m.

(a) What is the tension in the wire?

(b) What is the magnitude of the force exerted on each nail by the wire? (Assume that the tension is the same in each part of the wire.)

Compare your answers to the answers to Problem 3.18.

Problem 3.19

3.20 Assume that the 150-lb climber is in equilibrium. What are the tensions in the rope on the left and right sides?

3.21 If the mass of the climber shown in Problem 3.20 is 80 kg, what are the tensions in the rope on the left and right sides?

Problems 3.20/3.21

3.22 The construction worker exerts a 20-lb force on the rope to hold the crate in equilibrium in the position shown. What is the weight of the crate?

Problem 3.22

3.23 A construction worker on the moon, where the acceleration due to gravity is 1.62 m/s^2, holds the same crate described in Problem 3.22 in the position shown. What force must she exert on the cable to hold the crate in equilibrium (a) in newtons; (b) in pounds?

Problem 3.23

3.24 The person wants to cause the 200-lb crate to start sliding toward the right. To achieve this, the *horizontal component* of the force exerted on the crate by the rope must equal 0.35 times the normal force exerted on the crate by the floor. In Fig. a, the person pulls on the rope in the direction shown. In Fig. b, the person attaches the rope to a support as shown and pulls upward on the rope. What is the magnitude of the force he must exert on the rope in each case?

(a)

(b)

Problem 3.24

3.25 A traffic engineer wants to suspend a 200-lb traffic light above the center of the two right lanes of a four-lane thoroughfare as shown. Points *A* and *C* are at the same height. Determine the tensions in the cables *AB* and *BC*.

Problem 3.25

3.26 Cable *AB* is 3 m long and cable *BC* is 4 m long. Points *A* and *C* are at the same height. The mass of the suspended object is 350 kg. Determine the tensions in cables *AB* and *BC*.

3.27 The length of cable *AB* is adjustable. Cable *BC* is 4 m long. If you don't want the tension in either cable *AB* or cable *BC* to exceed 3 kN, what is the minimum acceptable length of cable *AB*?

Problems 3.26/3.27

3.28 What are the tensions in the upper and lower cables? (Your answers will be in terms of *W*. Neglect the weight of the pulley.)

Problem 3.28

3.29 Two tow trucks lift a 660-lb motorcycle out of a ravine following an accident. If the motorcycle is in equilibrium in the position shown, what are the tensions in cables *AB* and *AC*?

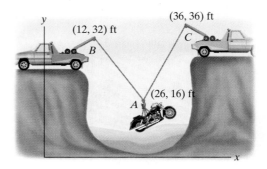

Problem 3.29

3.30 An astronaut candidate conducts experiments on an airbearing platform. While she carries out calibrations, the platform is held in place by the horizontal tethers AB, AC, and AD. The forces exerted by the tethers are the only horizontal forces acting on the platform. If the tension in tether AC is 2 N, what are the tensions in the other two tethers?

TOP VIEW

D

A

4.0 m

3.5 m

B

C

3.0 m   1.5 m

Problem 3.30

3.31 The bucket contains concrete and weighs 5800 lb. What are the tensions in the cables AB and AC?

(5, 34) ft   B

C   (20, 34) ft

y

(12, 16) ft   A

x

Problem 3.31

3.32 The slider A is in equilibrium and the bar is smooth. What is the mass of the slider?

20°

200 N

A

45°

Problem 3.32

3.33 The 20-kg mass is suspended from three cables. Cable AC is equipped with a turnbuckle so that its tension can be adjusted and a strain gauge that allows its tension to be measured. If the tension in cable AC is 40 N, what are the tensions in cables AB and AD?

|←0.4 m→|←0.4 m→|←0.48 m→|

B   C   D

0.64 m

A

Problem 3.33

3.34 The structural joint is in equilibrium. If $F_A = 1000$ lb and $F_D = 5000$ lb, what are F_B and F_C?

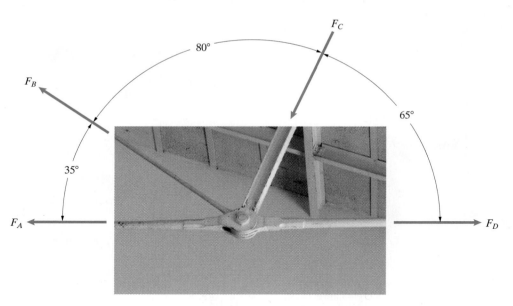

Problem 3.34

3.35 The collar A slides on the smooth vertical bar. The masses $m_A = 20$ kg and $m_B = 10$ kg. When $h = 0.1$ m, the spring is unstretched. When the system is in equilibrium, $h = 0.3$ m. Determine the spring constant k.

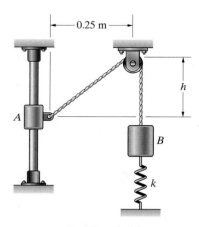

Problem 3.35

3.36* Suppose that you want to design a cable system to suspend an object of weight W from the ceiling. The two wires must be identical, and the dimension b is fixed. The ratio of the tension T in each wire to its cross-sectional area A must equal a specified value $T/A = \sigma$. The "cost" of your design is the total volume of material in the two wires, $V = 2A\sqrt{b^2 + h^2}$. Determine the value of h that minimizes the cost.

Problem 3.36

3.37 The system of cables suspends a 1000-lb bank of lights above a movie set. Determine the tensions in cables AB, CD, and CE.

3.38 A technician changes the position of the 1000-lb bank of lights by removing the cable CE. What is the tension in cable AB after the change?

3.39 While working on another exhibit, a curator at the Smithsonian Institution pulls the suspended *Voyager* aircraft to one side by attaching three horizontal cables as shown. The mass of the aircraft is 1250 kg. Determine the tensions in the cable segments AB, BC, and CD.

Problem 3.39

Problems 3.37/3.38

3.40 A truck dealer wants to suspend a 4000-kg truck as shown for advertising. The distance b = 15 m, and the sum of the lengths of the cables AB and BC is 42 m. Points A and C are at the same height. What are the tensions in the cables?

Problem 3.40

3.41 The distance $h = 12$ in, and the tension in cable AD is 200 lb. What are the tensions in cables AB and AC?

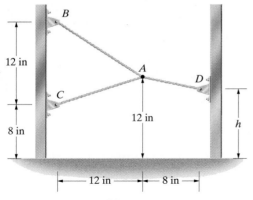

Problem 3.41

3.42 You are designing a cable system to support a suspended object of weight W. Because your design requires points A and B to be placed as shown, you have no control over the angle α, but you can choose the angle β by placing point C wherever you wish. Show that to minimize the tensions in cables AB and BC, you must choose $\beta = \alpha$ if the angle $\alpha \geq 45°$.

Strategy: Draw a diagram of the sum of the forces exerted by the three cables at A.

Problem 3.42

3.43* The length of the cable ABC is 1.4 m. The 2-kN force is applied to a small pulley. The system is stationary. What is the tension in the cable?

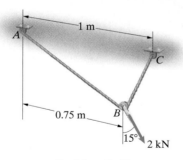

Problem 3.43

3.44 The masses $m_1 = 12$ kg and $m_2 = 6$ kg are suspended by the cable system shown. The cable BC is horizontal. Determine the angle α and the tensions in the cables AB, BC, and CD.

Problem 3.44

3.45 The weights $W_1 = 50$ lb and W_2 are suspended by the cable system shown. Determine the weight W_2 and the tensions in the cables AB, BC, and CD.

3.46 Assume that $W_2 = W_1/2$. If you don't want the tension anywhere in the supporting cable to exceed 200 lb, what is the largest acceptable value of W_1?

Problems 3.45/3.46

3.47 The hydraulic cylinder is subjected to three forces. An 8-kN force is exerted on the cylinder at B that is parallel to the cylinder and points from B toward C. The link AC exerts a force at C that is parallel to the line from A to C. The link CD exerts a force at C that is parallel to the line from C to D.

(a) Draw the free-body diagram of the cylinder. (The cylinder's weight is negligible.)

(b) Determine the magnitudes of the forces exerted by the links AC and CD.

Problem 3.47

3.48 The 50-lb cylinder rests on two smooth surfaces.

(a) Draw the free-body diagram of the cylinder.

(b) If $\alpha = 30°$, what are the magnitudes of the forces exerted on the cylinder by the left and right surfaces?

3.49 Obtain an equation for the force exerted on the 50-lb cylinder by the left surface in terms of the angle α in two ways: (a) using a coordinate system with the y axis vertical, (b) using a coordinate system with the y axis parallel to the right surface.

Problems 3.48/3.49

3.50 The two springs are identical, with unstretched length 0.4 m. When the 50-kg mass is suspended at B, the length of each spring increases to 0.6 m. What is the spring constant k?

Problem 3.50

3.51 The cable AB is 0.5 m in length. The *unstretched* length of the spring is 0.4 m. When the 50-kg mass is suspended at B, the length of the spring increases to 0.45 m. What is the spring constant k?

Problem 3.51

3.52* The small sphere of mass m is attached to a string of length L and rests on the smooth surface of a fixed sphere of radius R. The center of the sphere is directly below the point where the string is attached. Obtain an equation for the tension in the string in terms of m, L, h, and R.

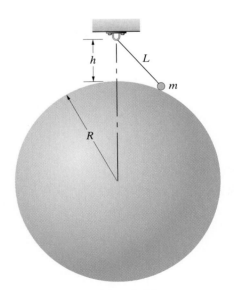

Problem 3.52

3.53 The inclined surface is smooth. Determine the force T that must be exerted on the cable to hold the 100-kg crate in equilibrium and compare your answer to the answer of Problem 3.11.

Problem 3.53

▶ **3.54** In Example 3.3, suppose that the mass of the suspended object is m_A and the masses of the pulleys are $m_B = 0.3m_A$, $m_C = 0.2m_A$, and $m_D = 0.2m_A$. Show that the force T necessary for the system to be in equilibrium is $0.275m_Ag$.

3.55 The mass of each pulley of the system is m and the mass of the suspended object A is m_A. Determine the force T necessary for the system to be in equilibrium.

Problem 3.55

3.56 The suspended mass $m_1 = 50$ kg. Neglecting the masses of the pulleys, determine the value of the mass m_2 necessary for the system to be in equilibrium.

Problem 3.56

3.57 The boy is lifting himself using the block and tackle shown. If the weight of the block and tackle is negligible, and the combined weight of the boy and the beam he is sitting on is 120 lb, what force does he have to exert on the rope to raise himself at a constant rate? (Neglect the deviation of the ropes from the vertical.)

Problem 3.57

3.58 Pulley systems containing one, two, and three pulleys are shown. Neglecting the weights of the pulleys, determine the force T required to support the weight W in each case.

3.59 The number of pulleys in the type of system shown could obviously be extended to an arbitrary number N.

(a) Neglecting the weights of the pulleys, determine the force T required to support the weight W as a function of the number of pulleys N in the system.

(b) Using the result of part (a), determine the force T required to support the weight W for a system with 10 pulleys.

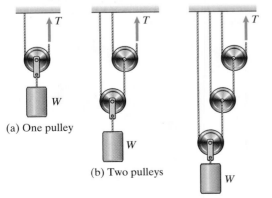

(a) One pulley

(b) Two pulleys

(c) Three pulleys

Problems 3.58/3.59

▶ **3.60** A 14,000-kg airplane is in steady flight in the vertical plane. The flight path angle is $\gamma = 10°$, the angle of attack is $\alpha = 4°$, and the thrust force exerted by the engine is $T = 60$ kN. What are the magnitudes of the lift and drag forces acting on the airplane? (See Example 3.4.)

▶ **3.61** An airplane is in steady flight, the angle of attack $\alpha = 0$, the thrust-to-drag ratio $T/D = 2$, and the lift-to-drag ratio $L/D = 4$. What is the flight path angle γ? (See Example 3.4.)

▶ **3.62** An airplane glides in steady flight ($T = 0$), and its lift-to-drag ratio is $L/D = 4$.

(a) What is the flight path angle γ?

(b) If the airplane glides from an altitude of 1000 m to zero altitude, what horizontal distance does it travel? (See Example 3.4.)

3.3 Three-Dimensional Force Systems

The equilibrium situations we have considered so far have involved only coplanar forces. When the system of external forces acting on an object in equilibrium is three dimensional, we can express the sum of the external forces as

$$\Sigma \mathbf{F} = (\Sigma F_x)\mathbf{i} + (\Sigma F_y)\mathbf{j} + (\Sigma F_z)\mathbf{k} = \mathbf{0}.$$

Each component of this equation must equal zero, resulting in three scalar equilibrium equations:

$$\Sigma F_x = 0, \quad \Sigma F_y = 0, \quad \Sigma F_z = 0. \tag{3.5}$$

The sums of the x, y, and z components of the external forces acting on an object in equilibrium must each equal zero.

Active Example 3.5 (▶ *Related Problem 3.63*)

The 100-kg cylinder is suspended from the ceiling by cables attached at points B, C, and D. What are the tensions in cables AB, AC, and AD?

Strategy

By isolating part of the cable system near point A, we will obtain a free-body diagram subjected to forces due to the tensions in the cables. Because the sums of the external forces in the x, y, and z directions must each equal zero, we can obtain three equilibrium equations for the three unknown tensions. To do so, we must express the forces exerted by the tensions in terms of their components.

Solution

Draw the Free-Body Diagram and Apply Equilibrium

Isolate part of the cable system near point A and show the forces exerted due to the tensions in the cables. The sum of the forces must equal zero:

$$\Sigma \mathbf{F} = \mathbf{T}_{AB} + \mathbf{T}_{AC} + \mathbf{T}_{AD} - (981\ \text{N})\mathbf{j} = \mathbf{0}.$$

Write the Forces in Terms of Their Components

$$\mathbf{r}_{AB} = (x_B - x_A)\mathbf{i} + (y_B - y_A)\mathbf{j} + (z_B - z_A)\mathbf{k}.$$
$$= 4\mathbf{i} + 4\mathbf{j} + 2\mathbf{k} \text{ (m).}$$
$$\mathbf{e}_{AB} = \frac{\mathbf{r}_{AB}}{|\mathbf{r}_{AB}|} = 0.667\mathbf{i} + 0.667\mathbf{j} + 0.333\mathbf{k}.$$

> Obtain a unit vector that has the same direction as the force $\mathbf{T}_{AB}$ by dividing the position vector $\mathbf{r}_{AB}$ from point A to point B by its magnitude.

$$\mathbf{T}_{AB} = T_{AB}\mathbf{e}_{AB}$$
$$= T_{AB}(0.667\mathbf{i} + 0.667\mathbf{j} + 0.333\mathbf{k}),$$
$$\mathbf{T}_{AC} = T_{AC}(-0.408\mathbf{i} + 0.816\mathbf{j} - 0.408\mathbf{k}),$$
$$\mathbf{T}_{AD} = T_{AD}(-0.514\mathbf{i} + 0.686\mathbf{j} + 0.514\mathbf{k}).$$

> Express the force $\mathbf{T}_{AB}$ in terms of its components by writing it as the product of the tension T_{AB} in cable AB and the unit vector $\mathbf{e}_{AB}$. Express the forces $\mathbf{T}_{AC}$ and $\mathbf{T}_{AD}$ in terms of their components using the same procedure.

> Substitute these expressions into the equilibrium equation
> $$\mathbf{T}_{AB} + \mathbf{T}_{AC} + \mathbf{T}_{AD} - (981 \text{ N})\mathbf{j} = \mathbf{0}.$$
> Because the $\mathbf{i}$, $\mathbf{j}$, and $\mathbf{k}$ components must each equal zero, this results in three equations:

$$0.667T_{AB} - 0.408T_{AC} - 0.514T_{AD} = 0,$$
$$0.667T_{AB} + 0.816T_{AC} + 0.686T_{AD} - 981 \text{ N} = 0,$$
$$0.333T_{AB} - 0.408T_{AC} + 0.514T_{AD} = 0.$$

Solving these three equations yields $T_{AB} = 519$ N, $T_{AC} = 636$ N, and $T_{AD} = 168$ N.

Practice Problem Suppose that cables AB, AC, and AD are lengthened so that the attachment point A is located at the point $(0, -6, 0)$ m. What are the tensions in the cables?

Answer: $T_{AB} = 432$ N, $T_{AC} = 574$ N, $T_{AD} = 141$ N.

Example 3.6   **Application of the Dot Product** (▶ *Related Problem 3.79*)

The 100-lb "slider" C is held in place on the smooth bar by the cable AC. Determine the tension in the cable and the force exerted on the slider by the bar.

Strategy
Because we want to determine forces that act on the slider, we need to draw its free-body diagram. The external forces acting on the slider are its weight and the forces exerted on it by the cable and the bar. If we approached this example as we did the previous one, our next step would be to express the forces in terms of their components. However, we don't know the direction of the force exerted on the slider by the bar. Since the smooth bar exerts negligible friction force, we do know that the force is normal to the bar's axis. Therefore we can eliminate this force from the equation $\Sigma \mathbf{F} = \mathbf{0}$ by taking the dot product of the equation with a unit vector that is parallel to the bar.

Solution
Draw the Free-Body Diagram   We isolate the slider (Fig. a) and complete the free-body diagram by showing the weight of the slider, the force $\mathbf{T}$ exerted by the tension in the cable, and the normal force $\mathbf{N}$ exerted by the bar (Fig. b).

Apply the Equilibrium Equations   The sum of the external forces acting on the free-body diagram is

$$\Sigma \mathbf{F} = \mathbf{T} + \mathbf{N} - (100\text{ lb})\mathbf{j} = \mathbf{0}. \tag{1}$$

Let $\mathbf{e}_{BD}$ be the unit vector pointing from point B toward point D. Since $\mathbf{N}$ is perpendicular to the bar, $\mathbf{e}_{BD} \cdot \mathbf{N} = 0$. Therefore,

$$\mathbf{e}_{BD} \cdot (\Sigma \mathbf{F}) = \mathbf{e}_{BD} \cdot [\mathbf{T} - (100\text{ lb})\mathbf{j}] = 0. \tag{2}$$

Determining $\mathbf{e}_{BD}$: We determine the vector from point B to point D,

$$\mathbf{r}_{BD} = (4 - 0)\mathbf{i} + (0 - 7)\mathbf{j} + (4 - 0)\mathbf{k} = 4\mathbf{i} - 7\mathbf{j} + 4\mathbf{k}\text{(ft)},$$

and divide it by its magnitude to obtain the unit vector $\mathbf{e}_{BD}$:

$$\mathbf{e}_{BD} = \frac{\mathbf{r}_{BD}}{|\mathbf{r}_{BD}|} = \frac{4}{9}\mathbf{i} - \frac{7}{9}\mathbf{j} + \frac{4}{9}\mathbf{k}.$$

Expressing **T** *in terms of components:* We need to determine the coordinates of the slider C. We can write the vector from B to C in terms of the unit vector $\mathbf{e}_{BD}$,

$$\mathbf{r}_{BC} = 6\mathbf{e}_{BD} = 2.67\mathbf{i} - 4.67\mathbf{j} + 2.67\mathbf{k} \text{ (ft)},$$

and then add it to the vector from the origin O to B to obtain the vector from O to C:

$$\mathbf{r}_{OC} = \mathbf{r}_{OB} + \mathbf{r}_{BC} = 7\mathbf{j} + (2.67\mathbf{i} - 4.67\mathbf{j} + 2.67\mathbf{k})$$
$$= 2.67\mathbf{i} + 2.33\mathbf{j} + 2.67\mathbf{k} \text{ (ft)}.$$

The components of this vector are the coordinates of point C. Now we can determine a unit vector with the same direction as **T**. The vector from C to A is

$$\mathbf{r}_{CA} = (0 - 2.67)\mathbf{i} + (7 - 2.33)\mathbf{j} + (4 - 2.67)\mathbf{k}$$
$$= -2.67\mathbf{i} + 4.67\mathbf{j} + 1.33\mathbf{k} \text{ (ft)},$$

and the unit vector that points from point C toward point A is

$$\mathbf{e}_{CA} = \frac{\mathbf{r}_{CA}}{|\mathbf{r}_{CA}|} = -0.482\mathbf{i} + 0.843\mathbf{j} + 0.241\mathbf{k}.$$

Let T be the tension in the cable AC. Then we can write the vector **T** as

$$\mathbf{T} = T\mathbf{e}_{CA} = T(-0.482\mathbf{i} + 0.843\mathbf{j} + 0.241\mathbf{k}).$$

Determining **T** *and* **N**: Substituting our expressions for $\mathbf{e}_{BD}$ and **T** in terms of their components into Eq. (2) yields

$$0 = \mathbf{e}_{BD} \cdot [\mathbf{T} - (100 \text{ lb})\mathbf{j}]$$

$$= \left(\frac{4}{9}\mathbf{i} - \frac{7}{9}\mathbf{j} + \frac{4}{9}\mathbf{k}\right) \cdot [-0.482T\mathbf{i} + (0.843T - 100 \text{ lb})\mathbf{j} + 0.241T\mathbf{k}]$$

$$= -0.762T + 77.8 \text{ lb},$$

and we obtain the tension $T = 102$ lb.

Now we can determine the force exerted on the slider by the bar by using Eq. (1):

$$\mathbf{N} = -\mathbf{T} + (100 \text{ lb})\mathbf{j}$$
$$= -(102 \text{ lb})(-0.482\mathbf{i} + 0.843\mathbf{j} + 0.241\mathbf{k}) + (100 \text{ lb})\mathbf{j}$$
$$= 49.1\mathbf{i} + 14.0\mathbf{j} - 24.6\mathbf{k} \text{ (lb)}.$$

(a)

$-100\,\mathbf{j}$ (lb)

(b)

(a) Isolating the slider.
(b) Free-body diagram of the slider showing the forces exerted by its weight, the cable, and the bar.

Critical Thinking

By taking the dot product of the equilibrium equation for the slider with a unit vector $\mathbf{e}_{BD}$ that is parallel to the smooth bar BD, we obtained Eq. (2), which does not contain the normal force **N**. Why does this happen? The formal answer is that $\mathbf{e}_{BD}$ is perpendicular to **N**, and so $\mathbf{e}_{BD} \cdot \mathbf{N} = 0$. But the physical interpretation of Eq. (2) provides a more compelling explanation: It states that *the component of the slider's weight parallel to the bar is balanced by the component of* **T** *parallel to the bar*. The normal force exerted on the slider by the smooth bar has no component parallel to the bar. We were therefore able to solve for the tension in the cable without knowing the normal force **N**.

Problems

▶ **3.63** In Active Example 3.5, suppose that the attachment point B is moved to the point $(5, 0, 0)$ m. What are the tensions in cables AB, AC, and AD?

3.64 The force $\mathbf{F} = 800\mathbf{i} + 200\mathbf{j}$ (lb) acts at point A where the cables AB, AC, and AD are joined. What are the tensions in the three cables?

3.65* Suppose that you want to apply a 1000-lb force $\mathbf{F}$ at point A in a direction such that the resulting tensions in cables AB, AC, and AD are equal. Determine the components of $\mathbf{F}$.

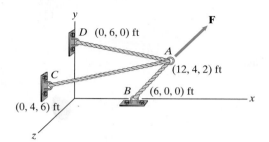

Problems 3.64/3.65

3.66 The 10-lb metal disk A is supported by the smooth inclined surface and the strings AB and AC. The disk is located at coordinates $(5, 1, 4)$ ft. What are the tensions in the strings?

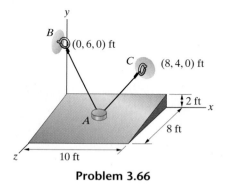

Problem 3.66

3.67 The bulldozer exerts a force $\mathbf{F} = 2\mathbf{i}$ (kip) at A. What are the tensions in cables AB, AC, and AD?

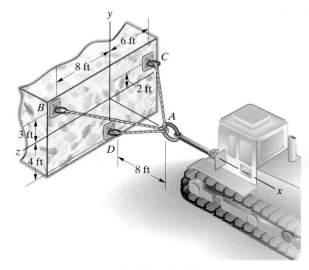

Problem 3.67

3.68 Prior to its launch, a balloon carrying a set of experiments to high altitude is held in place by groups of student volunteers holding the tethers at B, C, and D. The mass of the balloon, experiments package, and the gas it contains is 90 kg, and the buoyancy force on the balloon is 1000 N. The supervising professor conservatively estimates that each student can exert at least a 40-N tension on the tether for the necessary length of time. Based on this estimate, what minimum numbers of students are needed at B, C, and D?

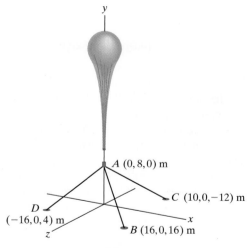

Problem 3.68

3.69 The 20-kg mass is suspended by cables attached to three vertical 2-m posts. Point A is at $(0, 1.2, 0)$ m. Determine the tensions in cables AB, AC, and AD.

3.71 The car in Fig. a and the pallet supporting it weigh 3000 lb. They are supported by four cables AB, AC, AD, and AE. The locations of the attachment points on the pallet are shown in Fig. b. The tensions in cables AB and AE are equal. Determine the tensions in the cables.

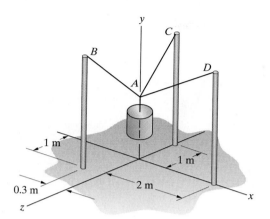

Problem 3.69

3.70 The weight of the horizontal wall section is $W = 20,000$ lb. Determine the tensions in the cables AB, AC, and AD.

Problem 3.70

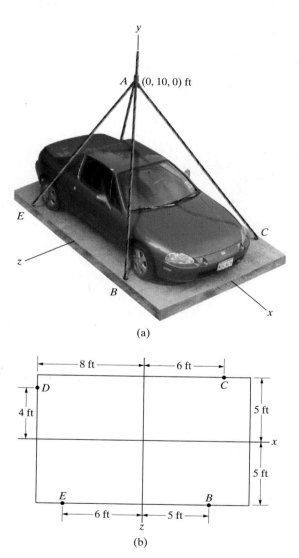

(a)

(b)

Problem 3.71

3.72 The 680-kg load suspended from the helicopter is in equilibrium. The aerodynamic drag force on the load is horizontal. The y axis is vertical, and cable OA lies in the x–y plane. Determine the magnitude of the drag force and the tension in cable OA.

3.73 The coordinates of the three cable attachment points B, C, and D are $(-3.3, -4.5, 0)$ m, $(1.1, -5.3, 1)$ m, and $(1.6, -5.4, -1)$ m, respectively. What are the tensions in cables OB, OC, and OD?

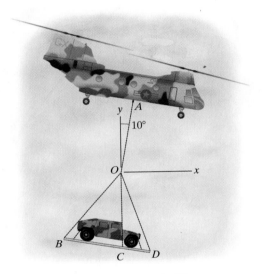

Problems 3.72/3.73

3.74 If the mass of the bar AB is negligible compared to the mass of the suspended object E, the bar exerts a force on the "ball" at B that points from A toward B. The mass of the object E is 200 kg. The y axis points upward. Determine the tensions in the cables BC and BD.

   Strategy: Draw a free-body diagram of the ball at B. (The weight of the ball is negligible.)

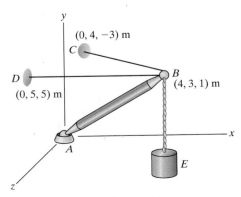

Problem 3.74

3.75* The 3400-lb car is at rest on the plane surface. The unit vector $e_n = 0.456i + 0.570j + 0.684k$ is perpendicular to the surface. Determine the magnitudes of the total normal force N and the total friction force f exerted on the surface by the car's wheels.

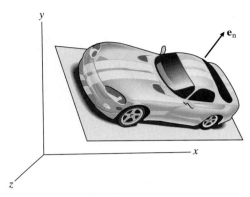

Problem 3.75

3.76 The system shown anchors a stanchion of a cable-suspended roof. If the tension in cable AB is 900 kN, what are the tensions in cables EF and EG?

3.77* The cables of the system will each safely support a tension of 1500 kN. Based on this criterion, what is the largest safe value of the tension in cable AB?

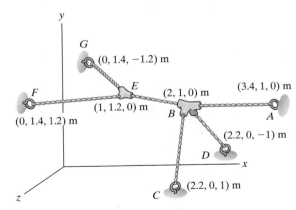

Problems 3.76/3.77

3.78 The 200-kg slider at A is held in place on the smooth vertical bar by the cable AB.

(a) Determine the tension in the cable.

(b) Determine the force exerted on the slider by the bar.

Problem 3.78

▶ **3.79** In Example 3.6, suppose that the cable AC is replaced by a longer one so that the distance from point B to the slider C increases from 6 ft to 8 ft. Determine the tension in the cable.

3.80 The cable AB keeps the 8-kg collar A in place on the smooth bar CD. The y axis points upward. What is the tension in the cable?

3.81* Determine the magnitude of the normal force exerted on the collar A by the smooth bar.

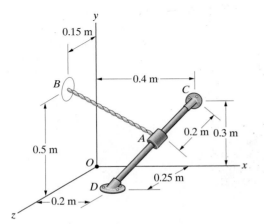

Problems 3.80/3.81

3.82* The 10-kg collar A and 20-kg collar B are held in place on the smooth bars by the 3-m cable from A to B and the force F acting on A. The force F is parallel to the bar. Determine F.

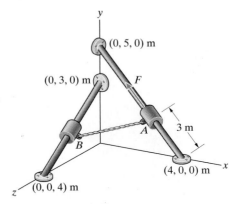

Problem 3.82

Review Problems

3.83 The 100-lb crate is held in place on the smooth surface by the rope *AB*. Determine the tension in the rope and the magnitude of the normal force exerted on the crate by the surface.

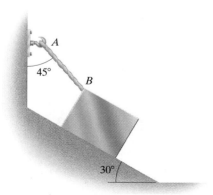

Problem 3.83

3.84 The system shown is called Russell's traction. If the sum of the downward forces exerted at *A* and *B* by the patient's leg is 32.2 lb, what is the weight *W*?

Problem 3.84

3.85 The 400-lb engine block is suspended by the cables *AB* and *AC*. If you don't want either T_{AB} or T_{AC} to exceed 400 lb, what is the smallest acceptable value of the angle α?

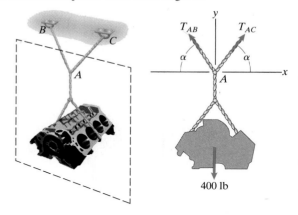

Problem 3.85

3.86 The cable *AB* is horizontal, and the box on the right weighs 100 lb. The surfaces are smooth.

(a) What is the tension in the cable?

(b) What is the weight of the box on the left?

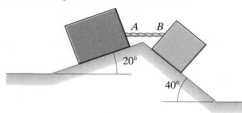

Problem 3.86

3.87 Assume that the forces exerted on the 170-lb climber by the slanted walls of the "chimney" are perpendicular to the walls. If he is in equilibrium and is exerting a 160-lb force on the rope, what are the magnitudes of the forces exerted on him by the left and right walls?

Problem 3.87

3.88 The mass of the suspended object A is m_A and the masses of the pulleys are negligible. Determine the force T necessary for the system to be in equilibrium.

Problem 3.88

3.89 The assembly A, including the pulley, weighs 60 lb. What force F is necessary for the system to be in equilibrium?

Problem 3.89

3.90 The mass of block A is 42 kg, and the mass of block B is 50 kg. The surfaces are smooth. If the blocks are in equilibrium, what is the force F?

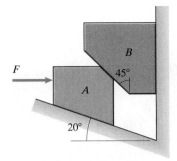

Problem 3.90

3.91 The climber A is being helped up an icy slope by two friends. His mass is 80 kg, and the direction cosines of the force exerted on him by the slope are $\cos \theta_x = -0.286$, $\cos \theta_y = 0.429$, and $\cos \theta_z = 0.857$. The y axis is vertical. If the climber is in equilibrium in the position shown, what are the tensions in the ropes AB and AC and the magnitude of the force exerted on him by the slope?

3.92 Consider the climber A being helped by his friends in Problem 3.91. To try to make the tensions in the ropes more equal, the friend at B moves to the position $(4, 2, 0)$ m. What are the new tensions in the ropes AB and AC and the magnitude of the force exerted on the climber by the slope?

Problems 3.91/3.92

3.93 A climber helps his friend up an icy slope. His friend is hauling a box of supplies. If the mass of the friend is 90 kg and the mass of the supplies is 22 kg, what are the tensions in the ropes AB and CD? Assume that the slope is smooth. That is, only normal forces are exerted on the man and the box by the slope.

Problem 3.93

3.94 The 2800-lb car is moving at constant speed on a road with the slope shown. The aerodynamics forces on the car are the drag $D = 270$ lb, which is parallel to the road, and the lift $L = 120$ lb, which is perpendicular to the road. Determine the magnitudes of the total normal and friction forces exerted on the car by the road.

Problem 3.94

3.95 An engineer doing preliminary design studies for a new radio telescope envisions a triangular receiving platform suspended by cables from three equally spaced 40-m towers. The receiving platform has a mass of 20 Mg (megagrams) and is 10 m below the tops of the towers. What tension would the cables be subjected to?

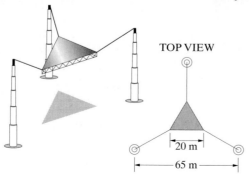

Problem 3.95

3.96 To support the tent, the tension in the rope AB must be 35 lb. What are the tensions in the ropes AC, AD, and AE?

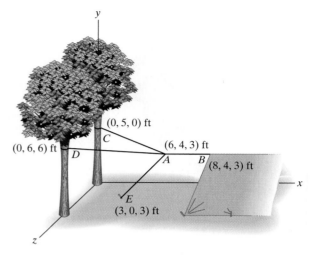

Problem 3.96

3.97 Cable AB is attached to the top of the vertical 3-m post, and its tension is 50 kN. What are the tensions in cables AO, AC, and AD?

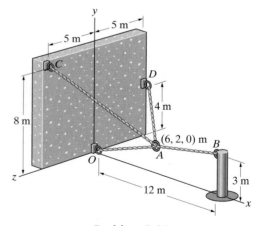

Problem 3.97

3.98* The 1350-kg car is at rest on a plane surface with its brakes locked. The unit vector $e_n = 0.231i + 0.923j + 0.308k$ is perpendicular to the surface. The y axis points upward. The direction cosines of the cable from A to B are $\cos \theta_x = -0.816$, $\cos \theta_y = 0.408$, $\cos \theta_z = -0.408$, and the tension in the cable is 1.2 kN. Determine the magnitudes of the normal and friction forces the car's wheels exert on the surface.

3.99* The brakes of the car are released, and the car is held in place on the plane surface by the cable AB. The car's front wheels are aligned so that the tires exert no friction forces parallel to the car's longitudinal axis. The unit vector $e_p = -0.941i + 0.131j + 0.314k$ is parallel to the plane surface and aligned with the car's longitudinal axis. What is the tension in the cable?

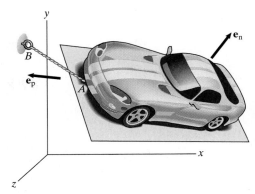

Problems 3.98/3.99

Design Project 1 A possible design for a simple scale to weigh objects is shown. The length of the string AB is 0.5 m. When an object is placed in the pan, the spring stretches and the string AB rotates. The object's weight can be determined by observing the change in the angle α.

(a) Assume that objects with masses in the range 0.2–2 kg are to be weighed. Choose the unstretched length and spring constant of the spring in order to obtain accurate readings for weights in the desired range. (Neglect the weights of the pan and spring. Notice that a significant change in the angle α is needed to determine the weight accurately.)

(b) Suppose that you can use the same components—the pan, protractor, a spring, string—and also one or more pulleys. Suggest another possible configuration for the scale. Use statics to analyze your proposed configuration and compare its accuracy with that of the configuration shown for objects with masses in the range 0.2–2 kg.

Design Project 2 Suppose that the positions of points A, C, and D of the system of cables suspending the 100-kg mass are fixed, but you are free to choose the x and z coordinates of point B. Investigate the effects of different choices of the location of point B on the tensions in the cables. If the cost of cable AB is proportional to the product of the tension in the cable and its length, investigate the effect of different choices of the location of point B on the cost of the cable. Write a brief report describing the results of your investigations and recommending a location for point B.

Systems of Forces and Moments

The effects of forces can depend not only on their magnitudes and directions but also on the moments, or torques, they exert. The rotations of objects such as the wheels of a vehicle, the crankshaft of an engine, and the rotor of an electric generator result from the moments of the forces exerted on them. If an object is in equilibrium, the moment about any point due to the forces acting on the object is zero. Before continuing our discussion of free-body diagrams and equilibrium, we must explain how to calculate moments and introduce the concept of equivalent systems of forces and moments.

◄ The counterweight of the building crane exerts a large moment that the crane's structure must support during assembly. In this chapter we calculate moments of forces and analyze systems of forces and moments.

4.1 Two-Dimensional Description of the Moment

BACKGROUND

Consider a force of magnitude F and a point P, and let's view them in the direction perpendicular to the plane containing the force vector and the point (Fig. 4.1a). The *magnitude of the moment* of the force about P is the product DF, where D is the perpendicular distance from P to the line of action of the force (Fig. 4.1b). In this example, the force would tend to cause counterclockwise rotation about point P. That is, if we imagine that the force acts on

(a)   (b)   (c)

Figure 4.1

(a) The force and point P.

(b) The perpendicular distance D from point P to the line of action of F.

(c) The direction of the moment is counterclockwise.

an object that can rotate about point P, the force would cause counterclockwise rotation (Fig. 4.1c). We say that the *direction of the moment* is counterclockwise. *We define counterclockwise moments to be positive and clockwise moments to be negative.* (This is the usual convention, although we occasionally encounter situations in which it is more convenient to define clockwise moments to be positive.) Thus, the moment of the force about P is

$$M_P = DF. \tag{4.1}$$

Notice that if the line of action of F passes through P, the perpendicular distance $D = 0$ and the moment of F about P is zero.

The dimensions of the moment are (distance) $\times$ (force). For example, moments can be expressed in newton-meters in SI units and in foot-pounds in U.S. Customary units.

Suppose that you want to place a television set on a shelf, and you aren't certain the attachment of the shelf to the wall is strong enough to support it. Intuitively, you place it near the wall (Fig. 4.2a), knowing that the attachment is more likely to fail if you place it away from the wall (Fig. 4.2b). What is the difference in the two cases? The magnitude and direction of the force exerted on the shelf by the weight of the television are the same in each case, but the moments exerted on the attachment are different. The moment exerted about P by its weight when it is near the wall, $M_P = -D_1 W$, is smaller in magnitude than the moment about P when it is placed away from the wall, $M_P = -D_2 W$.

The method we describe in this section can be used to determine the sum of the moments of a system of forces about a point if the forces are two-dimensional (coplanar) and the point lies in the same plane. For example, consider the construction crane shown in Fig. 4.3. The sum of the moments exerted about point P by the load W_1 and the counterweight W_2 is

$$\Sigma M_P = D_1 W_1 - D_2 W_2.$$

This moment tends to cause the top of the vertical tower to rotate and could cause it to collapse. If the distance D_2 is adjusted so that $D_1 W_1 = D_2 W_2$, the moment about point P due to the load and the counterweight is zero.

(a)

(b)

Figure 4.2

(a) Placing the television near the wall minimizes the moment exerted on the support of the shelf at P.

(b) Placing the television far from the wall exerts a large moment on the support at P and could cause it to fail.

Figure 4.3
A tower crane used in the construction of high-rise buildings.

If a force is expressed in terms of components, the moment of the force about a point P is equal to the sum of the moments of its components about P. We prove this very useful result in the next section.

RESULTS

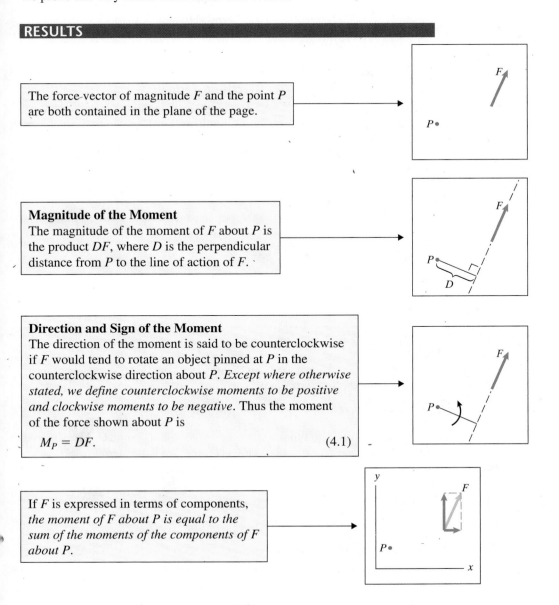

The force vector of magnitude F and the point P are both contained in the plane of the page.

Magnitude of the Moment
The magnitude of the moment of F about P is the product DF, where D is the perpendicular distance from P to the line of action of F.

Direction and Sign of the Moment
The direction of the moment is said to be counterclockwise if F would tend to rotate an object pinned at P in the counterclockwise direction about P. *Except where otherwise stated, we define counterclockwise moments to be positive and clockwise moments to be negative.* Thus the moment of the force shown about P is

$$M_P = DF. \qquad (4.1)$$

If F is expressed in terms of components, *the moment of F about P is equal to the sum of the moments of the components of F about P.*

Active Example 4.1 Determining a Moment (▶ *Related Problem 4.1*)

What is the moment of the 40-kN force about point *A*?

Strategy

We can calculate the magnitude of the moment by determining the perpendicular distance from point *A* to the line of action of the force.

Solution

The perpendicular distance from *A* to the line of action of the force is

$$D = (6 \text{ m})\sin 30° = 3 \text{ m}.$$

Therefore the magnitude of the moment is

$$(3 \text{ m})(40 \text{ kN}) = 120 \text{ kN-m}.$$

The direction of the moment is counterclockwise, so

$$M_A = 120 \text{ kN-m}.$$

Practice Problem Resolve the 40-kN force into horizontal and vertical components and calculate the sum of the moments of the components about *A*.

Answer: 120 kN-m.

Example 4.2 Moment of a System of Forces (▶ *Related Problem 4.12*)

Four forces act on the machine part. What is the sum of the moments of the forces about the origin *O*?

Strategy

We can determine the moments of the forces about point *O* directly from the given information except for the 4-kN force. We will determine its moment by expressing it in terms of components and summing the moments of the components.

Solution

Moment of the 3-kN Force The line of action of the 3-kN force passes through *O*. It exerts no moment about *O*.

Moment of the 5-kN Force   The line of action of the 5-kN force also passes through O. It too exerts no moment about O.

Moment of the 2-kN Force   The perpendicular distance from O to the line of action of the 2-kN force is 0.3 m, and the direction of the moment about O is clockwise. The moment of the 2-kN force about O is

$$-(0.3 \text{ m})(2 \text{ kN}) = -0.600 \text{ kN-m.}$$

(Notice that we converted the perpendicular distance from millimeters into meters, obtaining the result in terms of kilonewton-meters.)

Moment of the 4-kN Force   In Fig. a, we introduce a coordinate system and express the 4-kN force in terms of x and y components. The perpendicular distance from O to the line of action of the x component is 0.3 m, and the direction of the moment about O is clockwise. The moment of the x component about O is

$$-(0.3 \text{ m})(4 \cos 30° \text{ kN}) = -1.039 \text{ kN-m.}$$

(a) Resolving the 4-kN force into components.

The perpendicular distance from point O to the line of action of the y component is 0.7 m, and the direction of the moment about O is counterclockwise. The moment of the y component about O is

$$(0.7 \text{ m})(4 \sin 30° \text{ kN}) = 1.400 \text{ kN-m.}$$

The sum of the moments of the four forces about point O is

$$\Sigma M_0 = -0.600 - 1.039 + 1.400 = -0.239 \text{ kN-m.}$$

The four forces exert a 0.239 kN-m clockwise moment about point O.

Critical Thinking

If an object is subjected to a system of known forces, why is it useful to determine the sum of the moments of the forces about a given point? As we discuss in Chapter 5, the object is in equilibrium only if the sum of the moments about *any* point is zero, so calculating the sum of the moments provides a test for equilibrium. (Notice that the object in this example is not in equilibrium.) Furthermore, in dynamics the sum of the moments of the forces acting on objects must be determined in order to analyze their angular motions.

| Example 4.3 | Summing Moments to Determine an Unknown Force (▶ *Related Problem 4.23*) |

The weight $W = 300$ lb. The sum of the moments about C due to the weight W and the force exerted on the bar CA by the cable AB is zero. What is the tension in the cable?

Strategy

Let T be the tension in cable AB. Using the given dimensions, we can express the horizontal and vertical components of the force exerted on the bar by the cable in terms of T. Then by setting the sum of the moments about C due to the weight of the bar and the force exerted by the cable equal to zero, we can obtain an equation for T.

Solution

Using similar triangles, we express the force exerted on the bar by the cable in terms of horizontal and vertical components (Fig. a). The sum of the moments about C due to the weight of the bar and the force exerted by the cable AB is

$$\Sigma M_C = 4\left(\frac{4}{5}T\right) + 4\left(\frac{3}{5}T\right) - 2W = 0.$$

Solving for T, we obtain

$$T = 0.357W = 107.1 \text{ lb.}$$

Critical Thinking

This example is a preview of the applications we consider in Chapter 5 and demonstrates why you must know how to calculate moments of forces. If the bar is in equilibrium, the sum of the moments about C is zero. Applying this condition allowed us to determine the tension in the cable. Why didn't we need to consider the force exerted on the bar by its support at C? Because we know that the moment of that force about C is zero.

(a) Resolving the force exerted by the cable into horizontal and vertical components.

Problems

▶ **4.1** In Active Example 4.1, the 40-kN force points 30° above the horizontal. Suppose that the force points 30° below the horizontal instead. Draw a sketch of the beam with the new orientation of the force. What is the moment of the force about point A?

4.2 The mass $m_1 = 20$ kg. The magnitude of the total moment about B due to the forces exerted on bar AB by the weights of the two suspended masses is 170 N-m. What is the magnitude of the total moment due to the forces about point A?

Problem 4.2

4.3 The wheels of the overhead crane exert downward forces on the horizontal I-beam at B and C. If the force at B is 40 kip and the force at C is 44 kip, determine the sum of the moments of the forces on the beam about (a) point A, (b) point D.

Problem 4.3

4.4 What force F applied to the pliers is required to exert a 4 N-m moment about the center of the bolt at P?

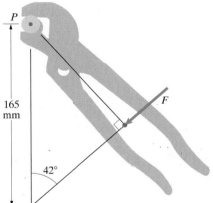

Problem 4.4

4.5 Two forces of equal magnitude F are applied to the wrench as shown. If a 50 N-m moment is required to loosen the nut, what is the necessary value of F?

Problem 4.5

4.6 The force $F = 8$ kN. What is the moment of the force about point P?

4.7 If the magnitude of the moment due to the force F about Q is 30 kN-m, what is F?

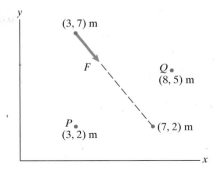

Problems 4.6/4.7

4.8 The support at the left end of the beam will fail if the moment about A of the 15-kN force F exceeds 18 kN-m. Based on this criterion, what is the largest allowable length of the beam?

Problem 4.8

4.9 The length of the bar AP is 650 mm. The radius of the pulley is 120 mm. Equal forces $T = 50$ N are applied to the ends of the cable. What is the sum of the moments of the forces (a) about A; (b) about P?

Problem 4.9

4.10 The force $F = 12$ kN. A structural engineer determines that the magnitude of the moment due to F about P should not exceed 5 kN-m. What is the acceptable range of the angle α? Assume that $0 \leq \alpha \leq 90°$.

Problem 4.10

4.11 The length of bar AB is 350 mm. The moments exerted about points B and C by the vertical force F are $M_B = -1.75$ kN-m and $M_C = -4.20$ kN-m. Determine the force F and the length of bar AC.

Problem 4.11

▶ **4.12** In Example 4.2, suppose that the 2-kN force points upward instead of downward. Draw a sketch of the machine part showing the orientations of the forces. What is the sum of the moments of the forces about the origin O?

4.13 Two equal and opposite forces act on the beam. Determine the sum of the moments of the two forces (a) about point P; (b) about point Q; (c) about the point with coordinates $x = 7$ m, $y = 5$ m.

Problem 4.13

4.14 The moment exerted about point E by the weight is 299 in-lb. What moment does the weight exert about point S?

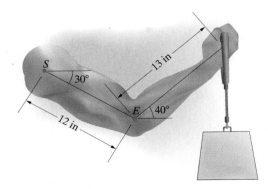

Problem 4.14

4.15 The magnitudes of the forces exerted on the pillar at D by the cables A, B, and C are equal: $F_A = F_B = F_C$. The magnitude of the total moment about E due to the forces exerted by the three cables at D is 1350 kN-m. What is F_A?

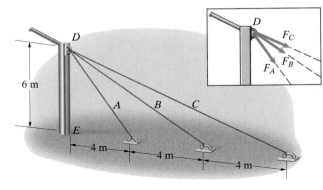

Problem 4.15

4.16 Three forces act on the piping. Determine the sum of the moments of the three forces about point P.

Problem 4.16

4.17 The forces $F_1 = 30$ N, $F_2 = 80$ N, and $F_3 = 40$ N. What is the sum of the moments of the forces about point A?

4.18 The force $F_1 = 30$ N. The vector sum of the three forces is zero. What is the sum of the moments of the forces about point A?

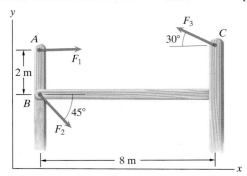

Problems 4.17/4.18

4.19 The forces $F_A = 30$ lb, $F_B = 40$ lb, $F_C = 20$ lb, and $F_D = 30$ lb. What is the sum of the moments of the forces about the origin of the coordinate system?

4.20 The force $F_A = 30$ lb. The vector sum of the forces on the beam is zero, and the sum of the moments of the forces about the origin of the coordinate system is zero. (a) Determine the forces F_B, F_C, and F_D. (b) Determine the sum of the moments of the forces about the right end of the beam.

Problems 4.19/4.20

4.21 Three forces act on the car. The sum of the forces is zero and the sum of the moments of the forces about point P is zero.

(a) Determine the forces A and B.

(b) Determine the sum of the moments of the forces about point Q.

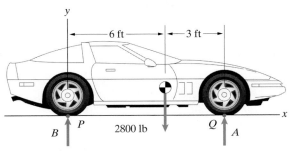

Problem 4.21

4.22 Five forces act on the piping. The vector sum of the forces is zero and the sum of the moments of the forces about point P is zero.

(a) Determine the forces A, B, and C.

(b) Determine the sum of the moments of the forces about point Q.

Problem 4.22

▶ **4.23** In Example 4.3, suppose that the attachment point B is moved upward and the cable is lengthened so that the vertical distance from C to B is 9 ft. (The positions of points C and A are unchanged.) Draw a sketch of the system with the cable in its new position. What is the tension in the cable?

4.24 The tension in the cable is the same on both sides of the pulley. The sum of the moments about point A due to the 800-lb force and the forces exerted on the bar by the cable at B and C is zero. What is the tension in the cable?

Problem 4.24

4.25 The 160-N weights of the arms AB and BC of the robotic manipulator act at their midpoints. Determine the sum of the moments of the three weights about A.

Problem 4.25

4.26 The space shuttle's attitude thrusters exert two forces of magnitude $F = 7.70$ kN. What moment do the thrusters exert about the center of mass G?

Problem 4.26

4.27 The force F exerts a 200 ft-lb counterclockwise moment about A and a 100 ft-lb clockwise moment about B. What are F and θ?

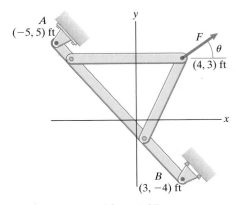

Problem 4.27

4.28 Five forces act on a link in the gear-shifting mechanism of a lawn mower. The vector sum of the five forces on the bar is zero. The sum of their moments about the point where the forces A_x and A_y act is zero.

(a) Determine the forces A_x, A_y, and B.

(b) Determine the sum of the moments of the forces about the point where the force B acts.

Problem 4.28

4.29 Five forces act on a model truss built by a civil engineering student as part of a design project. The dimensions are $b = 300$ mm and $h = 400$ mm and $F = 100$ N. The sum of the moments of the forces about the point where A_x and A_y act is zero. If the weight of the truss is negligible, what is the force B?

4.30 The dimensions are $b = 3$ ft and $h = 4$ ft and $F = 300$ lb. The vector sum of the forces acting on the truss is zero, and the sum of the moments of the forces about the point where A_x and A_y act is zero.

(a) Determine the forces A_x, A_y, and B.

(b) Determine the sum of the moments of the forces about the point where the force B acts.

Problems 4.29/4.30

4.31 The mass $m = 70$ kg. What is the moment about A due to the force exerted on the beam at B by the cable?

Problem 4.31

4.32 The weights W_1 and W_2 are suspended by the cable system shown. The weight $W_1 = 12$ lb. The cable BC is horizontal. Determine the moment about point P due to the force exerted on the vertical post at D by the cable CD.

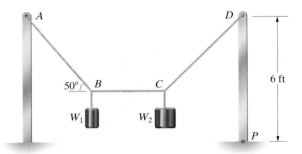

Problem 4.32

4.33 The bar AB exerts a force at B that helps support the vertical retaining wall. The force is parallel to the bar. The civil engineer wants the bar to exert a 38 kN-m moment about O. What is the magnitude of the force the bar must exert?

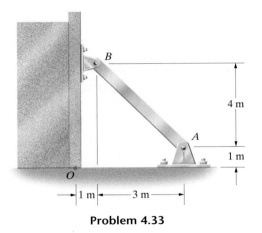

Problem 4.33

4.34 A contestant in a fly-casting contest snags his line in some grass. If the tension in the line is 5 lb, what moment does the force exerted on the rod by the line exert about point H, where he holds the rod?

Problem 4.34

4.35 The cables AB and AC help support the tower. The tension in cable AB is 5 kN. The points A, B, C, and O are contained in the same vertical plane. (a) What is the moment about O due to the force exerted on the tower by cable AB? (b) If the sum of the moments about O due to the forces exerted on the tower by the two cables is zero, what is the tension in cable AC?

Problem 4.35

4.36 The cable from B to A (the sailboat's forestay) exerts a 230-N force at B. The cable from B to C (the backstay) exerts a 660-N force at B. The bottom of the sailboat's mast is located at $x = 4$ m, $y = 0$. What is the sum of the moments about the bottom of the mast due to the forces exerted at B by the forestay and backstay?

Problem 4.36

4.37 The cable AB exerts a 290-kN force on the crane's boom at B. The cable AC exerts a 148-kN force on the boom at C. Determine the sum of the moments about P due to the forces the cables AB and AC exert on the boom.

4.38 The mass of the crane's boom is 9000 kg. Its weight acts at G. The sum of the moments about P due to the boom's weight, the force exerted at B by the cable AB, and the force exerted at C by the cable AC is zero. Assume that the tensions in cables AB and AC are equal. Determine the tension in the cables.

Problems 4.37/4.38

4.39 The mass of the luggage carrier and the suitcase combined is 12 kg. Their weight acts at A. The sum of the moments about the origin of the coordinate system due to the weight acting at A and the vertical force F applied to the handle of the luggage carrier is zero. Determine the force F (a) if $\alpha = 30°$; (b) if $\alpha = 50°$.

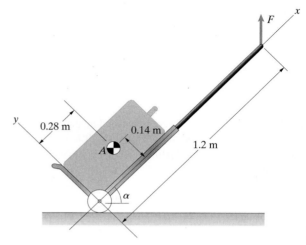

Problem 4.39

4.40 The hydraulic cylinder BC exerts a 300-kN force on the boom of the crane at C. The force is parallel to the cylinder. What is the moment of the force about A?

Problem 4.40

4.41 The hydraulic piston *AB* exerts a 400-lb force on the ladder at *B* in the direction parallel to the piston. The sum of the moments about *C* due to the force exerted on the ladder by the piston and the weight *W* of the ladder is zero. What is the weight of the ladder?

4.42 The hydraulic cylinder exerts an 8-kN force at *B* that is parallel to the cylinder and points from *C* toward *B*. Determine the moments of the force about points *A* and *D*.

Problem 4.42

Problem 4.41

4.43 The structure shown in the diagram is one of two identical structures that support the scoop of the excavator. The bar *BC* exerts a 700-N force at *C* that points from *C* toward *B*. What is the moment of this force about *K*?

4.44 The bar *BC* exerts a force at *C* that points from *C* toward *B*. The hydraulic cylinder *DH* exerts a 1550-N force at *D* that points from *D* toward *H*. The sum of the moments of these two forces about *K* is zero. What is the magnitude of the force that bar *BC* exerts at *C*?

Problems 4.43/4.44

(a)

(b)

(c)

Figure 4.4
(a) The force **F** and point *P*.
(b) A vector **r** from *P* to a point on the line
of action of **F**.
(c) The angle θ and the perpendicular
distance *D*.

4.2 The Moment Vector

BACKGROUND

The moment of a force about a point is a vector. In this section we define this vector and explain how it is evaluated. We then show that when we use the two-dimensional description of the moment described in Section 4.1, we are specifying the magnitude and direction of the moment vector.

Consider a force vector **F** and point *P* (Fig. 4.4a). The *moment* of **F** about *P* is the vector

$$\mathbf{M}_P = \mathbf{r} \times \mathbf{F}, \tag{4.2}$$

where **r** is a position vector from *P* to *any* point on the line of action of **F** (Fig. 4.4b).

Magnitude of the Moment

From the definition of the cross product, the magnitude of $\mathbf{M}_P$ is

$$|\mathbf{M}_P| = |\mathbf{r}||\mathbf{F}| \sin \theta,$$

where θ is the angle between the vectors **r** and **F** when they are placed tail to tail. The perpendicular distance from *P* to the line of action of **F** is $D = |\mathbf{r}| \sin \theta$ (Fig. 4.4c). Therefore the magnitude of the moment $\mathbf{M}_P$ equals the product of the perpendicular distance from *P* to the line of action of **F** and the magnitude of **F**:

$$|\mathbf{M}_P| = D|\mathbf{F}|. \tag{4.3}$$

Notice that if we know the vectors $\mathbf{M}_P$ and **F**, this equation can be solved for the perpendicular distance *D*.

Direction of the Moment

We know from the definition of the cross product that $\mathbf{M}_P$ is perpendicular to both **r** and **F**. That means that $\mathbf{M}_P$ is perpendicular to the plane containing *P* and **F** (Fig. 4.5a). Notice in this figure that we denote a moment by a circular arrow around the vector.

Figure 4.5
(a) $\mathbf{M}_P$ is perpendicular to the plane
containing *p* and **F**.
(b) The direction of $\mathbf{M}_P$ indicates the
direction of the moment.

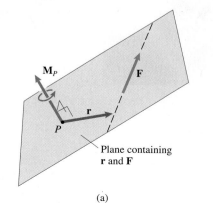

Plane containing
r and **F**

(a)

(b)

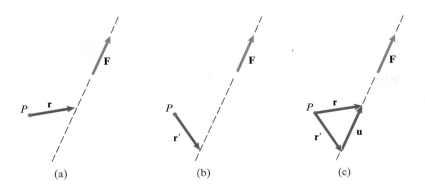

Figure 4.6
(a) A vector **r** from P to the line of action of **F**.
(b) A different vector **r**′.
(c) **r** = **r**′ + **u**.

The direction of $\mathbf{M}_P$ also indicates the direction of the moment: Pointing the thumb of the right hand in the direction of $\mathbf{M}_P$, the "arc" of the fingers indicates the direction of the rotation that **F** tends to cause about P (Fig. 4.5b).

The result obtained from Eq. (4.2) doesn't depend on where the vector **r** intersects the line of action of **F**. Instead of using the vector **r** in Fig. 4.6a, we could use the vector **r**′ in Fig. 4.6b. The vector **r** = **r**′ + **u**, where **u** is parallel to **F** (Fig. 4.6c). Therefore,

$$\mathbf{r} \times \mathbf{F} = (\mathbf{r}' + \mathbf{u}) \times \mathbf{F} = \mathbf{r}' \times \mathbf{F}$$

because the cross product of the parallel vectors **u** and **F** is zero.

In summary, the moment of a force **F** about a point P has three properties:

1. The magnitude of $\mathbf{M}_P$ is equal to the product of the magnitude of **F** and the perpendicular distance from P to the line of action of **F**. If the line of action of **F** passes through P, $\mathbf{M}_P = \mathbf{0}$.
2. $\mathbf{M}_P$ is perpendicular to the plane containing P and **F**.
3. The direction of $\mathbf{M}_P$ indicates the direction of the moment through a right-hand rule (Fig. 4.5b). Since the cross product is not commutative, it is essential to maintain the correct sequence of the vectors in the equation $\mathbf{M}_P = \mathbf{r} \times \mathbf{F}$.

Let us determine the moment of the force **F** in Fig. 4.7a about the point P. Since the vector **r** in Eq. (4.2) can be a position vector to any point on the line of action of **F**, we can use the vector from P to the point of application of **F** (Fig. 4.7b):

$$\mathbf{r} = (12 - 3)\mathbf{i} + (6 - 4)\mathbf{j} + (-5 - 1)\mathbf{k} = 9\mathbf{i} + 2\mathbf{j} - 6\mathbf{k} \text{ (ft)}.$$

The moment is

$$\mathbf{M}_P = \mathbf{r} \times \mathbf{F} = \begin{vmatrix} \mathbf{i} & \mathbf{j} & \mathbf{k} \\ 9 & 2 & -6 \\ 4 & 4 & 7 \end{vmatrix} = 38\mathbf{i} - 87\mathbf{j} + 28\mathbf{k} \text{ (ft-lb)}.$$

The magnitude of $\mathbf{M}_P$,

$$|\mathbf{M}_P| = \sqrt{(38)^2 + (-87)^2 + (28)^2} = 99.0 \text{ ft-lb},$$

(a)   (b)   (c)

Figure 4.7
(a) A force **F** and point P.
(b) The vector **r** from P to the point of application of **F**.
(c) $\mathbf{M}_P$ is perpendicular to the plane containing P and **F**.
  The right-hand rule indicates the direction of the moment.

equals the product of the magnitude of **F** and the perpendicular distance D from point P to the line of action of **F**. Therefore,

$$D = \frac{|\mathbf{M}_P|}{|\mathbf{F}|} = \frac{99.0 \text{ ft-lb}}{9 \text{ lb}} = 11.0 \text{ ft.}$$

The direction of $\mathbf{M}_P$ tells us both the orientation of the plane containing P and **F** and the direction of the moment (Fig. 4.7c).

Relation to the Two-Dimensional Description

If our view is perpendicular to the plane containing the point P and the force **F**, the two-dimensional description of the moment we used in Section 4.1 specifies both the magnitude and direction of the vector $\mathbf{M}_P$. In this situation, $\mathbf{M}_P$ is perpendicular to the page, and the right-hand rule indicates whether it points out of or into the page.

For example, in Fig. 4.8a, the view is perpendicular to the x–y plane and the 10-N force is contained in the x–y plane. Suppose that we want to determine the

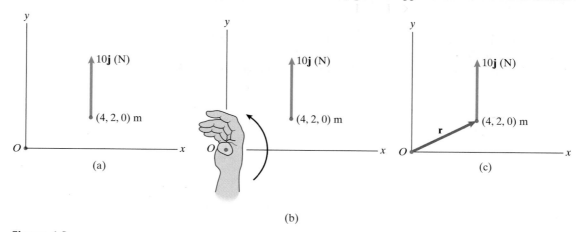

(b)

Figure 4.8
(a) The force is contained in the x–y plane.
(b) The counterclockwise direction of the moment indicates that $\mathbf{M}_O$ points out of the page.
(c) The vector **r** from O to the point of application of **F**.

moment of the force about the origin O. The perpendicular distance from O to the line of action of the force is 4 m. The two-dimensional description of the moment of the force about O is that its magnitude is $(4 \text{ m})(10 \text{ N}) = 40$ N-m and its direction is counterclockwise, or

$$M_O = 40 \text{ N-m.}$$

That tells us that the magnitude of the vector $\mathbf{M}_O$ is 40 N-m, and the right-hand rule (Fig. 4.8b) indicates that it points out of the page. Therefore,

$$\mathbf{M}_O = 40\mathbf{k} \text{ (N-m).}$$

We can confirm this result by using Eq. (4.2). If we let $\mathbf{r}$ be the vector from O to the point of application of the force (Fig. 4.8c),

$$\mathbf{M}_O = \mathbf{r} \times \mathbf{F} = (4\mathbf{i} + 2\mathbf{j}) \times 10\mathbf{j} = 40\mathbf{k} \text{ (N-m).}$$

As this example illustrates, the two-dimensional description of the moment determines the moment vector. The converse is also true. The magnitude of $\mathbf{M}_O$ equals the product of the magnitude of the force and the perpendicular distance from O to the line of action of the force, 40 N-m, and the direction of the vector $\mathbf{M}_O$ indicates that the moment is counterclockwise (Fig. 4.8b).

Varignon's Theorem

Let $\mathbf{F}_1, \mathbf{F}_2, \ldots, \mathbf{F}_N$ be a concurrent system of forces whose lines of action intersect at a point Q. The moment of the system about a point P is

$$(\mathbf{r}_{PQ} \times \mathbf{F}_1) + (\mathbf{r}_{PQ} \times \mathbf{F}_2) + \cdots + (\mathbf{r}_{PQ} \times \mathbf{F}_N)$$

$$= \mathbf{r}_{PQ} \times (\mathbf{F}_1 + \mathbf{F}_2 + \cdots + \mathbf{F}_N),$$

where $\mathbf{r}_{PQ}$ is the vector from P to Q (Fig. 4.9). This result, known as *Varignon's theorem*, follows from the distributive property of the cross product, Eq. (2.31). It confirms that the moment of a force about a point P is equal to the sum of the moments of its components about P.

Figure 4.9
A system of concurrent forces and a point P.

RESULTS

Moment
The moment of a force $\mathbf{F}$ about a point P is defined by

$$\mathbf{M}_P = \mathbf{r} \times \mathbf{F}, \qquad (4.2)$$

where $\mathbf{r}$ is a position vector from P to *any* point on the line of action of $\mathbf{F}$.

Magnitude of the Moment
The magnitude of the vector $\mathbf{M}_P$ is

$$|\mathbf{M}_P| = D|\mathbf{F}|, \qquad (4.3)$$

where D is the perpendicular distance from P to the line of action of $\mathbf{F}$.

Direction of the Moment
The vector $\mathbf{M}_P$ is perpendicular to the plane containing the point P and the vector $\mathbf{F}$. Pointing the thumb of the right hand in the direction of $\mathbf{M}_P$, the fingers point in the direction of the rotation that $\mathbf{F}$ tends to cause about P.

Active Example 4.4   **Determining a Moment** (▶ *Related Problem 4.45*)

Determine the moment of the 90-lb force $\mathbf{F}$ about point A.

Strategy
To apply Eq. (4.2), we must express the force $\mathbf{F}$ in terms of its components. The vector $\mathbf{r}$ is a vector from point A to any point on the line of action of $\mathbf{F}$, so we can use the vector from point A to point B.

Solution –

$$\mathbf{r}_{BC} = (x_C - x_B)\mathbf{i} + (y_C - y_B)\mathbf{j} + (z_C - z_B)\mathbf{k}$$
$$= -4\mathbf{i} + 7\mathbf{j} - 4\mathbf{k} \text{ (ft)}.$$

$$\mathbf{e}_{BC} = \frac{\mathbf{r}_{BC}}{|\mathbf{r}_{BC}|} = -\frac{4}{9}\mathbf{i} + \frac{7}{9}\mathbf{j} - \frac{4}{9}\mathbf{k}.$$

> Obtain a unit vector that has the same direction as the force **F** by dividing the position vector from point B to point C by its magnitude.

$$\mathbf{F} = (90 \text{ lb})\mathbf{e}_{BC}$$
$$= (90 \text{ lb})\left(-\frac{4}{9}\mathbf{i} + \frac{7}{9}\mathbf{j} - \frac{4}{9}\mathbf{k}\right)$$
$$= -40\mathbf{i} + 70\mathbf{j} - 40\mathbf{k} \text{ (lb)}.$$

> Express the force **F** in terms of its components by writing it as the product of its magnitude and the unit vector $\mathbf{e}_{BC}$.

$$\mathbf{r}_{AB} = (x_B - x_A)\mathbf{i} + (y_B - y_A)\mathbf{j} + (z_B - z_A)\mathbf{k}$$
$$= 11\mathbf{i} - 6\mathbf{j} - \mathbf{k} \text{ (ft)}.$$

$$\mathbf{M}_A = \mathbf{r}_{AB} \times \mathbf{F}$$
$$= \begin{vmatrix} \mathbf{i} & \mathbf{j} & \mathbf{k} \\ 11 & -6 & -1 \\ -40 & 70 & -40 \end{vmatrix}$$
$$= 310\mathbf{i} + 480\mathbf{j} + 530\mathbf{k} \text{ (ft-lb)}.$$

> Apply Eq. (4.2) to determine the moment of **F** about point A.

Practice Problem (a) Use Eq. (4.2) to determine the moment of **F** about point A, letting the vector **r** be the position vector from point A to point C. (b) Determine the perpendicular distance from point A to the line of action of **F**.

Answer: (a) $\mathbf{M}_A = 310\mathbf{i} + 480\mathbf{j} + 530\mathbf{k}$ (ft-lb). (b) 8.66 ft.

| Example 4.5 | Applying the Moment Vector (▶ *Related Problem 4.57*) |

The cables AB and AC extend from an attachment point A on the floor to attachment points B and C in the walls. The tension in cable AB is 10 kN, and the tension in cable AC is 20 kN. What is the sum of the moments about O due to the forces exerted on the attachment point A by the two cables?

Strategy

We must express the forces exerted on the attachment point A by the two cables in terms of their components. Then we can use Eq. (4.2) to determine the moments the forces exert about O.

Solution

Let $\mathbf{F}_{AB}$ and $\mathbf{F}_{AC}$ be the forces exerted on the attachment point A by the two cables (Fig. a). To express $\mathbf{F}_{AB}$ in terms of its components, we determine the position vector from A to B,

$$(0 - 4)\mathbf{i} + (4 - 0)\mathbf{j} + (8 - 6)\mathbf{k} = -4\mathbf{i} + 4\mathbf{j} + 2\mathbf{k} \text{ (m)},$$

and divide it by its magnitude to obtain a unit vector $\mathbf{e}_{AB}$ with the same direction as $\mathbf{F}_{AB}$ (Fig. b):

$$\mathbf{e}_{AB} = \frac{-4\mathbf{i} + 4\mathbf{j} + 2\mathbf{k} \text{ (m)}}{\sqrt{(-4 \text{ m})^2 + (4 \text{ m})^2 + (2 \text{ m})^2}} = -\frac{2}{3}\mathbf{i} + \frac{2}{3}\mathbf{j} + \frac{1}{3}\mathbf{k}.$$

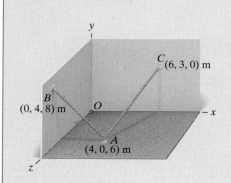

Now we write $\mathbf{F}_{AB}$ as

$$\mathbf{F}_{AB} = 10\mathbf{e}_{AB} = -6.67\mathbf{i} + 6.67\mathbf{j} + 3.33\mathbf{k} \text{ (kN)}.$$

We express the force $\mathbf{F}_{AC}$ in terms of its components in the same way:

$$\mathbf{F}_{AC} = 5.71\mathbf{i} + 8.57\mathbf{j} - 17.14\mathbf{k} \text{ (kN)}.$$

(a) The forces $\mathbf{F}_{AB}$ and $\mathbf{F}_{AC}$ exerted at A by the cables.

Choose the Vector r　Since the lines of action of both forces pass through point A, we can use the vector from O to A to determine the moments of both forces about point O (Fig. a):

$$\mathbf{r} = 4\mathbf{i} + 6\mathbf{k} \text{ (m)}.$$

Evaluate r × F　The sum of the moments is

$$\Sigma\mathbf{M}_O = (\mathbf{r} \times \mathbf{F}_{AB}) + (\mathbf{r} \times \mathbf{F}_{AC})$$

$$= \begin{vmatrix} \mathbf{i} & \mathbf{j} & \mathbf{k} \\ 4 & 0 & 6 \\ -6.67 & 6.67 & 3.33 \end{vmatrix} + \begin{vmatrix} \mathbf{i} & \mathbf{j} & \mathbf{k} \\ 4 & 0 & 6 \\ 5.71 & 8.57 & -17.14 \end{vmatrix}$$

$$= -91.4\mathbf{i} + 49.5\mathbf{j} + 61.0\mathbf{k} \text{ (kN-m)}.$$

(b) The unit vector $\mathbf{e}_{AB}$ has the same direction as $\mathbf{F}_{AB}$.

Critical Thinking

The lines of action of the forces $\mathbf{F}_{AB}$ and $\mathbf{F}_{AC}$ intersect at A. Notice that, according to Varignon's theorem, we could have summed the forces first, obtaining

$$\mathbf{F}_{AB} + \mathbf{F}_{AC} = -0.952\mathbf{i} + 15.24\mathbf{j} - 13.81\mathbf{k} \text{ (kN)},$$

and then determined the sum of the moments of the two forces about O by calculating the moment of the sum of the two forces about O:

$$\Sigma \mathbf{M}_O = \mathbf{r} \times (\mathbf{F}_{AB} + \mathbf{F}_{AC})$$

$$= \begin{vmatrix} \mathbf{i} & \mathbf{j} & \mathbf{k} \\ 4 & 0 & 6 \\ -0.952 & 15.24 & -13.81 \end{vmatrix}$$

$$= -91.4\mathbf{i} + 49.5\mathbf{j} + 61.0\mathbf{k} \ (\text{kN-m}).$$

Problems

▶ **4.45** In Active Example 4.4, what is the moment of $\mathbf{F}$ about the origin of the coordinate system?

4.46 Use Eq. (4.2) to determine the moment of the 80-N force about the origin O letting $\mathbf{r}$ be the vector (a) from O to A; (b) from O to B.

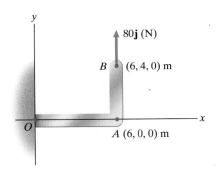

Problem 4.46

4.47 A bioengineer studying an injury sustained in throwing the javelin estimates that the magnitude of the maximum force exerted was $|\mathbf{F}| = 360$ N and the perpendicular distance from O to the line of action of $\mathbf{F}$ was 550 mm. The vector $\mathbf{F}$ and point O are contained in the x–y plane. Express the moment of $\mathbf{F}$ about the shoulder joint at O as a vector.

Problem 4.47

4.48 Use Eq. (4.2) to determine the moment of the 100-kN force (a) about A; (b) about B.

Problem 4.48

4.49 The cable AB exerts a 200-N force on the support at A that points from A toward B. Use Eq. (4.2) to determine the moment of this force about point P, (a) letting $\mathbf{r}$ be the vector from P to A; (b) letting $\mathbf{r}$ be the vector from P to B.

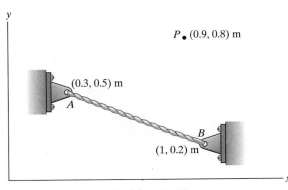

Problem 4.49

4.50 The line of action of **F** is contained in the x–y plane. The moment of **F** about O is 140**k** (N-m), and the moment of **F** about A is 280**k** (N-m). What are the components of **F**?

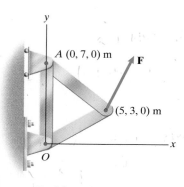

A (0, 7, 0) m

F

(5, 3, 0) m

O

x

y

Problem 4.50

4.51 Use Eq. (4.2) to determine the sum of the moments of the three forces (a) about A; (b) about B.

y

6 kN

3 kN

3 kN

A

B

x

0.2 m — 0.2 m — 0.2 m — 0.2 m

Problem 4.51

4.52 Three forces are applied to the plate. Use Eq. (4.2) to determine the sum of the moments of the three forces about the origin O.

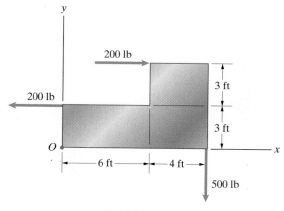

y

200 lb

3 ft

200 lb

3 ft

O

x

6 ft — 4 ft

500 lb

Problem 4.52

4.53 Three forces act on the plate. Use Eq. (4.2) to determine the sum of the moments of the three forces about point P.

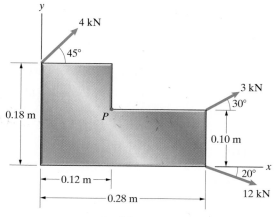

y

4 kN

45°

3 kN

30°

0.18 m

P

0.10 m

20° x

0.12 m

12 kN

0.28 m

Problem 4.53

4.54 (a) Determine the magnitude of the moment of the 150-N force about A by calculating the perpendicular distance from A to the line of action of the force.

(b) Use Eq. (4.2) to determine the magnitude of the moment of the 150-N force about A.

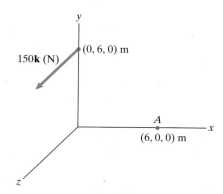

y

150**k** (N)

(0, 6, 0) m

A

(6, 0, 0) m

x

z

Problem 4.54

4.55 (a) Determine the magnitude of the moment of the 600-N force about A by calculating the perpendicular distance from A to the line of action of the force.

(b) Use Eq. (4.2) to determine the magnitude of the moment of the 600-N force about A.

Problem 4.55

4.56 What is the magnitude of the moment of **F** about point B?

Problem 4.56

▶ **4.57** In Example 4.5, suppose that the attachment point C is moved to the location $(8, 2, 0)$ m and the tension in cable AC changes to 25 kN. What is the sum of the moments about O due to the forces exerted on the attachment point A by the two cables?

4.58 The rope exerts a force of magnitude $|\mathbf{F}| = 200$ lb on the top of the pole at B. Determine the magnitude of the moment of **F** about A.

Problem 4.58

4.59 The force $\mathbf{F} = 30\mathbf{i} + 20\mathbf{j} - 10\mathbf{k}$ (N).

(a) Determine the magnitude of the moment of **F** about A.

(b) Suppose that you can change the direction of **F** while keeping its magnitude constant, and you want to choose a direction that maximizes the moment of **F** about A. What is the magnitude of the resulting maximum moment?

Problem 4.59

4.60 The direction cosines of the force **F** are $\cos \theta_x = 0.818$, $\cos \theta_y = 0.182$, and $\cos \theta_z = -0.545$. The support of the beam at O will fail if the magnitude of the moment of **F** about O exceeds 100 kN-m. Determine the magnitude of the largest force **F** that can safely be applied to the beam.

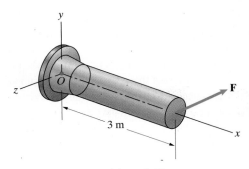

Problem 4.60

4.61 The force **F** exerted on the grip of the exercise machine points in the direction of the unit vector $\mathbf{e} = \frac{2}{3}\mathbf{i} - \frac{2}{3}\mathbf{j} + \frac{1}{3}\mathbf{k}$ and its magnitude is 120 N. Determine the magnitude of the moment of **F** about the origin O.

4.62 The force **F** points in the direction of the unit vector $\mathbf{e} = \frac{2}{3}\mathbf{i} - \frac{2}{3}\mathbf{j} + \frac{1}{3}\mathbf{k}$. The support at O will safely support a moment of 560 N-m magnitude. (a) Based on this criterion, what is the largest safe magnitude of **F**? (b) If the force **F** may be exerted in any direction, what is its largest safe magnitude?

Problems 4.61/4.62

4.63 A civil engineer in Boulder, Colorado, estimates that under the severest expected Chinook winds, the total force on the highway sign will be $\mathbf{F} = 2.8\mathbf{i} - 1.8\mathbf{j}$ (kN). Let $\mathbf{M}_O$ be the moment due to **F** about the base O of the cylindrical column supporting the sign. The y component of $\mathbf{M}_O$ is called the *torsion* exerted on the cylindrical column at the base, and the component of $\mathbf{M}_O$ parallel to the x–z plane is called the *bending moment*. Determine the magnitudes of the torsion and bending moment.

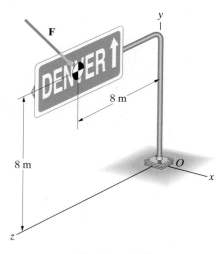

Problem 4.63

4.64 The weights of the arms OA and AB of the robotic manipulator act at their midpoints. The direction cosines of the centerline of arm OA are $\cos \theta_x = 0.500$, $\cos \theta_y = 0.866$, and $\cos \theta_z = 0$, and the direction cosines of the centerline of arm AB are $\cos \theta_x = 0.707$, $\cos \theta_y = 0.619$, and $\cos \theta_z = -0.342$. What is the sum of the moments about O due to the two forces?

Problem 4.64

4.65 The tension in cable AB is 100 lb. If you want the magnitude of the moment due to the forces exerted on the tree by the two ropes about the base O of the tree to be 1500 ft-lb, what is the necessary tension in rope AC?

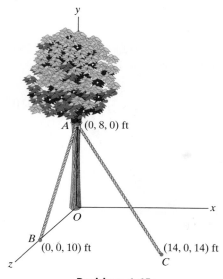

Problem 4.65

4.66* A force **F** acts at the top end *A* of the pole. Its magnitude is $|\mathbf{F}| = 6$ kN and its *x* component is $F_x = 4$ kN. The coordinates of point *A* are shown. Determine the components of **F** so that the magnitude of the moment due to **F** about the base *P* of the pole is as large as possible.

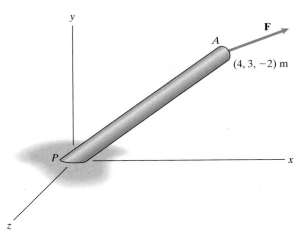

(4, 3, −2) m

Problem 4.66

4.67 The force $\mathbf{F} = 5\mathbf{i}$ (kN) acts on the ring *A* where the cables *AB*, *AC*, and *AD* are joined. What is the sum of the moments about point *D* due to the force **F** and the three forces exerted on the ring by the cables?

   Strategy: The ring is in equilibrium. Use what you know about the four forces acting on it.

4.68 In Problem 4.67, determine the moment about point *D* due to the force exerted on the ring *A* by the cable *AB*.

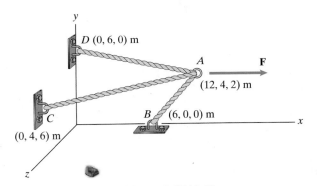

D (0, 6, 0) m
A
F
(12, 4, 2) m
B   (6, 0, 0) m
C
(0, 4, 6) m

Problems 4.67/4.68

4.69 The tower is 70 m tall. The tensions in cables *AB*, *AC*, and *AD* are 4 kN, 2 kN, and 2 kN, respectively. Determine the sum of the moments about the origin *O* due to the forces exerted by the cables at point *A*.

4.70 Suppose that the tension in cable *AB* is 4 kN, and you want to adjust the tensions in cables *AC* and *AD* so that the sum of the moments about the origin *O* due to the forces exerted by the cables at point *A* is zero. Determine the tensions.

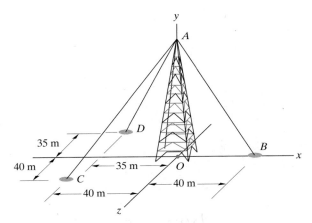

35 m
40 m
D
35 m
O
C
40 m
B
40 m

Problems 4.69/4.70

4.71 The tension in cable *AB* is 150 N. The tension in cable *AC* is 100 N. Determine the sum of the moments about *D* due to the forces exerted on the wall by the cables.

4.72 The total force exerted by the two cables in the direction perpendicular to the wall is 2 kN. The magnitude of the sum of the moments about *D* due to the forces exerted on the wall by the cables is 18 kN-m. What are the tensions in the cables?

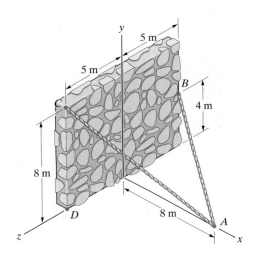

y
5 m
5 m
C
B
4 m
8 m
D
8 m
A

Problems 4.71/4.72

4.73 The tension in the cable BD is 1 kN. As a result, cable BD exerts a 1-kN force on the "ball" at B that points from B toward D. Determine the moment of this force about point A.

4.74* Suppose that the mass of the suspended object E is 100 kg and the mass of the bar AB is 20 kg. Assume that the weight of the bar acts at its midpoint. If the sum of the moments about point A due to the weight of the bar and the forces exerted on the "ball" at B by the three cables BC, BD, and BE is zero, determine the tensions in the cables BC and BD.

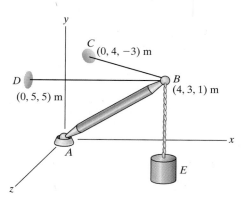

Problems 4.73/4.74

4.75 The 200-kg slider at A is held in place on the smooth vertical bar by the cable AB. Determine the moment about the bottom of the bar (point C with coordinates $x = 2$ m, $y = z = 0$) due to the force exerted on the slider by the cable.

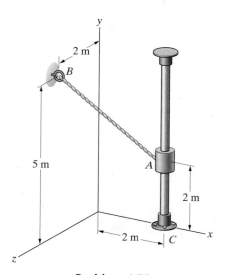

Problem 4.75

4.76 To evaluate the adequacy of the design of the vertical steel post, you must determine the moment about the bottom of the post due to the force exerted on the post at B by the cable AB. A calibrated strain gauge mounted on cable AC indicates that the tension in cable AC is 22 kN. What is the moment?

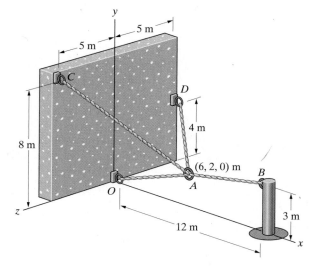

Problem 4.76

4.3 Moment of a Force About a Line

BACKGROUND

The device in Fig. 4.10, called a *capstan*, was used in the days of square-rigged sailing ships. Crewmen turned it by pushing on the handles as shown in Fig. 4.10a, providing power for such tasks as raising anchors and hoisting yards. A vertical force **F** applied to one of the handles as shown in Fig. 4.10b does not cause the capstan to turn, even though the magnitude of the moment about point P is $d|\mathbf{F}|$ in both cases.

The measure of the tendency of a force to cause rotation about a line, or axis, is called the moment of the force about the line. Suppose that a force **F** acts on an object such as a turbine that rotates about an axis L, and we resolve **F** into components in terms of the coordinate system shown in Fig. 4.11. The components F_x and F_z do not tend to rotate the turbine, just as the force parallel to the axis of the capstan did not cause it to turn. It is the component F_y that tends to cause rotation, by exerting a moment of magnitude aF_y about the turbine's axis. In this example we can determine the moment of **F** about L easily because the coordinate system is conveniently placed. We now introduce an expression that determines the moment of a force about any line.

(a)   (b)

Figure 4.10
(a) Turning a capstan.
(b) A vertical force does not turn the capstan.

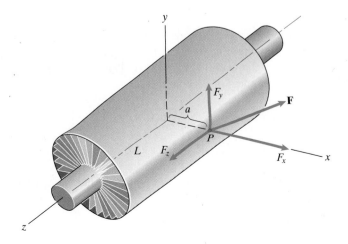

Figure 4.11
Applying a force to a turbine with axis of rotation L.

(a)

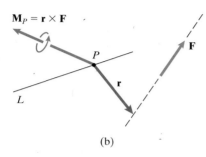

$\mathbf{M}_P = \mathbf{r} \times \mathbf{F}$

(b)

(c)

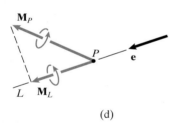

(d)

Figure 4.12
(a) The line L and force $\mathbf{F}$.
(b) $\mathbf{M}_P$ is the moment of $\mathbf{F}$ about any point P on L.
(c) The component $\mathbf{M}_L$ is the moment of $\mathbf{F}$ about L.
(d) A unit vector $\mathbf{e}$ along L.

Definition

Consider a line L and force $\mathbf{F}$ (Fig. 4.12a). Let $\mathbf{M}_P$ be the moment of $\mathbf{F}$ about an arbitrary point P on L (Fig. 4.12b). The moment of $\mathbf{F}$ about L is the component of $\mathbf{M}_P$ parallel to L, which we denote by $\mathbf{M}_L$ (Fig. 4.12c). The magnitude of the moment of $\mathbf{F}$ about L is $|\mathbf{M}_L|$, and when the thumb of the right hand is pointed in the direction of $\mathbf{M}_L$, the arc of the fingers indicates the direction of the moment about L. In terms of a unit vector $\mathbf{e}$ along L (Fig. 4.12d), $\mathbf{M}_L$ is given by

$$\mathbf{M}_L = (\mathbf{e} \cdot \mathbf{M}_P)\mathbf{e}. \qquad (4.4)$$

(The unit vector $\mathbf{e}$ can point in either direction. See our discussion of vector components parallel and normal to a line in Section 2.5.) The moment $\mathbf{M}_P = \mathbf{r} \times \mathbf{F}$, so we can also express $\mathbf{M}_L$ as

$$\mathbf{M}_L = [\mathbf{e} \cdot (\mathbf{r} \times \mathbf{F})]\mathbf{e}. \qquad (4.5)$$

The mixed triple product in this expression is given in terms of the components of the three vectors by

$$\mathbf{e} \cdot (\mathbf{r} \times \mathbf{F}) = \begin{vmatrix} e_x & e_y & e_z \\ r_x & r_y & r_z \\ F_x & F_y & F_z \end{vmatrix}. \qquad (4.6)$$

Notice that the value of the scalar $\mathbf{e} \cdot \mathbf{M}_P = \mathbf{e} \cdot (\mathbf{r} \times \mathbf{F})$ determines both the magnitude and direction of $\mathbf{M}_L$. The absolute value of $\mathbf{e} \cdot \mathbf{M}_P$ is the magnitude of $\mathbf{M}_L$. If $\mathbf{e} \cdot \mathbf{M}_P$ is positive, $\mathbf{M}_L$ points in the direction of $\mathbf{e}$, and if $\mathbf{e} \cdot \mathbf{M}_P$ is negative, $\mathbf{M}_L$ points in the direction opposite to $\mathbf{e}$.

The result obtained with Eq. (4.4) or (4.5) doesn't depend on which point on L is chosen to determine $\mathbf{M}_P = \mathbf{r} \times \mathbf{F}$. If we use point P in Fig. 4.13 to determine the moment of $\mathbf{F}$ about L, we get the result given by Eq. (4.5). If we use P' instead, we obtain the same result,

$$[\mathbf{e} \cdot (\mathbf{r}' \times \mathbf{F})]\mathbf{e} = \{\mathbf{e} \cdot [(\mathbf{r} + \mathbf{u}) \times \mathbf{F}]\}\mathbf{e}$$
$$= [\mathbf{e} \cdot (\mathbf{r} \times \mathbf{F}) + \mathbf{e} \cdot (\mathbf{u} \times \mathbf{F})]\mathbf{e}$$
$$= [\mathbf{e} \cdot (\mathbf{r} \times \mathbf{F})]\mathbf{e},$$

because $\mathbf{u} \times \mathbf{F}$ is perpendicular to $\mathbf{e}$.

Applications

To demonstrate that $\mathbf{M}_L$ is the measure of the tendency of $\mathbf{F}$ to cause rotation about L, we return to the turbine in Fig. 4.11. Let Q be a point on L at an arbitrary distance b from the origin (Fig. 4.14a). The vector $\mathbf{r}$ from Q to P is $\mathbf{r} = a\mathbf{i} - b\mathbf{k}$, so the moment of $\mathbf{F}$ about Q is

$$\mathbf{M}_Q = \mathbf{r} \times \mathbf{F} = \begin{vmatrix} \mathbf{i} & \mathbf{j} & \mathbf{k} \\ a & 0 & -b \\ F_x & F_y & F_z \end{vmatrix} = bF_y\mathbf{i} - (aF_z + bF_x)\mathbf{j} + aF_y\mathbf{k}.$$

Since the z axis is coincident with L, the unit vector $\mathbf{k}$ is along L. Therefore the moment of $\mathbf{F}$ about L is

$$\mathbf{M}_L = (\mathbf{k} \cdot \mathbf{M}_Q)\mathbf{k} = aF_y\mathbf{k}.$$

Figure 4.13
Using different points P and P' to determine the moment of $\mathbf{F}$ about L.

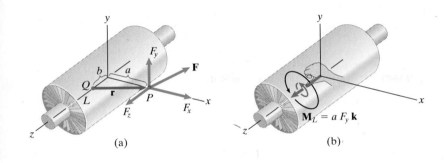

(a)

(b)

Figure 4.14
(a) An arbitrary point Q on L and the vector $\mathbf{r}$ from Q to P.
(b) $\mathbf{M}_L$ and the direction of the moment about L.

The components F_x and F_z exert no moment about L. If we assume that F_y is positive, it exerts a moment of magnitude aF_y about the turbine's axis in the direction shown in Fig. 4.14b.

Now let us determine the moment of a force about an arbitrary line L (Fig. 4.15a). The first step is to choose a point on the line. If we choose point A (Fig. 4.15b), the vector $\mathbf{r}$ from A to the point of application of $\mathbf{F}$ is

$$\mathbf{r} = (8 - 2)\mathbf{i} + (6 - 0)\mathbf{j} + (4 - 4)\mathbf{k} = 6\mathbf{i} + 6\mathbf{j} \; (\text{m}).$$

The moment of $\mathbf{F}$ about A is

$$\mathbf{M}_A = \mathbf{r} \times \mathbf{F} = \begin{vmatrix} \mathbf{i} & \mathbf{j} & \mathbf{k} \\ 6 & 6 & 0 \\ 10 & 60 & -20 \end{vmatrix}$$

$$= -120\mathbf{i} + 120\mathbf{j} + 300\mathbf{k} \; (\text{N-m}).$$

The next step is to determine a unit vector along L. The vector from A to B is

$$(-7 - 2)\mathbf{i} + (6 - 0)\mathbf{j} + (2 - 4)\mathbf{k} = -9\mathbf{i} + 6\mathbf{j} - 2\mathbf{k} \; (\text{m}).$$

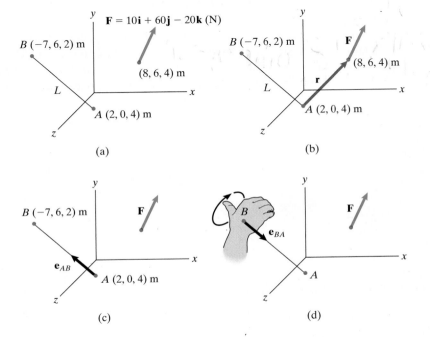

(a)

(b)

(c)

(d)

Figure 4.15
(a) A force $\mathbf{F}$ and line L.
(b) The vector $\mathbf{r}$ from A to the point of application of $\mathbf{F}$.
(c) $\mathbf{e}_{AB}$ points from A toward B.
(d) The right-hand rule indicates the direction of the moment.

Dividing this vector by its magnitude, we obtain a unit vector $\mathbf{e}_{AB}$ that points from A toward B (Fig. 4.15c):

$$\mathbf{e}_{AB} = -\frac{9}{11}\mathbf{i} + \frac{6}{11}\mathbf{j} - \frac{2}{11}\mathbf{k}.$$

The moment of $\mathbf{F}$ about L is

$$\mathbf{M}_L = (\mathbf{e}_{AB} \cdot \mathbf{M}_A)\mathbf{e}_{AB}$$

$$= \left[\left(-\frac{9}{11}\right)(-120 \text{ N-m}) + \left(\frac{6}{11}\right)(120 \text{ N-m}) + \left(-\frac{2}{11}\right)(300 \text{ N-m})\right]\mathbf{e}_{AB}$$

$$= 109\mathbf{e}_{AB} \text{ (N-m)}.$$

The magnitude of $\mathbf{M}_L$ is 109 N-m; pointing the thumb of the right hand in the direction of $\mathbf{e}_{AB}$ indicates the direction.

If we calculate $\mathbf{M}_L$ using the unit vector $\mathbf{e}_{BA}$ that points from B toward A instead, we obtain

$$\mathbf{M}_L = -109\mathbf{e}_{BA} \text{ (N-m)}.$$

We obtain the same magnitude, and the minus sign indicates that $\mathbf{M}_L$ points in the direction opposite to $\mathbf{e}_{BA}$, so the direction of $\mathbf{M}_L$ is the same. Therefore the right-hand rule indicates the same direction (Fig. 4.15d).

The preceding examples demonstrate three useful results that we can state in more general terms:

- When the line of action of $\mathbf{F}$ is perpendicular to a plane containing L (Fig. 4.16a), the magnitude of the moment of $\mathbf{F}$ about L is equal to the product of the magnitude of $\mathbf{F}$ and the perpendicular distance D from L to the point where the line of action intersects the plane: $|\mathbf{M}_L| = |\mathbf{F}|D$.
- When the line of action of $\mathbf{F}$ is parallel to L (Fig. 4.16b), the moment of $\mathbf{F}$ about L is zero: $\mathbf{M}_L = 0$. Since $\mathbf{M}_P = \mathbf{r} \times \mathbf{F}$ is perpendicular to $\mathbf{F}$, $\mathbf{M}_P$ is perpendicular to L and the vector component of $\mathbf{M}_P$ parallel to L is zero.
- When the line of action of $\mathbf{F}$ intersects L (Fig. 4.16c), the moment of $\mathbf{F}$ about L is zero. Since we can choose any point on L to evaluate $\mathbf{M}_P$, we can use the point where the line of action of $\mathbf{F}$ intersects L. The moment $\mathbf{M}_P$ about that point is zero, so its vector component parallel to L is zero.

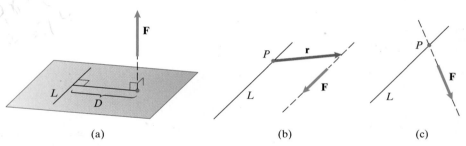

(a)   (b)   (c)

Figure 4.16
(a) $\mathbf{F}$ is perpendicular to a plane containing L.
(b) $\mathbf{F}$ is parallel to L.
(c) The line of action of $\mathbf{F}$ intersects L at P.

RESULTS

Determining the Moment of a Force F About a Line *L*

Choose *any* point *P* on the line and determine the moment $\mathbf{M}_P$ of **F** about *P*.

$$\mathbf{M}_P = \mathbf{r} \times \mathbf{F}$$

The component of $\mathbf{M}_P$ parallel to *L*, denoted by $\mathbf{M}_L$, is the moment of **F** about the line. (Pointing the thumb of the right hand in the direction of $\mathbf{M}_L$, the fingers point in the direction of the moment about the line.)

If **e** is a unit vector parallel to *L*,

$$\mathbf{M}_L = (\mathbf{e} \cdot \mathbf{M}_P)\, \mathbf{e}. \qquad (4.4)$$

Special Cases

When the line of action of **F** is perpendicular to a plane containing *L*, $|\mathbf{M}_L| = |\mathbf{F}|D$, where *D* is the perpendicular distance from *L* to the point where the line of action intersects the plane.

When the line of action of **F** is parallel to *L*, $\mathbf{M}_L = \mathbf{0}$.

When the line of action of **F** intersects *L*, $\mathbf{M}_L = \mathbf{0}$.

Active Example 4.6   **Moment of a Force About a Line** (▶ *Related Problem 4.87*)

What is the moment of the force **F** about the axis of the bar *BC*?

Strategy

Because we know the coordinates of points *A*, *B*, and *C*, we can determine the moment due to **F** about a point on the axis of the bar. We will determine the moment about point *B*. The component of that moment parallel to the axis *BC* is the moment of **F** about the axis. By obtaining a unit vector parallel to the axis, we can use Eq. (4.4) to determine the parallel component.

Solution

$$\mathbf{r} = (x_A - x_B)\mathbf{i} + (y_A - y_B)\mathbf{j} + (z_A - z_B)\mathbf{k}$$
$$= 4\mathbf{i} + 2\mathbf{j} - \mathbf{k} \text{ (m)}.$$

Determine the components of the vector from point *B* to the point of application of **F**.

$$\mathbf{M}_B = \mathbf{r} \times \mathbf{F}$$

$$= \begin{vmatrix} \mathbf{i} & \mathbf{j} & \mathbf{k} \\ 4 & 2 & -1 \\ -2 & 6 & 3 \end{vmatrix}$$

$$= 12\mathbf{i} - 10\mathbf{j} + 28\mathbf{k} \text{ (kN-m)}.$$

Calculate the moment of **F** about point *B*.

$$\mathbf{r}_{BC} = (x_C - x_B)\mathbf{i} + (y_C - y_B)\mathbf{j} + (z_C - z_B)\mathbf{k}$$
$$= 4\mathbf{j} - 3\mathbf{k} \text{ (m)}.$$

$$\mathbf{e}_{BC} = \frac{\mathbf{r}_{BC}}{|\mathbf{r}_{BC}|} = 0.8\mathbf{j} - 0.6\mathbf{k}.$$

> Obtain a unit vector parallel to the axis BC by dividing the position vector from point B to point C by its magnitude.

$$\mathbf{M}_{BC} = (\mathbf{e}_{BC} \cdot \mathbf{M}_B)\,\mathbf{e}_{BC}$$
$$= [(0)(12) + (0.8)(-10) + (-0.6)(28)]\mathbf{e}_{BC}$$
$$= -24.8\mathbf{e}_{BC} \text{ (kN-m)}.$$

> Apply Eq. (4.4) to determine the moment of $\mathbf{F}$ about the axis BC. Notice the negative result. Pointing the thumb of the right hand *opposite* to the direction of the unit vector $\mathbf{e}_{BC}$, the fingers point in the direction of the moment of $\mathbf{F}$ about the axis BC.

Practice Problem Determine the moment $\mathbf{M}_C$ of the force $\mathbf{F}$ about point C. Use it to calculate the moment of $\mathbf{F}$ about the axis BC by determining the component of $\mathbf{M}_C$ parallel to the axis.

Answer: $\mathbf{M}_{BC} = -24.8\mathbf{e}_{BC}$ (kN-m).

Example 4.7   Moment of a Force About the *x* Axis (▶ *Related Problem 4.77*)

What is the moment of the 50-lb force about the *x* axis?

Strategy

We can determine the moment in two ways.

First Method   We can use Eqs. (4.5) and (4.6). Since **r** can extend from any point on the *x* axis to the line of action of the force, we can use the vector from *O* to the point of application of the force. The vector **e** must be a unit vector along the *x* axis, so we can use either **i** or −**i**.

Second Method   This example is the first of the special cases we discussed, because the 50-lb force is perpendicular to the *x–z* plane. We can determine the magnitude and direction of the moment directly from the given information.

Solution

First Method   *Determine a vector* **r**. The vector from *O* to the point of application of the force is (Fig. a)

$$\mathbf{r} = 4\mathbf{i} + 3\mathbf{k} \ (\text{ft}).$$

(a) The vector **r** from *O* to the point of application of the force.

Determine a vector **e**. We can use the unit vector **i**.

Evaluate $\mathbf{M}_L$. From Eq. (4.6), the mixed triple product is

$$\mathbf{i} \cdot (\mathbf{r} \times \mathbf{F}) = \begin{vmatrix} 1 & 0 & 0 \\ 4 & 0 & 3 \\ 0 & 50 & 0 \end{vmatrix} = -150 \text{ ft-lb}.$$

Then from Eq. (4.5), the moment of the force about the *x* axis is

$$\mathbf{M}_{x\,\text{axis}} = [\mathbf{i} \cdot (\mathbf{r} \times \mathbf{F})]\mathbf{i} = -150\mathbf{i} \ (\text{ft-lb}).$$

The magnitude of the moment is 150 ft-lb, and its direction is as shown in Fig. b.
Second Method   Since the 50-lb force is perpendicular to a plane (the *x–z* plane) containing the *x* axis, the magnitude of the moment about the *x* axis is equal to the perpendicular distance from the *x* axis to the point where the line of action of the force intersects the *x–z* plane (Fig. c):

$$\left| \mathbf{M}_{x\,\text{axis}} \right| = (3 \text{ ft})(50 \text{ lb}) = 150 \text{ ft-lb}.$$

(b) The direction of the moment.

Pointing the arc of the fingers in the direction of the moment about the *x* axis (Fig. c), we find that the right-hand rule indicates that $\mathbf{M}_{x\,\text{axis}}$ points in the negative *x* axis direction. Therefore,

$$\mathbf{M}_{x\,\text{axis}} = -150\mathbf{i} \ (\text{ft-lb}).$$

(c) The distance from the *x* axis to the point where the line of action of the force intersects the *x–z* plane is 3 ft. The arrow indicates the direction of the moment about the *x* axis.

Critical Thinking

The hinged door in this example is designed to rotate about the *x* axis. If no other forces act on the door, you can see that the 50-lb upward force would tend to cause the door to rotate upward. It is the moment of the force about the *x* axis, and *not* the moment of the force about some point, that measures the tendency of the force to cause the door to rotate on its hinges. Furthermore, the direction of the moment of the force about the *x* axis indicates the direction in which the force tends to cause the door to rotate. (See Fig. b.)

Example 4.8   Rotating Machines (▶ *Related Problem 4.100*)

The crewman exerts the forces shown on the handles of the coffee grinder winch, where $\mathbf{F} = 4\mathbf{j} + 32\mathbf{k}$ N. Determine the total moment he exerts (a) about point O; (b) about the axis of the winch, which coincides with the x axis.

Strategy
(a) To obtain the total moment about point O, we must sum the moments of the two forces about O. Let the sum be denoted by $\Sigma\mathbf{M}_O$.

(b) Because point O is on the x axis, the total moment about the x axis is the component of $\Sigma\mathbf{M}_O$ parallel to the x axis, which is the x component of $\Sigma\mathbf{M}_O$.

Solution
(a) The total moment about point O is

$$\Sigma\mathbf{M}_O = \begin{vmatrix} \mathbf{i} & \mathbf{j} & \mathbf{k} \\ -0.18 & 0.28 & 0.1 \\ 0 & 4 & 32 \end{vmatrix} + \begin{vmatrix} \mathbf{i} & \mathbf{j} & \mathbf{k} \\ 0.18 & -0.28 & -0.1 \\ 0 & -4 & -32 \end{vmatrix}$$

$$= 17.1\mathbf{i} + 11.5\mathbf{j} - 1.4\mathbf{k} \text{ (N-m)}.$$

(a) The total moment about the x axis.

(b) The total moment about the x axis is the x component of $\Sigma\mathbf{M}_O$ (Fig. a):

$$\Sigma\mathbf{M}_{x \text{ axis}} = 17.1 \text{ (N-m)}.$$

Notice that this is the result given by Eq. (4.4): Since $\mathbf{i}$ is a unit vector parallel to the x axis,

$$\Sigma\mathbf{M}_{x \text{ axis}} = (\mathbf{i} \cdot \Sigma\mathbf{M}_O)\mathbf{i} = 17.1 \text{ (N-m)}.$$

Problems

▶ **4.77** The force $\mathbf{F} = 20\mathbf{i} + 40\mathbf{j} - 10\mathbf{k}$ (N). Use both of the procedures described in Example 4.7 to determine the moment due to $\mathbf{F}$ about the z axis.

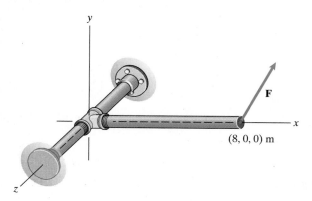

(8, 0, 0) m

Problem 4.77

4.78 Use Eqs. (4.5) and (4.6) to determine the moment of the 20-N force about (a) the x axis, (b) the y axis, (c) the z axis. (First see if you can write down the results without using the equations.)

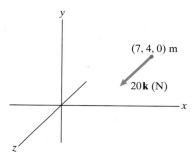

(7, 4, 0) m

20**k** (N)

Problem 4.78

4.79 Three forces parallel to the y axis act on the rectangular plate. Use Eqs. (4.5) and (4.6) to determine the sum of the moments of the forces about the x axis. (First see if you can write down the result without using the equations.)

4.80 The three forces are parallel to the y axis. Determine the sum of the moments of the forces (a) about the y axis; (b) about the z axis.

3 kN

2 kN

6 kN

600 mm

900 mm

Problems 4.79/4.80

4.81 The person exerts a force $\mathbf{F} = 0.2\mathbf{i} - 0.4\mathbf{j} + 1.2\mathbf{k}$ (lb) on the gate at C. Point C lies in the x–y plane. What moment does the person exert about the gate's hinge axis, which is coincident with the y axis?

A

C

3.5 ft

B ⟵ 2 ft ⟶

Problem 4.81

4.82 Four forces act on the plate. Their components are

$$\mathbf{F}_A = -2\mathbf{i} + 4\mathbf{j} + 2\mathbf{k} \ (\text{kN}),$$

$$\mathbf{F}_B = 3\mathbf{j} - 3\mathbf{k} \ (\text{kN}),$$

$$\mathbf{F}_C = 2\mathbf{j} + 3\mathbf{k} \ (\text{kN}),$$

$$\mathbf{F}_D = 2\mathbf{i} + 6\mathbf{j} + 4\mathbf{k} \ (\text{kN}).$$

Determine the sum of the moments of the forces (a) about the x axis; (b) about the z axis.

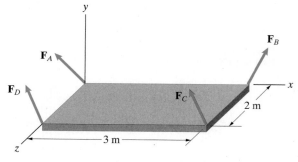

$\mathbf{F}_B$

$\mathbf{F}_A$

$\mathbf{F}_D$

$\mathbf{F}_C$

2 m

3 m

Problem 4.82

4.83 The force $\mathbf{F} = 30\mathbf{i} + 20\mathbf{j} - 10\mathbf{k}$ (lb).

(a) What is the moment of $\mathbf{F}$ about the y axis?

(b) Suppose that you keep the magnitude of $\mathbf{F}$ fixed, but you change its direction so as to make the moment of $\mathbf{F}$ about the y axis as large as possible. What is the magnitude of the resulting moment?

4.84 The moment of the force $\mathbf{F}$ about the x axis is $-80\mathbf{i}$ (ft-lb), the moment about the y axis is zero, and the moment about the z axis is $160\mathbf{k}$ (ft-lb). If $F_y = 80$ lb, what are F_x and F_z?

Problems 4.83/4.84

4.85 The robotic manipulator is stationary. The weights of the arms AB and BC act at their midpoints. The direction cosines of the centerline of arm AB are $\cos \theta_x = 0.500$, $\cos \theta_y = 0.866$, $\cos \theta_z = 0$, and the direction cosines of the centerline of arm BC are $\cos \theta_x = 0.707$, $\cos \theta_y = 0.619$, $\cos \theta_z = -0.342$. What total moment is exerted about the z axis by the weights of the arms?

4.86 In Problem 4.85, what total moment is exerted about the x axis by the weights of the arms?

Problems 4.85/4.86

4.87 In Active Example 4.6, suppose that the force changes to $\mathbf{F} = -2\mathbf{i} + 3\mathbf{j} + 6\mathbf{k}$ (kN). Determine the magnitude of the moment of the force about the axis of the bar BC.

4.88 Determine the moment of the 20-N force about the line AB. Use Eqs. (4.5) and (4.6), letting the unit vector $\mathbf{e}$ point (a) from A toward B; (b) from B toward A.

Problem 4.88

4.89 The force $\mathbf{F} = -10\mathbf{i} + 5\mathbf{j} - 5\mathbf{k}$ (kip). Determine the moment of $\mathbf{F}$ about the line AB. Draw a sketch to indicate the direction of the moment.

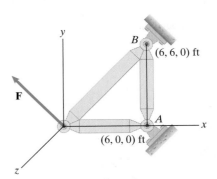

Problem 4.89

4.90 The force $\mathbf{F} = 10\mathbf{i} + 12\mathbf{j} - 6\mathbf{k}$ (N). What is the moment of $\mathbf{F}$ about the line AO? Draw a sketch to indicate the direction of the moment.

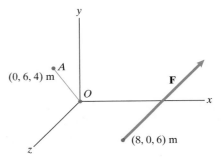

Problem 4.90

4.91 The tension in the cable AB is 1 kN. Determine the moment about the x axis due to the force exerted on the hatch by the cable at point B. Draw a sketch to indicate the direction of the moment.

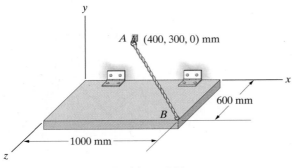

Problem 4.91

4.92 Determine the moment of the force applied at D about the straight line through the hinges A and B. (The line through A and B lies in the y–z plane.)

4.93 The tension in the cable CE is 160 lb. Determine the moment of the force exerted by the cable on the hatch at C about the straight line through the hinges A and B.

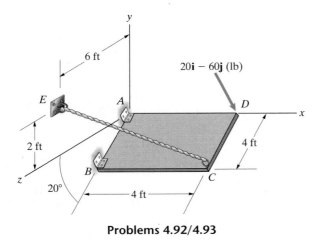

Problems 4.92/4.93

4.94 The coordinates of A are $(-2.4, 0, -0.6)$ m, and the coordinates of B are $(-2.2, 0.7, -1.2)$ m. The force exerted at B by the sailboat's main sheet AB is 130 N. Determine the moment of the force about the centerline of the mast (the y axis). Draw a sketch to indicate the direction of the moment.

Problem 4.94

4.95 The tension in cable AB is 200 lb. Determine the moments about each of the coordinate axes due to the force exerted on point B by the cable. Draw sketches to indicate the direction of the moments.

Problem 4.95

4.96 The total force exerted on the blades of the turbine by the steam nozzle is $\mathbf{F} = 20\mathbf{i} - 120\mathbf{j} + 100\mathbf{k}$ (N), and it effectively acts at the point (100, 80, 300) mm. What moment is exerted about the axis of the turbine (the x axis)?

Fixed
Rotating

Problem 4.96

4.97 The pneumatic support AB holds a trunk lid in place. It exerts a 35-N force on the fixture at B that points in the direction from A toward B. Determine the magnitude of the moment of the force about the hinge axis of the lid, which is the z axis.

B (60, 100, −30) mm

O

A
(480, −40, 40) mm

Problem 4.97

4.98 The tension in cable AB is 80 lb. What is the moment about the line CD due to the force exerted by the cable on the wall at B?

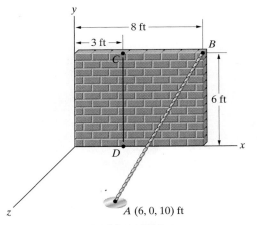

8 ft
3 ft
C
B
6 ft
D
x
A (6, 0, 10) ft

Problem 4.98

4.99 The magnitude of the force $\mathbf{F}$ is 0.2 N and its direction cosines are $\cos \theta_x = 0.727$, $\cos \theta_y = -0.364$, and $\cos \theta_z = 0.582$. Determine the magnitude of the moment of $\mathbf{F}$ about the axis AB of the spool.

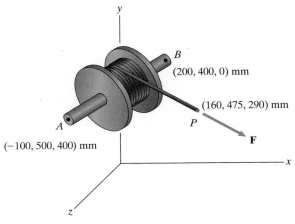

B
(200, 400, 0) mm
(160, 475, 290) mm
A
P
(−100, 500, 400) mm
$\mathbf{F}$
x

Problem 4.99

▶ **4.100** A motorist applies the two forces shown to loosen a lug nut. The direction cosines of **F** are $\cos \theta_x = \frac{4}{13}$, $\cos \theta_y = \frac{12}{13}$, and $\cos \theta_z = \frac{3}{13}$. If the magnitude of the moment about the x axis must be 32 ft-lb to loosen the nut, what is the magnitude of the forces the motorist must apply? (See Example 4.8.)

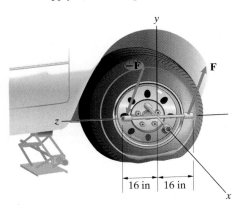

Problem 4.100

4.101 The tension in cable AB is 2 kN. What is the magnitude of the moment about the shaft CD due to the force exerted by the cable at A? Draw a sketch to indicate the direction of the moment about the shaft.

Problem 4.101

4.102 The axis of the car's wheel passes through the origin of the coordinate system and its direction cosines are $\cos \theta_x = 0.940$, $\cos \theta_y = 0$, $\cos \theta_z = 0.342$. The force exerted on the tire by the road effectively acts at the point $x = 0$, $y = -0.36$ m, $z = 0$ and has components $\mathbf{F} = -720\mathbf{i} + 3660\mathbf{j} + 1240\mathbf{k}$ (N). What is the moment of **F** about the wheel's axis?

Problem 4.102

4.103 The direction cosines of the centerline OA are $\cos \theta_x = 0.500$, $\cos \theta_y = 0.866$, and $\cos \theta_z = 0$, and the direction cosines of the line AG are $\cos \theta_x = 0.707$, $\cos \theta_y = 0.619$, and $\cos \theta_z = -0.342$. What is the moment about OA due to the 250-N weight? Draw a sketch to indicate the direction of the moment about the shaft.

Problem 4.103

4.104 The radius of the steering wheel is 200 mm. The distance from O to C is 1 m. The center C of the steering wheel lies in the x–y plane. The driver exerts a force $\mathbf{F} = 10\mathbf{i} + 10\mathbf{j} - 5\mathbf{k}$ (N) on the wheel at A. If the angle $\alpha = 0$, what is the magnitude of the moment about the shaft OC? Draw a sketch to indicate the direction of the moment about the shaft.

Problem 4.104

4.105* The magnitude of the force $\mathbf{F}$ is 10 N. Suppose that you want to choose the direction of the force $\mathbf{F}$ so that the magnitude of its moment about the line L is a maximum. Determine the components of $\mathbf{F}$ and the magnitude of its moment about L. (There are two solutions for $\mathbf{F}$.)

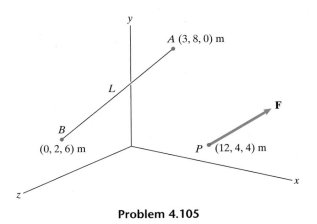

Problem 4.105

4.106 The weight W causes a tension of 100 lb in cable CD. If $d = 2$ ft, what is the moment about the z axis due to the force exerted by the cable CD at point C?

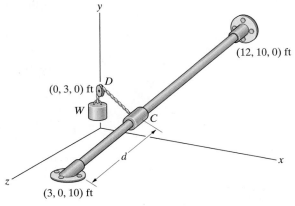

Problem 4.106

4.107* The y axis points upward. The weight of the 4-kg rectangular plate acts at the midpoint G of the plate. The sum of the moments about the straight line through the supports A and B due to the weight of the plate and the force exerted on the plate by the cable CD is zero. What is the tension in the cable?

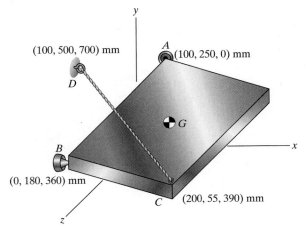

Problem 4.107

4.4 Couples

BACKGROUND

Now that we have described how to calculate the moment due to a force, consider this question: Is it possible to exert a moment on an object without subjecting it to a net force? The answer is yes, and it occurs when a compact disk begins rotating or a screw is turned by a screwdriver. Forces are exerted on these objects, but in such a way that the net force is zero while the net moment is not zero.

Two forces that have equal magnitudes, opposite directions, and different lines of action are called a *couple* (Fig. 4.17a). A couple tends to cause rotation of an object even though the vector sum of the forces is zero, and it has the remarkable property that *the moment it exerts is the same about any point*.

The moment of a couple is simply the sum of the moments of the forces about a point P (Fig. 4.17b):

$$\mathbf{M} = [\mathbf{r}_1 \times \mathbf{F}] + [\mathbf{r}_2 \times (-\mathbf{F})] = (\mathbf{r}_1 - \mathbf{r}_2) \times \mathbf{F}.$$

The vector $\mathbf{r}_1 - \mathbf{r}_2$ is equal to the vector $\mathbf{r}$ shown in Fig. 4.17c, so we can express the moment as

$$\mathbf{M} = \mathbf{r} \times \mathbf{F}.$$

Since $\mathbf{r}$ doesn't depend on the position of P, the moment $\mathbf{M}$ is the same for *any* point P.

Because a couple exerts a moment but the sum of the forces is zero, it is often represented in diagrams simply by showing the moment (Fig. 4.17d). Like the Cheshire cat in *Alice's Adventures in Wonderland,* which vanished except for its grin, the forces don't appear; only the moment they exert is visible. But we recognize the origin of the moment by referring to it as a *moment of a couple,* or simply a *couple*.

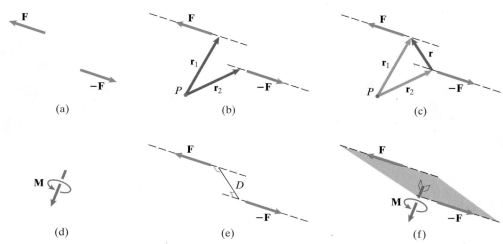

Figure 4.17
(a) A couple.
(b) Determining the moment about P.
(c) The vector $\mathbf{r} = \mathbf{r}_1 - \mathbf{r}_2$.
(d) Representing the moment of the couple.
(e) The distance D between the lines of action.
(f) $\mathbf{M}$ is perpendicular to the plane containing $\mathbf{F}$ and $-\mathbf{F}$.

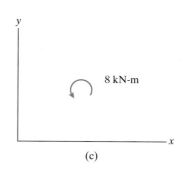

Figure 4.18
(a) A couple consisting of 2-kN forces.
(b) Determining the sum of the moments of the forces about O.
(c) Representing a couple in two dimensions.

Notice in Fig. 4.17c that $\mathbf{M} = \mathbf{r} \times \mathbf{F}$ is the moment of $\mathbf{F}$ about a point on the line of action of the force $-\mathbf{F}$. The magnitude of the moment of a force about a point equals the product of the magnitude of the force and the perpendicular distance from the point to the line of action of the force, so $|\mathbf{M}| = D|\mathbf{F}|$, where D is the perpendicular distance between the lines of action of the two forces (Fig. 4.17e). The cross product $\mathbf{r} \times \mathbf{F}$ is perpendicular to $\mathbf{r}$ and $\mathbf{F}$, which means that $\mathbf{M}$ is perpendicular to the plane containing $\mathbf{F}$ and $-\mathbf{F}$ (Fig. 4.17f). Pointing the thumb of the right hand in the direction of $\mathbf{M}$, the arc of the fingers indicates the direction of the moment.

In Fig. 4.18a, our view is perpendicular to the plane containing the two forces. The distance between the lines of action of the forces is 4 m, so the magnitude of the moment of the couple is $|\mathbf{M}| = (4\ \mathrm{m})(2\ \mathrm{kN}) = 8\ \mathrm{kN\text{-}m}$. The moment $\mathbf{M}$ is perpendicular to the plane containing the two forces. Pointing the arc of the fingers of the right hand counterclockwise, we find that the right-hand rule indicates that $\mathbf{M}$ points out of the page. Therefore, the moment of the couple is

$$\mathbf{M} = 8\mathbf{k}\ (\mathrm{kN\text{-}m}).$$

We can also determine the moment of the couple by calculating the sum of the moments of the two forces about *any* point. The sum of the moments of the forces about the origin O is (Fig. 4.18b)

$$\mathbf{M} = [\mathbf{r}_1 \times (2\mathbf{j})] + [\mathbf{r}_2 \times (-2\mathbf{j})]$$
$$= [(7\mathbf{i} + 2\mathbf{j}) \times (2\mathbf{j})] + [(3\mathbf{i} + 7\mathbf{j}) \times (-2\mathbf{j})]$$
$$= 8\mathbf{k}\ (\mathrm{kN\text{-}m}).$$

In a two-dimensional situation like this example, it isn't convenient to represent a couple by showing the moment vector, because the vector is perpendicular to the page. Instead, we represent the couple by showing its magnitude and a circular arrow that indicates its direction (Fig. 4.18c).

By grasping a bar and twisting it (Fig. 4.19a), a moment can be exerted about its axis (Fig. 4.19b). Although the system of forces exerted is distributed over the surface of the bar in a complicated way, the effect is the same as if two equal and opposite forces are exerted (Fig. 4.19c). When we represent a couple as in Fig. 4.19b, or by showing the moment vector $\mathbf{M}$, we imply that some system of forces exerts that moment. The system of forces (such as the forces exerted in twisting the bar, or the forces on the crankshaft that exert a moment on

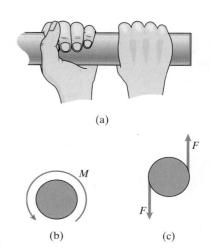

Figure 4.19
(a) Twisting a bar.
(b) The moment about the axis of the bar.
(c) The same effect is obtained by applying two equal and opposite forces.

the drive shaft of a car) is nearly always more complicated than two equal and opposite forces, but the effect is the same. For this reason, we can *model* the actual system as a simple system of two forces.

RESULTS

Two forces with equal magnitudes, opposite directions, and different lines of action are called a *couple*.

The moment about a point due to a couple is the sum of the moments of the two forces about that point. *The moment **M** due to a couple is the same about any point.* Its magnitude is $D|\mathbf{F}|$, where D is the perpendicular distance between the lines of action of the forces. The vector **M** is perpendicular to the plane containing the lines of action.

Because the total force exerted by a couple is zero, a couple is often represented by the moment it exerts.

When the lines of action of the forces of a couple lie in the *x–y* plane, the couple can be represented by its magnitude and a circular arrow that indicates its direction.

Active Example 4.9   **Moment of a Couple** (▶ *Related Problem 4.108*)

The force $\mathbf{F} = 10\mathbf{i} - 4\mathbf{j}$ (N). Determine the moment due to the couple. Represent the moment by its magnitude and a circular arrow indicating its direction.

Strategy

We will determine the moment in two ways. In the first method, we will choose a point and calculate the sum of the moments of the two forces about that point. Because the moment due to a couple is the same about any point, we can choose any convenient point. In the second method, we will sum the moments of the two couples formed by the *x* and *y* components of the forces.

Solution

First Method

$$\mathbf{M} = \mathbf{r} \times (-\mathbf{F})$$
$$= (-2\mathbf{i} + 3\mathbf{j}) \times (-10\mathbf{i} + 4\mathbf{j})$$
$$= 22\mathbf{k} \text{ (N-m)}.$$

Calculate the sum of the moments of the two forces about the point of application of the force **F**.

The magnitude of the moment is 22 N-m. Pointing the thumb of the right hand in the direction of the unit vector **k**, the direction of the moment in the x–y plane is counterclockwise.

22 N-m

Second Method

The components of the two forces form two couples.

The magnitude of the moment due to the 10-N couple is $(3 \text{ m})(10 \text{ N}) = 30$ N-m, and the moment is counterclockwise. The magnitude of the moment due to the 4-N couple is $(2 \text{ m})(4 \text{ N}) = 8$ N-m, and the moment is clockwise. Therefore the total counterclockwise moment is $30 - 8 = 22$ N-m.

Practice Problem Use the cross product to calculate the sum of the moments of the forces **F** and −**F** about the point P with coordinates $(10, 7, 3)$ m. Represent the moment by its magnitude and a circular arrow indicating its direction.

Answer: 22**k** (N-m), or 22 N-m counterclockwise.

Example 4.10 | **Determining Unknown Forces** (▶ *Related Problem 4.113*)

Two forces A and B and a 200 ft-lb couple act on the beam. The sum of the forces is zero, and the sum of the moments about the left end of the beam is zero. What are the forces A and B?

Strategy

By summing the two forces (the couple exerts no net force on the beam) and summing the moments due to the forces and the couple about the left end of the beam, we will obtain two equations in terms of the two unknown forces.

Solution

The sum of the forces is

$$\Sigma F_y = A + B = 0.$$

The moment of the couple (200 ft-lb clockwise) is the same about any point, so the sum of the moments about the left end of the beam is

$$\Sigma M_{\text{left end}} = (4 \text{ ft}) B - 200 \text{ ft-lb} = 0.$$

The forces are $B = 50$ lb and $A = -50$ lb.

The forces on the beam form a couple.

Critical Thinking

Notice that the total moment about the left end of the beam is the sum of the moment due to the force B and the moment due to the 200 ft-lb couple. As we observe in Chapter 5, if an object subjected to forces and couples is in equilibrium, the sum of the forces is zero and the sum of the moments about any point, *including moments due to couples*, is zero. In this example we needed both these conditions to determine the unknown forces A and B.

Example 4.11	Sum of the Moments Due to Two Couples (▶ *Related Problem 4.119*)

Determine the sum of the moments exerted on the pipe by the two couples.

Strategy

We will express the moment exerted by each couple as a vector. To express the 30-N couple in terms of a vector, we will express the forces in terms of their components. We can then sum the moment vectors to determine the sum of the moments exerted by the couples.

Solution

Consider the 20-N couple. The magnitude of the moment of the couple is $(2 \text{ m})(20 \text{ N}) = 40$ N-m. The direction of the moment vector is perpendicular to the y–z plane, and the right-hand rule indicates that it points in the positive x axis direction. The moment of the 20-N couple is $40\mathbf{i}$ (N-m).

By resolving the 30-N forces into y and z components, we obtain the two couples in Fig. a. The moment of the couple formed by the y components is $-(30 \sin 60°)(4)\mathbf{k}$ (N-m), and the moment of the couple formed by the z components is $(30 \cos 60°)(4)\mathbf{j}$ (N-m).

The sum of the moments is therefore

$$\Sigma\mathbf{M} = 40\mathbf{i} + (30 \cos 60°)(4)\mathbf{j} - (30 \sin 60°)(4)\mathbf{k} \text{ (N-m)}$$

$$= 40\mathbf{i} + 60\mathbf{j} - 104\mathbf{k} \text{ (N-m)}.$$

(a) Resolving the 30-N forces into y and z components.

Critical Thinking

Although the method we used in this example helps you recognize the contributions of the individual couples to the sum of the moments, it is convenient only when the orientations of the forces and their points of application relative to the coordinate system are fairly simple. When that is not the case, you can determine the sum of the moments by choosing any point and calculating the sum of the moments of the forces about that point.

Problems

▶ **4.108** In Active Example 4.9, suppose that the point of application of the force **F** is moved from (8, 3, 0) m to (8, 8, 0) m. Draw a sketch showing the new position of the force. From your sketch, will the moment due to the couple be clockwise or counterclockwise? Calculate the moment due to the couple. Represent the moment by its magnitude and a circular arrow indicating its direction.

4.109 The forces are contained in the x–y plane.

(a) Determine the moment of the couple and represent it as shown in Fig. 4.18c.

(b) What is the sum of the moments of the two forces about the point (10, −40, 20) ft?

Problem 4.109

4.110 The moment of the couple is 600**k** (N-m). What is the angle α?

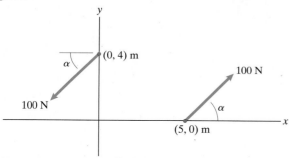

Problem 4.110

4.111 Point P is contained in the x–y plane, $|\mathbf{F}| = 100$ N, and the moment of the couple is −500**k** (N-m). What are the coordinates of P?

Problem 4.111

4.112 Three forces of equal magnitude are applied parallel to the sides of an equilateral triangle. (a) Show that the sum of the moments of the forces is the same about any point. (b) Determine the magnitude of the sum of the moments.

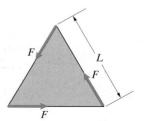

Problem 4.112

▶ **4.113** In Example 4.10, suppose that the 200 ft-lb couple is counterclockwise instead of clockwise. Draw a sketch of the beam showing the forces and couple acting on it. What are the forces A and B?

4.114 The moments of two couples are shown. What is the sum of the moments about point P?

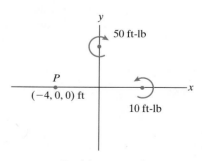

Problem 4.114

4.115 Determine the sum of the moments exerted on the plate by the two couples.

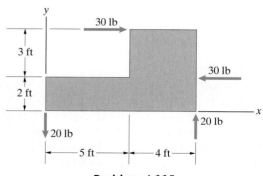

Problem 4.115

4.116 Determine the sum of the moments exerted about A by the couple and the two forces.

Problem 4.116

4.117 Determine the sum of the moments exerted about A by the couple and the two forces.

Problem 4.117

4.118 The sum of the moments about point A due to the forces and couples acting on the bar is zero.

(a) What is the magnitude of the couple C?

(b) Determine the sum of the moments about point B due to the forces and couples acting on the bar.

Problem 4.118

▶ **4.119** In Example 4.11, suppose that instead of acting in the positive z axis direction, the upper 20-N force acts in the positive x axis direction. Instead of acting in the negative z axis direction, let the lower 20-N force act in the negative x axis direction. Draw a sketch of the pipe showing the forces acting on it. Determine the sum of the moments exerted on the pipe by the two couples.

4.120 (a) What is the moment of the couple?

(b) Determine the perpendicular distance between the lines of action of the two forces.

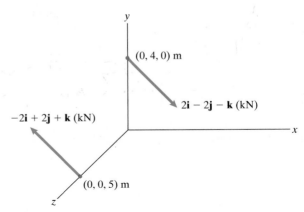

Problem 4.120

4.121 Determine the sum of the moments exerted on the plate by the three couples. (The 80-lb forces are contained in the x–z plane.)

Problem 4.121

4.122 What is the magnitude of the sum of the moments exerted on the T-shaped structure by the two couples?

Problem 4.122

4.123 The tension in cables AB and CD is 500 N.

(a) Show that the two forces exerted by the cables on the rectangular hatch at B and C form a couple.

(b) What is the moment exerted on the plate by the cables?

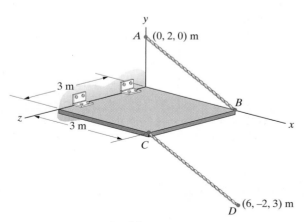

Problem 4.123

4.124 The cables AB and CD exert a couple on the vertical pipe. The tension in each cable is 8 kN. Determine the magnitude of the moment the cables exert on the pipe.

Problem 4.124

4.125 The bar is loaded by the forces

$$\mathbf{F}_B = 2\mathbf{i} + 6\mathbf{j} + 3\mathbf{k} \text{ (kN)},$$
$$\mathbf{F}_C = \mathbf{i} - 2\mathbf{j} + 2\mathbf{k} \text{ (kN)},$$

and the couple

$$\mathbf{M}_C = 2\mathbf{i} + \mathbf{j} - 2\mathbf{k} \text{ (kN-m)}.$$

Determine the sum of the moments of the two forces and the couple about A.

4.126 The forces

$$\mathbf{F}_B = 2\mathbf{i} + 6\mathbf{j} + 3\mathbf{k} \text{ (kN)},$$
$$\mathbf{F}_C = \mathbf{i} - 2\mathbf{j} + 2\mathbf{k} \text{ (kN)},$$

and the couple

$$\mathbf{M}_C = M_{Cy}\mathbf{j} + M_{Cz}\mathbf{k} \text{ (kN-m)}.$$

Determine the values of M_{Cy} and M_{Cz} so that the sum of the moments of the two forces and the couple about A is zero.

Problems 4.125/4.126

4.127 Two wrenches are used to tighten an elbow fitting. The force $\mathbf{F} = 10\mathbf{k}$ (lb) on the right wrench is applied at $(6, -5, -3)$ in, and the force $-\mathbf{F}$ on the left wrench is applied at $(4, -5, 3)$ in.

(a) Determine the moment about the x axis due to the force exerted on the right wrench.

(b) Determine the moment of the couple formed by the forces exerted on the two wrenches.

(c) Based on the results of (a) and (b), explain why two wrenches are used.

Problem 4.127

4.5 Equivalent Systems

A *system of forces and moments* is simply a particular set of forces and moments of couples. The systems of forces and moments dealt with in engineering can be complicated. This is especially true in the case of distributed forces, such as the pressure forces exerted by water on a dam. Fortunately, if we are concerned only with the total force and moment exerted, we can represent complicated systems of forces and moments by much simpler systems.

Conditions for Equivalence

We define two systems of forces and moments, designated as system 1 and system 2, to be *equivalent* if the sums of the forces are equal, or

$$(\Sigma \mathbf{F})_1 = (\Sigma \mathbf{F})_2, \tag{4.7}$$

and the sums of the moments about a point P are equal, or

$$(\Sigma \mathbf{M}_P)_1 = (\Sigma \mathbf{M}_P)_2. \tag{4.8}$$

To see what the conditions for equivalence mean, consider the systems of forces and moments in Fig. 4.20a. In system 1, an object is subjected to two forces $\mathbf{F}_A$ and $\mathbf{F}_B$ and a couple $\mathbf{M}_C$. In system 2, the object is subjected to a force $\mathbf{F}_D$ and two couples $\mathbf{M}_E$ and $\mathbf{M}_F$. The first condition for equivalence is

$$(\Sigma \mathbf{F})_1 = (\Sigma \mathbf{F})_2:$$
$$\mathbf{F}_A + \mathbf{F}_B = \mathbf{F}_D. \tag{4.9}$$

(a)

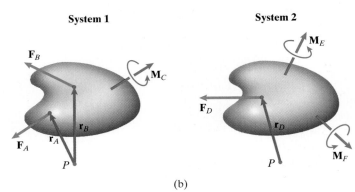

(b)

Figure 4.20
(a) Different systems of forces and moments applied to an object.
(b) Determining the sum of the moments about a point P for each system.

System 1

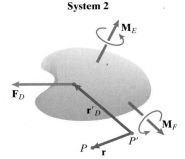

System 2

Figure 4.21
Determining the sums of the moments about
a different point P'.

System 1

System 2

Figure 4.22
(a) An arbitrary system of forces and
moments.
(b) A force acting at P and a couple.

If we determine the sums of the moments about the point P in Fig. 4.20b, the
second condition for equivalence is

$$(\Sigma \mathbf{M}_P)_1 = (\Sigma \mathbf{M}_P)_2:$$

$$(\mathbf{r}_A \times \mathbf{F}_A) + (\mathbf{r}_B \times \mathbf{F}_B) + \mathbf{M}_C = (\mathbf{r}_D \times \mathbf{F}_D) + \mathbf{M}_E + \mathbf{M}_F. \quad (4.10)$$

If these conditions are satisfied, systems 1 and 2 are equivalent.

We will use this example to demonstrate that *if the sums of the forces are
equal for two systems of forces and moments and the sums of the moments about
one point P are equal, then the sums of the moments about any point are equal.*
Suppose that Eq. (4.9) is satisfied, and Eq. (4.10) is satisfied for the point P in
Fig. 4.20b. For a different point P' (Fig. 4.21), we will show that

$$(\Sigma \mathbf{M}_{P'})_1 = (\Sigma \mathbf{M}_{P'})_2:$$

$$(\mathbf{r}'_A \times \mathbf{F}_A) + (\mathbf{r}'_B \times \mathbf{F}_B) + \mathbf{M}_C = (\mathbf{r}'_D \times \mathbf{F}_D) + \mathbf{M}_E + \mathbf{M}_F. \quad (4.11)$$

In terms of the vector $\mathbf{r}$ from P' to P, the relations between the vectors $\mathbf{r}'_A$, $\mathbf{r}'_B$,
and $\mathbf{r}'_D$ in Fig. 4.21 and the vectors $\mathbf{r}_A$, $\mathbf{r}_B$, and $\mathbf{r}_D$ in Fig. 4.20b are

$$\mathbf{r}'_A = \mathbf{r} + \mathbf{r}_A, \qquad \mathbf{r}'_B = \mathbf{r} + \mathbf{r}_B, \qquad \mathbf{r}'_D = \mathbf{r} + \mathbf{r}_D.$$

Substituting these expressions into Eq. (4.11), we obtain

$$[(\mathbf{r} + \mathbf{r}_A) \times \mathbf{F}_A] + [(\mathbf{r} + \mathbf{r}_B) \times \mathbf{F}_B] + \mathbf{M}_C$$
$$= [(\mathbf{r} + \mathbf{r}_D) \times \mathbf{F}_D] + \mathbf{M}_E + \mathbf{M}_F.$$

Rearranging terms, we can write this equation as

$$[\mathbf{r} \times (\Sigma \mathbf{F})_1] + (\Sigma \mathbf{M}_P)_1 = [\mathbf{r} \times (\Sigma \mathbf{F})_2] + (\Sigma \mathbf{M}_P)_2,$$

which holds in view of Eqs. (4.9) and (4.10). The sums of the moments of the
two systems about any point are equal.

Representing Systems by Equivalent Systems

If we are concerned only with the total force and total moment exerted on an
object by a given system of forces and moments, we can *represent* the system
by an equivalent one. By this we mean that instead of showing the actual forces
and couples acting on an object, we would show a different system that exerts
the same total force and moment. In this way, we can replace a given system
by a less complicated one to simplify the analysis of the forces and moments
acting on an object and to gain a better intuitive understanding of their effects
on the object.

Representing a System by a Force and a Couple   Let us consider an
arbitrary system of forces and moments and a point P (system 1 in Fig. 4.22). We
can represent this system by one consisting of a single force acting at P and a
single couple (system 2). The conditions for equivalence are

$$(\Sigma \mathbf{F})_2 = (\Sigma \mathbf{F})_1:$$

$$\mathbf{F} = (\Sigma \mathbf{F})_1$$

and

$$(\Sigma \mathbf{M}_P)_2 = (\Sigma \mathbf{M}_P)_1:$$

$$\mathbf{M} = (\Sigma \mathbf{M}_P)_1.$$

These conditions are satisfied if **F** equals the sum of the forces in system 1 and **M** equals the sum of the moments about P in system 1.

Thus *no matter how complicated a system of forces and moments may be, it can be represented by a single force acting at a given point and a single couple.*

Representing a Force by a Force and a Couple   A force $\mathbf{F}_P$ acting at a point P (system 1 in Fig. 4.23a) can be represented by a force **F** acting at a different point Q and a couple **M** (system 2). The moment of system 1 about point Q is $\mathbf{r} \times \mathbf{F}_P$, where **r** is the vector from Q to P (Fig. 4.23b). The conditions for equivalence are

$$(\Sigma \mathbf{F})_2 = (\Sigma \mathbf{F})_1:$$

$$\mathbf{F} = \mathbf{F}_P$$

and

$$(\Sigma \mathbf{M}_Q)_2 = (\Sigma \mathbf{M}_Q)_1:$$

$$\mathbf{M} = \mathbf{r} \times \mathbf{F}_P.$$

The systems are equivalent if the force **F** equals the force $\mathbf{F}_P$ and the couple **M** equals the moment of $\mathbf{F}_P$ about Q.

Concurrent Forces Represented by a Force   A system of concurrent forces whose lines of action intersect at a point P (system 1 in Fig. 4.24) can be represented by a single force whose line of action intersects P (system 2). The sums of the forces in the two systems are equal if

$$\mathbf{F} = \mathbf{F}_1 + \mathbf{F}_2 + \cdots + \mathbf{F}_N.$$

The sum of the moments about P equals zero for each system, so the systems are equivalent if the force **F** equals the sum of the forces in system 1.

Parallel Forces Represented by a Force   A system of parallel forces whose sum is not zero can be represented by a single force **F** (Fig. 4.25). We demonstrate this result in Example 4.14.

Representing a System by a Wrench

We have shown that *any* system of forces and moments can be represented by a single force acting at a given point and a single couple. This raises an interesting question: What is the simplest system that can be equivalent to any system of forces and moments?

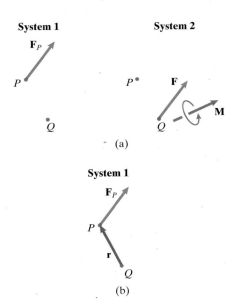

Figure 4.23
(a) System 1 is a force $\mathbf{F}_P$ acting at point P. System 2 consists of a force **F** acting at point Q and a couple **M**.
(b) Determining the moment of system 1 about point Q.

Figure 4.24
A system of concurrent forces and a system consisting of a single force **F**.

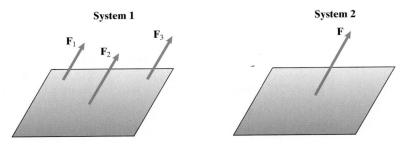

Figure 4.25
A system of parallel forces and a system consisting of a single force **F**.

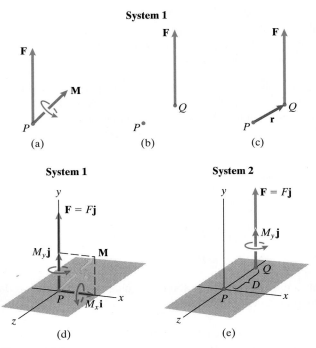

Figure 4.26
(a) System 1 is a single force and a single couple.
(b) Can system 1 be represented by a single force and no couple?
(c) The moment of **F** about P is $\mathbf{r} \times \mathbf{F}$.
(d) **F** is along the y axis, and **M** is contained in the x–y plane.
(e) System 2 is the force **F** and the component of **M** parallel to **F**.

To consider this question, let us begin with an arbitrary force **F** acting at a point P and an arbitrary couple **M** (system 1 in Fig. 4.26a) and see whether we can represent this system by a simpler one. For example, can we represent it by the force **F** acting at a different point Q and no couple (Fig. 4.26b)? The sum of the forces is the same as in system 1. If we can choose the point Q so that $\mathbf{r} \times \mathbf{F} = \mathbf{M}$, where **r** is the vector from P to Q (Fig. 4.26c), the sum of the moments about P is the same as in system 1 and the systems are equivalent. But the vector $\mathbf{r} \times \mathbf{F}$ is perpendicular to **F**, so it can equal **M** only if **M** is perpendicular to **F**. That means that, in general, we can't represent system 1 by the force **F** alone.

However, we can represent system 1 by the force **F** acting at a point Q and the component of **M** that is parallel to **F**. Figure 4.26d shows system 1 with a coordinate system placed so that **F** is along the y axis and **M** is contained in the x–y plane. In terms of this coordinate system, we can express the force and couple as $\mathbf{F} = F\mathbf{j}$ and $\mathbf{M} = M_x\mathbf{i} + M_y\mathbf{j}$. System 2 in Fig. 4.26e consists of the force **F** acting at a point on the z axis and the component of **M** parallel to **F**. If we choose the distance D so that $D = M_x/F$, system 2 is equivalent to system 1. The sum of the forces in each system is **F**. The sum of the moments about P in system 1 is **M**, and the sum of the moments about P in system 2 is

$$(\Sigma\mathbf{M}_P)_2 = [(-D\mathbf{k}) \times (F\mathbf{j})] + M_y\mathbf{j} = M_x\mathbf{i} + M_y\mathbf{j} = \mathbf{M}.$$

A force **F** and a couple $\mathbf{M}_p$ that is parallel to **F** is called a *wrench. It is the simplest system that can be equivalent to an arbitrary system of forces and moments.*

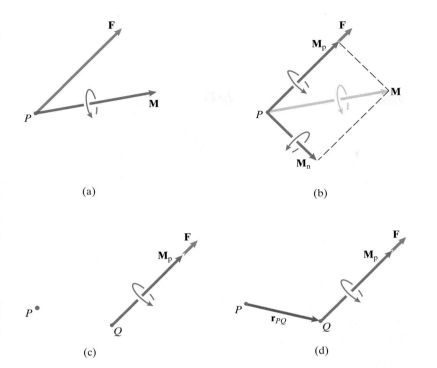

(a)

(b)

(c)

(d)

Figure 4.27
(a) If necessary, first represent the system by a single force and a single couple.
(b) The components of **M** parallel and normal to **F**.
(c) The wrench.
(d) Choose Q so that the moment of **F** about P equals the normal component of **M**.

How can we represent a given system of forces and moments by a wrench? If the system is a single force or a single couple or if it consists of a force **F** and a couple that is parallel to **F**, it is a wrench, and we can't simplify it further. If the system is more complicated than a single force and a single couple, we can begin by choosing a convenient point P and representing the system by a force **F** acting at P and a couple **M** (Fig. 4.27a). Then representing this system by a wrench requires two steps:

1. Determine the components of **M** parallel and normal to **F** (Fig. 4.27b).
2. The wrench consists of the force **F** acting at a point Q and the parallel component $\mathbf{M}_P$ (Fig. 4.27c). To achieve equivalence, the point Q must be chosen so that the moment of **F** about P equals the normal component $\mathbf{M}_n$ (Fig. 4.27d)—that is, so that $\mathbf{r}_{PQ} \times \mathbf{F} = \mathbf{M}_n$.

RESULTS

Equivalent Systems of Forces and Moments

A *system of forces and moments* is simply a particular set of forces and moments due to couples. We define two systems of forces and moments, designated as system 1 and system 2, to be *equivalent* if two conditions are satisfied:
1. The sum of the forces in system 1 is equal to the sum of the forces in system 2.
2. The sum of the moments about any point P due to the forces and moments in system 1 is equal to the sum of the moments about *the same point P* due to the forces and moments in system 2.

Representing Systems of Forces and Moments by Equivalent Systems

System 1

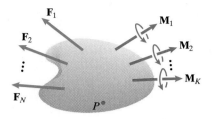

Representing an Arbitrary System by a Force and Couple

Any system of forces and moments (system 1) can be represented by an equivalent system consisting of a force **F** acting at any point *P* and a couple **M** (system 2). The systems are equivalent if **F** equals the sum of the forces in system 1 and **M** equals the sum of the moments about *P* due to the forces and moments in system 1.

System 2

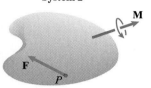

Representing a Force by a Force and Couple

A force **F** acting at a point *P* (system 1) can be represented by an equivalent system consisting of the force **F** acting at a different point *Q* and a couple **M** (system 2). The systems are equivalent if **M** equals the moment about point *Q* due to system 1.

System 1   **System 2**

Representing Concurrent Forces by a Force

A system of concurrent forces whose lines of action intersect at a point *P* (system 1) can be represented by an equivalent system consisting of a force **F** whose line of action passes through *P* (system 2). The systems are equivalent if **F** equals the sum of the forces in system 1.

System 1   **System 2**

Representing Parallel Forces by a Force

A system of parallel forces whose sum is not zero (system 1) can be represented by an equivalent system consisting of a force **F** acting at a point *P* (system 2). The systems are equivalent if **F** equals the sum of the forces in system 1 and the sum of the moments about any point due to the forces in system 1 is equal to the sum of the moments about the same point due to the forces in system 2.

System 1   **System 2**

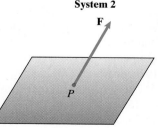

Active Example 4.12   (▶ *Related Problem 4.151*)

System 1 consists of the following forces and couples:

$$\mathbf{F}_A = -10\mathbf{i} + 10\mathbf{j} - 15\mathbf{k} \ (\text{kN}),$$
$$\mathbf{F}_B = 30\mathbf{i} + 5\mathbf{j} + 10\mathbf{k} \ (\text{kN}),$$
$$\mathbf{M}_C = -90\mathbf{i} + 150\mathbf{j} + 60\mathbf{k} \ (\text{kN-m}).$$

Suppose that you want to represent system 1 by an equivalent system consisting of a force **F** acting at the point P with coordinates $(4, 3, -2)$ m and a couple **M** (system 2). Determine **F** and **M**.

Strategy
The conditions for equivalence are satisfied if **F** equals the sum of the forces in system 1 and **M** equals the sum of the moments about point P due to the forces and moments in system 1. We can use these conditions to determine **F** and **M**.

Solution

$$\begin{aligned}\mathbf{F} &= \mathbf{F}_A + \mathbf{F}_B \\ &= 20\mathbf{i} + 15\mathbf{j} - 5\mathbf{k} \ (\text{kN}).\end{aligned}$$

> The force **F** must equal the sum of the forces in system 1.

$$\mathbf{M} = \begin{vmatrix} \mathbf{i} & \mathbf{j} & \mathbf{k} \\ -4 & -3 & 2 \\ -10 & 10 & -15 \end{vmatrix} + \begin{vmatrix} \mathbf{i} & \mathbf{j} & \mathbf{k} \\ 2 & -3 & 2 \\ 30 & 5 & 10 \end{vmatrix}$$
$$+ (-90\mathbf{i} + 150\mathbf{j} + 60\mathbf{k})$$
$$= -105\mathbf{i} + 110\mathbf{j} + 90\mathbf{k} \ (\text{kN-m}).$$

> The couple **M** must equal the sum of the moments about point P due to the forces and moments in system 1.

Practice Problem Suppose that you want to represent system 2 by an equivalent system consisting of a force **F**′ acting at the origin of the coordinate system and a couple **M**′ (system 3). Determine **F**′ and **M**′.

Answer: $\mathbf{F}' = 20\mathbf{i} + 15\mathbf{j} - 5\mathbf{k} \ (\text{kN})$, $\mathbf{M}' = -90\mathbf{i} + 90\mathbf{j} + 90\mathbf{k} \ (\text{kN-m})$.

Example 4.13	Representing a System by a Simpler Equivalent System (▶ *Related Problem 4.137*)

System 1

System 1 consists of two forces and a couple acting on a pipe. Represent system 1 by (a) a single force acting at the origin O of the coordinate system and a single couple and (b) a single force.

Strategy

(a) We can represent system 1 by a force $\mathbf{F}$ acting at the origin and a couple M (system 2 in Fig. a) and use the conditions for equivalence to determine $\mathbf{F}$ and $\mathbf{M}$.
(b) Suppose that we place the force $\mathbf{F}$ with its point of application a distance D along the x axis (system 3 in Fig. b). The sums of the forces in systems 2 and 3 are equal. If we can choose the distance D so that the moment about O in system 3 equals $\mathbf{M}$, system 3 will be equivalent to system 2 and therefore equivalent to system 1.

System 2

System 3

(a) A force $\mathbf{F}$ acting at O and a couple M.   **(b)** A system consisting of the force $\mathbf{F}$ acting at a point on the x axis.

Solution

(a) The conditions for equivalence are

$$(\Sigma \mathbf{F})_2 = (\Sigma \mathbf{F})_1:$$

$$\mathbf{F} = 30\mathbf{j} + (20\mathbf{i} + 20\mathbf{j}) \ (kN) = 20\mathbf{i} + 50\mathbf{j} \ (kN),$$

and

$$(\Sigma M_O)_2 = (\Sigma M_O)_1:$$

$$M = (30 \ kN)(3 \ m) + (20 \ kN)(5 \ m) + 210 \ kN\text{-}m$$

$$= 400 \ kN\text{-}m.$$

(b) The sums of the forces in systems 2 and 3 are equal. Equating the sums of the moments about O yields

$$(\Sigma M_O)_3 = (\Sigma M_O)_2:$$

$$(50 \ kN)D = 400 \ kN\text{-}m,$$

and we find that system 3 is equivalent to system 2 if $D = 8$ m.

Critical Thinking

In part (b), why did we assume that the point of application of the force is on the x axis? In order to represent the system in Fig. a by a single force, we needed to place the line of action of the force so that the force would exert a 400 kN-m counterclockwise moment about O. Placing the point of application of the force a distance D along the x axis was simply a convenient way to accomplish that.

| **Example 4.14** | **Representing Parallel Forces by a Single Force** (▶ *Related Problem 4.154*) |

System 1 consists of parallel forces. Suppose you want to represent it by a force **F** (system 2). What is **F**, and where does its line of action intersect the *x–z* plane?

System 1

Strategy

We can determine **F** from the condition that the sums of the forces in the two systems must be equal. For the two systems to be equivalent, we must choose the point of application P so that the sums of the moments about a point are equal. This condition will tell us where the line of action intersects the *x–z* plane.

Solution

The sums of the forces must be equal.

$$(\Sigma \mathbf{F})_2 = (\Sigma \mathbf{F})_1:$$
$$\mathbf{F} = 30\mathbf{j} + 20\mathbf{j} - 10\mathbf{j} \text{ (lb)} = 40\mathbf{j} \text{ (lb)}.$$

The sums of the moments about an arbitrary point must be equal: Let the coordinates of point P be (x, y, z). The sums of the moments about the origin O must be equal.

$$(\Sigma \mathbf{M}_O)_2 = (\Sigma \mathbf{M}_O)_1:$$

$$\begin{vmatrix} \mathbf{i} & \mathbf{j} & \mathbf{k} \\ x & y & z \\ 0 & 40 & 0 \end{vmatrix} = \begin{vmatrix} \mathbf{i} & \mathbf{j} & \mathbf{k} \\ 6 & 0 & 2 \\ 0 & 30 & 0 \end{vmatrix} + \begin{vmatrix} \mathbf{i} & \mathbf{j} & \mathbf{k} \\ 2 & 0 & 4 \\ 0 & -10 & 0 \end{vmatrix} + \begin{vmatrix} \mathbf{i} & \mathbf{j} & \mathbf{k} \\ -3 & 0 & -2 \\ 0 & 20 & 0 \end{vmatrix}$$

Expanding the determinants, we obtain

$$[20 \text{ ft-lb} + (40 \text{ lb})z]\mathbf{i} + [100 \text{ ft-lb} - (40 \text{ lb})x]\mathbf{k} = \mathbf{0}.$$

The sums of the moments about the origin are equal if

$$x = 2.5 \text{ ft},$$
$$z = -0.5 \text{ ft}.$$

The systems are equivalent if $\mathbf{F} = 40\mathbf{j}$ (lb) and its line of action intersects the *x–z* plane at $x = 2.5$ ft and $z = -0.5$ ft. Notice that we did not obtain an equation for the y coordinate of P. The systems are equivalent if **F** is applied at any point along the line of action.

System 2

Critical Thinking

In this example we could have determined the x and z coordinates of point P in a simpler way. Since the sums of the moments about any point must be equal for the systems to be equivalent, the sums of the moments about any *line* must also be equal. Equating the sums of the moments about the x axis yields

$$(\Sigma M_{x \text{ axis}})_2 = (\Sigma M_{x \text{ axis}})_1:$$
$$-(40 \text{ lb})z = -(30 \text{ lb})(2 \text{ ft}) + (10 \text{ lb})(4 \text{ ft}) + (20 \text{ lb})(2 \text{ ft}),$$

and we obtain $z = -0.5$ ft. Also, equating the sums of the moments about the z axis gives

$$(\Sigma M_{z \text{ axis}})_2 = (\Sigma M_{z \text{ axis}})_1:$$
$$(40 \text{ lb})x = (30 \text{ lb})(6 \text{ ft}) - (10 \text{ lb})(2 \text{ ft}) - (20 \text{ lb})(3 \text{ ft}),$$

and we obtain $x = 2.5$ ft.

Example 4.15 | **Representing a Force and Couple by a Wrench** (▶ *Related Problems 4.170, 4.171*)

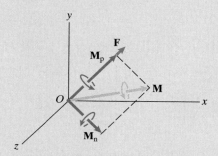

(a) Resolving **M** into components parallel and normal to **F**.

(b) The wrench acting at a point in the x–z plane.

The system consists of the force and couple

$$\mathbf{F} = 3\mathbf{i} + 6\mathbf{j} + 2\mathbf{k} \ (\text{N}),$$
$$\mathbf{M} = 12\mathbf{i} + 4\mathbf{j} + 6\mathbf{k} \ (\text{N-m}).$$

Represent it by a wrench, and determine where the line of action of the wrench's force intersects the x–z plane.

Strategy

The wrench is the force **F** and the component of **M** parallel to **F** (Figs. a, b). We must choose the point of application P so that the moment of **F** about O equals the normal component $\mathbf{M_n}$. By letting P be an arbitrary point of the x–z plane, we can determine where the line of action of **F** intersects that plane.

Solution

Dividing **F** by its magnitude, we obtain a unit vector **e** with the same direction as **F**:

$$\mathbf{e} = \frac{\mathbf{F}}{|\mathbf{F}|} = \frac{3\mathbf{i} + 6\mathbf{j} + 2\mathbf{k} \ (\text{N})}{\sqrt{(3\,\text{N})^2 + (6\,\text{N})^2 + (2\,\text{N})^2}} = 0.429\mathbf{i} + 0.857\mathbf{j} + 0.286\mathbf{k}.$$

We can use **e** to calculate the component of **M** parallel to **F**:

$$\mathbf{M_p} = (\mathbf{e}\cdot\mathbf{M})\mathbf{e} = [(0.429)(12\,\text{N-m}) + (0.857)(4\,\text{N-m}) + (0.286)(6\,\text{N-m})]\mathbf{e}$$
$$= 4.408\mathbf{i} + 8.816\mathbf{j} + 2.939\mathbf{k} \ (\text{N-m}).$$

The component of **M** normal to **F** is

$$\mathbf{M_n} = \mathbf{M} - \mathbf{M_p} = 7.592\mathbf{i} - 4.816\mathbf{j} + 3.061\mathbf{k} \ (\text{N-m}).$$

The wrench is shown in Fig. b. Let the coordinates of P be $(x, 0, z)$. The moment of **F** about O is

$$\mathbf{r}_{OP} \times \mathbf{F} = \begin{vmatrix} \mathbf{i} & \mathbf{j} & \mathbf{k} \\ x & 0 & z \\ 3 & 6 & 2 \end{vmatrix} = -6z\mathbf{i} - (2x - 3z)\mathbf{j} + 6x\mathbf{k} \ (\text{N-m}).$$

By equating this moment to $\mathbf{M_n}$, or

$$-6z\mathbf{i} - (2x - 3z)\mathbf{j} + 6x\mathbf{k} \ (\text{N-m}) = 7.592\mathbf{i} - 4.816\mathbf{j} + 3.061\mathbf{k} \ (\text{N-m}),$$

we obtain the equations

$$-6z = 7.592,$$
$$-2x + 3z = -4.816,$$
$$6x = 3.061.$$

Solving these equations, we find the coordinates of point P are $x = 0.510$ m, $z = -1.265$ m.

Critical Thinking

Why did we place point P at an arbitrary point $(x, 0, z)$ in the x–z plane? Our objective was to place the line of action of the force **F** of the wrench so as to satisfy the condition that the moment of **F** about O would equal $\mathbf{M_n}$. Placing the point of application of **F** at a point $(x, 0, z)$ and then using this condition to determine x and z was a convenient way to determine the necessary location of the line of action. The point $(x, 0, z) = (0.510, 0, -1.265)$ m is the intersection of the line of action with the x–z plane.

Problems

4.128 Two systems of forces act on the beam. Are they equivalent?

Strategy: Check the two conditions for equivalence. The sums of the forces must be equal, and the sums of the moments about an arbitrary point must be equal.

Problem 4.128

4.129 Two systems of forces and moments act on the beam. Are they equivalent?

Problem 4.129

4.130 Four systems of forces and moments act on an 8-m beam. Which systems are equivalent?

4.131 The four systems can be made equivalent by adding a couple to one of the systems. Which system is it, and what couple must be added?

Problems 4.130/4.131

4.132 System 1 is a force **F** acting at a point O. System 2 is the force **F** acting at a different point O' along the same line of action. Explain why these systems are equivalent. (This simple result is called the *principle of transmissibility.*)

Problem 4.132

4.133 The vector sum of the forces exerted on the log by the cables is the same in the two cases. Show that the systems of forces exerted on the log are equivalent.

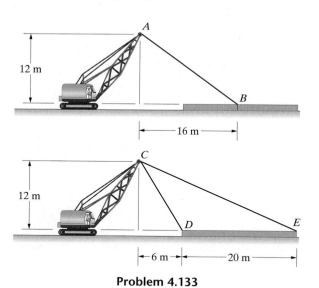

Problem 4.133

4.134 Systems 1 and 2 each consist of a couple. If they are equivalent, what is F?

Problem 4.134

4.135 Two equivalent systems of forces and moments act on the L-shaped bar. Determine the forces F_A and F_B and the couple M.

Problem 4.135

4.136 Two equivalent systems of forces and moments act on the plate. Determine the force F and the couple M.

Problem 4.136

▶ **4.137** In Example 4.13, suppose that the 30-kN vertical force in system 1 is replaced by a 230-kN vertical force. Draw a sketch of the new system 1. If you represent system 1 by a single force $\mathbf{F}$ as in system 3, at what position D on the x axis must the force be placed?

4.138 Three forces and a couple are applied to a beam (system 1).
(a) If you represent system 1 by a force applied at A and a couple (system 2), what are $\mathbf{F}$ and M?
(b) If you represent system 1 by the force $\mathbf{F}$ (system 3), what is the distance D?

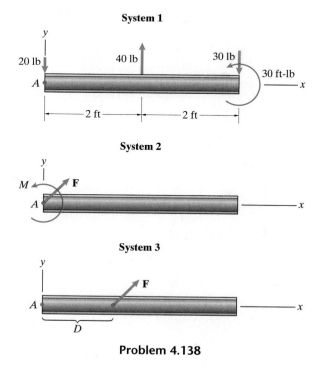

Problem 4.138

4.139 Represent the two forces and couple acting on the beam by a force **F**. Determine **F** and determine where its line of action intersects the *x* axis.

Problem 4.139

4.140 The bracket is subjected to three forces and a couple. If you represent this system by a force **F**, what is **F** and where does its line of action intersect the *x* axis?

Problem 4.140

4.141 The vector sum of the forces acting on the beam is zero, and the sum of the moments about the left end of the beam is zero.

(a) Determine the forces A_x and A_y, and the couple M_A.

(b) Determine the sum of the moments about the right end of the beam.

(c) If you represent the 600-N force, the 200-N force, and the 30 N-m couple by a force **F** acting at the left end of the beam and a couple *M*, what are **F** and *M*?

Problem 4.141

4.142 The vector sum of the forces acting on the truss is zero, and the sum of the moments about the origin *O* is zero.

(a) Determine the forces A_x, A_y, and *B*.

(b) If you represent the 2-kip, 4-kip, and 6-kip forces by a force **F**, what is **F**, and where does its line of action intersect the *y* axis?

(c) If you replace the 2-kip, 4-kip, and 6-kip forces by the force you determined in (b), what are the vector sum of the forces acting on the truss and the sum of the moments about *O*?

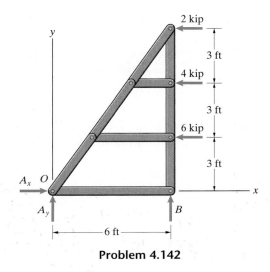

Problem 4.142

4.143 The distributed force exerted on part of a building foundation by the soil is represented by five forces. If you represent them by a force **F**, what is **F**, and where does its line of action intersect the *x* axis?

Problem 4.143

4.144 At a particular instant, aerodynamic forces distributed over the airplane's surface exert the 88-kN and 16-kN vertical forces and the 22 kN-m counterclockwise couple shown. If you represent these forces and couple by a system consisting of a force **F** acting at the center of mass G and a couple M, what are **F** and M?

4.145 If you represent the two forces and couple acting on the airplane by a force **F**, what is **F**, and where does its line of action intersect the x axis?

88 kN

16 kN

G

x

5 m

22 kN-m

5.7 m

9 m

Problems 4.144/4.145

4.146 The system is in equilibrium. If you represent the forces F_{AB} and F_{AC} by a force **F** acting at A and a couple M, what are **F** and M?

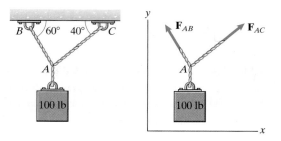

B 60°   40° C

A

100 lb

y

F_{AB} F_{AC}

A

100 lb

x

Problem 4.146

4.147 Three forces act on the beam.

(a) Represent the system by a force **F** acting at the origin O and a couple M.

(b) Represent the system by a single force. Where does the line of action of the force intersect the x axis?

y

30 N

5 m

O x

30 N   6 m   4 m   50 N

Problem 4.147

4.148 The tension in cable AB is 400 N, and the tension in cable CD is 600 N.

(a) If you represent the forces exerted on the left post by the cables by a force **F** acting at the origin O and a couple M, what are **F** and M?

(b) If you represent the forces exerted on the left post by the cables by the force **F** alone, where does its line of action intersect the y axis?

4.149 The tension in each of the cables AB and CD is 400 N. If you represent the forces exerted on the right post by the cables by a force **F**, what is **F**, and where does its line of action intersect the y axis?

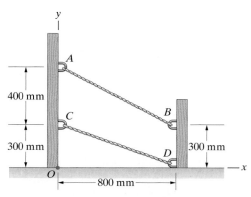

y

A

400 mm

C

B

300 mm

D 300 mm

O x

800 mm

Problems 4.148/4.149

4.150 If you represent the three forces acting on the beam cross section by a force **F**, what is **F**, and where does its line of action intersect the x axis?

y

500 lb

800 lb

6 in

x

6 in

z

500 lb

Problem 4.150

▶ **4.151** In Active Example 4.12, suppose that the force F_B is changed to $F_B = 20\mathbf{i} - 15\mathbf{j} + 30\mathbf{k}$ (kN), and you want to represent system 1 by an equivalent system consisting of a force **F** acting at the point P with coordinates $(4, 3, -2)$ m and a couple **M** (system 2). Determine **F** and **M**.

4.152 The wall bracket is subjected to the force shown.

(a) Determine the moment exerted by the force about the z axis.

(b) Determine the moment exerted by the force about the y axis.

(c) If you represent the force by a force **F** acting at O and a couple **M**, what are **F** and **M**?

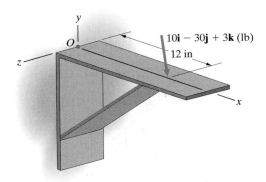

Problem 4.152

4.153 A basketball player executes a "slam dunk" shot, then hangs momentarily on the rim, exerting the two 100-1b forces shown. The dimensions are $h = 14\frac{1}{2}$ in and $r = 9\frac{1}{2}$ in, and the angle $\alpha = 120°$.

(a) If you represent the forces he exerts by a force **F** acting at O and a couple **M**, what are **F** and **M**?

(b) The glass backboard will shatter if $|\mathbf{M}| > 4000$ in-lb. Does it break?

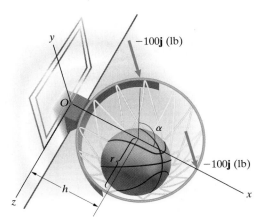

Problem 4.153

▶ **4.154** In Example 4.14, suppose that the 30-lb upward force in system 1 is changed to a 25-lb upward force. If you want to represent system 1 by a single force **F** (system 2), where does the line of action of **F** intersect the x–z plane?

4.155 The normal forces exerted on the car's tires by the road are

$$\mathbf{N}_A = 5104\mathbf{j} \ (\text{N}),$$
$$\mathbf{N}_B = 5027\mathbf{j} \ (\text{N}),$$
$$\mathbf{N}_C = 3613\mathbf{j} \ (\text{N}),$$
$$\mathbf{N}_D = 3559\mathbf{j} \ (\text{N}).$$

If you represent these forces by a single equivalent force **N**, what is **N** and where does its line of action intersect the x–z plane?

Problem 4.155

4.156 Two forces act on the beam. If you represent them by a force **F** acting at C and a couple **M**, what are **F** and **M**?

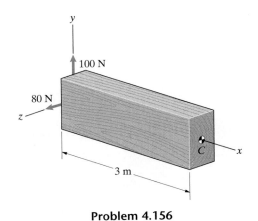

Problem 4.156

4.157 An axial force of magnitude P acts on the beam. If you represent it by a force $\mathbf{F}$ acting at the origin O and a couple $\mathbf{M}$, what are $\mathbf{F}$ and $\mathbf{M}$?

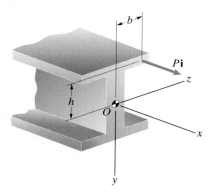

Problem 4.157

4.158 The brace is being used to remove a screw.

(a) If you represent the forces acting on the brace by a force $\mathbf{F}$ acting at the origin O and a couple $\mathbf{M}$, what are $\mathbf{F}$ and $\mathbf{M}$?

(b) If you represent the forces acting on the brace by a force $\mathbf{F}'$ acting at a point P with coordinates (x_P, y_P, z_P) and a couple $\mathbf{M}'$, what are $\mathbf{F}'$ and $\mathbf{M}'$?

Problem 4.158

4.159 Two forces and a couple act on the cube. If you represent them by a force $\mathbf{F}$ acting at point P and a couple $\mathbf{M}$, what are $\mathbf{F}$ and $\mathbf{M}$?

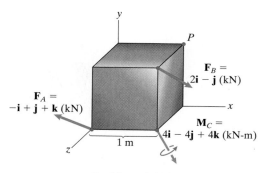

Problem 4.159

4.160 The two shafts are subjected to the torques (couples) shown.

(a) If you represent the two couples by a force $\mathbf{F}$ acting at the origin O and a couple $\mathbf{M}$, what are $\mathbf{F}$ and $\mathbf{M}$?

(b) What is the magnitude of the total moment exerted by the two couples?

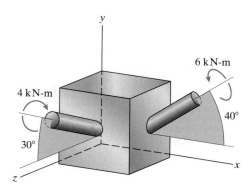

Problem 4.160

4.161 The two systems of forces and moments acting on the bar are equivalent. If

$$\mathbf{F}_A = 30\mathbf{i} + 30\mathbf{j} - 20\mathbf{k} \ (\text{kN}),$$
$$\mathbf{F}_B = 40\mathbf{i} - 20\mathbf{j} + 25\mathbf{k} \ (\text{kN}),$$
$$\mathbf{M}_B = 10\mathbf{i} + 40\mathbf{j} - 10\mathbf{k} \ (\text{kN-m}),$$

what are $\mathbf{F}$ and $\mathbf{M}$?

System 1

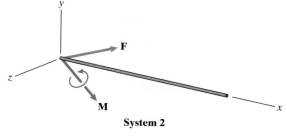

System 2

Problem 4.161

4.162 Point G is at the center of the block. The forces are

$$\mathbf{F}_A = -20\mathbf{i} + 10\mathbf{j} + 20\mathbf{k} \text{ (lb)},$$

$$\mathbf{F}_B = 10\mathbf{j} - 10\mathbf{k} \text{ (lb)}.$$

If you represent the two forces by a force $\mathbf{F}$ acting at G and a couple $\mathbf{M}$, what are $\mathbf{F}$ and $\mathbf{M}$?

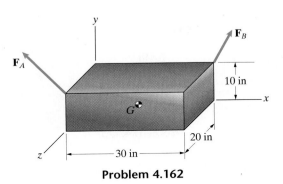

Problem 4.162

4.163 The engine above the airplane's fuselage exerts a thrust $T_0 = 16$ kip, and each of the engines under the wings exerts a thrust $T_U = 12$ kip. The dimensions are $h = 8$ ft, $c = 12$ ft, and $b = 16$ ft. If you represent the three thrust forces by a force $\mathbf{F}$ acting at the origin O and a couple $\mathbf{M}$, what are $\mathbf{F}$ and $\mathbf{M}$?

4.164 Consider the airplane described in Problem 4.163 and suppose that the engine under the wing to the pilot's right loses thrust.

(a) If you represent the two remaining thrust forces by a force $\mathbf{F}$ acting at the origin O and a couple $\mathbf{M}$, what are $\mathbf{F}$ and $\mathbf{M}$?

(b) If you represent the two remaining thrust forces by the force $\mathbf{F}$ alone, where does its line of action intersect the x–y plane?

Problems 4.163/4.164

4.165 The tension in cable AB is 100 lb, and the tension in cable CD is 60 lb. Suppose that you want to replace these two cables by a single cable EF so that the force exerted on the wall at E is equivalent to the two forces exerted by cables AB and CD on the walls at A and C. What is the tension in cable EF, and what are the coordinates of points E and F?

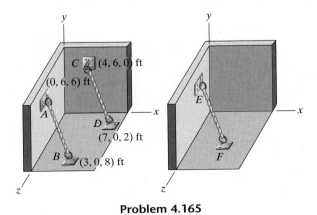

Problem 4.165

4.166 The distance $s = 4$ m. If you represent the force and the 200-N-m couple by a force $\mathbf{F}$ acting at the origin O and a couple $\mathbf{M}$, what are $\mathbf{F}$ and $\mathbf{M}$?

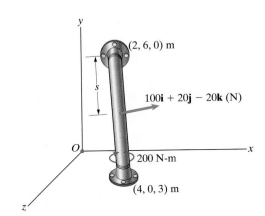

Problem 4.166

4.167 The force **F** and couple **M** in system 1 are

$$\mathbf{F} = 12\mathbf{i} + 4\mathbf{j} - 3\mathbf{k} \text{ (lb)},$$

$$\mathbf{M} = 4\mathbf{i} + 7\mathbf{j} + 4\mathbf{k} \text{ (ft-lb)}.$$

Suppose you want to represent system 1 by a wrench (system 2). Determine the couple $\mathbf{M}_p$ and the coordinates x and z where the line of action of the force intersects the x–z plane.

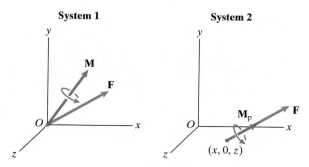

Problem 4.167

4.168 A system consists of a force **F** acting at the origin O and a couple **M**, where

$$\mathbf{F} = 10\mathbf{i} \text{ (lb)}, \qquad \mathbf{M} = 20\mathbf{j} \text{ (ft-lb)}.$$

If you represent the system by a wrench consisting of the force **F** and a parallel couple $\mathbf{M}_p$, what is $\mathbf{M}_p$, and where does the line of action of **F** intersect the y–z plane?

4.169 A system consists of a force **F** acting at the origin O and a couple **M**, where

$$\mathbf{F} = \mathbf{i} + 2\mathbf{j} + 5\mathbf{k} \text{ (N)}, \qquad \mathbf{M} = 10\mathbf{i} + 8\mathbf{j} - 4\mathbf{k} \text{ (N-m)}.$$

If you represent it by a wrench consisting of the force **F** and a parallel couple $\mathbf{M}_p$, (a) determine $\mathbf{M}_p$, and determine where the line of action of **F** intersects (b) the x–z plane, (c) the y–z plane.

▶ **4.170** Consider the force **F** acting at the origin O and the couple **M** given in Example 4.15. If you represent this system by a wrench, where does the line of action of the force intersect the x–y plane?

▶ **4.171** Consider the force **F** acting at the origin O and the couple **M** given in Example 4.15. If you represent this system by a wrench, where does the line of action of the force intersect the plane $y = 3$ m?

4.172 A wrench consists of a force of magnitude 100 N acting at the origin O and a couple of magnitude 60 N-m. The force and couple point in the direction from O to the point $(1, 1, 2)$ m. If you represent the wrench by a force **F** acting at the point $(5, 3, 1)$ m and a couple **M**, what are **F** and **M**?

4.173 System 1 consists of two forces and a couple. Suppose that you want to represent it by a wrench (system 2). Determine the force **F**, the couple $\mathbf{M}_p$, and the coordinates x and z where the line of action of **F** intersects the x–z plane.

Problem 4.173

4.174 A plumber exerts the two forces shown to loosen a pipe.
(a) What total moment does he exert about the axis of the pipe?
(b) If you represent the two forces by a force **F** acting at O and a couple **M**, what are **F** and **M**?
(c) If you represent the two forces by a wrench consisting of the force **F** and a parallel couple $\mathbf{M}_p$, what is $\mathbf{M}_p$, and where does the line of action of **F** intersect the x–y plane?

Problem 4.174

Review Problems

4.175 The Leaning Tower of Pisa is approximately 55 m tall and 7 m in diameter. The horizontal displacement of the top of the tower from the vertical is approximately 5 m. Its mass is approximately 3.2×10^6 kg. If you model the tower as a cylinder and assume that its weight acts at the center, what is the magnitude of the moment exerted by the weight about the point at the center of the tower's base?

Problem 4.175

4.176 The cable AB exerts a 300-N force on the support A that points from A toward B. Determine the magnitude of the moment the force exerts about point P.

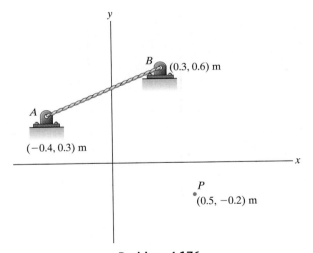

Problem 4.176

4.177 Three forces act on the structure. The sum of the moments due to the forces about A is zero. Determine the magnitude of the force F.

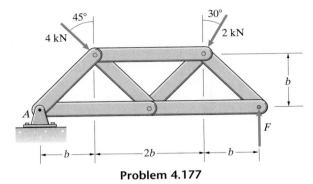

Problem 4.177

4.178 Determine the moment of the 400-N force (a) about A, (b) about B.

Problem 4.178

4.179 Determine the sum of the moments exerted about A by the three forces and the couple.

4.180 If you represent the three forces and the couple by an equivalent system consisting of a force $\mathbf{F}$ acting at A and a couple $\mathbf{M}$, what are the magnitudes of $\mathbf{F}$ and $\mathbf{M}$?

Problems 4.179/4.180

4.181 The vector sum of the forces acting on the beam is zero, and the sum of the moments about A is zero.

(a) What are the forces A_x, A_y, and B?

(b) What is the sum of the moments about B?

Problem 4.181

4.182 The hydraulic piston BC exerts a 970-lb force on the boom at C in the direction parallel to the piston. The angle $\alpha = 40°$. The sum of the moments about A due to the force exerted on the boom by the piston and the weight of the suspended load is zero. What is the weight of the suspended load?

Problem 4.182

4.183 The force $\mathbf{F} = -60\mathbf{i} + 60\mathbf{j}$ (lb).

(a) Determine the moment of $\mathbf{F}$ about point A.

(b) What is the perpendicular distance from point A to the line of action of $\mathbf{F}$?

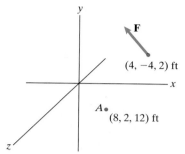

Problem 4.183

4.184 The 20-kg mass is suspended by cables attached to three vertical 2-m posts. Point A is at (0, 1.2, 0) m. Determine the moment about the base E due to the force exerted on the post BE by the cable AB.

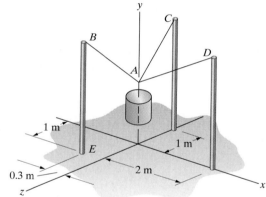

Problem 4.184

4.185 What is the total moment due to the two couples?

(a) Express the answer by giving the magnitude and stating whether the moment is clockwise or counterclockwise.

(b) Express the answer as a vector.

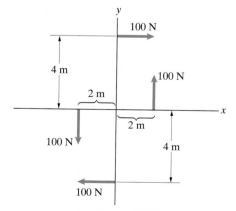

Problem 4.185

4.186 The bar AB supporting the lid of the grand piano exerts a force $\mathbf{F} = -6\mathbf{i} + 35\mathbf{j} - 12\mathbf{k}$ (lb) at B. The coordinates of B are $(3, 4, 3)$ ft. What is the moment of the force about the hinge line of the lid (the x axis)?

Problem 4.186

4.187 Determine the moment of the vertical 800-lb force about point C.

4.188 Determine the moment of the vertical 800-lb force about the straight line through points C and D.

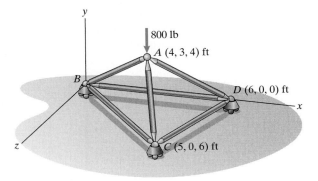

Problems 4.187/4.188

4.189 The system of cables and pulleys supports the 300-lb weight of the work platform. If you represent the upward force exerted at E by cable EF and the upward force exerted at G by cable GH by a single equivalent force $\mathbf{F}$, what is $\mathbf{F}$, and where does its line of action intersect the x axis?

4.190 The system of cables and pulleys supports the 300-lb weight of the work platform.

(a) What are the tensions in cables AB and CD?

(b) If you represent the forces exerted on the work platform by the cables at A and C by a single equivalent force $\mathbf{F}$, what is $\mathbf{F}$ and where does its line of action intersect the x axis?

Problems 4.189/4.190

4.191 The two systems are equivalent. Determine the forces A_x and A_y, and the couple M_A.

4.192 If you represent the equivalent systems in Problem 4.191 by a force $\mathbf{F}$ acting at the origin and a couple M, what are $\mathbf{F}$ and M?

4.193 If you represent the equivalent systems in Problem 4.191 by a force $\mathbf{F}$, what is $\mathbf{F}$, and where does its line of action intersect the x axis?

4.195 The tugboats A and B exert forces $F_A = 1$ kN and $F_B = 1.2$ kN on the ship. The angle $\theta = 30°$. If you represent the two forces by a force $\mathbf{F}$ acting at the origin O and a couple M, what are $\mathbf{F}$ and M?

4.196 The tugboats A and B exert forces $F_A = 600$ N and $F_B = 800$ N on the ship. The angle $\theta = 45°$. If you represent the two forces by a force $\mathbf{F}$, what is $\mathbf{F}$, and where does its line of action intersect the y axis?

4.197 The tugboats A and B want to exert two forces on the ship that are equivalent to a force $\mathbf{F}$ acting at the origin O of 2-kN magnitude. If $F_A = 800$ N, determine the necessary values of F_B and θ.

System 1

System 2

Problems 4.191–4.193

Problems 4.195–4.197

4.194 The two systems are equivalent. If

$$\mathbf{F} = -100\mathbf{i} + 40\mathbf{j} + 30\mathbf{k} \ (\text{lb}),$$
$$\mathbf{M}' = -80\mathbf{i} + 120\mathbf{j} + 40\mathbf{k} \ (\text{in-lb}),$$

determine $\mathbf{F}'$ and $\mathbf{M}$.

Problem 4.194

4.198 If you represent the forces exerted by the floor on the table legs by a force **F** acting at the origin *O* and a couple **M**, what are **F** and **M**?

4.199 If you represent the forces exerted by the floor on the table legs by a force **F**, what is **F**, and where does its line of action intersect the *x–z* plane?

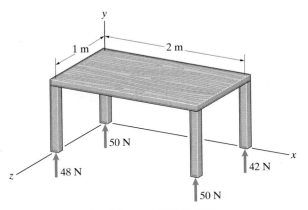

Problems 4.198/4.199

4.200 Two forces are exerted on the crankshaft by the connecting rods. The direction cosines of **F**$_A$ are $\cos \theta_x = -0.182$, $\cos \theta_y = 0.818$, and $\cos \theta_z = 0.545$, and its magnitude is 4 kN. The direction cosines of **F**$_B$ are $\cos \theta_x = 0.182$, $\cos \theta_y = 0.818$, and $\cos \theta_z = -0.545$, and its magnitude is 2 kN. If you represent the two forces by a force **F** acting at the origin *O* and a couple **M**, what are **F** and **M**?

4.201 If you represent the two forces exerted on the crankshaft in Problem 4.200 by a wrench consisting of a force **F** and a parallel couple **M**$_p$, what are **F** and **M**$_p$, and where does the line of action of **F** intersect the *x–z* plane?

Problems 4.200/4.201

Design Project   A relatively primitive device for exercising the biceps muscle is shown. Suggest an improved configuration for the device. You can use elastic cords (which behave like linear springs), weights, and pulleys. Seek a design such that the variation of the moment about the elbow joint as the device is used is small in comparison to the design shown. Give consideration to the safety of your device, its reliability, and the requirement to accommodate users having a range of dimensions and strengths. Choosing specific dimensions, determine the range of the magnitude of the moment exerted about the elbow joint as your device is used.

Objects in Equilibrium

Building on concepts developed in Chapters 3 and 4, we first state the general equilibrium equations. We describe various ways that structural members can be supported, or held in place. Using free-body diagrams and equilibrium equations, we then show how to determine unknown forces and couples exerted on structural members by their supports. The principal motivation for this procedure is that it is the initial step in answering an essential question in structural analysis: How do engineers design structural elements so that they will support the loads to which they are subjected?

◄ The beam is in equilibrium under the actions of its weight and the forces exerted by the chains. In this chapter we apply the equilibrium equations to determine unknown forces and couples acting on objects.

5.1 Two-Dimensional Applications

BACKGROUND

When an object acted upon by a system of forces and moments is in equilibrium, the following conditions are satisfied:

1. The sum of the forces is zero:

$$\Sigma \mathbf{F} = \mathbf{0}. \tag{5.1}$$

2. The sum of the moments about any point is zero:

$$\Sigma \mathbf{M}_{\text{any point}} = \mathbf{0}. \tag{5.2}$$

From our discussion of equivalent systems of forces and moments in Chapter 4, Eqs. (5.1) and (5.2) imply that the system of forces and moments acting on an object in equilibrium is equivalent to a system consisting of no forces and no couples. This provides insight into the nature of equilibrium. From the standpoint of the total force and total moment exerted on an object in equilibrium, the effects are the same as if no forces or couples acted on the object. This observation also makes it clear that if the sum of the forces on an object is zero and the sum of the moments about one point is zero, then the sum of the moments about every point is zero.

The Scalar Equilibrium Equations

When the loads and reactions on an object in equilibrium form a two-dimensional system of forces and moments, they are related by three scalar equilibrium equations:

$$\Sigma F_x = 0, \tag{5.3}$$

$$\Sigma F_y = 0, \tag{5.4}$$

$$\Sigma M_{\text{any point}} = 0. \tag{5.5}$$

A natural question is whether more than one equation can be obtained from Eq. (5.5) by evaluating the sum of the moments about more than one point. The answer is yes, and in some cases it is convenient to do so. But there is a catch—the additional equations will not be independent of Eqs. (5.3)–(5.5). In other words, *more than three independent equilibrium equations cannot be obtained from a two-dimensional free-body diagram, which means we can solve for at most three unknown forces or couples.* We discuss this point further in Section 5.2.

Supports

When you are standing, the floor supports you. When you sit in a chair with your feet on the floor, the chair and floor support you. In this section we are concerned with the ways objects can be supported, or held in place. Forces and couples exerted on an object by its supports are called *reactions,* expressing the fact that the supports "react" to the other forces and couples, or *loads,* acting on the object. For example, a bridge is held up by the reactions exerted by its supports, and the loads are the forces exerted by the weight of the bridge itself, the traffic crossing it, and the wind.

Some very common kinds of supports are represented by stylized models called *support conventions.* Actual supports often closely resemble the support

Figure 5.1
(a) A pin support.
(b) Side view showing the pin passing through the beam.
(c) Holding a supported bar.
(d) The pin support is capable of exerting two components of force.

conventions, but even when they don't, we represent them by these conventions if the actual supports exert the same (or approximately the same) reactions as the models.

The Pin Support   Figure 5.1a shows a *pin support*. The diagram represents a bracket to which an object (such as a beam) is attached by a smooth pin that passes through the bracket and the object. The side view is shown in Fig. 5.1b.

   To understand the reactions that a pin support can exert, it's helpful to imagine holding a bar attached to a pin support (Fig. 5.1c). If you try to move the bar without rotating it (that is, translate the bar), the support exerts a reactive force that prevents this movement. However, you can rotate the bar about the axis of the pin. The support cannot exert a couple about the pin axis to prevent rotation. Thus a pin support can't exert a couple about the pin axis, but it can exert a force on an object in any direction, which is usually expressed by representing the force in terms of components (Fig. 5.1d). The arrows indicate the directions of the reactions if A_x and A_y are positive. If you determine A_x or A_y to be negative, the reaction is in the direction opposite to that of the arrow.

   The pin support is used to represent any real support capable of exerting a force in any direction but not exerting a couple. Pin supports are used in many common devices, particularly those designed to allow connected parts to rotate relative to each other (Fig. 5.2).

Figure 5.2
Pin supports in a pair of scissors and a stapler.

The Roller Support   The convention called a *roller support* (Fig. 5.3a) represents a pin support mounted on wheels. Like the pin support, it cannot exert a couple about the axis of the pin. Since it can move freely in the direction parallel to the surface on which it rolls, it can't exert a force parallel to the surface but can only exert a force normal (perpendicular) to this surface (Fig. 5.3b). Figures 5.3c–e are other commonly used conventions equivalent to the roller support. The wheels of vehicles and wheels supporting parts of machines are roller supports if the friction forces exerted on them are negligible in comparison to the normal forces. A plane smooth surface can also be modeled by a roller

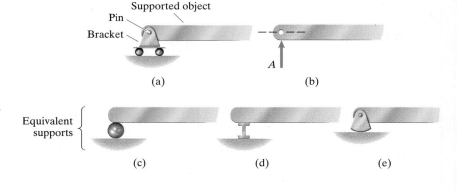

Figure 5.3
(a) A roller support.
(b) The reaction consists of a force normal to the surface.
(c)–(e) Supports equivalent to the roller support.

Figure 5.4
Supporting an object with a plane smooth surface.

support (Fig. 5.4). Beams and bridges are sometimes supported in this way so that they will be free to undergo thermal expansion and contraction.

The supports shown in Fig. 5.5 are similar to the roller support in that they cannot exert a couple and can only exert a force normal to a particular direction. (Friction is neglected.) In these supports, the supported object is attached to a pin or slider that can move freely in one direction but is constrained in the perpendicular direction. Unlike the roller support, these supports can exert a normal force in either direction.

The Fixed Support The *fixed support* shows the supported object literally built into a wall (Fig. 5.6a). This convention is also called a *built-in* support. To understand the reactions, imagine holding a bar attached to a fixed support (Fig. 5.6b). If you try to translate the bar, the support exerts a reactive force that

Figure 5.5
Supports similar to the roller support except that the normal force can be exerted in either direction.

(a) Pin in a slot.   (b) Slider in a slot.   (c) Slider on a shaft.

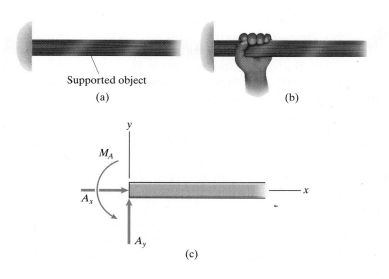

Supported object
(a)   (b)

Figure 5.6
(a) Fixed support.
(b) Holding a supported bar.
(c) The reactions a fixed support is capable of exerting.

prevents translation, and if you try to rotate the bar, the support exerts a reactive couple that prevents rotation. A fixed support can exert two components of force and a couple (Fig. 5.6c). The term M_A is the couple exerted by the support, and the curved arrow indicates its direction. Fence posts and lampposts have fixed supports. The attachments of parts connected so that they cannot move or rotate relative to each other, such as the head of a hammer and its handle, can be modeled as fixed supports.

Table 5.1 summarizes the support conventions commonly used in two-dimensional applications, including those we discussed in Chapter 3. Although

Table 5.1   Supports used in two-dimensional applications.

Supports	Reactions
Rope or Cable Spring	A Collinear Force
Contact with a Smooth Surface	A Force Normal to the Supporting Surface
Contact with a Rough Surface	Two Force Components
Pin Support	Two Force Components
Roller Support Equivalents	A Force Normal to the Supporting Surface
Constrained Pin or Slider	A Normal Force
Fixed (Built-in) Support	Two Force Components and a Couple

the number of conventions may appear daunting, the examples and problems will help you become familiar with them. You should also observe how various objects you see in your everyday experience are supported and think about whether each support could be represented by one of the conventions.

Free-Body Diagrams

We introduced free-body diagrams in Chapter 3 and used them to determine forces acting on simple objects in equilibrium. By using the support conventions, we can model more elaborate objects and construct their free-body diagrams in a systematic way.

For example, the beam in Fig. 5.7a has a pin support at the left end and a roller support at the right end and is loaded by a force F. The roller support rests on a surface inclined at $30°$ to the horizontal. To obtain the free-body diagram of the beam, we first isolate it from its supports (Fig. 5.7b), since the free-body diagram must contain no object other than the beam. We complete the free-body diagram by showing the reactions that may be exerted on the beam by the supports (Fig. 5.7c). Notice that the reaction B exerted by the roller support is normal to the surface on which the support rests.

The object in Fig. 5.8a has a fixed support at the left end. A cable passing over a pulley is attached to the object at two points. We isolate it from its supports (Fig. 5.8b) and complete the free-body diagram by showing the reactions at the fixed support and the forces exerted by the cable (Fig. 5.8c). *Don't forget the couple at a fixed support.* Since we assume the tension in the cable is the same on both sides of the pulley, the two forces exerted by the cable have the same magnitude T.

Once you have obtained the free-body diagram of an object in equilibrium to identify the loads and reactions acting on it, you can apply the equilibrium equations.

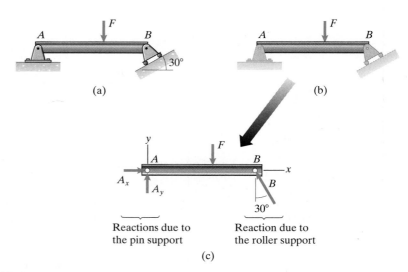

Figure 5.7
(a) A beam with pin and roller supports.
(b) Isolating the beam from its supports.
(c) The completed free-body diagram.

Figure 5.8
(a) An object with a fixed support.
(b) Isolating the object.
(c) The completed free-body diagram.

RESULTS

Equilibrium Equations

When an object is in equilibrium, the system of forces and moments acting on it satisfies two conditions.

> The sum of the forces is zero:
> $$\Sigma \mathbf{F} = \mathbf{0}. \qquad (5.1)$$

> The sum of the moments about any point is zero:
> $$\Sigma \mathbf{M}_{\text{any point}} = \mathbf{0}. \qquad (5.2)$$

When the system of forces and moments acting on an object in equilibrium is two dimensional, it satisfies three scalar equilibrium equations.

$$\Sigma F_x = 0, \qquad (5.3)$$
$$\Sigma F_y = 0, \qquad (5.4)$$
$$\Sigma M_{\text{any point}} = 0. \qquad (5.5)$$

Supports

To draw the free-body diagram of an object, isolate it from its supports and show the *reactions*, the forces and moments that the supports may exert (Table 5.1).

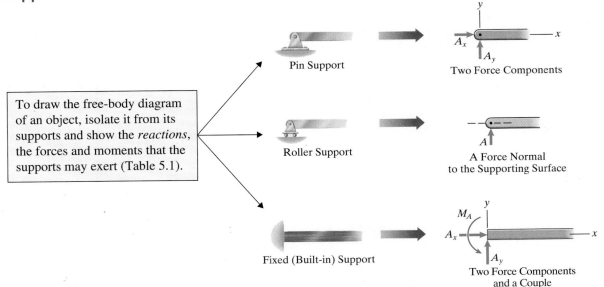

Pin Support — Two Force Components

Roller Support — A Force Normal to the Supporting Surface

Fixed (Built-in) Support — Two Force Components and a Couple

Active Example 5.1 | **Reactions at a Fixed Support** (▶ *Related Problem 5.1*)

The beam has a fixed support at *A* and is subjected to a 4-kN force. (a) Draw the free-body diagram of the beam. (b) Determine the reactions at the fixed support.

Strategy

To draw the free-body diagram of the beam we must isolate it from the built-in support and show the reactions that the support may exert. Then we can apply the equilibrium equations to determine the unknown reactions.

Solution

(a) Draw a diagram of the beam isolated from its fixed support and show the reactions due to the support.

(b) Write the equilibrium equations,

$$\Sigma F_x = A_x = 0,$$

$$\Sigma F_y = A_y - 4 \text{ kN} = 0,$$

$$\Sigma M_{\text{left end}} = M_A - (2 \text{ m})(4 \text{ kN}) = 0,$$

and solve them, obtaining

$$A_x = 0, \quad A_y = 4 \text{ kN}, \quad M_A = 8 \text{ kN-m}.$$

Practice Problem The beam has pin and roller supports and is subjected to a 4-kN force. (a) Draw the free-body diagram of the beam. (b) Determine the reactions at the supports.

Answer: $A_x = 0$, $A_y = 1.33$ kN, $B = 2.67$ kN.

Example 5.2	Reactions at a Fixed Support (▶ *Related Problem 5.9*)

The object has a fixed support at A and is subjected to two forces and a couple. What are the reactions at the support?

Strategy

We will obtain a free-body diagram by isolating the object from the fixed support at A and showing the reactions exerted at A, *including the couple that may be exerted by a fixed support*. Then we can determine the unknown reactions by applying the equilibrium equations.

Solution

Draw the Free-Body Diagram We isolate the object from its support and show the reactions at the fixed support (Fig. a). There are three unknown reactions: two force components A_x and A_y and a couple M_A. (Remember that we can choose the directions of these arrows arbitrarily.) We also resolve the 100-lb force into its components.

(a) Drawing the free-body diagram.

Apply the Equilibrium Equations Summing the moments about point A, the equilibrium equations are

$$\Sigma F_x = A_x + 100 \cos 30° \text{ lb} = 0,$$
$$\Sigma F_y = A_y - 200 \text{ lb} + 100 \sin 30° \text{ lb} = 0,$$
$$\Sigma M_{\text{point } A} = M_A + 300 \text{ ft-lb} - (2\text{ft})(200 \text{ lb}) - (2 \text{ ft})(100 \cos 30° \text{ lb})$$
$$+ (4 \text{ ft})(100 \sin 30° \text{ lb}) = 0.$$

Solving these equations, we obtain the reactions $A_x = -86.6$ lb, $A_y = 150$ lb, and $M_A = 73.2$ ft-lb.

Critical Thinking

Why don't the 300 ft-lb couple and the couple M_A exerted by the fixed support appear in the first two equilibrium equations? Remember that a couple exerts no net force. Also, because the moment due to a couple is the same about any point, the moment about A due to the 300 ft-lb counterclockwise couple is 300 ft-lb counterclockwise.

Example 5.3 — Choosing the Point About Which to Evaluate Moments (▶ *Related Problem 5.15*)

The structure AB supports a suspended 2-Mg (megagram) mass. The structure is attached to a slider in a vertical slot at A and has a pin support at B. What are the reactions at A and B?

Strategy
We will draw the free-body diagram of the structure and the suspended mass by removing the supports at A and B. Notice that the support at A can exert only a horizontal reaction. Then we can use the equilibrium equations to determine the reactions at A and B.

Solution

(a) Drawing the free-body diagram.

Draw the Free-Body Diagram We isolate the structure and mass from the supports and show the reactions at the supports and the force exerted by the weight of the 2000-kg mass (Fig. a). The slot at A can exert only a horizontal force on the slider.

Apply the Equilibrium Equations Summing moments about point B, we find that the equilibrium equations are

$$\Sigma F_x = A + B_x = 0,$$

$$\Sigma F_y = B_y - (2000)(9.81)\text{ N} = 0,$$

$$\Sigma M_{\text{point } B} = (3\text{ m})A + (2\text{ m})[(2000)(9.81)\text{ N}] = 0.$$

The reactions are $A = -13.1$ kN, $B_x = 13.1$ kN, and $B_y = 19.6$ kN.

Critical Thinking
Although the point about which moments are evaluated in writing equilibrium equations can be chosen arbitrarily, a careful choice can often simplify your solution. In this example, point B lies on the lines of action of the two unknown reactions B_x and B_y. By evaluating moments about B, we obtained an equation containing only one unknown, the reaction at A.

Example 5.4 — Analysis of a Luggage Carrier (▶ *Related Problems 5.65–5.68*)

The figure shows an airport luggage carrier and its free-body diagram when it is held in equilibrium in the tilted position. If the luggage carrier supports a weight $W = 50$ lb, the angle $\alpha = 30°$, $a = 8$ in, $b = 16$ in, and $d = 48$ in, what force F must the user exert?

Strategy
The unknown reactions on the free-body diagram are the force F and the normal force N exerted by the floor. If we sum moments about the center of the wheel C, we obtain an equation in which F is the only unknown reaction.

Solution
Summing moments about C,

$$\Sigma M_{(\text{point } C)} = d(F \cos \alpha) + a(W \sin \alpha) - b(W \cos \alpha) = 0,$$

and solving for F, we obtain

$$F = \frac{(b - a \tan \alpha)W}{d}.$$

Substituting the values of W, α, a, b, and d yields the solution $F = 11.9$ lb.

Problems

Assume that objects are in equilibrium. In the statements of the answers, x components are positive to the right and y components are positive upward.

▶ **5.1** In Active Example 5.1, suppose that the beam is subjected to a 6 kN-m counterclockwise couple at the right end in addition to the 4-kN downward force. Draw a sketch of the beam showing its new loading. Draw the free-body diagram of the beam and apply the equilibrium equations to determine the reactions at A.

5.2 The beam has a fixed support at A and is loaded by two forces and a couple. Draw the free-body diagram of the beam and apply equilibrium to determine the reactions at A.

Problem 5.2

5.3 The beam is subjected to a load $F = 400$ N and is supported by the rope and the smooth surfaces at A and B.

(a) Draw the free-body diagram of the beam.

(b) What are the magnitudes of the reactions at A and B?

Problem 5.3

5.4 (a) Draw the free-body diagram of the beam.

(b) Determine the tension in the rope and the reactions at B.

Problem 5.4

5.5 (a) Draw the free-body diagram of the 60-lb drill press, assuming that the surfaces at A and B are smooth.

(b) Determine the reactions at A and B.

Problem 5.5

5.6 The masses of the person and the diving board are 54 kg and 36 kg, respectively. Assume that they are in equilibrium.

(a) Draw the free-body diagram of the diving board.

(b) Determine the reactions at the supports A and B.

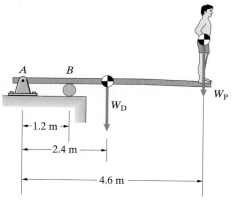

Problem 5.6

5.7 The ironing board has supports at A and B that can be modeled as roller supports.

(a) Draw the free-body diagram of the ironing board.

(b) Determine the reactions at A and B.

Problem 5.7

5.8 The distance $x = 9$ m.

(a) Draw the free-body diagram of the beam.

(b) Determine the reactions at the supports.

Problem 5.8

▶ **5.9** In Example 5.2, suppose that the 200-lb downward force and the 300 ft-lb counterclockwise couple change places; the 200-lb downward force acts at the right end of the horizontal bar, and the 300 ft-lb counterclockwise couple acts on the horizontal bar 2 ft to the right of the support A. Draw a sketch of the object showing the new loading. Draw the free-body diagram of the object and apply the equilibrium equations to determine the reactions at A.

5.10 (a) Draw the free-body diagram of the beam.

(b) Determine the reactions at the supports.

Problem 5.10

5.11 The person exerts 20-N forces on the pliers. The free-body diagram of one part of the pliers is shown. Notice that the pin at *C* connecting the two parts of the pliers behaves like a pin support. Determine the reactions at *C* and the force *B* exerted on the pliers by the bolt.

Problem 5.11

5.12 (a) Draw the free-body diagram of the beam.
(b) Determine the reactions at the pin support *A*.

Problem 5.12

5.13 (a) Draw the free-body diagram of the beam.
(b) Determine the reactions at the supports.

Problem 5.13

5.14 (a) Draw the free-body diagram of the beam.
(b) If $F = 4$ kN, what are the reactions at *A* and *B*?

Problem 5.14

▶ **5.15** In Example 5.3, suppose that the attachment point for the suspended mass is moved toward point *B* such that the horizontal distance from *A* to the attachment point increases from 2 m to 3 m. Draw a sketch of the beam *AB* showing the new geometry. Draw the free-body diagram of the beam and apply the equilibrium equations to determine the reactions at *A* and *B*.

5.16 A person doing push-ups pauses in the position shown. His 180-lb weight *W* acts at the point shown. The dimensions $a = 15$ in, $b = 42$ in, and $c = 16$ in. Determine the normal force exerted by the floor on each of his hands and on each of his feet.

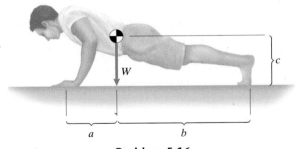

Problem 5.16

5.17 The hydraulic piston AB exerts a 400-lb force on the ladder at B in the direction parallel to the piston. Determine the weight of the ladder and the reactions at C.

Problem 5.17

5.18 Draw the free-body diagram of the structure by isolating it from its supports at A and E. Determine the reactions at A and E.

Problem 5.18

5.19 (a) Draw the free-body diagram of the beam.

(b) Determine the tension in the cable and the reactions at A.

Problem 5.19

5.20 The unstretched length of the spring CD is 350 mm. Suppose that you want the lever ABC to exert a 120-N normal force on the smooth surface at A. Determine the necessary value of the spring constant k and the resulting reactions at B.

Problem 5.20

5.21 The mobile is in equilibrium. The fish B weighs 27 oz. Determine the weights of the fish A, C, and D. (The weights of the crossbars are negligible.)

Problem 5.21

5.22 The car's wheelbase (the distance between the wheels) is 2.82 m. The mass of the car is 1760 kg and its weight acts at the point $x = 2.00$ m, $y = 0.68$ m. If the angle $\alpha = 15°$, what is the total normal force exerted on the two rear tires by the sloped ramp?

Problem 5.22

5.23 The link *AB* exerts a force on the bucket of the excavator at *A* that is parallel to the link. The weight $W = 1500$ lb. Draw the free-body diagram of the bucket and determine the reactions at *C*. (The connection at *C* is equivalent to a pin support of the bucket.)

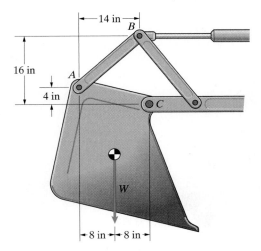

Problem 5.23

5.24 The 14.5-lb chain saw is subjected to the loads at *A* by the log it cuts. Determine the reactions *R*, B_x, and B_y that must be applied by the person using the saw to hold it in equilibrium.

Problem 5.24

5.25 The mass of the trailer is 2.2 Mg (megagrams). The distances $a = 2.5$ m and $b = 5.5$ m. The truck is stationary, and the wheels of the trailer can turn freely, which means the road exerts no horizontal force on them. The hitch at *B* can be modeled as a pin support.

(a) Draw the free-body diagram of the trailer.

(b) Determine the total normal force exerted on the rear tires at *A* and the reactions exerted on the trailer at the pin support *B*.

Problem 5.25

5.26 The total weight of the wheelbarrow and its load is $W = 100$ lb. (a) What is the magnitude of the upward force *F* necessary to lift the support at *A* off the ground? (b) What is the magnitude of the downward force necessary to raise the wheel off the ground?

Problem 5.26

5.27 The airplane's weight is $W = 2400$ lb. Its brakes keep the rear wheels locked. The front (nose) wheel can turn freely, and so the ground exerts no horizontal force on it. The force *T* exerted by the airplane's propeller is horizontal.

(a) Draw the free-body diagram of the airplane. Determine the reaction exerted on the nose wheel and the total normal reaction exerted on the rear wheels

(b) when $T = 0$;

(c) when $T = 250$ lb.

Problem 5.27

5.28 A safety engineer establishing limits on the loads that can be carried by a forklift analyzes the situation shown. The dimensions are $a = 32$ in, $b = 30$ in, and $c = 26$ in. The combined weight of the forklift and operator is $W_F = 1200$ lb. As the weight W_L supported by the forklift increases, the normal force exerted on the floor by the rear wheels at B decreases. The forklift is on the verge of tipping forward when the normal force at B is zero. Determine the value of W_L that will cause this condition.

Problem 5.28

5.29 Paleontologists speculate that the stegosaur could stand on its hind limbs for short periods to feed. Based on the free-body diagram shown and assuming that $m = 2000$ kg, determine the magnitudes of the forces B and C exerted by the ligament–muscle brace and vertebral column, and determine the angle α.

Problem 5.29

5.30 The weight of the fan is $W = 20$ lb. Its base has four equally spaced legs of length $b = 12$ in. Each leg has a pad near the end that contacts the floor and supports the fan. The height $h = 32$ in. If the fan's blade exerts a thrust $T = 2$ lb, what total normal force is exerted on the two legs at A?

5.31 The weight of the fan is $W = 20$ lb. Its base has four equally spaced legs of length $b = 12$ in. Each leg has a pad near the end that contacts the floor and supports the fan. The height $h = 32$ in. As the thrust T of the fan increases, the normal force supported by the two legs at A decreases. When the normal force at A is zero, the fan is on the verge of tipping over. Determine the value of T that will cause this condition.

5.33 A force $F = 400$ N acts on the bracket. What are the reactions at A and B?

Problem 5.33

5.34 The sign's weight $W_s = 32$ lb acts at the point shown. The 10-lb weight of bar AD acts at the midpoint of the bar. Determine the tension in the cable AE and the reactions at D.

Side View   Top View

Problems 5.30/5.31

5.32 In a measure to decrease costs, the manufacturer of the fan described in Problem 5.31 proposes to support the fan with three equally spaced legs instead of four. An engineer is assigned to analyze the safety implications of the change. The weight of the fan decreases to $W = 19.6$ lb. The dimensions b and h are unchanged. What thrust T will cause the fan to be on the verge of tipping over in this case? Compare your answer to the answer to Problem 5.31.

Problem 5.32

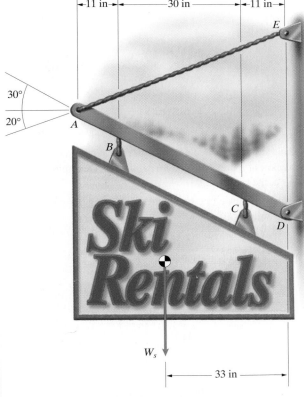

Problem 5.34

5.35 The device shown, called a *swape* or *shadoof*, helps a person lift a heavy load. (Devices of this kind were used in Egypt at least as early as 1550 B.C. and are still in use in various parts of the world.) The dimensions $a = 3.6$ m and $b = 1.2$ m. The mass of the bar and counterweight is 90 kg, and their weight W acts at the point shown. The mass of the load being lifted is 45 kg. Determine the vertical force the person must exert to support the stationary load (a) when the load is just above the ground (the position shown); (b) when the load is 1 m above the ground. Assume that the rope remains vertical.

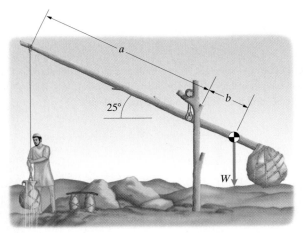

Problem 5.35

5.36 This structure, called a *truss*, has a pin support at A and a roller support at B and is loaded by two forces. Determine the reactions at the supports.

    Strategy: Draw a free-body diagram treating the entire truss as a single object.

Problem 5.36

5.37 An Olympic gymnast is stationary in the "iron cross" position. The weight of his left arm and the weight of his body *not including his arms* are shown. The distances are $a = b = 9$ in and $c = 13$ in. Treat his shoulder S as a fixed support, and determine the magnitudes of the reactions at his shoulder. That is, determine the force and couple his shoulder must support.

Problem 5.37

5.38 Determine the reactions at A.

Problem 5.38

5.39 The car's brakes keep the rear wheels locked, and the front wheels are free to turn. Determine the forces exerted on the front and rear wheels by the road when the car is parked (a) on an upslope with $\alpha = 15°$; (b) on a downslope with $\alpha = -15°$.

Problem 5.39

5.40 The length of the bar is $L = 4$ ft. Its weight $W = 6$ lb acts at the midpoint of the bar. The floor and wall are smooth. The spring is unstretched when the angle $\alpha = 0$. If the bar is in equilibrium when $\alpha = 40°$, what is the spring constant k?

5.41 The weight W of the bar acts at its midpoint. The floor and wall are smooth. The spring is unstretched when the angle $\alpha = 0$. Determine the angle α at which the bar is in equilibrium in terms of W, k, and L.

Problems 5.40/5.41

5.42 The plate is supported by a pin in a smooth slot at B. What are the reactions at the supports?

Problem 5.42

5.43 Determine the reactions at the fixed support A.

Problem 5.43

5.44 Suppose that you want to represent the two forces and couple acting on the beam in Problem 5.43 by an equivalent force **F** as shown. (a) Determine **F** and the distance D at which its line of action crosses the x axis. (b) Assume that **F** is the only load acting on the beam and determine the reactions at the fixed support A. Compare your answers to the answers to Problem 5.43.

Problem 5.44

5.45 The bicycle brake on the right is pinned to the bicycle's frame at A. Determine the force exerted by the brake pad on the wheel rim at B in terms of the cable tension T.

Problem 5.45

5.46 The mass of each of the suspended weights is 80 kg. Determine the reactions at the supports at A and E.

5.47 The suspended weights are each of mass m. The supports at A and E will each safely support a force of 6 kN magnitude. Based on this criterion, what is the largest safe value of m?

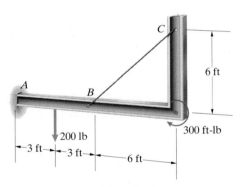

Problems 5.46/5.47

5.48 The tension in cable BC is 100 lb. Determine the reactions at the fixed support.

Problem 5.48

5.49 The tension in cable AB is 2 kN. What are the reactions at C in the two cases?

(a)   (b)

Problem 5.49

5.50 Determine the reactions at the supports.

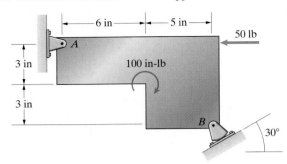

Problem 5.50

5.51 The weight $W = 2$ kN. Determine the tension in the cable and the reactions at A.

5.52 The cable will safely support a tension of 6 kN. Based on this criterion, what is the largest safe value of the weight W?

Problems 5.51/5.52

5.53 The blocks being compressed by the clamp exert a 200-N force on the pin at D that points from A toward D. The threaded shaft BE exerts a force on the pin at E that points from B toward E.

(a) Draw a free-body diagram of the arm DCE of the clamp, assuming that the pin at C behaves like a pin support.

(b) Determine the reactions at C.

5.54 The blocks being compressed by the clamp exert a 200-N force on the pin at A that points from D toward A. The threaded shaft BE exerts a force on the pin at B that points from E toward B.

(a) Draw a free-body diagram of the arm ABC of the clamp, assuming that the pin at C behaves like a pin support.

(b) Determine the reactions at C.

Problems 5.53/5.54

5.55 Suppose that you want to design the safety valve to open when the difference between the pressure p in the circular pipe (diameter = 150 mm) and atmospheric pressure is 10 MPa (megapascals; a pascal is 1 N/m²). The spring is compressed 20 mm when the valve is closed. What should the value of the spring constant be?

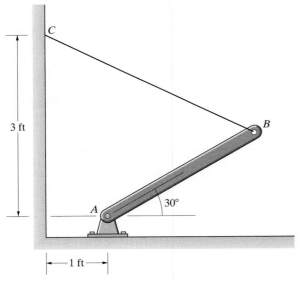

Problem 5.55

5.56 The 10-lb weight of the bar AB acts at the midpoint of the bar. The length of the bar is 3 ft. Determine the tension in the string BC and the reactions at A.

Problem 5.56

5.57 The crane's arm has a pin support at A. The hydraulic cylinder BC exerts a force on the arm at C in the direction parallel to BC. The crane's arm has a mass of 200 kg, and its weight can be assumed to act at a point 2 m to the right of A. If the mass of the suspended box is 800 kg and the system is in equilibrium, what is the magnitude of the force exerted by the hydraulic cylinder?

5.58 In Problem 5.57, what is the magnitude of the force exerted on the crane's arm by the pin support at A?

Problems 5.57/5.58

5.59 A speaker system is suspended by the cables attached at D and E. The mass of the speaker system is 130 kg, and its weight acts at G. Determine the tensions in the cables and the reactions at A and C.

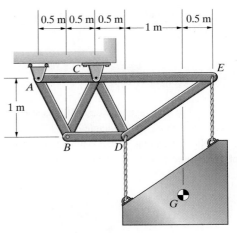

Problem 5.59

5.60 The weight $W_1 = 1000$ lb. Neglect the weight of the bar AB. The cable goes over a pulley at C. Determine the weight W_2 and the reactions at the pin support A.

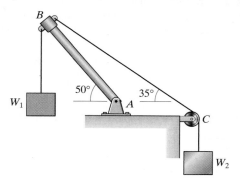

Problem 5.60

5.61 The dimensions $a = 2$ m and $b = 1$ m. The couple $M = 2400$ N-m. The spring constant is $k = 6000$ N/m, and the spring would be unstretched if $h = 0$. The system is in equilibrium when $h = 2$ m and the beam is horizontal. Determine the force F and the reactions at A.

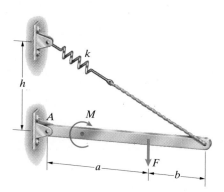

Problem 5.61

5.62 The bar is 1 m long, and its weight W acts at its midpoint. The distance $b = 0.75$ m, and the angle $\alpha = 30°$. The spring constant is $k = 100$ N/m, and the spring is unstretched when the bar is vertical. Determine W and the reactions at A.

Problem 5.62

5.63 The boom derrick supports a suspended 15-kip load. The booms BC and DE are each 20 ft long. The distances are $a = 15$ ft and $b = 2$ ft, and the angle $\theta = 30°$. Determine the tension in cable AB and the reactions at the pin supports C and D.

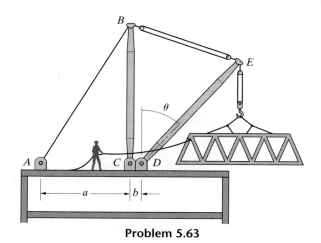

Problem 5.63

5.64 The arrangement shown controls the elevators of an airplane. (The elevators are the horizontal control surfaces in the airplane's tail.) The elevators are attached to member EDG. Aerodynamic pressures on the elevators exert a clockwise couple of 120 in-lb. Cable BG is slack, and its tension can be neglected. Determine the force F and the reactions at the pin support A.

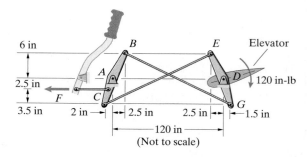

Problem 5.64

▶ **5.65** In Example 5.4, suppose that $\alpha = 40°$, $d = 1$ m, $a = 200$ mm, $b = 500$ mm, $R = 75$ mm, and the mass of the luggage is 40 kg. Determine F and N.

▶ **5.66** In Example 5.4, suppose that $\alpha = 35°$, $d = 46$ in, $a = 10$ in, $b = 14$ in, $R = 3$ in, and you don't want the user to have to exert a force F larger than 20 lb. What is the largest luggage weight that can be placed on the carrier?

▶ **5.67** One of the difficulties in making design decisions is that you don't know how the user will place the luggage on the carrier in Example 5.4. Suppose you assume that the point where the weight acts may be anywhere within the "envelope" $R \le a \le 0.75c$ and $0 \le b \le 0.75d$. If $\alpha = 30°$, $c = 14$ in, $d = 48$ in, $R = 3$ in, and $W = 80$ lb, what is the largest force F the user will have to exert for any luggage placement?

▶ **5.68** In Example 5.4, assume a user that would hold the carrier's handle at $h = 36$ in above the floor. Assume that $R = 3$ in, $a = 6$ in, $b = 12$ in, and $d = 4$ ft. The resulting ratio of the force the user must exert to the weight of the luggage is $F/W = 0.132$. Suppose that people with a range of heights use this carrier. Obtain a graph of F/W as a function of h for $24 \le h \le 36$ in.

5.2 Statically Indeterminate Objects

BACKGROUND

In Section 5.1 we discussed examples in which we were able to use the equilibrium equations to determine unknown forces and couples acting on objects in equilibrium. It is important to be aware of two common situations in which this procedure doesn't lead to a solution. First, the free-body diagram of an object can have more unknown forces or couples than the number of independent equilibrium equations that can be obtained. For example, because no more than three independent equilibrium equations can be obtained from a given free-body diagram in a two-dimensional problem, if there are more than three unknowns they can't all be determined from the equilibrium equations alone. This occurs, for example, when an object has more supports than the minimum number necessary to maintain it in equilibrium. Such an object is said to have *redundant supports*. The second situation is when the supports of an object are improperly designed such that they cannot maintain equilibrium under the loads acting on it. The object is said to have *improper supports*. In either situation, the object is said to be *statically indeterminate*.

Engineers use redundant supports whenever possible for strength and safety. Some designs, however, require that the object be incompletely supported so that it is free to undergo certain motions. These two situations—more supports than necessary for equilibrium or not enough—are so common that we consider them in detail.

Redundant Supports

Consider a beam with a fixed support (Fig. 5.9a). From its free-body diagram (Fig. 5.9b), we obtain the equilibrium equations

$$\Sigma F_x = A_x = 0,$$
$$\Sigma F_y = A_y - F = 0,$$
$$\Sigma M_{\text{point } A} = M_A - \left(\frac{L}{2}\right)F = 0.$$

Assuming we know the load F, we have three equations and three unknown reactions, for which we obtain the solutions $A_x = 0$, $A_y = F$, and $M_A = FL/2$.

(a)

(b)

Figure 5.9
(a) A beam with a fixed support.
(b) The free-body diagram has three unknown reactions.

Figure 5.10
(a) A beam with fixed and roller supports.
(b) The free-body diagram has four
unknown reactions.

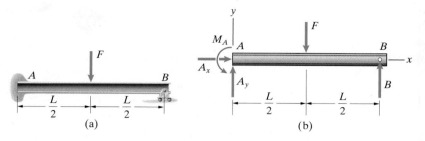

(a)

(b)

Now suppose we add a roller support at the right end of the beam (Fig. 5.10a). From the new free-body diagram (Fig. 5.10b), we obtain the equilibrium equations

$$\Sigma F_x = A_x = 0, \tag{5.6}$$

$$\Sigma F_y = A_y - F + B = 0, \tag{5.7}$$

$$\Sigma M_{\text{point } A} = M_A - \left(\frac{L}{2}\right) F + LB = 0. \tag{5.8}$$

Now we have three equations and four unknown reactions. Although the first equation tells us that $A_x = 0$, we can't solve the two equations (5.7) and (5.8) for the three reactions A_y, B, and M_A.

When faced with this situation, students often attempt to sum the moments about another point, such as point B, to obtain an additional equation:

$$\Sigma M_{\text{point } B} = M_A + \left(\frac{L}{2}\right) F - LA_y = 0.$$

Unfortunately, this doesn't help. This is not an independent equation, but is a linear combination of Eqs. (5.7) and (5.8):

$$\Sigma M_{\text{point } B} = M_A + \left(\frac{L}{2}\right) F - LA_y$$

$$= \underbrace{M_A - \left(\frac{L}{2}\right) F + LB}_{\text{Eq. (5.8)}} - \underbrace{L(A_y - F + B)}_{\text{Eq. (5.7)}}.$$

As this example demonstrates, each support added to an object results in additional reactions. The difference between the number of reactions and the number of independent equilibrium equations is called the *degree of redundancy*. Even if an object is statically indeterminate due to redundant supports, it may be possible to determine some of the reactions from the equilibrium equations. Notice that in our previous example we were able to determine the reaction A_x even though we could not determine the other reactions.

Since redundant supports are so ubiquitous, you may wonder why we devote so much effort to teaching you how to analyze objects whose reactions can be determined with the equilibrium equations. We want to develop your understanding of equilibrium and give you practice writing equilibrium equations. The reactions on an object with redundant supports *can* be determined by supplementing the equilibrium equations with additional equations that relate the forces and couples acting on the object to its deformation, or change in shape. Thus obtaining the equilibrium equations is the first step of the solution.

Improper Supports

We say that an object has improper supports if it will not remain in equilibrium under the action of the loads exerted on it. Thus an object with improper supports will move when the loads are applied. In two-dimensional problems, this can occur in two ways:

1. *The supports can exert only parallel forces.* This leaves the object free to move in the direction perpendicular to the support forces. If the loads exert a component of force in that direction, the object is not in equilibrium. Figure 5.11a shows an example of this situation. The two roller supports can exert only vertical forces, while the force F has a horizontal component. The beam will move horizontally when F is applied. This is particularly apparent from the free-body diagram (Fig. 5.11b). The sum of the forces in the horizontal direction cannot be zero because the roller supports can exert only vertical reactions.

2. *The supports can exert only concurrent forces.* If the loads exert a moment about the point where the lines of action of the support forces intersect, the object is not in equilibrium. For example, consider the beam in Fig. 5.12a. From its free-body diagram (Fig. 5.12b) we see that the reactions A and B exert no moment about the point P, where their lines of action intersect, but the load F does. The sum of the moments about point P is not zero, and the beam will rotate when the load is applied.

Except for problems that deal explicitly with improper supports, objects in our examples and problems have proper supports. You should develop the habit of examining objects in equilibrium and thinking about why they are properly supported for the loads acting on them.

(a)

(b)

Figure 5.11

(a) A beam with two roller supports is not in equilibrium when subjected to the load shown.

(b) The sum of the forces in the horizontal direction is not zero.

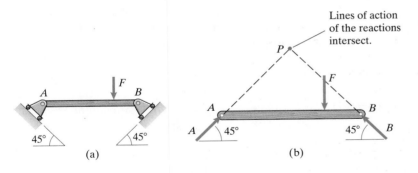

(a)

(b)

Figure 5.12

(a) A beam with roller supports on sloped surfaces.

(b) The sum of the moments about point P is not zero.

RESULTS

A supported object is said to be *statically indeterminate* in two circumstances:

Redundant Supports

The object has more supports than the minimum number necessary to maintain equilibrium. The difference between the number of reactions due to the supports and the number of independent equilibrium equations is called the *degree of redundancy*.

Improper Supports

The supports cannot maintain the object in equilibrium under the loads acting on it.

Active Example 5.5   **Recognizing a Statically Indeterminate Object** (▶ *Related Problem 5.69*)

The beam has two pin supports and is loaded by a 2-kN force.

(a) Show that the beam is statically indeterminate and determine the degree of redundancy.

(b) Determine as many reactions as possible.

Strategy

The beam is statically indeterminate if its free-body diagram has more unknown reactions than the number of independent equilibrium equations we can obtain. The difference between the number of reactions and the number of equilibrium equations is the degree of redundancy. Even if the beam is statically indeterminate, it may be possible to solve the equilibrium equation for some of the reactions.

Solution

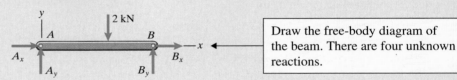

Draw the free-body diagram of the beam. There are four unknown reactions.

$$\Sigma F_x = A_x + B_x = 0,$$
$$\Sigma F_y = A_y + B_y - 2 \text{ kN} = 0,$$
$$\Sigma M_{\text{point } A} = (5 \text{ m})B_y - (3 \text{ m})(2 \text{ kN}) = 0.$$

Write the equilibrium equations.

There are three independent equilibrium equations, so the beam is statically indeterminate and the degree of redundancy is $4 - 3 = 1$. We cannot determine A_x or B_x from the equilibrium equations, but we can determine A_y and B_y.

$$B_y = \frac{(3 \text{ m})(2 \text{ kN})}{(5 \text{ m})} = 1.2 \text{ kN},$$
$$A_y = 2 \text{ kN} - B_y = 0.8 \text{ kN}.$$

Determine the reactions A_y and B_y.

Practice Problem Suppose that the pin support at point A of the beam is replaced by a fixed support. (a) Show that the beam is statically indeterminate and determine the degree of redundancy. (b) Determine as many reactions as possible.

Answer: (a) Degree of redundancy is 2. (b) No reactions can be determined.

| Example 5.6 | **Proper and Improper Supports** (▶ *Related Problems 5.75, 5.76*) |

State whether each L-shaped bar is properly or improperly supported. If a bar is properly supported, determine the reactions at its supports.

Strategy

By drawing the free-body diagram of each bar, we can determine whether the reactions of the supports can exert only parallel or concurrent forces on it. If so, we can then recognize whether the applied load results in the bar not being in equilibrium.

(a)   (b)   (c)

Solution

Consider the free-body diagrams of the bars (shown below):

Bar (a)   The lines of action of the reactions due to the two roller supports intersect at P, and the load F exerts a moment about P. This bar is improperly supported.

Bar (b)   The lines of action of the reactions intersect at A, and the load F exerts a moment about A. This bar is also improperly supported.

Bar (c)   The three support forces are neither parallel nor concurrent. This bar is properly supported. The equilibrium equations are

$$\Sigma F_x = A_x - B = 0,$$
$$\Sigma F_y = A_y - F = 0,$$
$$\Sigma M_{\text{point } A} = BL - FL = 0.$$

Solving these equations, the reactions are $A_x = F$, $A_y = F$, and $B = F$.

(a)   (b)   (c)

Critical Thinking

An essential part of learning mechanics is developing your intuition about the behaviors of the physical systems we study. In this example, think about the effects of the loads on the three systems, and see if you can predict whether they are properly supported. Will the loads cause the bars to move or not? Then see if your judgment is confirmed by the analyses given in the example.

Problems

▶ **5.69** (a) Draw the free-body diagram of the beam and show that it is statically indeterminate. (See Active Example 5.5.)

(b) Determine as many of the reactions as possible.

5.70 Choose supports at A and B so that the beam is not statically indeterminate. Determine the reactions at the supports.

Problems 5.69/5.70

5.71 (a) Draw the free-body diagram of the beam and show that it is statically indeterminate. (The external couple M_0 is known.)

(b) By an analysis of the beam's deflection, it is determined that the vertical reaction B exerted by the roller support is related to the couple M_0 by $B = 2M_0/L$. What are the reactions at A?

5.72 Choose supports at A and B so that the beam is not statically indeterminate. Determine the reactions at the supports.

Problems 5.71/5.72

5.73 Draw the free-body diagram of the L-shaped pipe assembly and show that it is statically indeterminate. Determine as many of the reactions as possible.

Strategy: Place the coordinate system so that the x axis passes through points A and B.

5.74 Choose supports at A and B so that the pipe assembly is not statically indeterminate. Determine the reactions at the supports.

Problems 5.73/5.74

▶ **5.75** State whether each of the L-shaped bars shown is properly or improperly supported. If a bar is properly supported, determine the reactions at its supports. (See Active Example 5.6.)

Problem 5.75

▶ **5.76** State whether each of the L-shaped bars shown is properly or improperly supported. If a bar is properly supported, determine the reactions at its supports. (See Active Example 5.6.)

Problem 5.76

5.3 Three-Dimensional Applications

BACKGROUND

We have seen that when an object in equilibrium is subjected to a two-dimensional system of forces and moments, no more than three independent equilibrium equations can be obtained. In the case of a three-dimensional system of forces and moments, up to six independent equilibrium equations can be obtained. The three components of the sum of the forces must equal zero and the three components of the sum of the moments about a point must equal zero. The procedure for determining the reactions on an object subjected to a three-dimensional system of forces and moments—drawing a free-body diagram and applying the equilibrium equations—is the same as in two dimensions.

The Scalar Equilibrium Equations

When an object is in equilibrium, the system of forces and couples acting on it satisfy Eqs. (5.1) and (5.2). The sum of the forces is zero and the sum of the moments about any point is zero. Expressing these equations in terms of cartesian components in three dimensions yields the six scalar equilibrium equations.

$$\Sigma F_x = 0, \tag{5.9}$$

$$\Sigma F_y = 0, \tag{5.10}$$

$$\Sigma F_z = 0, \tag{5.11}$$

$$\Sigma M_x = 0, \tag{5.12}$$

$$\Sigma M_y = 0, \tag{5.13}$$

$$\Sigma M_z = 0. \tag{5.14}$$

The sums of the moments can be evaluated about any point. Although more equations can be obtained by summing moments about other points, they will not be independent of these equations. *More than six independent equilibrium equations cannot be obtained from a given free-body diagram, so at most six unknown forces or couples can be determined.*

The steps required to determine reactions in three dimensions are familiar from the two-dimensional applications we have discussed. First obtain a free-body diagram by isolating an object and showing the loads and reactions acting on it, then use Eqs. (5.9)–(5.14) to determine the reactions.

Supports

We present five conventions frequently used in three-dimensional problems. Even when actual supports do not physically resemble these models, we represent them by the models if they exert the same (or approximately the same) reactions.

The Ball and Socket Support   In the *ball and socket support*, the supported object is attached to a ball enclosed within a spherical socket (Fig. 5.13a). The socket permits the ball to rotate freely (friction is neglected) but prevents it from translating in any direction.

Imagine holding a bar attached to a ball and socket support (Fig. 5.13b). If you try to translate the bar (move it without rotating it) in any direction, the support exerts a reactive force to prevent the motion. However, you can rotate

(a)   (b)   (c)

Figure 5.13
(**a**) A ball and socket support.
(**b**) Holding a supported bar.
(**c**) The ball and socket support can exert three components of force.

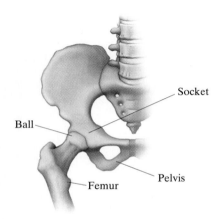

Figure 5.14
The human femur is attached to the pelvis by a ball and socket support.

the bar about the support. The support cannot exert a couple to prevent rotation. Thus a ball and socket support can't exert a couple but can exert three components of force (Fig. 5.13c). It is the three-dimensional analog of the two-dimensional pin support.

The human hip joint is an example of a ball and socket support (Fig. 5.14). The support of the gear shift lever of a car can be modeled as a ball and socket support within the lever's range of motion.

The Roller Support   The *roller support* (Fig. 5.15a) is a ball and socket support that can roll freely on a supporting surface. A roller support can exert only a force normal to the supporting surface (Fig. 5.15b). The rolling "casters" sometimes used to support furniture legs are supports of this type.

The Hinge   The hinge support is the familiar device used to support doors. It permits the supported object to rotate freely about a line, the *hinge axis*. An object is attached to a hinge in Fig. 5.16a. The z axis of the coordinate system is aligned with the hinge axis.

If you imagine holding a bar attached to a hinge (Fig. 5.16b), notice that you can rotate the bar about the hinge axis. The hinge cannot exert a couple about the hinge axis (the z axis) to prevent rotation. However, you can't rotate the bar about the x or y axis because the hinge can exert couples about those

(a)   (b)

Figure 5.15
(**a**) A roller support.
(**b**) The reaction is normal to the supporting surface.

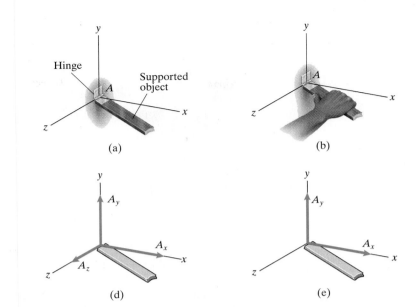

Figure 5.16
(a) A hinge. The z axis is aligned with the hinge axis.
(b) Holding a supported bar.
(c) In general, a hinge can exert five reactions: three force
    components and two couple components.
(d) The reactions when the hinge exerts no couples.
(e) The reactions when the hinge exerts neither couples nor
    a force parallel to the hinge axis.

axes to resist the motion. In addition, you can't translate the bar in any direction. The reactions a hinge can exert on an object are shown in Fig. 5.16c. There are three components of force, A_x, A_y, and A_z, and couples about the x and y axes, M_{Ax} and M_{Ay}.

   In some situations, either a hinge exerts no couples on the object it supports, or they are sufficiently small to neglect. An example of the latter case is when the axes of the hinges supporting a door are properly aligned (the axes of the individual hinges coincide). In these situations the hinge exerts only forces on an object (Fig. 5.16d). Situations also arise in which a hinge exerts no couples on an object and exerts no force in the direction of the hinge axis. (The hinge may actually be designed so that it cannot support a force parallel to the hinge axis.) Then the hinge exerts forces only in the directions perpendicular to the hinge axis (Fig. 5.16e). In examples and problems, we indicate when a hinge does not exert all five of the reactions in Fig. 5.16c.

The Bearing   The type of bearing shown in Fig. 5.17a supports a circular shaft while permitting it to rotate about its axis. The reactions are identical to those exerted by a hinge. In the most general case (Fig. 5.17b), the bearing can exert a force on the supported shaft in each coordinate direction and can exert couples about axes perpendicular to the shaft but cannot exert a couple about the axis of the shaft.

   As in the case of the hinge, situations can occur in which the bearing exerts no couples (Fig. 5.17c) or exerts no couples and no force parallel to the shaft axis (Fig. 5.17d). Some bearings are designed in this way for specific

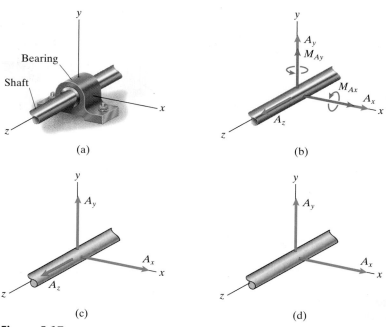

(a)

(b)

(c)

(d)

Figure 5.17
(a) A bearing. The z axis is aligned with the axis of the shaft.
(b) In general, a bearing can exert five reactions: three force components and two couple components.
(c) The reactions when the bearing exerts no couples.
(d) The reactions when the bearing exerts neither couples nor a force parallel to the axis of the shaft.

applications. In examples and problems, we indicate when a bearing does not exert all of the reactions in Fig. 5.17b.

The Fixed Support You are already familiar with the fixed, or built-in, support (Fig. 5.18a). Imagine holding a bar with a fixed support (Fig. 5.18b). You cannot translate it in any direction, and you cannot rotate it about any axis.

(a)

(b)

(c)

Figure 5.18
(a) A fixed support.
(b) Holding a supported bar.
(c) A fixed support can exert six reactions: three force components and three couple components.

The support is capable of exerting forces A_x, A_y, and A_z in each coordinate direction and couples M_{Ax}, M_{Ay}, and M_{Az} about each coordinate axis (Fig. 5.18c).

Table 5.2 summarizes the support conventions commonly used in three-dimensional applications.

Table 5.2 Supports used in three-dimensional applications.

Supports	Reactions
Rope or Cable	A Collinear Force
Contact with a Smooth Surface	A Normal Force
Contact with a Rough Surface	Three Force Components
Ball and Socket Support	Three Force Components
Roller Support	A Normal Force

Table 5.2   *continued*

Supports	Reactions
 Hinge (The z axis is parallel to the hinge axis.)	 Three Force Components, Two Couple Components
 Bearing (The z axis is parallel to the axis of the supported shaft.)	 (When no couples are exerted) (When no couples and no axial force are exerted)
 Fixed (Built-in) Support	 Three Force Components, Three Couple Components

Equilibrium Equations

If an object is in equilibrium, the sum of the external forces acting on it equals zero,

$$\Sigma \mathbf{F} = \mathbf{0}$$

$$\Sigma F_x = 0, \qquad (5.9)$$
$$\Sigma F_y = 0, \qquad (5.10)$$
$$\Sigma F_z = 0, \qquad (5.11)$$

and the sum of the moments about any point due to the forces and couples acting on it is zero,

$$\Sigma \mathbf{M}_{\text{any point}} = \mathbf{0}$$

$$\Sigma M_x = 0, \qquad (5.12)$$
$$\Sigma M_y = 0, \qquad (5.13)$$
$$\Sigma M_z = 0. \qquad (5.14)$$

Supports

Examples of supports used in three-dimensional applications. (See Table 5.2.)

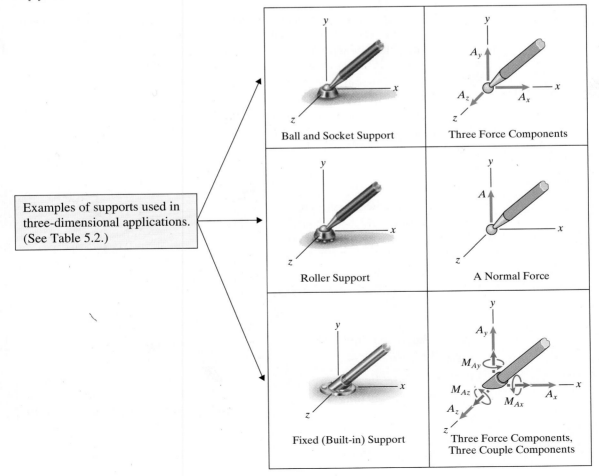

Ball and Socket Support

Three Force Components

Roller Support

A Normal Force

Fixed (Built-in) Support

Three Force Components, Three Couple Components

Active Example 5.7 | **Determining Reactions in Three Dimensions** (▶ *Related Problem 5.86*)

The bar *AB* is supported by the cables *BC* and *BD* and a ball and socket support at *A*. Cable *BC* is parallel to the *z* axis and cable *BD* is parallel to the *x* axis. The 200-N force acts at the midpoint of the bar. Determine the tensions in the cables and the reactions at *A*.

Strategy

We must obtain the free-body diagram of the bar by isolating it and showing the reactions exerted by the cables and the ball and socket support. Then we can apply the equilibrium equations to determine the reactions.

Solution

Draw the Free-Body Diagram of the Bar

| Isolate the bar and show the reactions exerted by the cables and the ball and socket support. |

Forces exerted by the cables

Reactions due to the ball and socket support

Apply the Equilibrium Equations

$$\Sigma F_x = A_x - T_{BD} = 0,$$
$$\Sigma F_y = A_y - 200 \text{ N} = 0,$$
$$\Sigma F_z = A_z - T_{BC} = 0.$$

The sums of the forces in each coordinate direction equal zero.

$$\Sigma \mathbf{M}_{\text{point } A} = [\mathbf{r}_{AB} \times (-T_{BC}\mathbf{k})] + [\mathbf{r}_{AB} \times (-T_{BD}\mathbf{i})] + \left[\frac{1}{2}\,\mathbf{r}_{AB} \times (-200\mathbf{j})\right]$$

$$= \begin{vmatrix} \mathbf{i} & \mathbf{j} & \mathbf{k} \\ 1 & 0.6 & 0.4 \\ 0 & 0 & -T_{BC} \end{vmatrix} + \begin{vmatrix} \mathbf{i} & \mathbf{j} & \mathbf{k} \\ 1 & 0.6 & 0.4 \\ -T_{BD} & 0 & 0 \end{vmatrix} + \begin{vmatrix} \mathbf{i} & \mathbf{j} & \mathbf{k} \\ 0.5 & 0.3 & 0.2 \\ 0 & -200 & 0 \end{vmatrix}$$

$$= (-0.6T_{BC} + 40)\mathbf{i} + (T_{BC} - 0.4T_{BD})\mathbf{j} + (0.6T_{BD} - 100)\mathbf{k}.$$

> The sum of the moments about any point equals zero.

> The components of this vector (the sums of the moments about the three coordinate axes) must each equal zero.

$$\Sigma M_x = -(0.6 \text{ m})T_{BC} + 40 \text{ N-m} = 0,$$
$$\Sigma M_y = (1 \text{ m})T_{BC} - (0.4 \text{ m})T_{BD} = 0,$$
$$\Sigma M_z = (0.6 \text{ m})T_{BD} - 100 \text{ N-m} = 0.$$

> Solving the six scalar equilibrium equations yields
> $A_x = 166.7$ N, $A_y = 200$ N, $A_z = 66.7$ N, $T_{BC} = 66.7$ N,
> and $T_{BD} = 166.7$ N.

Practice Problem Suppose that the cables BC and BD are removed and the ball and socket joint at A is replaced by a fixed support. Determine the reactions at A.

Answer: $A_x = 0$, $A_y = 200$ N, $A_z = 0$, $M_{Ax} = -40$ N-m, $M_{Ay} = 0$, $M_{Az} = 100$ N-m.

| Example 5.8 | Reactions at a Hinge Support (▶ *Related Problem 5.104*) |

The bar AC is 4 ft long and is supported by a hinge at A and the cable BD. The hinge axis is along the z axis. The centerline of the bar lies in the x–y plane, and the cable attachment point B is the midpoint of the bar. Determine the tension in the cable and the reactions exerted on the bar by the hinge.

Strategy

We will obtain a free-body diagram of bar AC by isolating it from the cable and hinge. (The reactions the hinge can exert on the bar are shown in Table 5.2.) Then we can determine the reactions by applying the equilibrium equations.

Solution

Draw the Free-Body Diagram   We isolate the bar from the hinge support and the cable and show the reactions they exert (Fig. a). The terms A_x, A_y, and A_z are the components of force exerted by the hinge, and the terms M_{Ax} and M_{Ay} are the couples exerted by the hinge about the x and y axes. (Remember that the hinge cannot exert a couple on the bar about the hinge axis.) The term T is the tension in the cable.

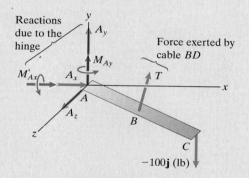

(a) The free-body diagram of the bar.

Apply the Equilibrium Equations   To write the equilibrium equations, we must first express the cable force in terms of its components. The coordinates of point B are $(2 \cos 30°, -2 \sin 30°, 0)$ ft, so the position vector from B to D is

$$\mathbf{r}_{BD} = (2 - 2 \cos 30°)\mathbf{i} + [2 - (-2 \sin 30°)]\mathbf{j} + (-1 - 0)\mathbf{k}$$

$$= 0.268\mathbf{i} + 3\mathbf{j} - \mathbf{k} \ (\text{ft}).$$

We divide this vector by its magnitude to obtain a unit vector $\mathbf{e}_{BD}$ that points from point B toward point D:

$$\mathbf{e}_{BD} = \frac{\mathbf{r}_{BD}}{|\mathbf{r}_{BD}|} = 0.084\mathbf{i} + 0.945\mathbf{j} - 0.315\mathbf{k}.$$

Now we can write the cable force as the product of its magnitude and $\mathbf{e}_{BD}$:

$$T\mathbf{e}_{BD} = T(0.084\mathbf{i} + 0.945\mathbf{j} - 0.315\mathbf{k}).$$

The sums of the forces in each coordinate direction must equal zero:

$$\Sigma F_x = A_x + 0.084T = 0,$$

$$\Sigma F_y = A_y + 0.945T - 100 \ \text{lb} = 0, \tag{1}$$

$$\Sigma F_z = A_z - 0.315T = 0.$$

If we sum moments about A, the resulting equations do not contain the unknown reactions A_x, A_y, and A_z. The position vectors from A to B and from A to C are

$$\mathbf{r}_{AB} = 2 \cos 30°\mathbf{i} - 2 \sin 30°\mathbf{j} \text{ (ft)},$$

$$\mathbf{r}_{AC} = 4 \cos 30°\mathbf{i} - 4 \sin 30°\mathbf{j} \text{ (ft)}.$$

The sum of the moments about A, with forces in lb and distances in ft, is

$$\Sigma\mathbf{M}_{\text{point } A} = M_{Ax}\mathbf{i} + M_{Ay}\mathbf{j} + [\mathbf{r}_{AB} \times (Te_{BD})] + [\mathbf{r}_{AC} \times (-100\mathbf{j})]$$

$$= M_{Ax}\mathbf{i} + M_{Ay}\mathbf{j} + \begin{vmatrix} \mathbf{i} & \mathbf{j} & \mathbf{k} \\ 1.732 & -1 & 0 \\ 0.084T & 0.945T & -0.315T \end{vmatrix}$$

$$+ \begin{vmatrix} \mathbf{i} & \mathbf{j} & \mathbf{k} \\ 3.464 & -2 & 0 \\ 0 & -100 & 0 \end{vmatrix}$$

$$= (M_{Ax} + 0.315T)\mathbf{i} + (M_{Ay} + 0.546T)\mathbf{j}$$

$$+ (1.72T - 346)\mathbf{k} = 0.$$

From this vector equation, we obtain the scalar equations

$$\Sigma\dot{M}_x = M_{Ax} + (0.315 \text{ ft})T = 0,$$

$$\Sigma M_y = M_{Ay} + (0.546 \text{ ft})T = 0,$$

$$\Sigma M_z = (1.72 \text{ ft})T_{BD} - 346 \text{ ft-lb} = 0.$$

Solving these equations yields the reactions

$$T = 201 \text{ lb}, \qquad M_{Ax} = -63.4 \text{ ft-lb}, \qquad M_{Ay} = -109.8 \text{ ft-lb}.$$

Then from Eqs. (1) we obtain the forces exerted on the bar by the hinge:

$$A_x = -17.0 \text{ lb}, \qquad A_y = -90.2 \text{ lb}, \qquad A_z = 63.4 \text{ lb}.$$

Critical Thinking

Notice in Table 5.2 that there are three possibilities for the reactions exerted by a hinge or bearing. How do you know which one to choose? Under certain circumstances, a hinge may not exert significant couples on the object to which it is connected, and it also may not exert a significant force in the direction of the hinge axis. For example, when an object has two hinge supports and their axes are aligned (see Example 5.9), you can often assume that each individual hinge does not exert couples on the object. But in general, it requires experience to make such judgments. In upcoming examples and problems, we will indicate the reactions that you can assume are exerted by a hinge. Whenever you are in doubt, you should assume that a hinge may exert the most general set of reactions shown in Table 5.2 (three force components and two couple components).

| Example 5.9 | Reactions at Properly Aligned Hinges (▶ *Related Problem 5.112*) |

The plate is supported by hinges at A and B and the cable CE. The properly aligned hinges do not exert couples on the plate, and the hinge at A does not exert a force on the plate in the direction of the hinge axis. Determine the reactions at the hinges and the tension in the cable.

Strategy

We will draw the free-body diagram of the plate, using the given information about the reactions exerted by the hinges at A and B. Before the equilibrium equations can be applied, we must express the force exerted on the plate by the cable in terms of its components.

Solution

Draw the Free-Body Diagram   We isolate the plate and show the reactions at the hinges and the force exerted by the cable (Fig. a). The term T is the force exerted on the plate by cable CE.

Apply the Equilibrium Equations   Since we know the coordinates of points C and E, we can express the cable force as the product of its magnitude T and a unit vector directed from C toward E. The result is

$$T(-0.842\mathbf{i} + 0.337\mathbf{j} + 0.421\mathbf{k}).$$

The sums of the forces in each coordinate direction equal zero:

$$\Sigma F_x = A_x + B_x - 0.842T = 0,$$
$$\Sigma F_y = A_y + B_y + 0.337T - 400 = 0, \qquad (1)$$
$$\Sigma F_z = B_z + 0.421T = 0.$$

(a) The free-body diagram of the plate.

If we sum the moments about B, the resulting equations will not contain the three unknown reactions at B. The sum of the moments about B, with forces in N and distances in m, is

$$\Sigma\mathbf{M}_{\text{point }B} = \begin{vmatrix} \mathbf{i} & \mathbf{j} & \mathbf{k} \\ 0.2 & 0 & 0 \\ -0.842T & 0.337T & 0.421T \end{vmatrix} + \begin{vmatrix} \mathbf{i} & \mathbf{j} & \mathbf{k} \\ 0 & 0 & 0.2 \\ A_x & A_y & 0 \end{vmatrix}$$

$$+ \begin{vmatrix} \mathbf{i} & \mathbf{j} & \mathbf{k} \\ 0.2 & 0 & 0.2 \\ 0 & -400 & 0 \end{vmatrix}$$

$$= (-0.2A_y + 80)\mathbf{i} + (-0.0842T + 0.2A_x)\mathbf{j}$$
$$+ (0.0674T - 80)\mathbf{k} = 0.$$

The scalar equations are

$$\Sigma M_x = -(0.2\text{ m})A_y + 80\text{ N-m} = 0,$$
$$\Sigma M_y = -(0.0842\text{ m})T + (0.2\text{ m})A_x = 0,$$
$$\Sigma M_z = (0.0674\text{ m})T - 80\text{ N-m} = 0.$$

Solving these equations, we obtain the reactions

$$T = 1187\text{ N}, \qquad A_x = 500\text{ N}, \qquad A_y = 400\text{ N}.$$

Then from Eqs. (1), the reactions at B are

$$B_x = 500\text{ N}, \qquad B_y = -400\text{ N}, \qquad B_z = -500\text{ N}.$$

Critical Thinking

"Properly aligned hinges" means hinges that are mounted on an object so that their axes are aligned. When this is the case, as in this example, it can usually be assumed that each individual hinge does not exert couples on the object. Notice that it is also assumed in this example that the hinge at A exerts no reaction parallel to the hinge axis but the hinge at B does. The hinges can be intentionally designed so that this is the case, or it can result from the way they are installed.

If our only objective in this example had been to determine the tension T, we could have done so easily by evaluating the sum of the moments about the line AB (the z axis). Because the reactions at the hinges exert no moment about the z axis, we obtain the equation

$$(0.2 \text{ m})(0.337T) - (0.2 \text{ m})(400 \text{ N}) = 0,$$

which yields $T = 1187$ N.

Problems

5.77 The bar AB has a fixed support at A and is loaded by the forces

$$\mathbf{F}_B = 2\mathbf{i} + 6\mathbf{j} + 3\mathbf{k} \text{ (kN)},$$
$$\mathbf{F}_C = \mathbf{i} - 2\mathbf{j} + 2\mathbf{k} \text{ (kN)}.$$

(a) Draw the free-body diagram of the bar.

(b) Determine the reactions at A.

Strategy: (a) Draw a diagram of the bar isolated from its supports. Complete the free-body diagram of the bar by adding the two external forces and the reactions due to the fixed support (see Table 5.2). (b) Use the scalar equilibrium equations (5.9)–(5.14) to determine the reactions.

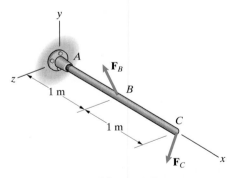

Problem 5.77

5.78 The bar AB has a fixed support at A. The tension in cable BC is 8 kN. Determine the reactions at A.

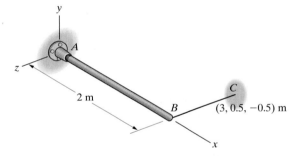

Problem 5.78

5.79 The bar AB has a fixed support at A. The collar at B is fixed to the bar. The tension in the rope BC is 300 lb. (a) Draw the free-body diagram of the bar. (b) Determine the reactions at A.

5.80 The bar AB has a fixed support at A. The collar at B is fixed to the bar. Suppose that you don't want the support at A to be subjected to a couple of magnitude greater than 3000 ft-lb. What is the largest allowable tension in the rope BC?

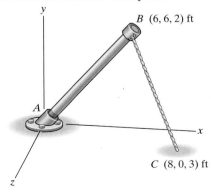

Problems 5.79/5.80

5.81 The total force exerted on the highway sign by its weight and the most severe anticipated winds is $\mathbf{F} = 2.8\mathbf{i} - 1.8\mathbf{j}$ (kN). Determine the reactions at the fixed support.

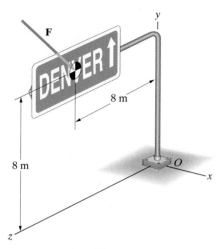

Problem 5.81

5.82 The tension in cable AB is 800 lb. Determine the reactions at the fixed support C.

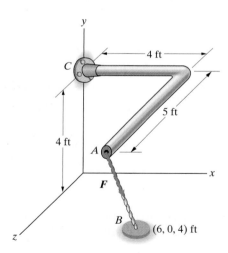

Problem 5.82

5.83 The tension in cable AB is 24 kN. Determine the reactions at the fixed support D.

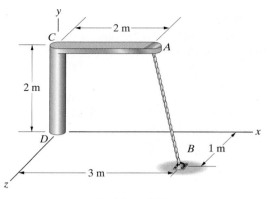

Problem 5.83

5.84 The robotic manipulator is stationary and the y axis is vertical. The weights of the arms AB and BC act at their midpoints. The direction cosines of the centerline of arm AB are $\cos \theta_x = 0.174$, $\cos \theta_y = 0.985$, $\cos \theta_z = 0$, and the direction cosines of the centerline of arm BC are $\cos \theta_x = 0.743$, $\cos \theta_y = 0.557$, $\cos \theta_z = -0.371$. The support at A behaves like a fixed support.

(a) What is the sum of the moments about A due to the weights of the two arms?

(b) What are the reactions at A?

Problem 5.84

5.85 The force exerted on the grip of the exercise machine is $F = 260i - 130j$ (N). What are the reactions at the fixed support at O?

Problem 5.85

▶ **5.86** In Active Example 5.7, suppose that cable BD is lengthened and the attachment point D moved from (0, 600, 400) mm to (0, 600, 600) mm. (The end B of bar AB remains where it is.) Draw a sketch of the bar and its supports showing cable BD in its new position. Draw the free-body diagram of the bar and apply equilibrium to determine the tensions in the cables and the reactions at A.

5.87 The force F acting on the boom ABC at C points in the direction of the unit vector $0.512i - 0.384j + 0.768k$ and its magnitude is 8 kN. The boom is supported by a ball and socket at A and the cables BD and BE. The collar at B is fixed to the boom.

(a) Draw the free-body diagram of the boom.

(b) Determine the tensions in the cables and the reactions at A.

5.88 The cables BD and BE in Problem 5.87 will each safely support a tension of 25 kN. Based on this criterion, what is the largest acceptable magnitude of the force F?

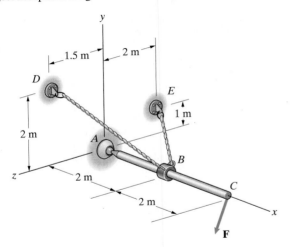

Problems 5.87/5.88

5.89 The suspended load exerts a force $F = 600$ lb at A, and the weight of the bar OA is negligible. Determine the tensions in the cables and the reactions at the ball and socket support O.

5.90 The suspended load exerts a force $F = 600$ lb at A and bar OA weighs 200 lb. Assume that the bar's weight acts at its midpoint. Determine the tensions in the cables and the reactions at the ball and socket support O.

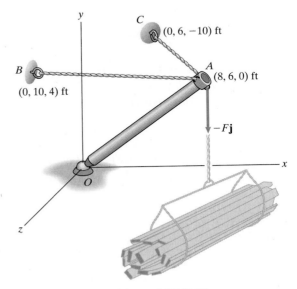

Problems 5.89/5.90

5.91 The 158,000-kg airplane is at rest on the ground ($z = 0$ is ground level). The landing gear carriages are at A, B, and C. The coordinates of the point G at which the weight of the plane acts are (3, 0.5, 5) m. What are the magnitudes of the normal reactions exerted on the landing gear by the ground?

Problem 5.91

5.92 The horizontal triangular plate is suspended by the three vertical cables A, B, and C. The tension in each cable is 80 N. Determine the x and z coordinates of the point where the plate's weight effectively acts.

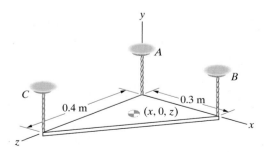

Problem 5.92

5.93 The 800-kg horizontal wall section is supported by the three vertical cables A, B, and C. What are the tensions in the cables?

Problem 5.93

5.94 The bar AC is supported by the cable BD and a bearing at A that can rotate about the z axis. The person exerts a force $\mathbf{F} = 10\mathbf{j}$ (lb) at C. Determine the tension in the cable and the reactions at A.

Problem 5.94

5.95 The L-shaped bar is supported by a bearing at A and rests on a smooth horizontal surface at B. The vertical force $F = 4$ kN and the distance $b = 0.15$ m. Determine the reactions at A and B.

5.96 The vertical force $F = 4$ kN and the distance $b = 0.15$ m. If you represent the reactions at A and B by an equivalent system consisting of a single force, what is the force and where does its line of action intersect the x–z plane?

5.97 The vertical force $F = 4$ kN. The bearing at A will safely support a force of 2.5-kN magnitude and a couple of 0.5 kN-m magnitude. Based on these criteria, what is the allowable range of the distance b?

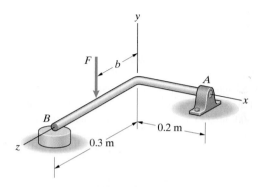

Problems 5.95–5.97

5.98 The 1.1-m bar is supported by a ball and socket support at A and the two smooth walls. The tension in the vertical cable CD is 1 kN.

(a) Draw the free-body diagram of the bar.

(b) Determine the reactions at A and B.

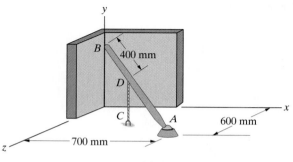

Problem 5.98

5.99 The 8-ft bar is supported by a ball and socket support at A, the cable BD, and a roller support at C. The collar at B is fixed to the bar at its midpoint. The force $\mathbf{F} = -50\mathbf{k}$ (lb). Determine the tension in cable BD and the reactions at A and C.

5.100 The bar is 8 ft in length. The force $\mathbf{F} = F_y\mathbf{j} - 50\mathbf{k}$ (lb). What is the largest value of F_y for which the roller support at C will remain on the floor?

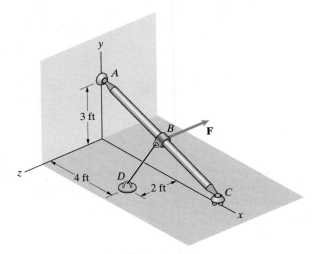

Problems 5.99/5.100

5.101 The tower is 70 m tall. The tension in each cable is 2 kN. Treat the base of the tower A as a fixed support. What are the reactions at A?

5.102 The tower is 70 m tall. If the tension in cable BC is 2 kN, what must the tensions in cables BD and BE be if you want the couple exerted on the tower by the fixed support at A to be zero? What are the resulting reactions at A?

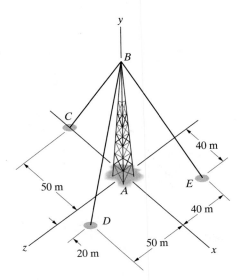

Problems 5.101/5.102

5.103 The space truss has roller supports at B, C, and D and is subjected to a vertical force $F = 20$ kN at A. What are the reactions at the roller supports?

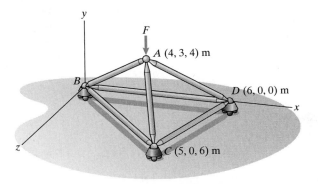

Problem 5.103

▶ **5.104** In Example 5.8, suppose that the cable BD is lengthened and the attachment point B is moved to the end of the bar at C. The positions of the attachment point D and the bar are unchanged. Draw a sketch of the bar showing cable BD in its new position. Draw the free-body diagram of the bar and apply equilibrium to determine the tension in the cable and the reactions at A.

5.105 The 40-lb door is supported by hinges at A and B. The y axis is vertical. The hinges do not exert couples on the door, and the hinge at B does not exert a force parallel to the hinge axis. The weight of the door acts at its midpoint. What are the reactions at A and B?

Problem 5.105

5.106 The vertical cable is attached at A. Determine the tension in the cable and the reactions at the bearing B due to the force $F = 10i - 30j - 10k$ (N).

5.107 Suppose that the z component of the force F is zero, but otherwise F is unknown. If the couple exerted on the shaft by the bearing at B is $M_B = 6j - 6k$ N-m, what are the force F and the tension in the cable?

Problems 5.106/5.107

5.108 The device in Problem 5.106 is badly designed because of the couples that must be supported by the bearing at B, which would cause the bearing to "bind." (Imagine trying to open a door supported by only one hinge.) In this improved design, the bearings at B and C support no couples, and the bearing at C does not exert a force in the x direction. If the force $F = 10i - 30j - 10k$ (N), what are the tension in the vertical cable and the reactions at the bearings B and C?

Problem 5.108

5.109 The rocket launcher is supported by the hydraulic jack DE and the bearings A and B. The bearings lie on the x axis and support shafts parallel to the x axis. The hydraulic cylinder DE exerts a force on the launcher that points along the line from D to E. The coordinates of D are $(7, 0, 7)$ ft, and the coordinates of E are $(9, 6, 4)$ ft. The weight $W = 30$ kip acts at $(4.5, 5, 2)$ ft. What is the magnitude of the reaction on the launcher at E?

5.110 Consider the rocket launcher described in Problem 5.109. The bearings at A and B do not exert couples, and the bearing B does not exert a force in the x direction. Determine the reactions at A and B.

Problems 5.109/5.110

5.111 The crane's cable CD is attached to a stationary object at D. The crane is supported by the bearings E and F and the horizontal cable AB. The tension in cable AB is 8 kN. Determine the tension in the cable CD.

Strategy: Since the reactions exerted on the crane by the bearings do not exert moments about the z axis, the sum of the moments about the z axis due to the forces exerted on the crane by the cables AB and CD equals zero.

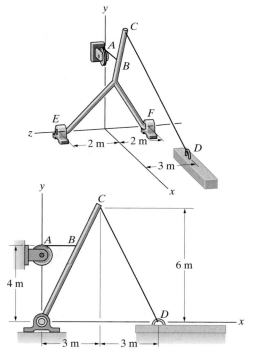

Problem 5.111

▶ **5.112** In Example 5.9, suppose that the cable CE is shortened and its attachment point E is moved to the point (0, 80, 0) mm. The plate remains in the same position. Draw a sketch of the plate and its supports showing the new position of cable CE. Draw the free-body diagram of the plate and apply equilibrium to determine the reactions at the hinges and the tension in the cable.

5.113 The plate is supported by hinges at A and B and the cable CE, and it is loaded by the force at D. The edge of the plate to which the hinges are attached lies in the y–z plane, and the axes of the hinges are parallel to the line through points A and B. The hinges do not exert couples on the plate. What is the tension in cable CE?

5.114 In Problem 5.113, the hinge at B does not exert a force on the plate in the direction of the hinge axis. What are the magnitudes of the forces exerted on the plate by the hinges at A and B?

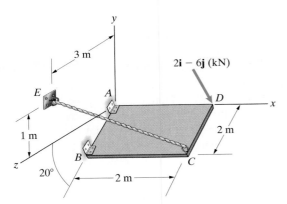

Problems 5.113/5.114

5.115 The bar ABC is supported by ball and socket supports at A and C and the cable BD. The suspended mass is 1800 kg. Determine the tension in the cable.

5.116* In Problem 5.115, assume that the ball and socket support at A is designed so that it exerts no force parallel to the straight line from A to C. Determine the reactions at A and C.

Problems 5.115/5.116

5.117 The bearings at A, B, and C do not exert couples on the bar and do not exert forces in the direction of the axis of the bar. Determine the reactions at the bearings due to the two forces on the bar.

Problem 5.117

5.118 The support that attaches the sailboat's mast to the deck behaves like a ball and socket support. The line that attaches the spinnaker (the sail) to the top of the mast exerts a 200-lb force on the mast. The force is in the horizontal plane at 15° from the centerline of the boat. (See the top view.) The spinnaker pole exerts a 50-lb force on the mast at P. The force is in the horizontal plane at 45° from the centerline. (See the top view.) The mast is supported by two cables, the backstay AB and the port shroud ACD. (The forestay AE and the starboard shroud AFG are slack, and their tensions can be neglected.) Determine the tensions in the cables AB and CD and the reactions at the bottom of the mast.

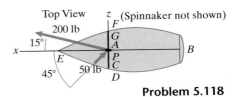

Problem 5.118

5.119* The bar AC is supported by the cable BD and a bearing at A that can rotate about the axis AE. The person exerts a force $\mathbf{F} = 50\mathbf{j}$ (N) at C. Determine the tension in the cable.

Strategy: Use the fact that the sum of the moments about the axis AE due to the forces acting on the free-body diagram of the bar must equal zero.

5.120* In Problem 5.119, determine the reactions at the bearing A.

Strategy: Write the couple exerted on the free-body diagram of the bar by the bearing as $\mathbf{M}_A = M_{Ax}\mathbf{i} + M_{Ay}\mathbf{j} + M_{Az}\mathbf{k}$. Then, in addition to the equilibrium equations, obtain an equation by requiring the component of $\mathbf{M}_A$ parallel to the axis AE to equal zero.

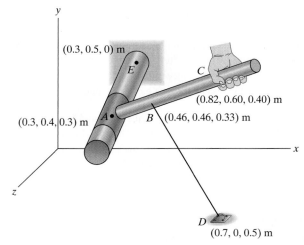

Problems 5.119/5.120

5.4 Two-Force and Three-Force Members

BACKGROUND

We have shown how the equilibrium equations are used to analyze objects that are supported and loaded in different ways. Here we discuss two particular types of loading that occur so frequently they deserve particular attention. The first type, the two-force member, is especially important and plays an important role in our analysis of structures in Chapter 6.

Two-Force Members

If the system of forces and moments acting on an object is equivalent to two forces acting at different points, we refer to the object as a *two-force member*. For example, the object in Fig. 5.19a is subjected to two sets of concurrent forces whose lines of action intersect at A and B. Since we can represent them by single forces acting at A and B (Fig. 5.19b), where $\mathbf{F} = \mathbf{F}_1 + \mathbf{F}_2 + \cdots + \mathbf{F}_N$ and $\mathbf{F}' = \mathbf{F}'_1 + \mathbf{F}'_2 + \cdots + \mathbf{F}'_M$, this object is a two-force member.

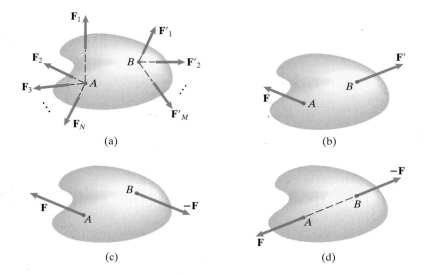

Figure 5.19
(a) An object subjected to two sets of concurrent forces.
(b) Representing the concurrent forces by two forces $\mathbf{F}$ and $\mathbf{F}'$.
(c) If the object is in equilibrium, the forces must be equal and opposite.
(d) The forces form a couple unless they have the same line of action.

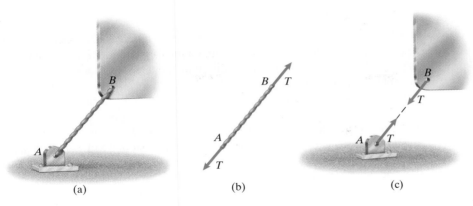

Figure 5.20
(a) A cable attached at A and B.
(b) The cable is a two-force member.
(c) The forces exerted by the cable.

If the object is in equilibrium, what can we infer about the forces **F** and **F'**? The sum of the forces equals zero only if $\mathbf{F}' = -\mathbf{F}$ (Fig. 5.19c). Furthermore, the forces **F** and $-\mathbf{F}$ form a couple, so the sum of the moments is not zero unless the lines of action of the forces lie along the line through the points A and B (Fig. 5.19d). Thus equilibrium tells us that *the two forces are equal in magnitude, are opposite in direction, and have the same line of action.* However, without additional information, we cannot determine their magnitude.

A cable attached at two points (Fig. 5.20a) is a familiar example of a two-force member (Fig. 5.20b). The cable exerts forces on the attachment points that are directed along the line between them (Fig. 5.20c).

A bar that has two supports that exert only forces on it (no couples) and is not subjected to any loads is a two-force member (Fig. 5.21a). Such bars are often used as supports for other objects. Because the bar is a two-force member, the lines of action of the forces exerted on the bar must lie along the line between the supports (Fig. 5.21b). Notice that, unlike the cable, the bar can exert forces at A and B either in the directions shown in Fig. 5.21c or in the opposite directions. (In other words, the cable can only pull on its supports, while the bar can either pull or push.)

In these examples we assumed that the weights of the cable and the bar could be neglected in comparison with the forces exerted on them by their supports. When that is not the case, they are clearly not two-force members.

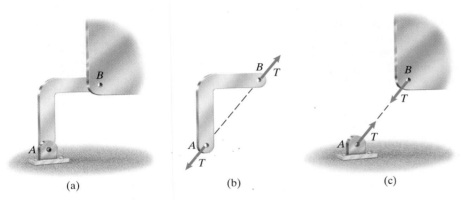

Figure 5.21
(a) The bar AB attaches the object to the pin support.
(b) The bar AB is a two-force member.
(c) The force exerted on the supported object by the bar AB.

(a)

(b)

(c)

Figure 5.22
(a) The three forces and the plane P.
(b) Determining the moment due to force $\mathbf{F}_3$ about L.
(c) If the forces are not parallel, they must be concurrent.

Three-Force Members

If the system of forces and moments acting on an object is equivalent to three forces acting at different points, we call it a *three-force member*. We can show that if a three-force member is in equilibrium, the three forces are coplanar and are either parallel or concurrent.

We first prove that the forces are coplanar. Let them be called $\mathbf{F}_1$, $\mathbf{F}_2$, and $\mathbf{F}_3$, and let P be the plane containing the three points of application (Fig. 5.22a). Let L be the line through the points of application of $\mathbf{F}_1$ and $\mathbf{F}_2$. Since the moments due to $\mathbf{F}_1$ and $\mathbf{F}_2$ about L are zero, the moment due to $\mathbf{F}_3$ about L must equal zero (Fig. 5.22b):

$$[\mathbf{e} \cdot (\mathbf{r} \times \mathbf{F}_3)]\mathbf{e} = [\mathbf{F}_3 \cdot (\mathbf{e} \times \mathbf{r})]\mathbf{e} = \mathbf{0}.$$

This equation requires that $\mathbf{F}_3$ be perpendicular to $\mathbf{e} \times \mathbf{r}$, which means that $\mathbf{F}_3$ is contained in P. The same procedure can be used to show that $\mathbf{F}_1$ and $\mathbf{F}_2$ are contained in P, so the forces are coplanar. (A different proof is required if the points of application lie on a straight line, but the result is the same.)

If the three coplanar forces are not parallel, there will be points where their lines of action intersect. Suppose that the lines of action of two of the forces intersect at a point Q. Then the moments of those two forces about Q are zero, and the sum of the moments about Q is zero only if the line of action of the third force also passes through Q. Therefore, either the forces are parallel or they are concurrent (Fig. 5.22c).

The analysis of an object in equilibrium can often be simplified by recognizing that it is a two-force or three-force member. However, in doing so we are not getting something for nothing. Once the free-body diagram of a two-force member is drawn, as shown in Figs. 5.20b and 5.21b, no further information can be obtained from the equilibrium equations. And when we require that the lines of action of nonparallel forces acting on a three-force member be coincident, we have used the fact that the sum of the moments about a point must be zero and cannot obtain further information from that condition.

RESULTS

Two-Force Member
If an object in equilibrium is subjected to two forces acting at different points and *no other forces or couples*, it is called a two-force member. Equilibrium requires that the two forces be equal and opposite and parallel to the line between the two points.

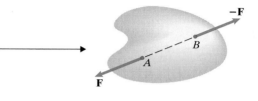

Three-Force Member
If an object in equilibrium is subjected to three forces acting at different points and *no other forces or couples*, it is called a three-force member. Equilibrium requires that the three forces be coplanar and either parallel or concurrent.

Active Example 5.10 **Two- and Three-Force Members** (▶ *Related Problem 5.121*)

The 100-lb weight of the rectangular plate acts at its midpoint. Neglect the weight of the link *AB*. Determine the reactions exerted on the plate at *B* and *C*.

Strategy

The plate is subjected to its weight and the reactions exerted by the pin supports at *B* and *C*, so it is a three-force member. The link *BC* is a two-force member, so the line of action of the reaction it exerts on the plate at *B* must be directed along the line between *A* and *B*. We can use this information to simplify the free-body diagram of the plate.

The reaction exerted on the plate by the two-force member *AB* must be directed along the line between *A* and *B*.

Solution

> The force exerted on the plate by the bar *AB* must be directed along the line between *A* and *B*, and the line of action of the weight of the plate is vertical, so the three forces on the plate are not parallel. Therefore they must be concurrent.

$$\Sigma F_x = B \sin 45° - C \sin 45° = 0,$$
$$\Sigma F_y = B \cos 45° + C \cos 45° - 100 \text{ lb} = 0.$$
Solving yields the reactions $B = C = 70.7$ lb.

> Apply the equilibrium equations.

Practice Problem Suppose that the plate is replaced with a 100-lb plate whose thickness (the dimension perpendicular to the page) is not uniform. The line of action of the weight of the nonuniform plate is 3 ft to the right of point *B*. Determine the reactions exerted on the plate at *B* and *C*.

Answer: $B = 35.4$ lb, $C = 79.1$ lb.

Example 5.11 **A Two-Force Member** (▶ *Related Problem 5.122*)

The L-shaped bar has a pin support at A and is loaded by a 6-kN force at B. Neglect the weight of the bar. Determine the angle α and the reactions at A.

Strategy

The bar is a two-force member because it is subjected only to the 6-kN force at B and the force exerted by the pin support. (If we could not neglect the weight of the bar, it would not be a two-force member.) We will determine the angle α and the reactions at A in two ways, first by applying the equilibrium equations in the usual way and then by using the fact that the bar is a two-force member.

Solution

Applying the Equilibrium Equations We draw the free-body diagram of the bar in Fig. a, showing the reactions at the pin support. Summing moments about point A, the equilibrium equations are

$$\Sigma F_x = A_x + 6 \cos \alpha \text{ kN} = 0,$$
$$\Sigma F_y = A_y + 6 \sin \alpha \text{ kN} = 0,$$
$$\Sigma M_{\text{point } A} = (0.7 \text{ m})(6 \sin \alpha \text{ kN}) - (0.4 \text{ m})(6 \cos \alpha \text{ kN}) = 0.$$

From the third equation we see that $\alpha = \arctan(0.4/0.7)$. In the range $0 \leq \alpha \leq 360°$, this equation has the two solutions $\alpha = 29.7°$ and $\alpha = 209.7°$. Knowing α, we can determine A_x and A_y from the first two equilibrium equations. The solutions for the two values of α are

$$\alpha = 29.7°, \quad A_x = -5.21 \text{ kN}, \quad A_y = -2.98 \text{ kN},$$

and

$$\alpha = 209.7°, \quad A_x = 5.21 \text{ kN}, \quad A_y = 2.98 \text{ kN}.$$

(a) The free-body diagram of the bar.

Treating the Bar as a Two-Force Member We know that the 6-kN force at B and the force exerted by the pin support must be equal in magnitude, opposite in direction, and directed along the line between points A and B. The two possibilities are shown in Figs. b and c. Thus by recognizing that the bar is a two-force member, we immediately know the possible directions of the forces and the magnitude of the reaction at A.

In Fig. b we can see that $\tan \alpha = 0.4/0.7$, so $\alpha = 29.7°$ and the components of the reaction at A are

$$A_x = -6 \cos 29.7° \text{ kN} = -5.21 \text{ kN},$$
$$A_y = -6 \sin 29.7° \text{ kN} = -2.98 \text{ kN}.$$

In Fig. c, $\alpha = 180° + 29.7° = 209.7°$, and the components of the reaction at A are

$$A_x = 6 \cos 29.7° \text{ kN} = 5.21 \text{ kN}$$
$$A_y = 6 \sin 29.7° \text{ kN} = 2.98 \text{ kN}.$$

(b)

(c)

(b), (c) The possible directions of the forces.

Critical Thinking

Why is it worthwhile to recognize that an object is a two-force member? Doing so tells you the directions of the forces acting on the object and also that the forces are equal and opposite. As this example demonstrates, such information frequently simplifies the solution of a problem.

Problems

▶ **5.121** In Active Example 5.10, suppose that the support at *A* is moved so that the angle between the bar *AB* and the vertical decreases from 45° to 30°. The position of the rectangular plate does not change. Draw the free-body diagram of the plate showing the point *P* where the lines of action of the three forces acting on the plate intersect. Determine the magnitudes of the reactions on the plate at *B* and *C*.

▶ **5.122** The magnitude of the reaction exerted on the L-shaped bar at *B* is 60 lb. (See Example 5.11.)

(a) What is the magnitude of the reaction exerted on the bar by the support at *A*?

(b) What are the *x* and *y* components of the reaction exerted on the bar by the support at *A*?

Problem 5.122

5.123 The suspended load weighs 1000 lb. The structure is a three-force member if its weight is neglected. Use this fact to determine the magnitudes of the reactions at *A* and *B*.

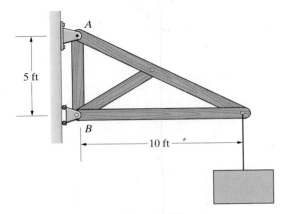

Problem 5.123

5.124 The weight *W* = 50 lb acts at the center of the disk. Use the fact that the disk is a three-force member to determine the tension in the cable and the magnitude of the reaction at the pin support.

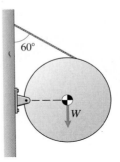

Problem 5.124

5.125 The weight *W* = 40 N acts at the center of the disk. The surfaces are rough. What force *F* is necessary to lift the disk off the floor?

Problem 5.125

5.126 Use the fact that the horizontal bar is a three-force member to determine the angle *α* and the magnitudes of the reactions at *A* and *B*. Assume that $0 \leq \alpha \leq 90°$.

Problem 5.126

5.127 The suspended load weighs 600 lb. Use the fact that ABC is a three-force member to determine the magnitudes of the reactions at A and B.

Problem 5.127

5.128 (a) Is the L-shaped bar a three-force member?
(b) Determine the magnitudes of the reactions at A and B.
(c) Are the three forces acting on the L-shaped bar concurrent?

Problem 5.128

5.129 The hydraulic piston exerts a horizontal force at B to support the weight $W = 1500$ lb of the bucket of the excavator. Determine the magnitude of the force the hydraulic piston must exert. (The vector sum of the forces exerted at B by the hydraulic piston, the two-force member AB, and the two-force member BD must equal zero.)

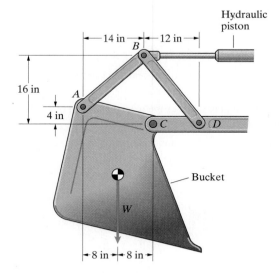

Problem 5.129

5.130 The member ACG of the front-end loader is subjected to a load $W = 2$ kN and is supported by a pin support at A and the hydraulic cylinder BC. Treat the hydraulic cylinder as a two-force member.

(a) Draw the free-body diagrams of the hydraulic cylinder and the member ACG.

(b) Determine the reactions on the member ACG.

5.131 In Problem 5.130, determine the reactions on the member ACG by using the fact that it is a three-force member.

Problems 5.130/5.131

5.132 A rectangular plate is subjected to two forces A and B (Fig. a). In Fig. b, the two forces are resolved into components. By writing equilibrium equations in terms of the components A_x, A_y, B_x, and B_y, show that the two forces A and B are equal in magnitude, opposite in direction, and directed along the line between their points of application.

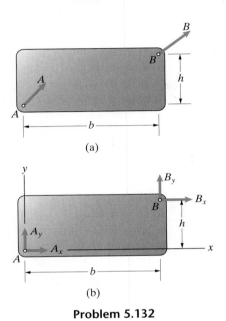

(a)

(b)

Problem 5.132

5.133 An object in equilibrium is subjected to three forces whose points of application lie on a straight line. Prove that the forces are coplanar.

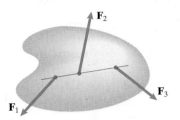

Problem 5.133

Review Problems

5.134 The suspended cable weighs 12 lb.

(a) Draw the free-body diagram of the cable. (The tensions in the cable at A and B are *not* equal.)

(b) Determine the tensions in the cable at A and B.

(c) What is the tension in the cable at its lowest point?

Problem 5.134

5.135 Determine the reactions at the fixed support.

Problem 5.135

5.136 (a) Draw the free-body diagram of the 50-lb plate, and explain why it is statically indeterminate.

(b) Determine as many of the reactions at A and B as possible.

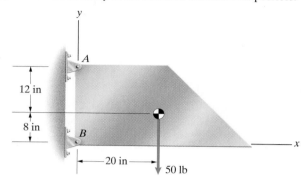

Problem 5.136

5.137 The mass of the truck is 4000 kg. Its wheels are locked, and the tension in its cable is $T = 10$ kN.

(a) Draw the free-body diagram of the truck.

(b) Determine the normal forces exerted on the truck's wheels at A and B by the road.

Problem 5.137

5.138 Assume that the force exerted on the head of the nail by the hammer is vertical, and neglect the hammer's weight.

(a) Draw the free-body diagram of the hammer.

(b) If $F = 10$ lb, what are the magnitudes of the force exerted on the nail by the hammer and the normal and friction forces exerted on the floor by the hammer?

Problem 5.138

5.139 The spring constant is $k = 9600$ N/m and the unstretched length of the spring is 30 mm. Treat the bolt at A as a pin support and assume that the surface at C is smooth. Determine the reactions at A and the normal force at C.

5.140 The engineer designing the release mechanism wants the normal force exerted at C to be 120 N. If the unstretched length of the spring is 30 mm, what is the necessary value of the spring constant k?

Problems 5.139/5.140

5.141 The truss supports a 90-kg suspended object. What are the reactions at the supports A and B?

Problem 5.141

5.142 The trailer is parked on a 15° slope. Its wheels are free to turn. The hitch *H* behaves like a pin support. Determine the reactions at *A* and *H*.

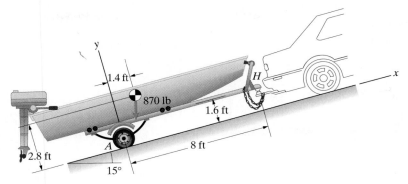

Problem 5.142

5.143 To determine the location of the point where the weight of a car acts (the *center of mass*), an engineer places the car on scales and measures the normal reactions at the wheels for two values of α, obtaining the following results.

α	A_y (kN)	*B* (kN)
10°	10.134	4.357
20°	10.150	3.677

What are the distances *b* and *h*?

Problem 5.143

5.144 The bar is attached by pin supports to collars that slide on the two fixed bars. Its mass is 10 kg, it is 1 m in length, and its weight acts at its midpoint. Neglect friction and the masses of the collars. The spring is unstretched when the bar is vertical (α = 0), and the spring constant is *k* = 100 N/m. Determine the values of α in the range 0 ≤ α ≤ 60° at which the bar is in equilibrium.

Problem 5.144

5.145 With each of the devices shown you can support a load R by applying a force F. They are called levers of the first, second, and third class.

(a) The ratio R/F is called the *mechanical advantage*. Determine the mechanical advantage of each lever.

(b) Determine the magnitude of the reaction at A for each lever. (Express your answers in terms of F.)

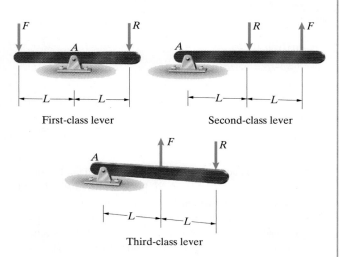

First-class lever   Second-class lever

Third-class lever

Problem 5.145

5.146 The force exerted by the weight of the horizontal rectangular plate is 800 N. The weight of the rectangular plate acts at its midpoint. If you represent the reactions exerted on the plate by the three cables by a single equivalent force, what is the force, and where does its line of action intersect the plate?

Problem 5.146

5.147 The 20-kg mass is suspended by cables attached to three vertical 2-m posts. Point A is at $(0, 1.2, 0)$ m. Determine the reactions at the fixed support at E.

5.148 In Problem 5.147, the fixed support of each vertical post will safely support a couple of 800 N-m magnitude. Based on this criterion, what is the maximum safe value of the suspended mass?

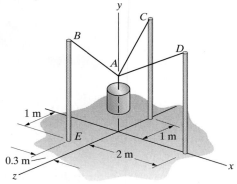

Problems 5.147/148

5.149 The 80-lb bar is supported by a ball and socket support at A, the smooth wall it leans against, and the cable BC. The weight of the bar acts at its midpoint.

(a) Draw the free-body diagram of the bar.

(b) Determine the tension in cable BC and the reactions at A.

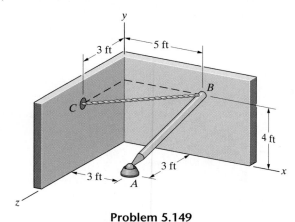

Problem 5.149

5.150 The horizontal bar of weight W is supported by a roller support at A and the cable BC. Use the fact that the bar is a three-force member to determine the angle α, the tension in the cable, and the magnitude of the reaction at A.

Problem 5.150

Design Project 1 The traditional wheelbarrow shown is designed to transport a load W while being supported by an upward force F applied to the handles by the user. (a) Use statics to analyze the effects of a range of choices of the dimensions a and b on the size of load that could be carried. Also consider the implications of these dimensions on the wheelbarrow's ease and practicality of use. (b) Suggest a different design for this classic device that achieves the same function. Use statics to compare your design to the wheelbarrow with respect to load-carrying ability and ease of use.

Design Project 2 The figure shows an example of the popular devices called "mobiles," which were introduced as an art form by American artist Alexander Calder (1898–1976). Suppose that you want to design a mobile representing the solar system, and have chosen colored spheres to represent the planets. The masses of the spheres that represent Mercury, Venus, Earth, Mars, Jupiter, Saturn, Uranus, Neptune, and Pluto are 10 g, 25 g, 25 g, 10 g, 50 g, 40 g, 40 g, 40 g, and 10 g. Assume that the cross bars and string you use are of negligible mass. Design your mobile so that the planets are in their correct order relative to the sun. Write a brief report including a drawing of your design and the analysis proving that your mobile is balanced.

Design Project 3 The bed of the dump truck (Fig. a) is raised by two tandem hydraulic cylinders AB (Fig. b). The mass of the truck's bed and load is 16,000 kg and its weight acts at point G. (Assume that the position of point G *relative to the bed* does not change when the bed is raised.)

(a) Draw a graph of the magnitude of the total force the hydraulic cylinders must exert to support the stationary bed for values of the angle α from zero to 30°.

(b) Consider other choices for the locations of the attachment points A and B that appear to be feasible and investigate how your choices affect the magnitude of the total force the hydraulic cylinders must exert as α varies from zero to 30°. Also compare the costs of your choices of the attachment points to the choices shown in Fig. a, assuming that the cost of the hydraulic cylinders is proportional to the product of the maximum force they must exert as α varies from zero to 30° and their length when $\alpha = 30°$.

(c) Write a brief report presenting your investigations and making a recommendation for the locations of points A and B.

(a)

(b)

Structures in Equilibrium

In engineering, the term *structure* can refer to any object that has the capacity to support and exert loads. In this chapter we consider structures composed of interconnected parts, or *members*. To design such a structure, or to determine whether an existing one is adequate, it is necessary to determine the forces and couples acting on the structure as a whole as well as on its individual members. We first demonstrate how this is done for the structures called trusses, which are composed entirely of two-force members. The familiar frameworks of steel members that support some highway bridges are trusses. We then consider other structures, called *frames* if they are designed to remain stationary and support loads and *machines* if they are designed to move and exert loads.

◄ The Neolithic engineers who built Stonehenge set an example for the design of enduring structures. In this chapter we describe techniques for determining the forces and couples acting on individual members of structures.

6.1 Trusses

BACKGROUND

We can explain the nature of truss structures such as the beams supporting a house (Fig. 6.1) by starting with very simple examples. Suppose we pin three bars together at their ends to form a triangle. If we add supports as shown in Fig. 6.2a, we obtain a structure that will support a load F. We can construct more elaborate structures by adding more triangles (Figs. 6.2b and c). The bars are the members of these structures, and the places where the bars are pinned together are called the *joints*. Even though these examples are quite simple, you can see that Fig. 6.2c, which is called a Warren truss, begins to resemble the structures used to support bridges and the roofs of houses (Fig. 6.3). If these structures are supported and loaded at their joints and we neglect the weights of the bars, each bar is a two-force member. We call such a structure a *truss*.

We draw the free-body diagram of a member of a truss in Fig. 6.4a. Because it is a two-force member, the forces at the ends, which are the sums of the forces exerted on the member at its joints, must be equal in magnitude, opposite in direction, and directed along the line between the joints. We call the force T the *axial force* in the member. When T is positive in the direction shown (that is, when the forces are directed away from each other), the member is in *tension*. When the forces are directed toward each other, the member is in *compression*.

In Fig. 6.4b, we "cut" the member by a plane and draw the free-body diagram of the part of the member on one side of the plane. We represent the system of internal forces and moments exerted by the part not included in the free-body diagram by a force **F** acting at the point P where the plane intersects

Figure 6.1
A typical house is supported by trusses made of wood beams.

(a)

(b)

(c)

Figure 6.2
Making structures by pinning bars together to form triangles.

Figure 6.3
Simple examples of bridge and roof structures. (The lines represent members, the circles represent joints.)

Figure 6.4
(a) Each member of a truss is a two-force member.
(b) Obtaining the free-body diagram of part of the member.
(c) The internal force is equal and opposite to the force acting at the joint, and the internal couple is zero.

the axis of the member and a couple **M**. The sum of the moments about P must equal zero, so **M** $= \mathbf{0}$. Therefore we have a two-force member, which means that **F** must be equal in magnitude and opposite in direction to the force T acting at the joint (Fig. 6.4c). The internal force is a tension or compression equal to the tension or compression exerted at the joint. Notice the similarity to a rope or cable, in which the internal force is a tension equal to the tension applied at the ends.

Although many actual structures, including "roof trusses" and "bridge trusses," consist of bars connected at the ends, very few have pinned joints. For example, a joint of a bridge truss is shown in Fig. 6.5. The ends of the members are welded at the joint and are not free to rotate. It is obvious that such a joint can exert couples on the members. Why are these structures called trusses?

The reason is that they are designed to function as trusses, meaning that they support loads primarily by subjecting their members to axial forces. They can usually be *modeled* as trusses, treating the joints as pinned connections under the assumption that couples they exert on the members are small in comparison to axial forces. When we refer to structures with riveted joints as trusses in problems, we mean that you can model them as trusses.

Figure 6.5
A joint of a bridge truss.

RESULTS

Trusses
Structures that consist of straight bars pinned at the ends and are supported and loaded only at the *joints* where the members are connected are called *trusses*. It is assumed that the weights of the members are negligible in comparison to the applied loads.

Free-Body Diagram of an Individual Member
Because *each member of a truss is a two-force member*, it is subjected only to equal and opposite axial loads. We call the force T the *axial force* in a member. When T is positive in the direction shown (that is, when the forces are directed away from each other), the member is in *tension* (T). When the forces are directed toward each other, the member is in *compression* (C).

6.2 The Method of Joints

BACKGROUND

The method of joints involves drawing free-body diagrams of the joints of a truss one by one and using the equilibrium equations to determine the axial forces in the members. Before beginning, it is usually necessary to draw a free-body diagram of the entire truss (that is, treat the truss as a single object) and determine the reactions at its supports. For example, let's consider the Warren truss in Fig. 6.6a, which has members 2 m in length and supports

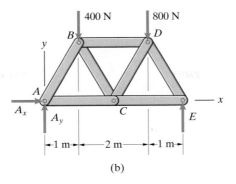

Figure 6.6
(a) A Warren truss supporting two loads.
(b) Free-body diagram of the truss.

loads at B and D. We draw its free-body diagram in Fig. 6.6b. From the equilibrium equations,

$$\Sigma F_x = A_x = 0,$$
$$\Sigma F_y = A_y + E - 400\text{ N} - 800\text{ N} = 0,$$
$$\Sigma M_{\text{point }A} = -(1\text{ m})(400\text{ N}) - (3\text{ m})(800\text{ N}) + (4\text{ m})E = 0,$$

we obtain the reactions $A_x = 0, A_y = 500$ N, and $E = 700$ N.

Our next step is to choose a joint and draw its free-body diagram. In Fig. 6.7a, we isolate joint A by cutting members AB and AC. The terms T_{AB} and T_{AC} are the axial forces in members AB and AC, respectively. Although the directions of the arrows representing the unknown axial forces can be chosen arbitrarily, notice that we have chosen them so that a member is in tension if we obtain a positive value for the axial force. Consistently choosing the directions in this way helps avoid errors.

The equilibrium equations for joint A are

$$\Sigma F_x = T_{AC} + T_{AB}\cos 60° = 0,$$
$$\Sigma F_y = T_{AB}\sin 60° + 500\text{ N} = 0.$$

Solving these equations, we obtain the axial forces $T_{AB} = -577$ N and $T_{AC} = 289$ N. Member AB is in compression, and member AC is in tension (Fig. 6.7b).

Although we use a realistic figure for the joint in Fig. 6.7a to help you understand the free-body diagram, in your own work you can use a simple figure showing only the forces acting on the joint (Fig. 6.7c).

We next obtain a free-body diagram of joint B by cutting members AB, BC, and BD (Fig. 6.8a). From the equilibrium equations for joint B,

$$\Sigma F_x = T_{BD} + T_{BC}\cos 60° + 577\cos 60°\text{ N} = 0,$$
$$\Sigma F_y = -400\text{ N} + 577\sin 60°\text{ N} - T_{BC}\sin 60° = 0,$$

we obtain $T_{BC} = 115$ N and $T_{BD} = -346$ N. Member BC is in tension, and member BD is in compression (Fig. 6.8b). By continuing to draw free-body diagrams of the joints, we can determine the axial forces in all of the members.

In two dimensions, you can obtain only two independent equilibrium equations from the free-body diagram of a joint. Summing the moments about a point does not result in an additional independent equation because the forces are concurrent. Therefore when applying the method of joints, you should choose joints to analyze that are subjected to no more than two unknown forces. In our

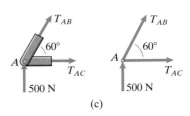

Figure 6.7
(a) Obtaining the free-body diagram of joint A.
(b) The axial forces on members AB and AC.
(c) Realistic and simple free-body diagrams of joint A.

Figure 6.8
(a) Obtaining the free-body diagram of joint B.
(b) Axial forces in members BD and BC.

Figure 6.9
(a) A joint with two collinear members and no load.
(b) Free-body diagram of the joint.

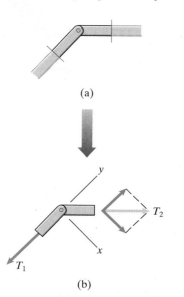

Figure 6.10
(a) A joint with two noncollinear members and no load.
(b) Free-body diagram of the joint.

example, we analyzed joint A first because it was subjected to the known reaction exerted by the pin support and two unknown forces, the axial forces T_{AB} and T_{AC} (Fig. 6.7a). We could then analyze joint B because it was subjected to two known forces and two unknown forces, T_{BC} and T_{BD} (Fig. 6.8a). If we had attempted to analyze joint B first, there would have been three unknown forces.

When you determine the axial forces in the members of a truss, your task will often be simpler if you are familiar with three particular types of joints.

- **Truss joints with two collinear members and no load** (Fig. 6.9). The sum of the forces must equal zero, $T_1 = T_2$. The axial forces are equal.
- **Truss joints with two noncollinear members and no load** (Fig. 6.10). Because the sum of the forces in the x direction must equal zero, $T_2 = 0$. Therefore T_1 must also equal zero. The axial forces are zero.
- **Truss joints with three members, two of which are collinear, and no load** (Fig. 6.11). Because the sum of the forces in the x direction must equal zero, $T_3 = 0$. The sum of the forces in the y direction must equal zero, so $T_1 = T_2$. The axial forces in the collinear members are equal, and the axial force in the third member is zero.

Figure 6.11
(a) A joint with three members, two of which are collinear, and no load.
(b) Free-body diagram of the joint.

RESULTS

Method of Joints

Before beginning, it is usually necessary to draw the free-body diagram of the entire truss considered as a single object and apply the equilibrium equations to determine the reactions at the supports.

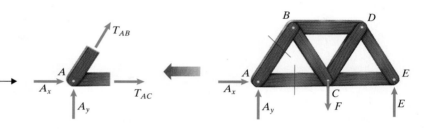

Isolate an individual joint by passing planes through the connected members. Complete the free-body diagram by showing the axial forces in the members. Apply the equilibrium equations $\Sigma F_x = 0$ and $\Sigma F_y = 0$ to the free-body diagram of the joint. Repeat this process for other joints until the desired axial loads have been determined.

Special Joints

If a joint consists of two collinear members and no external load is applied to the joint, the axial forces in the members are equal.

If a joint consists of two noncollinear members and no external load is applied to the joint, there is no axial force in either member.

If a joint consists of three members, two of which are collinear, and no external load is applied to the joint, the axial forces in the collinear members are equal and the axial force in the third member is zero.

Active Example 6.1	The Method of Joints (▶ *Related Problem 6.1*)

Determine the axial forces in members AB and AC of the truss.

Strategy

We will first draw a free-body diagram of the entire truss, treating it as a single object, and determine the reactions at the supports. Then we can determine the axial forces in members AB and AC by drawing the free-body diagram of joint A.

Solution

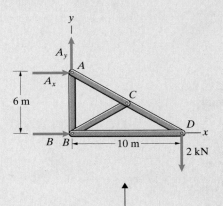

$\Sigma F_x = A_x + B = 0,$

$\Sigma F_y = A_y - 2 \text{ kN} = 0,$

$\Sigma M_{\text{point } B} = -(6 \text{ m})A_x - (10 \text{ m})(2 \text{ kN}) = 0.$

Solving yields $A_x = -3.33$ kN, $A_y = 2$ kN, and $B = 3.33$ kN.

> Draw the free-body diagram of the entire truss and apply the equilibrium equations.

The angle $\alpha = \arctan(5/3) = 59.0°$.

$\Sigma F_x = T_{AC} \sin\alpha - 3.33 \text{ kN} = 0,$

$\Sigma F_y = 2 \text{ kN} - T_{AB} - T_{AC} \cos\alpha = 0.$

Solving yields $T_{AB} = 0$ and $T_{AC} = 3.89$ kN. The axial force in member AB is zero and the axial force in member AC is 3.89 kN in tension, which we write as

    AB: zero, AC: 3.89 kN (T).

> Draw the free-body diagram of joint A and apply the equilibrium equations.

Practice Problem Determine the axial forces in members *BC* and *BD* of the truss. In doing so, use the fact that it is already known from the analysis of joint *A* that the axial force in member *AB* is zero.

Answer: *BC*: zero, *BD*: 3.33 kN (C).

Example 6.2   A Bridge Truss (▶ *Related Problem 6.31*)

The loads a bridge structure must support and pin supports where the structure is to be attached are shown in Fig. 1. Assigned to design the structure, a civil engineering student proposes the structure shown in Fig. 2. What are the axial forces in the members?

Strategy

The vertical members *AG*, *BH*, *CI*, *DJ*, and *EK* are subjected to compressive forces of magnitude *F*. Because of the symmetry of the structure, we can determine the axial loads in the remaining members by analyzing joints *C* and *B*.

(1)   (2)

Solution

We will leave it as an exercise to show by drawing the free-body diagram of joint *C* that members *BC* and *CD* are subjected to equal compressive loads of magnitude 1.93*F*. We draw the free-body diagram of joint *B* in Fig. a, where $T_{BC} = -1.93F$.

From the equilibrium equations

$$\Sigma F_x = -T_{AB} \cos \alpha + T_{BC} \cos 15° = 0,$$

$$\Sigma F_y = -T_{AB} \sin \alpha + T_{BC} \sin 15° - F = 0,$$

we obtain $T_{AB} = -2.39F$ and $\alpha = 38.8°$. By symmetry, $T_{DE} = T_{AB}$. The axial forces in the members are shown in the table.

(a) Free-body diagram of joint *B*.

Axial forces in the members of the bridge structure

Members	Axial Force
AG, BH, CI, DJ, EK	F (C)
AB, DE	2.39F (C)
BC, CD	1.93F (C)

Problems

▶ **6.1** In Active Example 6.1, suppose that in addition to the 2-kN downward force acting at point D, a 2-kN downward force acts at point C. Draw a sketch of the truss showing the new loading. Determine the axial forces in members AB and AC of the truss.

6.2 Determine the axial forces in the members of the truss and indicate whether they are in tension (T) or compression (C).

Problem 6.2

6.3 Member AB of the truss is subjected to a 1000-lb tensile force. Determine the weight W and the axial force in member AC.

Problem 6.3

6.4 Determine the axial forces in members BC and CD of the truss.

Problem 6.4

6.5 Each suspended weight has mass $m = 20$ kg. Determine the axial forces in the members of the truss and indicate whether they are in tension (T) or compression (C).

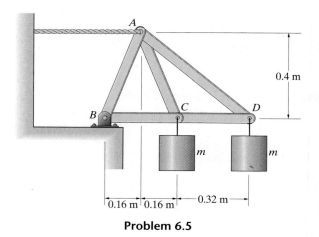

Problem 6.5

6.6 Determine the largest tensile and compressive forces that occur in the members of the truss, and indicate the members in which they occur if

(a) the dimension $h = 0.1$ m;

(b) the dimension $h = 0.5$ m.

Observe how a simple change in design affects the maximum axial loads.

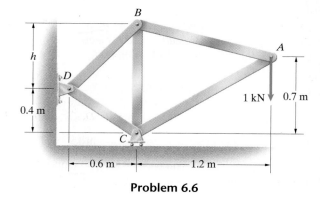

Problem 6.6

6.7 This steel truss bridge is in the Gallatin National Forest south of Bozeman, Montana. Suppose that one of the tandem trusses supporting the bridge is loaded as shown. Determine the axial forces in members *AB*, *BC*, *BD*, and *BE*.

6.8 Determine the largest tensile and compressive forces that occur in the members of the bridge truss, and indicate the members in which they occur.

Problems 6.7/6.8

6.9 The trusses supporting the bridge in Problems 6.7 and 6.8 are called Pratt trusses. Suppose that the bridge designers had decided to use the truss shown instead, which is called a Howe truss. Determine the largest tensile and compressive forces that occur in the members, and indicate the members in which they occur. Compare your answers to the answers to Problem 6.8.

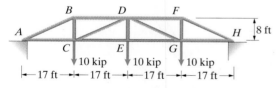

Problem 6.9

6.10 Determine the axial forces in members *BD*, *CD*, and *CE* of the truss.

Problem 6.10

6.11 The loads $F_1 = F_2 = 8$ kN. Determine the axial forces in members *BD*, *BE*, and *BG*.

Problem 6.11

6.12 Determine the largest tensile and compressive forces that occur in the members of the truss, and indicate the members in which they occur if

(a) the dimension $h = 5$ in;

(b) the dimension $h = 10$ in.

Observe how a simple change in design affects the maximum axial loads.

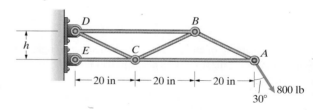

Problem 6.12

6.13 The truss supports loads at *C* and *E*. If $F = 3$ kN, what are the axial forces in members *BC* and *BE*?

Problem 6.13

6.14 If you don't want the members of the truss to be subjected to an axial load (tension or compression) greater than 20 kN, what is the largest acceptable magnitude of the downward force F?

Problem 6.14

6.15 The truss is a preliminary design for a structure to attach one end of a stretcher to a rescue helicopter. Based on dynamic simulations, the design engineer estimates that the downward forces the stretcher will exert will be no greater than 1.6 kN at A and at B. What are the resulting axial forces in members CF, DF, and FG?

6.16 Upon learning of an upgrade in the helicopter's engine, the engineer designing the truss does new simulations and concludes that the downward forces the stretcher will exert at A and at B may be as large as 1.8 kN. What are the resulting axial forces in members DE, DF, and DG?

Problems 6.15/6.16

6.17 Determine the axial forces in the members in terms of the weight W.

Problem 6.17

6.18 The lengths of the members of the truss are shown. The mass of the suspended crate is 900 kg. Determine the axial forces in the members.

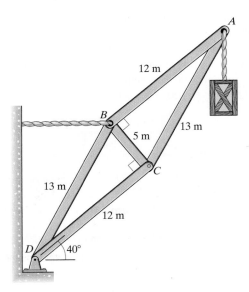

Problem 6.18

6.19 The loads F_1 = 600 lb and F_2 = 300 lb. Determine the axial forces in members AE, BD, and CD.

6.20 The loads F_1 = 450 lb and F_2 = 150 lb. Determine the axial forces in members AB, AC, and BC.

Problems 6.19/6.20

6.21 Determine the axial forces in members BC, CD, and CE of the truss.

Problem 6.21

6.22 The Warren truss supporting the walkway is designed to support vertical 50-kN loads at B, D, F, and H. If the truss is subjected to these loads, what are the resulting axial forces in members BC, CD, and CE?

6.23 For the Warren truss in Problem 6.22, determine the axial forces in members DF, EF, and FG.

Problems 6.22/6.23

6.24 The Pratt bridge truss supports five forces (F = 300 kN). The dimension L = 8 m. Determine the axial forces in members BC, BI, and BJ.

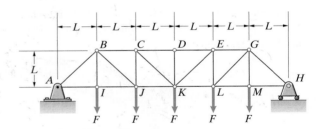

Problem 6.24

6.25 For the roof truss shown, determine the axial forces in members AD, BD, DE, and DG. Model the supports at A and I as roller supports.

Problem 6.25

6.26 The Howe truss helps support a roof. Model the supports at A and G as roller supports. Determine the axial forces in members AB, BC, and CD.

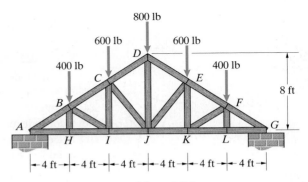

Problem 6.26

6.27 The plane truss forms part of the supports of a crane on an offshore oil platform. The crane exerts vertical 75-kN forces on the truss at B, C, and D. You can model the support at A as a pin support and model the support at E as a roller support that can exert a force normal to the dashed line but cannot exert a force parallel to it. The angle $\alpha = 45°$. Determine the axial forces in the members of the truss.

Problem 6.27

6.28 (a) Design a truss attached to the supports A and B that supports the loads applied at points C and D. (b) Determine the axial forces in the members of the truss you designed in (a).

Problem 6.28

6.29 (a) Design a truss attached to the supports A and B that goes over the obstacle and supports the load applied at C.

(b) Determine the axial forces in the members of the truss you designed in (a).

Problem 6.29

6.30 Suppose that you want to design a truss supported at A and B (Fig. a) to support a 3-kN downward load at C. The simplest design (Fig. b) subjects member AC to a 5-kN tensile force. Redesign the truss so that the largest tensile force is less than 3 kN.

Problem 6.30

▶ **6.31** The bridge structure shown in Example 6.2 can be given a higher arch by increasing the 15° angles to 20°. If this is done, what are the axial forces in members AB, BC, CD, and DE?

6.3 The Method of Sections

BACKGROUND

When we need to know the axial forces only in certain members of a truss, we often can determine them more quickly using the method of sections than using the method of joints. For example, let's reconsider the Warren truss we used to introduce the method of joints (Fig. 6.12a). It supports loads at B and D, and each member is 2 m in length. Suppose that we need to determine only the axial force in member BC.

(a)

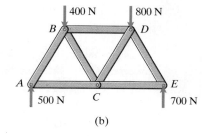

(b)

Figure 6.12
(a) A Warren truss supporting two loads.
(b) Free-body diagram of the truss, showing the reactions at the supports.

Just as in the method of joints, we begin by drawing a free-body diagram of the entire truss and determining the reactions at the supports. The results of this step are shown in Fig. 6.12b. Our next step is to cut the members AC, BC, and BD to obtain a free-body diagram of a part, or *section*, of the truss (Fig. 6.13). Summing moments about point B, the equilibrium equations for the section are

$$\Sigma F_x = T_{AC} + T_{BD} + T_{BC} \cos 60° = 0,$$
$$\Sigma F_y = 500 \text{ N} - 400 \text{ N} - T_{BC} \sin 60° = 0,$$
$$\Sigma M_{\text{point } B} = (2 \sin 60° \text{ m})T_{AC} - (2 \cos 60° \text{ m})(500 \text{ N}) = 0.$$

Solving them, we obtain $T_{AC} = 289$ N, $T_{BC} = 115$ N, and $T_{BD} = -346$ N.

Notice how similar this method is to the method of joints. Both methods involve cutting members to obtain free-body diagrams of parts of a truss. In the method of joints, we move from joint to joint, drawing free-body diagrams of the joints and determining the axial forces in the members as we go. In the method of sections, we try to obtain a single free-body diagram that allows us to determine the axial forces in specific members. In our example, we obtained a free-body diagram by cutting three members, including the one (member BC) whose axial force we wanted to determine.

In contrast to the free-body diagrams of joints, the forces on the free-body diagrams used in the method of sections are not usually concurrent, and as in our example, we can obtain three independent equilibrium equations. Although there are exceptions, it is usually necessary to choose a section that requires cutting no more than three members, or there will be more unknown axial forces than equilibrium equations.

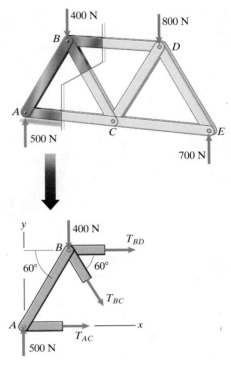

Figure 6.13
Obtaining a free-body diagram of a section of the truss.

The Method of Sections

When the axial forces in particular members of a truss must be determined, the method of sections can often provide the needed results more efficiently than the method of joints.

Before beginning, it is usually advantageous to draw the free-body diagram of the entire truss considered as a single object and apply the equilibrium equations to determine the reactions at the supports.

Pass planes through enough members to isolate a part, or *section*, of the truss. In doing so, attempt to pass planes through members whose axial forces are to be determined. Complete the free-body diagram of the section by showing the axial forces in the members. Apply the equilibrium equations to the free-body diagram of the section.

Active Example 6.3   **The Method of Sections** (▶ *Related Problem 6.32*)

The horizontal members of the truss are each 1 m in length. Determine the axial forces in members *CD*, *CJ*, and *IJ*.

Strategy

By passing planes through members *CD*, *CJ*, and *IJ*, we will obtain a section from which we can obtain the desired axial forces.

Solution

Pass planes through members *CD*, *CJ*, and *IJ* and draw the free-body diagram of the section.

$$\Sigma F_x = -T_{CD} - T_{CJ} \cos 45° - T_{IJ} = 0,$$

$$\Sigma F_y = T_{CJ} \sin 45° - 100 \text{ kN} = 0,$$

$$\Sigma M_{\text{point } J} = (1 \text{ m})T_{CD} - (3 \text{ m})(100 \text{ kN}) = 0.$$

Solving yields $T_{CD} = 300$ kN, $T_{CJ} = 141$ kN, and $T_{IJ} = -400$ kN. The axial loads are *CD*: 300 kN (T), *CJ*: 141 kN (T), *IJ*: 400 kN (C).

Apply the equilibrium equations.

Practice Problem Use the method of sections to determine the axial forces in members *DE*, *DK*, and *JK* of the truss.

Answer: *DE*: 200 kN (T), *DK*: 141 kN (T), *JK*: 300 kN (C).

| Example 6.4 | Choosing an Appropriate Section (▶ *Related Problem 6.33*) |

Determine the axial forces in members *DG* and *BE* of the truss.

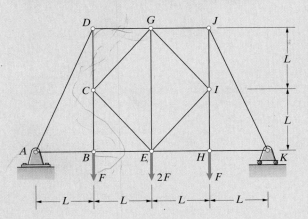

Strategy
We can't obtain a section that involves cutting members *DG* and *BE* without cutting more than three members. However, cutting members *DG*, *BE*, *CD*, and *BC* results in a section with which we can determine the axial forces in members *DG* and *BE*.

Solution
Determine the Reactions at the Supports We draw the free-body diagram of the entire truss in Fig. a. From the equilibrium equations,

$$\Sigma F_x = A_x = 0,$$

$$\Sigma F_y = A_y + K - F - 2F - F = 0,$$

$$\Sigma M_{\text{point } A} = -LF - (2L)(2F) - (3L)F + (4L)K = 0,$$

we obtain the reactions $A_x = 0$, $A_y = 2F$, and $K = 2F$.

Choose a Section
In Fig. b, we obtain a section by cutting members *DG*, *CD*, *BC*, and *BE*. Because the lines of action of T_{BE}, T_{BC}, and T_{CD} pass through point *B*, we can determine T_{DG} by summing moments about *B*:

$$\Sigma M_{\text{point } B} = -L(2F) - (2L)T_{DG} = 0.$$

The axial force $T_{DG} = -F$. Then, from the equilibrium equation

$$\Sigma F_x = T_{DG} + T_{BE} = 0,$$

we see that $T_{BE} = -T_{DG} = F$. Member *DG* is in compression, and member *BE* is in tension.

Critical Thinking
This is a clever example, but not one that is typical of problems faced in practice. The section used to solve it might not be obvious even to a person with experience analyzing structures. Notice that the free-body diagram in Fig. b is statically indeterminate, although it can be used to determine the axial forces in members *DG* and *BE*.

(a) Free-body diagram of the entire truss.

(b) A section of the truss obtained by passing planes through members *DG*, *CD*, *BC*, and *BE*.

Problems

▶ **6.32** In Active Example 6.3, use the method of sections to determine the axial forces in members *BC*, *BI*, and *HI*.

▶ **6.33** In Example 6.4, obtain a section of the truss by passing planes through members *BE*, *CE*, *CG*, and *DG*. Using the fact that the axial forces in members *DG* and *BE* have already been determined, use your section to determine the axial forces in members *CE* and *CG*.

6.34 The truss supports a 100-kN load at *J*. The horizontal members are each 1 m in length.

(a) Use the method of joints to determine the axial force in member *DG*.

(b) Use the method of sections to determine the axial force in member *DG*.

6.35 The horizontal members are each 1 m in length. Use the method of sections to determine the axial forces in members *BC*, *CF*, and *FG*.

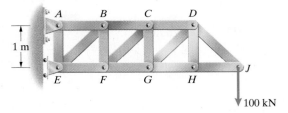

Problems 6.34/6.35

6.36 Use the method of sections to determine the axial forces in members *AB*, *BC*, and *CE*.

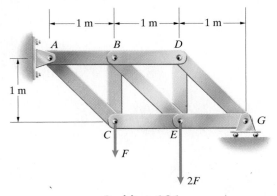

Problem 6.36

6.37 Use the method of sections to determine the axial forces in members *DF*, *EF*, and *EG*.

Problem 6.37

6.38 The Pratt bridge truss is loaded as shown. Use the method of sections to determine the axial forces in members *BD*, *BE*, and *CE*.

Problem 6.38

6.39 The Howe bridge truss is loaded as shown. Use the method of sections to determine the axial forces in members *BD*, *CD*, and *CE*.

6.40 For the Howe bridge truss, use the method of sections to determine the axial forces in members *DF*, *DG*, and *EG*.

Problems 6.39/6.40

6.41 The Pratt bridge truss supports five forces *F* = 340 kN. The dimension *L* = 8 m. Use the method of sections to determine the axial force in member *JK*.

6.42 For the Pratt bridge truss in Problem 6.41, use the method of sections to determine the axial force in member *EK*.

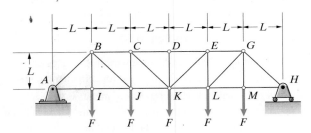

Problems 6.41/6.42

6.43 The walkway exerts vertical 50-kN loads on the Warren truss at *B*, *D*, *F*, and *H*. Use the method of sections to determine the axial force in member *CE*.

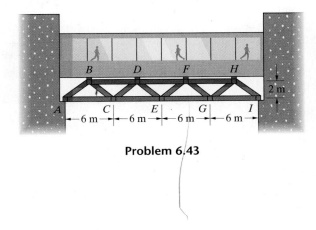

Problem 6.43

6.44 Use the method of sections to determine the axial forces in members *AC*, *BC*, and *BD*.

Problem 6.44

6.45 Use the method of sections to determine the axial forces in members *FH*, *GH*, and *GI*.

6.46 Use the method of sections to determine the axial forces in members *DF*, *DG*, and *EG*.

Problems 6.45/6.46

6.47 The Howe truss helps support a roof. Model the supports at *A* and *G* as roller supports.

(a) Use the method of joints to determine the axial force in member *BI*.

(b) Use the method of sections to determine the axial force in member *BI*.

6.48 Use the method of sections to determine the axial force in member *EJ*.

Problems 6.47/6.48

6.49 Use the method of sections to determine the axial forces in members *CE*, *DE*, and *DF*.

Problem 6.49

6.50 For the bridge truss shown, use the method of sections to determine the axial forces in members *CE*, *CF*, and *DF*.

Problem 6.50

6.51 The load $F = 20$ kN and the dimension $L = 2$ m. Use the method of sections to determine the axial force in member *HK*.

Strategy: Obtain a section by cutting members *HK*, *HI*, *IJ*, and *JM*. You can determine the axial forces in members *HK* and *JM* even though the resulting free-body diagram is statically indeterminate.

Problem 6.51

6.52 The weight of the bucket is $W = 1000$ lb. The cable passes over pulleys at A and D.

(a) Determine the axial forces in members FG and HI.

(b) By drawing free-body diagrams of sections, explain why the axial forces in members FG and HI are equal.

6.53 The weight of the bucket is $W = 1000$ lb. The cable passes over pulleys at A and D. Determine the axial forces in members IK and JL.

6.54 The truss supports loads at N, P, and R. Determine the axial forces in members IL and KM.

6.55 Determine the axial forces in members HJ and GI.

6.56 By drawing free-body diagrams of sections, explain why the axial forces in members DE, FG, and HI are zero.

Problems 6.52/6.53

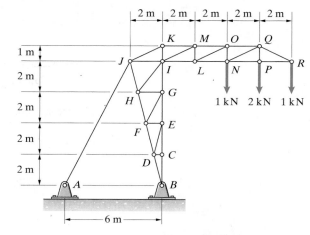

Problems 6.54–6.56

6.4 Space Trusses

BACKGROUND

We can form a simple three-dimensional structure by connecting six bars at their ends to obtain a tetrahedron, as shown in Fig. 6.14a. By adding members, we can obtain more elaborate structures (Figs. 6.14b and c). Three-dimensional structures such as these are called *space trusses* if they have joints that do not exert couples on the members (that is, the joints behave like ball and socket supports) and they are loaded and supported at their joints. Space trusses are analyzed by the same methods we described for two-dimensional trusses. The only difference is the need to cope with the more complicated geometry.

Consider the space truss in Fig. 6.15a. Suppose that the load $\mathbf{F} = -2\mathbf{i} - 6\mathbf{j} - \mathbf{k}$ (kN). The joints A, B, and C rest on the smooth floor. Joint

(a)

(b)

(c)

Figure 6.14
Space trusses with 6, 9, and 12 members.

(a)

(b)

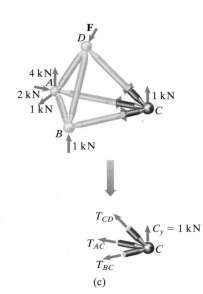

(c)

Figure 6.15
(a) A space truss supporting a load **F**.
(b) Free-body diagram of the entire truss.
(c) Obtaining the free-body diagram of
   joint C.

A is supported by the corner where the smooth walls meet, and joint C rests against the back wall. We can apply the method of joints to this truss.

First we must determine the reactions exerted by the supports (the floor and walls). We draw the free-body diagram of the entire truss in Fig. 6.15b. The corner can exert three components of force at A, the floor and wall can exert two components of force at C, and the floor can exert a normal force at B. Summing moments about A, we find that the equilibrium equations, with forces in kN and distances in m, are

$$\Sigma F_x = A_x - 2 = 0,$$

$$\Sigma F_y = A_y + B_y + C_y - 6 = 0,$$

$$\Sigma F_z = A_z + C_z - 1 = 0,$$

$$\Sigma M_{\text{point } A} = (\mathbf{r}_{AB} \times B_y\mathbf{j}) + [\mathbf{r}_{AC} \times (C_y\mathbf{j} + C_z\mathbf{k})] + (\mathbf{r}_{AD} \times \mathbf{F})$$

$$= \begin{vmatrix} \mathbf{i} & \mathbf{j} & \mathbf{k} \\ 2 & 0 & 3 \\ 0 & B_y & 0 \end{vmatrix} + \begin{vmatrix} \mathbf{i} & \mathbf{j} & \mathbf{k} \\ 4 & 0 & 0 \\ 0 & C_y & C_z \end{vmatrix} + \begin{vmatrix} \mathbf{i} & \mathbf{j} & \mathbf{k} \\ 2 & 3 & 1 \\ -2 & -6 & -1 \end{vmatrix}$$

$$= (-3B_y + 3)\mathbf{i} + (-4C_z)\mathbf{j}$$
$$+ (2B_y + 4C_y - 6)\mathbf{k} = 0.$$

Solving these equations, we obtain the reactions $A_x = 2$ kN, $A_y = 4$ kN, $A_z = 1$ kN, $B_y = 1$ kN, $C_y = 1$ kN, and $C_z = 0$.

In this example, we can determine the axial forces in members AC, BC, and CD from the free-body diagram of joint C (Fig. 6.15c). To write the equilibrium equations for the joint, we must express the three axial forces in terms of their components. Because member AC lies along the x axis, we express the force exerted on joint C by the axial force T_{AC} as the vector $-T_{AC}\mathbf{i}$. Let $\mathbf{r}_{CB}$ be the position vector from C to B:

$$\mathbf{r}_{CB} = (2 - 4)\mathbf{i} + (0 - 0)\mathbf{j} + (3 - 0)\mathbf{k} = -2\mathbf{i} + 3\mathbf{k} \text{ (m)}.$$

Dividing this vector by its magnitude to obtain a unit vector that points from C toward B yields

$$\mathbf{e}_{CB} = \frac{\mathbf{r}_{CB}}{|\mathbf{r}_{CB}|} = -0.555\mathbf{i} + 0.832\mathbf{k},$$

and we express the force exerted on joint C by the axial force T_{BC} as the vector

$$T_{BC}\,\mathbf{e}_{CB} = T_{BC}(-0.555\mathbf{i} + 0.832\mathbf{k}).$$

In the same way, we express the force exerted on joint C by the axial force T_{CD} as the vector

$$T_{CD}(-0.535\mathbf{i} + 0.802\mathbf{j} + 0.267\mathbf{k}).$$

Setting the sum of the forces on the joint equal to zero, we obtain

$$-T_{AC}\mathbf{i} + T_{BC}(-0.555\mathbf{i} + 0.832\mathbf{k})$$
$$+T_{CD}(-0.535\mathbf{i} + 0.802\mathbf{j} + 0.267\mathbf{k}) + (1\text{ kN})\mathbf{j} = 0,$$

and then get the three equilibrium equations

$$\Sigma F_x = -T_{AC} - 0.555T_{BC} - 0.535T_{CD} = 0,$$
$$\Sigma F_y = 0.802T_{CD} + 1\text{ kN} = 0,$$
$$\Sigma F_z = 0.832T_{BC} + 0.267T_{CD} = 0.$$

Solving these equations, the axial forces are $T_{AC} = 0.444$ kN, $T_{BC} = 0.401$ kN, and $T_{CD} = -1.247$ kN. Members AC and BC are in tension, and member CD is in compression. By continuing to draw free-body diagrams of the joints, we can determine the axial forces in all the members.

As our example demonstrates, three equilibrium equations can be obtained from the free-body diagram of a joint in three dimensions, so it is usually necessary to choose joints to analyze that are subjected to known forces and no more than three unknown forces.

RESULTS

A *space truss* is a truss whose members are not coplanar. Axial forces in the members of a statically determinate space truss can be determined by applying the method of joints.

Before beginning, it is usually necessary to draw the free-body diagram of the entire truss considered as a single object and apply the equilibrium equations to determine the reactions at the supports.

Isolate an individual joint by passing planes through the connected members. Complete the free-body diagram by showing the axial forces in the members. Apply the equilibrium equation $\Sigma\mathbf{F} = \mathbf{0}$ to the free-body diagram of the joint. Repeat this process for other joints until the desired axial loads have been determined.

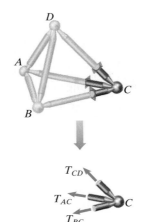

Active Example 6.5 **Space Truss** (▶ *Related Problem 6.57*)

The space truss has roller supports at B, C, and D and supports a vertical 1200-lb load at A. Determine the axial forces in members AD, BD, and CD.

Strategy

We will first draw a free-body diagram of the entire truss, treating it as a single object, and determine the reactions at the supports. Then we can determine the axial forces in members AD, BD, and CD by drawing the free-body diagram of joint D.

Solution

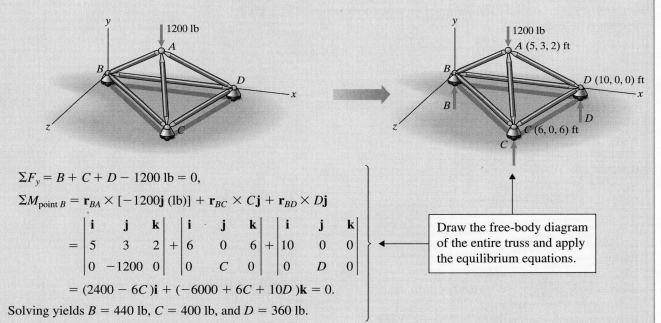

$$\Sigma F_y = B + C + D - 1200 \text{ lb} = 0,$$

$$\Sigma M_{\text{point } B} = \mathbf{r}_{BA} \times [-1200\mathbf{j} \text{ (lb)}] + \mathbf{r}_{BC} \times C\mathbf{j} + \mathbf{r}_{BD} \times D\mathbf{j}$$

$$= \begin{vmatrix} \mathbf{i} & \mathbf{j} & \mathbf{k} \\ 5 & 3 & 2 \\ 0 & -1200 & 0 \end{vmatrix} + \begin{vmatrix} \mathbf{i} & \mathbf{j} & \mathbf{k} \\ 6 & 0 & 6 \\ 0 & C & 0 \end{vmatrix} + \begin{vmatrix} \mathbf{i} & \mathbf{j} & \mathbf{k} \\ 10 & 0 & 0 \\ 0 & D & 0 \end{vmatrix}$$

$$= (2400 - 6C)\mathbf{i} + (-6000 + 6C + 10D)\mathbf{k} = 0.$$

Solving yields $B = 440$ lb, $C = 400$ lb, and $D = 360$ lb.

Draw the free-body diagram of the entire truss and apply the equilibrium equations.

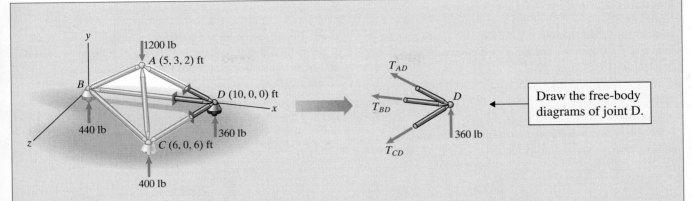

Draw the free-body diagrams of joint D.

$\mathbf{r}_{DA} = -5\mathbf{i} + 3\mathbf{j} + 2\mathbf{k} \text{ (ft)}.$

$\mathbf{e}_{DA} = \dfrac{\mathbf{r}_{DA}}{|\mathbf{r}_{DA}|} = -0.811\mathbf{i} + 0.487\mathbf{j} + 0.324\mathbf{k}.$

$T_{AD}\mathbf{e}_{DA} = T_{AD}(-0.811\mathbf{i} + 0.487\mathbf{j} + 0.324\mathbf{k}),$

$T_{BD}\mathbf{e}_{DB} = -T_{BD}\mathbf{i},$

$T_{CD}\mathbf{e}_{DC} = T_{CD}(-0.555\mathbf{i} + 0.832\mathbf{k}).$

Divide the position vector from D to A by its magnitude to obtain a unit vector $\mathbf{e}_{DA}$ that points from D toward A. Express the axial force in member AD in terms of its components by writing it as $T_{AD}\,\mathbf{e}_{DA}$. Express the axial forces in members BD and CD in terms of their components in the same way.

$T_{AD}\mathbf{e}_{DA} + T_{BD}\mathbf{e}_{DB} + T_{CD}\mathbf{e}_{DC} + (360 \text{ lb})\mathbf{j} = 0.$

The $\mathbf{i}$, $\mathbf{j}$, and $\mathbf{k}$ components of this equation must each equal zero, resulting in the three equations

$-0.811T_{AD} - T_{BD} - 0.555T_{CD} = 0,$

$0.487T_{AD} + 360 \text{ lb} = 0,$

$0.324T_{AD} + 0.832T_{CD} = 0.$

Solving yields $T_{AD} = -740$ lb, $T_{BD} = 440$ lb, and $T_{CD} = 288$ lb. The axial forces are AD: 740 lb (C), BD: 440 lb (T), CD: 288 lb (T).

Apply equilibrium.

Practice Problem Determine the axial forces in members AB and AC of the truss.

Answer: AB: 904 lb (C), AC: 680 lb (C).

Problems

▶ **6.57** In Active Example 6.5, draw the free-body diagram of joint B of the space truss and use it to determine the axial forces in members AB, BC, and BD.

6.58 The space truss supports a vertical 10-kN load at D. The reactions at the supports at joints A, B, and C are shown. What are the axial forces in members AD, BD, and CD?

6.59 The reactions at the supports at joints A, B, and C are shown. What are the axial forces in members AB, AC, and AD?

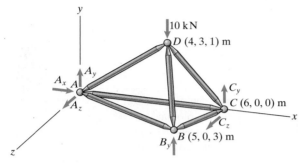

Problems 6.58/6.59

6.60 The space truss supports a vertical load F at A. Each member is of length L, and the truss rests on the horizontal surface on roller supports at B, C, and D. Determine the axial forces in members AB, AC, and AD.

6.61 For the truss in Problem 6.60, determine the axial forces in members AB, BC, and BD.

Problems 6.60/6.61

6.62 The space truss has roller supports at B, C, and D and supports a vertical 800-lb load at A. What are the axial forces in members AB, AC, and AD?

Problem 6.62

6.63 The space truss shown models an airplane's landing gear. It has ball and socket supports at C, D, and E. If the force exerted at A by the wheel is $\mathbf{F} = 40\mathbf{j}$ (kN), what are the axial forces in members AB, AC, and AD?

6.64 If the force exerted at point A of the truss in Problem 6.63 is $\mathbf{F} = 10\mathbf{i} + 60\mathbf{j} + 20\mathbf{k}$ (kN), what are the axial forces in members BC, BD, and BE?

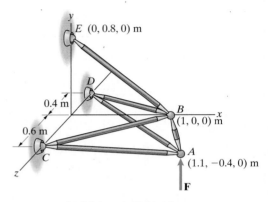

Problems 6.63/6.64

6.65 The space truss is supported by roller supports on the horizontal surface at C and D and a ball and socket support at E. The y axis points upward. The mass of the suspended object is 120 kg. The coordinates of the joints of the truss are A: (1.6, 0.4, 0) m, B: (1.0, 1.0, −0.2) m, C: (0.9, 0, 0.9) m, D: (0.9, 0, −0.6) m, and E: (0, 0.8, 0) m. Determine the axial forces in members AB, AC, and AD.

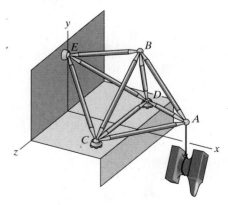

Problem 6.65

6.66 The free-body diagram of the part of the construction crane to the left of the plane is shown. The coordinates (in meters) of the joints A, B, and C are $(1.5, 1.5, 0)$, $(0, 0, 1)$, and $(0, 0, -1)$, respectively. The axial forces P_1, P_2, and P_3 are parallel to the x axis. The axial forces P_4, P_5, and P_6 point in the directions of the unit vectors

$$e_4 = 0.640i - 0.640j - 0.426k,$$
$$e_5 = 0.640i - 0.640j + 0.426k,$$
$$e_6 = 0.832i - 0.555k.$$

The total force exerted on the free-body diagram by the weight of the crane and the load it supports is $-Fj = -44j$ (kN) acting at the point $(-20, 0, 0)$ m. What is the axial force P_3?

Strategy: Use the fact that the moment about the line that passes through joints A and B equals zero.

6.67 In Problem 6.66, what are the axial forces P_1, P_4, and P_5?

Strategy: Write the equilibrium equations for the entire free-body diagram.

Problems 6.66/6.67

6.68 The mirror housing of the telescope is supported by a 6-bar space truss. The mass of the housing is 3 Mg (megagrams), and its weight acts at G. The distance from the axis of the telescope to points A, B, and C is 1 m, and the distance from the axis to points D, E, and F is 2.5 m. If the telescope axis is vertical ($\alpha = 90°$), what are the axial forces in the members of the truss?

6.69 Consider the telescope described in Problem 6.68. Determine the axial forces in the members of the truss if the angle α between the horizontal and the telescope axis is 20°.

END VIEW

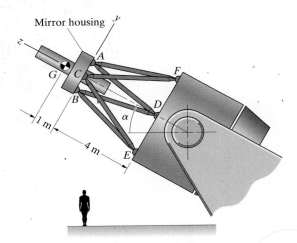

Problems 6.68/6.69

6.5 Frames and Machines

Many structures, such as the frame of a car and the human structure of bones, tendons, and muscles (Fig. 6.16), are not composed entirely of two-force members and thus cannot be modeled as trusses. In this section we consider structures of interconnected members that do not satisfy the definition of a truss. Such structures are called *frames* if they are designed to remain stationary and support loads and *machines* if they are designed to move and apply loads.

When trusses are analyzed by cutting members to obtain free-body diagrams of joints or sections, the internal forces acting at the "cuts" are simple axial forces (see Fig. 6.4). This is not generally true for frames or machines, and a different method of analysis is necessary. Instead of cutting members, we isolate entire members, or in some cases combinations of members, from the structure.

To begin analyzing a frame or machine, we draw a free-body diagram of the entire structure (that is, treat the structure as a single object) and determine the reactions at its supports. In some cases the entire structure will be statically indeterminate, but it is helpful to determine as many of the reactions as possible. We then draw free-body diagrams of individual members, or selected combinations of members, and apply the equilibrium equations to determine the forces and couples acting on them. For example, consider the stationary structure in

Figure 6.16
The internal structure of a person and a car's frame are not trusses.

Fig. 6.17. Member *BE* is a two-force member, but the other three members—*ABC*, *CD*, and *DEG*—are not. This structure is a frame. Our objective is to determine the forces on its members.

Analyzing the Entire Structure

We draw the free-body diagram of the entire frame in Fig. 6.18. It is statically indeterminate: There are four unknown reactions, A_x, A_y, G_x, and G_y, whereas we can write only three independent equilibrium equations. However, notice that the lines of action of three of the unknown reactions intersect at *A*. Summing moments about *A* yields

$$\Sigma M_{\text{point } A} = (2 \text{ m})G_x + (1 \text{ m})(8 \text{ kN}) - (3 \text{ m})(6 \text{ kN}) = 0,$$

and we obtain the reaction $G_x = 5$ kN. Then, from the equilibrium equation

$$\Sigma F_x = A_x + G_x + 8 \text{ kN} = 0,$$

we obtain the reaction $A_x = -13$ kN. Although we cannot determine A_y or G_y from the free-body diagram of the entire structure, we can do so by analyzing the individual members.

Analyzing the Members

Our next step is to draw free-body diagrams of the members. To do so, we treat the attachment of a member to another member just as if it were a support. Looked at in this way, we can think of each member as a supported object of the kind analyzed in Chapter 5. Furthermore, the forces and couples the members exert on one another are *equal in magnitude and opposite in direction*. A simple

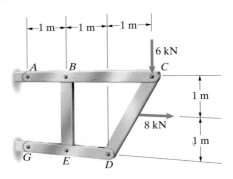

Figure 6.17
A frame supporting two loads.

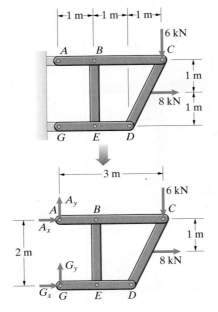

Figure 6.18
Obtaining the free-body diagram of the entire frame.

(a)   (b)

Figure 6.19
Demonstrating Newton's third law:
(a) Clasp your hands and pull on your left hand.
(b) Your hands exert equal and opposite forces.

demonstration is instructive. If you clasp your hands as shown in Fig. 6.19a and exert a force on your left hand with your right hand, your left hand exerts an equal and opposite force on your right hand (Fig. 6.19b). Similarly, if you exert a couple on your left hand, your left hand exerts an equal and opposite couple on your right hand.

In Fig. 6.20 we "disassemble" the frame and draw free-body diagrams of its members. Observe that the forces exerted on one another by the members are equal and opposite. For example, at point C on the free-body diagram of member ABC, the force exerted by member CD is denoted by the components

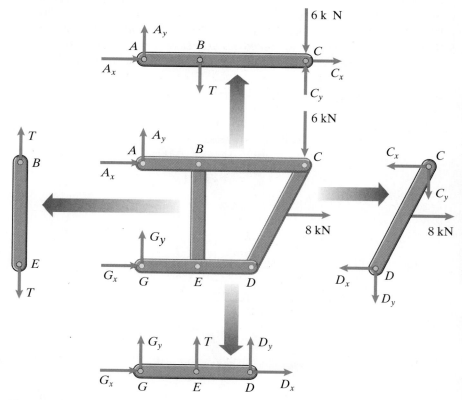

Figure 6.20
Obtaining the free-body diagrams of the members.

C_x and C_y. The forces exerted by member ABC on member CD at point C must be equal and opposite, as shown.

We need to discuss two important aspects of these free-body diagrams before completing the analysis.

Two-Force Members   Member BE is a two-force member, and we have taken this into account in drawing its free-body diagram in Fig. 6.20. The force T is the axial force in member BE, and an equal and opposite force is subjected on member ABC at B and on member GED at E.

Recognizing two-force members in frames and machines and drawing their free-body diagrams as we have done will reduce the number of unknowns and will greatly simplify the analysis. In our example, if we did not treat member BE as a two-force member, its free-body diagram would have four unknown forces (Fig. 6.21a). By treating it as a two-force member (Fig. 6.21b), we reduce the number of unknown forces by three.

Loads Applied at Joints   A question arises when a load is applied at a joint: Where does the load appear on the free-body diagrams of the individual members? The answer is that you can place the load on *any one* of the members attached at the joint. For example, in Fig. 6.17, the 6-kN load acts at the joint where members ABC and CD are connected. In drawing the free-body diagrams of the individual members (Fig. 6.20), we assumed that the 6-kN load acted on member ABC. The force components C_x and C_y on the free-body diagram of member ABC are the forces exerted by the member CD.

To explain why we can draw the free-body diagrams in this way, let us assume that the 6-kN force acts on the pin connecting members ABC and CD, and draw separate free-body diagrams of the pin and the two members (Fig. 6.22a). The force components C'_x and C'_y are the forces exerted by the pin on member ABC, and C_x and C_y are the forces exerted by the pin on member CD. If we superimpose the free-body diagrams of the pin and member ABC, we obtain the two free-body diagrams in Fig. 6.22b, which is the way we drew them in Fig. 6.20. Alternatively, by superimposing the free-body diagrams of the pin and member CD, we obtain the two free-body diagrams in Fig. 6.22c.

Thus if a load acts at a joint, it can be placed on any one of the members attached at the joint when drawing the free-body diagrams of the individual members. Just make sure not to place it on more than one member.

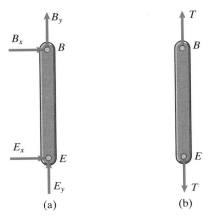

Figure 6.21
Free-body diagram of member BE:
(a) Not treating it as a two-force member.
(b) Treating it as a two-force member.

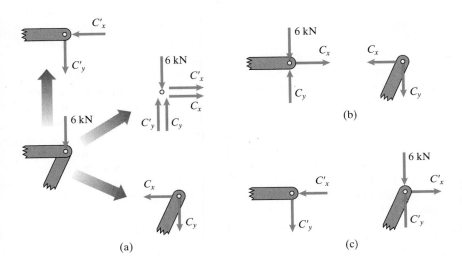

Figure 6.22
(a) Drawing free-body diagrams of the pin and the two members.
(b) Superimposing the pin on member ABC.
(c) Superimposing the pin on member CD.

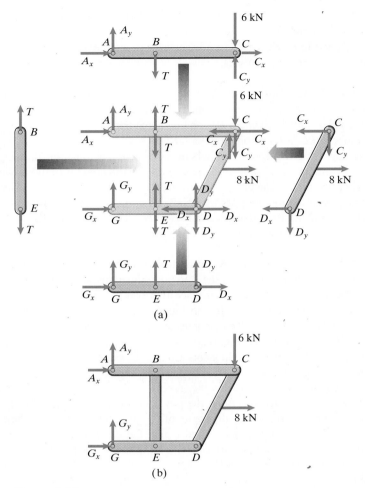

Figure 6.23
(a) "Reassembling" the free-body diagrams of the individual members.
(b) The free-body diagram of the entire frame is recovered.

To detect errors in the free-body diagrams of the members, it is helpful to "reassemble" them (Fig. 6.23a). The forces at the connections between the members cancel (they are internal forces once the members are reassembled), and the free-body diagram of the entire structure is recovered (Fig. 6.23b).

Our final step is to apply the equilibrium equations to the free-body diagrams of the members (Fig. 6.24). In two dimensions, we can obtain three independent equilibrium equations from the free-body diagram of each member of a structure that we do not treat as a two-force member. (By assuming that the forces on a two-force member are equal and opposite axial forces, we have already used the three equilibrium equations for that member.) In this example, there are three members in addition to the two-force member, so we can write $3 \times 3 = 9$ independent equilibrium equations, and there are nine unknown forces: A_x, A_y, C_x, C_y, D_x, D_y, G_x, G_y, and T.

Recall that we determined that $A_x = -13$ kN and $G_x = 5$ kN from our analysis of the entire structure. The equilibrium equations we obtained from the free-body diagram of the entire structure are not independent of the equilibrium

Figure 6.24
Free-body diagrams of the members.

(a)

(b)

(c)

equations obtained from the free-body diagrams of the members, but by using them to determine A_x and G_x, we get a head start on solving the equations for the members. Consider the free-body diagram of member ABC (Fig. 6.24a). Because we know A_x, we can determine C_x from the equation

$$\Sigma F_x = A_x + C_x = 0,$$

obtaining $C_x = -A_x = 13$ kN. Now consider the free-body diagram of GED (Fig. 6.24b). We can determine D_x from the equation

$$\Sigma F_x = G_x + D_x = 0,$$

obtaining $D_x = -G_x = -5$ kN. Now consider the free-body diagram of member CD (Fig. 6.24c). Because we know C_x, we can determine C_y by summing moments about D:

$$\Sigma M_{\text{point } D} = (2\text{ m})C_x - (1\text{ m})C_y - (1\text{ m})(8\text{ kN}) = 0.$$

We obtain $C_y = 18$ kN. Then, from the equation

$$\Sigma F_y = -C_y - D_y = 0,$$

we find that $D_y = -C_y = -18$ kN. Now we can return to the free-body diagrams of members ABC and GED to determine A_y and G_y. Summing moments about point B of member ABC yields

$$\Sigma M_{\text{point } B} = -(1\text{ m})A_y + (2\text{ m})C_y - (2\text{ m})(6\text{ kN}) = 0,$$

Figure 6.25
Forces on the members of the frame.

and we obtain $A_y = 2C_y - 12$ kN $= 24$ kN. Then, summing moments about point E of member GED, we have

$$\Sigma M_{\text{point } E} = (1 \text{ m})D_y - (1 \text{ m})G_y = 0,$$

from which we obtain $G_y = D_y = -18$ kN. Finally, from the free-body diagram of member GED, we use the equilibrium equation

$$\Sigma F_y = D_y + G_y + T = 0,$$

which gives us the result $T = -D_y - G_y = 36$ kN. The forces on the members are shown in Fig. 6.25. As this example demonstrates, determination of the forces on the members can often be simplified by carefully choosing the order in which the equations are solved.

We see that determining the forces and couples on the members of frames and machines involves two steps:

1. Determine the reactions at the supports—Draw the free-body diagram of the entire structure, and determine the reactions at its supports. Although this step is not essential, it can greatly simplify your analysis of the members. If the free-body diagram is statically indeterminant, determine as many of the reactions as possible.

2. Analyze the members—Draw free-body diagrams of the members, and apply the equilibrium equations to determine the forces acting on them. You can simplify this step by identifying two-force members. If a load acts at a joint of the structure, you can place the load on the free-body diagram of any one of the members attached at that joint.

RESULTS

A structure of interconnected members that cannot be modeled as a truss is called a *frame* if it is designed to remain stationary and support loads and a *machine* if it is designed to move and apply loads. The forces and couples acting on the individual members of a frame or machine in equilibrium can often be determined by applying the equilibrium equations to the individual members.

It is often advantageous to begin by drawing the free-body diagram of the entire structure considered as a single object and applying the equilibrium equations. *Even if the free-body diagram of the entire structure is statically indeterminate, it may be possible to determine the reactions from the subsequent analysis of the individual members.*

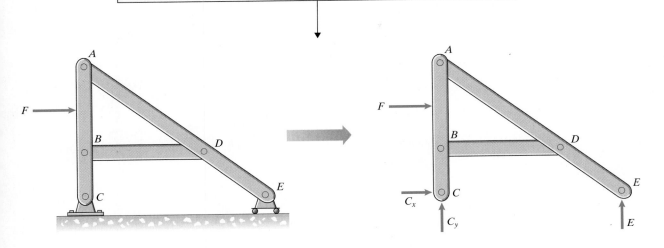

Draw the free-body diagrams of the individual members and apply the equilibrium equations to them. Notice that where two members are connected, *the reactions they exert on each other are equal and opposite*. Notice that member *BD* is a two-force member. Recognizing two-force members will simplify the analysis of a structure.

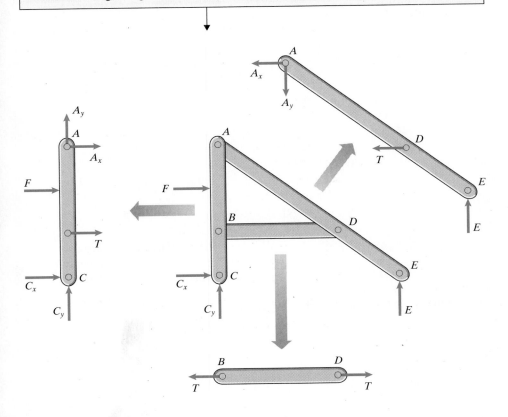

Active Example 6.6   **Analyzing a Frame** (▶ *Related Problem 6.70*)

Determine the forces and couples acting on the members of the frame.

Strategy

We will first draw a free-body diagram of the entire frame, treating it as a single object, and attempt to determine the reactions at the supports. We will then draw free-body diagrams of the individual members and apply the equilibrium equations to determine the forces and couples acting on them.

Solution

$\Sigma F_x = A_x = 0,$

$\Sigma F_y = A_y + C = 0,$

$\Sigma M_{\text{point } A} = M_A - 200 \text{ N-m} + (1.0 \text{ m})C = 0.$

The reaction $A_x = 0$, but A_y, C, and M_A cannot be determined from these equations. The free-body diagram of the entire frame is statically indeterminate.

Draw the free-body diagram of the entire frame and apply the equilibrium equations.

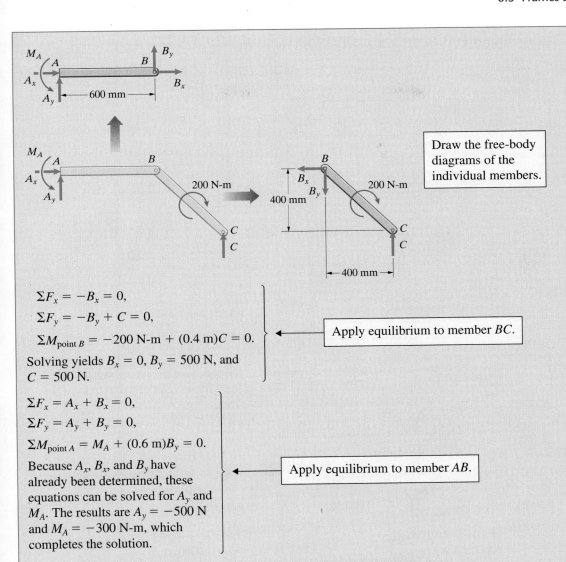

$\Sigma F_x = -B_x = 0,$

$\Sigma F_y = -B_y + C = 0,$

$\Sigma M_{\text{point } B} = -200 \text{ N-m} + (0.4 \text{ m})C = 0.$

Solving yields $B_x = 0$, $B_y = 500$ N, and $C = 500$ N.

Apply equilibrium to member BC.

$\Sigma F_x = A_x + B_x = 0,$

$\Sigma F_y = A_y + B_y = 0,$

$\Sigma M_{\text{point } A} = M_A + (0.6 \text{ m})B_y = 0.$

Because A_x, B_x, and B_y have already been determined, these equations can be solved for A_y and M_A. The results are $A_y = -500$ N and $M_A = -300$ N-m, which completes the solution.

Apply equilibrium to member AB.

Practice Problem The frame has pin supports at A and C. Determine the forces and couples acting on member BC at B and C.

Answer: $B_x = -500$ N, $B_y = 0$, $C_x = 500$ N, $C_y = 0$. (In the statements of the answers, x components are positive to the right and y components are positive upward.)

Example 6.7 Determining Forces on Members of a Frame (▶ *Related Problem 6.74*)

The frame supports a suspended weight $W = 40$ lb. Determine the forces on members *ABCD* and *CEG*.

Strategy

We will draw a free-body diagram of the entire frame and attempt to determine the reactions at the supports. We will then draw free-body diagrams of the individual members and use the equilibrium equations to determine the forces and couples acting on them. In doing so, we can take advantage of the fact that the bar *BE* is a two-force member.

Solution

Determine the Reactions at the Supports   We draw the free-body diagram of the entire frame in Fig. a. From the equilibrium equations

$$\Sigma F_x = A_x - D = 0,$$
$$\Sigma F_y = A_y - 40 \text{ lb} = 0,$$
$$\Sigma M_{\text{point } A} = (18 \text{ in})D - (19 \text{ in})(40 \text{ lb}) = 0,$$

we obtain the reactions $A_x = 42.2$ lb, $A_y = 40$ lb, and $D = 42.2$ lb.

Analyze the Members   We obtain the free-body diagrams of the members in Fig. b. Notice that *BE* is a two-force member. The angle $\alpha = \arctan(6/8) = 36.9°$.

The free-body diagram of the pulley has only two unknown forces. From the equilibrium equations

$$\Sigma F_x = G_x - 40 \text{ lb} = 0,$$
$$\Sigma F_y = G_y - 40 \text{ lb} = 0,$$

we obtain $G_x = 40$ lb and $G_y = 40$ lb. There are now only three unknown forces on the free-body diagram of member *CEG*. From the equilibrium equations

$$\Sigma F_x = -C_x - R \cos \alpha - 40 \text{ lb} = 0,$$
$$\Sigma F_y = -C_y - R \sin \alpha - 40 \text{ lb} = 0,$$
$$\Sigma M_{\text{point } C} = -(8 \text{ in})R \sin \alpha - (16 \text{ in})(40 \text{ lb}) = 0,$$

we obtain $C_x = 66.7$ lb, $C_y = 40$ lb, and $R = -133.3$ lb, completing the solution (Fig. c).

(a) Free-body diagram of the entire frame.

(b) Obtaining the free-body diagrams of the members.

Critical Thinking

In problems of this kind, the reactions on the individual members of the frame can be determined from the free-body diagrams of the members. Why did we draw the free-body diagram of the entire frame and solve the associated equilibrium equations? The reason is that it gave us a head start on solving the equilibrium equations for the members. In this example, when we drew the free-body diagrams of the members we already knew the reactions at A and D, which simplified the remaining analysis. Analyzing the entire frame can also provide a check on your work. Notice that we did not use the equilibrium equations for member $ABCD$. We can check our analysis by confirming that this member is in equilibrium (Fig. c):

$$\Sigma F_x = 42.2 \text{ lb} - 133.3 \cos 36.9° \text{ lb} + 66.7 \text{ lb} + 40 \text{ lb} - 42.2 \text{ lb} = 0,$$

$$\Sigma F_y = 40 \text{ lb} - 133.3 \sin 36.9° \text{ lb} + 40 \text{ lb} = 0,$$

$$\Sigma M_{\text{point } A} = (6 \text{ in})(133.3 \cos 36.9° \text{ lb}) - (12 \text{ in})(66.7 \text{ lb})$$

$$- (15 \text{ in})(40 \text{ lb}) + (18 \text{ in})(42.2 \text{ lb}) = 0.$$

(c) Forces on members $ABCD$ and CEG.

Example 6.8   Analyzing a Machine (▶ *Related Problem 6.103*)

What forces are exerted on the ball at *E* as a result of the 150-N forces on the pliers?

Strategy

A pair of pliers is a simple example of a machine, a structure designed to move and exert forces. The interconnections of the members are designed to create a mechanical advantage, subjecting an object to forces greater than the forces exerted by the user.

In this case there is no information to be gained from the free-body diagram of the entire structure. We must determine the forces exerted on the ball by drawing free-body diagrams of the members.

Solution

We "disassemble" the pliers in Fig. a to obtain the free-body diagrams of the members, labeled (1), (2), and (3). The force R on free-body diagrams (1) and (3) is exerted by the two-force member AB. The angle $\alpha = \arctan(30/70) = 23.2°$. Our objective is to determine the force E exerted by the ball.

The free-body diagram of member (3) has only three unknown forces and the 150-N load, so we can determine R, D_x, and D_y from this free-body diagram alone. The equilibrium equations are

$$\Sigma F_x = D_x + R \cos \alpha = 0,$$

$$\Sigma F_y = D_y - R \sin \alpha + 150 \text{ N} = 0,$$

$$\Sigma M_{\text{point } B} = (30 \text{ mm})D_y - (100 \text{ mm})(150 \text{ N}) = 0.$$

Solving these equations, we obtain $D_x = -1517$ N, $D_y = 500$ N, and $R = 1650$ N. Knowing D_x, we can determine E from the free-body diagram of member (2) by summing moments about C:

$$\Sigma M_{\text{point } C} = -(30 \text{ mm})E - (30 \text{ mm})D_x = 0.$$

The force exerted on the ball by the pliers is $E = -D_x = 1517$ N. The mechanical advantage of the pliers is $(1517 \text{ N})/(150 \text{ N}) = 10.1$.

(a) Obtaining the free-body diagrams of the members.

Critical Thinking

What is the motivation for determining the reactions on the members of the pliers? This process is essential for machine and tool design. To design the configuration of the pliers and choose the materials and dimensions of its members, it is necessary to determine all the forces acting on the members, as we have done in this example. Once the forces are known, the methods of mechanics of materials can be used to assess the adequacy of the members to support them.

Problems

Assume that objects are in equilibrium. In the statements of the answers, x components are positive to the right and y components are positive upward.

▶ **6.70** In Active Example 6.6, suppose that in addition to being loaded by the 200 N-m couple, the frame is subjected to a 400-N force at C that is horizontal and points toward the left. Draw a sketch of the frame showing the new loading. Determine the forces and couples acting on member AB of the frame.

6.71 The object suspended at E weighs 200 lb. Determine the reactions on member ACD at A and C.

Problem 6.71

6.72 The mass of the object suspended at *G* is 100 kg. Determine the reactions on member *CDE* at *C* and *E*.

Problem 6.72

6.73 The force $F = 10$ kN. Determine the forces on member *ABC*, presenting your answers as shown in Fig. 6.25.

Problem 6.73

▶ **6.74** In Example 6.7, suppose that the frame is redesigned so that the distance from point *C* to the attachment point *E* of the two-force member *BE* is increased from 8 in to 10 in. Determine the forces acting at *C* on member *ABCD*.

6.75 The tension in cable *BD* is 500 lb. Determine the reactions at *A* for cases (1) and (2).

Problem 6.75

6.76 Determine the reactions on member *ABCD* at *A*, *C*, and *D*.

Problem 6.76

6.77 Determine the forces exerted on member *ABC* at *A* and *C*.

Problem 6.77

6.78 An athlete works out with a squat thrust machine. To rotate the bar *ABD*, she must exert a vertical force at *A* that causes the magnitude of the axial force in the two-force member *BC* to be 1800 N. When the bar *ABD* is on the verge of rotating, what are the reactions on the vertical bar *CDE* at *D* and *E*?

Problem 6.78

6.79 The frame supports a 6-kN vertical load at *C*. The bars *ABC* and *DEF* are horizontal. Determine the reactions on the frame at *A* and *D*.

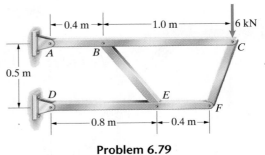

Problem 6.79

6.80 The mass *m* = 120 kg. Determine the forces on member *ABC*, presenting your answers as shown in Fig. 6.25.

Problem 6.80

6.81 Determine the reactions on member *BCD*.

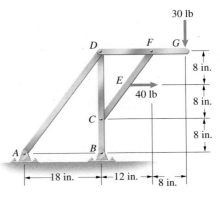

Problem 6.81

6.82 The weight of the suspended object is $W = 50$ lb. Determine the tension in the spring and the reactions at F. (The slotted member DE is vertical.)

Problem 6.82

6.83 The mass $m = 50$ kg. Bar DE is horizontal. Determine the forces on member $ABCD$, presenting your answers as shown in Fig. 6.25.

Problem 6.83

6.84 Determine the forces on member BCD.

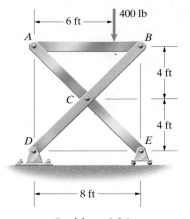

Problem 6.84

6.85 Determine the forces on member ABC.

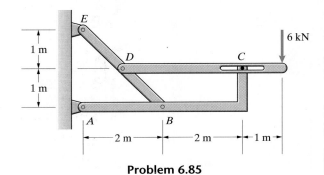

Problem 6.85

6.86 Determine the forces on member ABD.

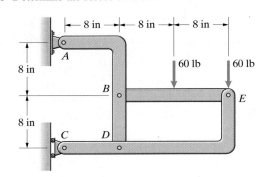

Problem 6.86

6.87 The mass $m = 12$ kg. Determine the forces on member CDE.

Problem 6.87

6.88 The weight $W = 80$ lb. Determine the forces on member $ABCD$.

Problem 6.88

6.89 The woman using the exercise machine is holding the 80-lb weight stationary in the position shown. What are the reactions at the fixed support E and the pin support F? (A and C are pinned connections.)

Problem 6.89

6.90 Determine the reactions on member ABC at A and B.

Problem 6.90

6.91 The mass of the suspended object is $m = 50$ kg. Determine the reactions on member ABC.

Problem 6.91

6.92 The unstretched length of the spring is L_0. Show that when the system is in equilibrium the angle α satisfies the relation $\sin \alpha = 2(L_0 - 2F/k)/L$.

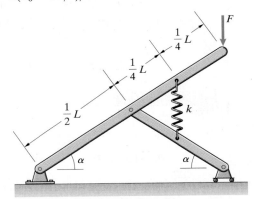

Problem 6.92

6.93 The pin support B will safely support a force of 24-kN magnitude. Based on this criterion, what is the largest mass m that the frame will safely support?

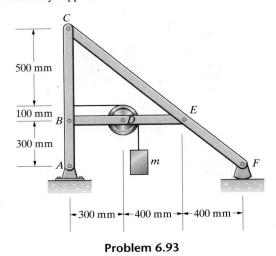

Problem 6.93

6.94 Determine the reactions at A and C.

Problem 6.94

6.95 Determine the forces on member AD.

Problem 6.95

6.96 The frame shown is used to support high-tension wires. If $b = 3$ ft, $\alpha = 30°$, and $W = 200$ lb, what is the axial force in member HJ?

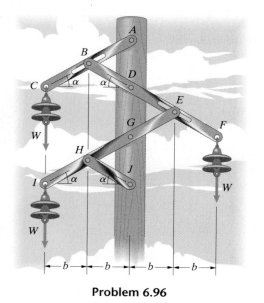

Problem 6.96

6.97 Determine the force exerted on the ball by the bolt cutters and the magnitude of the axial force in the two-force member AB.

Problem 6.97

6.98 The woman exerts 20-N forces to the pliers as shown.

(a) What is the magnitude of the forces the pliers exert on the bolt at B?

(b) Determine the magnitude of the force the members of the pliers exert on each other at the pinned connection C.

Problem 6.98

6.99 Figure a is a diagram of the bones and biceps muscle of a person's arm supporting a mass. Tension in the biceps muscle holds the forearm in the horizontal position, as illustrated in the simple mechanical model in Fig. b. The weight of the forearm is 9 N, and the mass $m = 2$ kg.

(a) Determine the tension in the biceps muscle AB.

(b) Determine the magnitude of the force exerted on the upper arm by the forearm at the elbow joint C.

Problem 6.99

6.100 The bones and tendons in a horse's rear leg are shown in Fig. a. A biomechanical model of the leg is shown in Fig. b. If the horse is stationary and the normal force exerted on its leg by the ground is $N = 1200$ N, determine the tensions in the superficial digital flexor BC and the patellar ligament DF.

Problem 6.100

6.101 The pressure force exerted on the piston is 2 kN toward the left. Determine the couple M necessary to keep the system in equilibrium.

6.102 In Problem 6.101, determine the forces on member AB at A and B.

Problems 6.101/6.102

▶ **6.103** In Example 6.8, suppose that the object being held by the pliers is moved to the left so that the horizontal distance from D to the object at E decreases from 30 mm to 20 mm. Draw a sketch of the pliers showing the new position of the object. What forces are exerted on the object at E as a result of the 150-N forces on the pliers?

6.104 The shovel of the excavator is supported by a pin support at E and the two-force member BC. The 300-lb weight W of the shovel acts at the point shown. Determine the reactions on the shovel at E and the magnitude of the axial force in the two-force member BC.

6.105 The shovel of the excavator has a pin support at E. The position of the shovel is controlled by the horizontal hydraulic piston AB, which is attached to the shovel through a linkage of the two-force members BC and BD. The 300-lb weight W of the shovel acts at the point shown. What is the magnitude of the force the hydraulic piston must exert to hold the shovel in equilibrium?

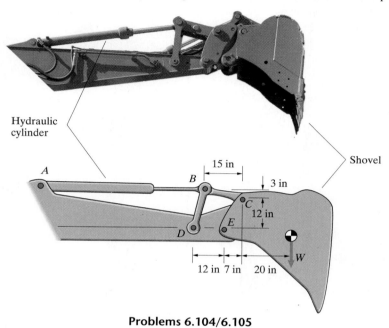

Problems 6.104/6.105

6.106 The woman exerts 20-N forces on the handles of the shears. Determine the magnitude of the forces exerted on the branch at A.

Problem 6.106

6.107 The person exerts 40-N forces on the handles of the locking wrench. Determine the magnitude of the forces the wrench exerts on the bolt at A.

6.108 Determine the magnitude of the force the members of the wrench exert on each other at B and the axial force in the two-force member DE.

Problems 6.107/6.108

6.109 This device is designed to exert a large force on the horizontal bar at A for a stamping operation. If the hydraulic cylinder DE exerts an axial force of 800 N and $\alpha = 80°$, what horizontal force is exerted on the horizontal bar at A?

Problem 6.109

6.110 This device raises a load W by extending the hydraulic actuator DE. The bars AD and BC are 4 ft long, and the distances $b = 2.5$ ft and $h = 1.5$ ft. If $W = 300$ lb, what force must the actuator exert to hold the load in equilibrium?

Problem 6.110

6.111 The four-bar linkage operates the forks of a fork lift truck. The force supported by the forks is $W = 8$ kN. Determine the reactions on member CDE.

Problem 6.111

6.112 If the horizontal force on the scoop is $F = 2000$ lb, what is the magnitude of the axial force in the hydraulic actuator AC?

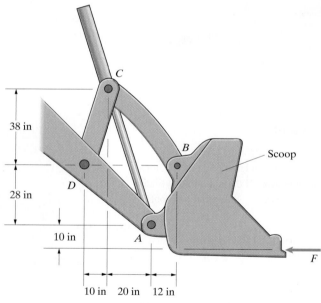

Problem 6.112

6.113 A 10-kip horizontal force acts on the bucket of the excavator. Determine the reactions on member *ACF* at *A* and *F*.

Problem 6.113

6.114 The structure shown in the diagram (one of the two identical structures that support the scoop of the excavator) supports a downward force *F* = 1800 N at *G*. Members *BC* and *DH* can be treated as two-force members. Determine the reactions on member *CDK* at *K*.

Problem 6.114

Review Problems

6.115 The loads $F_1 = 440$ N and $F_2 = 160$ N. Determine the axial forces in the members. Indicate whether they are in tension (T) or compression (C).

Problem 6.115

6.116 The truss supports a load $F = 10$ kN. Determine the axial forces in members $AB, AC,$ and BC.

6.117 Each member of the truss will safely support a tensile force of 40 kN and a compressive force of 32 kN. Based on this criterion, what is the largest downward load F that can safely be applied at C?

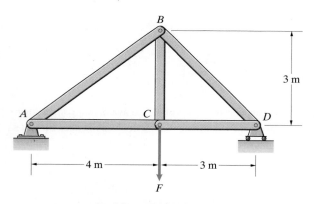

Problems 6.116/6.117

6.118 The Pratt bridge truss supports loads at $F, G,$ and H. Determine the axial forces in members $BC, BG,$ and FG.

6.119 Determine the axial forces in members $CD, GD,$ and GH.

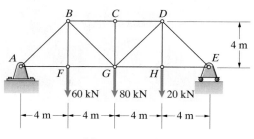

Problems 6.118/6.119

6.120 The truss supports loads at F and H. Determine the axial forces in members $AB, AC, BC, BD, CD,$ and CE.

6.121 Determine the axial forces in members EH and FH.

Problems 6.120/6.121

6.122 Determine the axial forces in members *BD*, *CD*, and *CE*.

6.123 Determine the axial forces in members *DF*, *EF*, and *EG*.

Problems 6.122/6.123

6.124 The truss supports a 400-N load at *G*. Determine the axial forces in members *AC*, *CD*, and *CF*.

6.125 Determine the axial forces in members *CE*, *EF*, and *EH*.

6.126 Which members have the largest tensile and compressive forces, and what are their values?

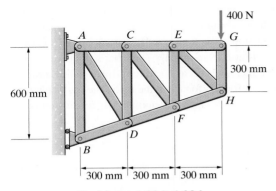

Problems 6.124–6.126

6.127 The Howe truss helps support a roof. Model the supports at *A* and *G* as roller supports. Use the method of joints to determine the axial forces in members *BC*, *CD*, *CI*, and *CJ*.

6.128 Use the method of sections to determine the axial forces in members *CD*, *CJ*, and *IJ*.

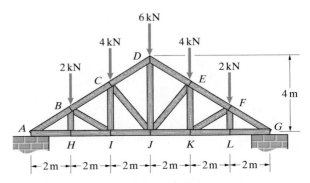

Problems 6.127/6.128

6.129 A speaker system is suspended from the truss by cables attached at *D* and *E*. The mass of the speaker system is 130 kg, and its weight acts at *G*. Determine the axial forces in members *BC* and *CD*.

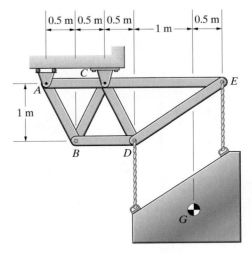

Problem 6.129

6.130 The mass of the suspended object is 900 kg. Determine the axial forces in the bars AB and AC.

Strategy: Draw the free-body diagram of joint A.

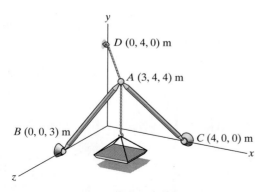

Problem 6.130

6.131 Determine the forces on member ABC, presenting your answers as shown in Fig. 6.25. Obtain the answers in two ways:

(a) When you draw the free-body diagrams of the individual members, place the 400-lb load on the free-body diagram of member ABC.

(b) When you draw the free-body diagrams of the individual members, place the 400-lb load on the free-body diagram of member CD.

Problem 6.131

6.132 The mass $m = 120$ kg. Determine the forces on member ABC.

Problem 6.132

6.133 Determine the reactions on member ABC at B and C.

Problem 6.133

6.134 The truck and trailer are parked on a 10° slope. The 14,000-lb weight of the truck and the 8000-lb weight of the trailer act at the points shown. The truck's brakes prevent its rear wheels at B from turning. The truck's front wheels at C and the trailer's wheels at A can turn freely, which means they do not exert friction forces on the road. The trailer hitch at D behaves like a pin support. Determine the forces exerted on the truck at B, C, and D.

Problem 6.134

6.135 The 600-lb weight of the scoop acts at a point 1 ft 6 in to the right of the vertical line *CE*. The line *ADE* is horizontal. The hydraulic actuator *AB* can be treated as a two-force member. Determine the axial force in the hydraulic actuator *AB* and the forces exerted on the scoop at *C* and *E*.

Problem 6.135

6.136 Determine the force exerted on the bolt by the bolt cutters.

6.137 Determine the magnitude of the force the members of the bolt cutters exert on each other at the pin connection *B* and the axial force in the two-force member *CD*.

Problems 6.136/6.137

Design Project 1 Design a truss structure to support a foot bridge with an unsupported span (width) of 8 m. Make conservative estimates of the loads the structure will need to support if the pathway supported by the truss is made of wood. Consider two options: (1) Your client wants the bridge to be supported by a truss below the bridge so that the upper surface will be unencumbered by structure. (2) The client wants the truss to be above the bridge and designed so that it can serve as handrails. For each option, use statics to estimate the maximum axial forces to which the members of the structure will be subjected. Investigate alternative designs and compare the resulting axial loads.

Design Project 2 The truss shown connects one end of a stretcher to a rescue helicopter. Consider alternative truss designs that support the stretcher at *A* and *B* and are supported at *E* and *G*. Compare the maximum tensile and compressive loads in the members of your designs to those in the truss shown. Assuming that the cost of a truss is proportional to the sum of the lengths of its members, compare the costs of your designs to that of the truss shown. Write a brief report describing your analysis and recommending the design you would choose.

Design Project 3 Go to a fitness center and choose an exercise device that seems mechanically interesting. (For example, it may employ weights, pulleys, and levers.) By measuring dimensions (while the device is not in use), drawing sketches, and perhaps taking photographs, gather the information necessary to analyze the device. Use statics to determine the range of forces a person must exert in using the device.

Suggest changes to the design of the device (other than simply increasing weights) that will increase the maximum force the user must exert.

Prepare a brief report that (1) describes the original device; (2) presents your model and analysis of the device; (3) describes your proposed changes and any analyses supporting them; and (4) recommends the design change you would choose to increase the maximum force the user must employ.

CHAPTER

7

Centroids and Centers of Mass

An object's weight does not act at a single point—it is distributed over the entire volume of the object. But the weight can be represented by a single equivalent force acting at a point called the center of mass. When the equilibrium equations are used to determine the reactions exerted on an object by its supports, the location of the center of mass must be known if the weight of the object is to be included in the analysis. The dynamic behaviors of objects also depend on the locations of their centers of mass. In this chapter we define the center of mass and show how it is determined for various kinds of objects. We also introduce definitions that can be interpreted as the average positions of areas, volumes, and lines. These average positions are called centroids. Centroids coincide with the centers of mass of particular classes of objects, and they also arise in many other engineering applications.

◀ To be balanced, the woman's center of mass—the point at which her weight effectively acts—must be directly above her hands. In this chapter we introduce the concept of an average position, or centroid, and show how to locate the centers of mass of objects.

7.1 Centroids of Areas

BACKGROUND

(a)

(b)

Figure 7.1
(a) A group of students in a classroom.
(b) Their average position.

Suppose that we want to determine the average position of a group of students sitting in a room. First, we introduce a coordinate system so that we can specify the position of each student. For example, we can align the axes with the walls of the room (Fig. 7.1a). We number the students from 1 to N and denote the position of student 1 by (x_1, y_1), the position of student 2 by (x_2, y_2), and so on. The average x coordinate, which we denote by $\bar{x}$, is the sum of their x coordinates divided by N; that is,

$$\bar{x} = \frac{x_1 + x_2 + \cdots + x_N}{N} = \frac{\sum_i x_i}{N}, \tag{7.1}$$

where the symbol $\sum_i$ means "sum over the range of i." The average y coordinate is

$$\bar{y} = \frac{\sum_i y_i}{N}. \tag{7.2}$$

We indicate the average position by the symbol shown in Fig. 7.1b.

Now suppose that we pass out some pennies to the students. Let the number of coins given to student 1 be c_1, the number given to student 2 be c_2, and so on. What is the average position of the coins in the room? Clearly, the average position of the coins may not be the same as the average position of the students. For example, if the students in the front of the room have more coins, the average position of the coins will be closer to the front of the room than the average position of the students.

To determine the x coordinate of the average position of the coins, we need to sum the x coordinates of the coins and divide by the number of coins. We can obtain the sum of the x coordinates of the coins by multiplying the number of coins each student has by his or her x coordinate and summing. We can obtain the number of coins by summing the numbers $c_1, c_2, \ldots$. Thus, the average x coordinate of the coins is

$$\bar{x} = \frac{\sum_i x_i c_i}{\sum_i c_i}. \tag{7.3}$$

We can determine the average y coordinate of the coins in the same way:

$$\bar{y} = \frac{\sum_i y_i c_i}{\sum_i c_i}. \tag{7.4}$$

By assigning other meanings to $c_1, c_2, \ldots$, we can determine the average positions of other measures associated with the students. For example, we could determine the average position of their age or the average position of their height.

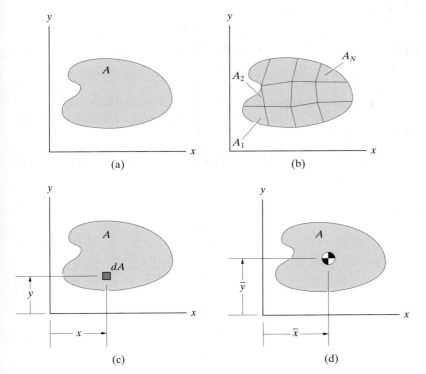

Figure 7.2
(a) The area A.
(b) Dividing A into N parts.
(c) A differential element of area dA with coordinates (x, y).
(d) The centroid of the area.

More generally, we can use Eqs. (7.3) and (7.4) to determine the average position of any set of quantities with which we can associate positions. An average position obtained from these equations is called a *weighted average position*, or *centroid*. The "weight" associated with position (x_1, y_1) is c_1, the weight associated with position (x_2, y_2) is c_2, and so on. In Eqs. (7.1) and (7.2), the weight associated with the position of each student is 1. When the census is taken, the centroid of the population of the United States—the average position of the population—is determined in this way.

Let us consider an arbitrary area A in the x–y plane (Fig. 7.2a). Divide the area into parts $A_1, A_2, \ldots, A_N$ (Fig. 7.2b) and denote the positions of the parts by $(x_1, y_1), (x_2, y_2), \ldots, (x_N, y_N)$. We can obtain the centroid, or average position of the area, by using Eqs. (7.3) and (7.4) with the areas of the parts as the weights:

$$\bar{x} = \frac{\sum_i x_i A_i}{\sum_i A_i}, \qquad \bar{y} = \frac{\sum_i y_i A_i}{\sum_i A_i}. \tag{7.5}$$

A question arises if we try to carry out this procedure: What are the exact positions of the areas $A_1, A_2, \ldots, A_N$? We could reduce the uncertainty in their positions by dividing A into smaller parts, but we would still obtain only approximate values for $\bar{x}$ and $\bar{y}$. To determine the exact location of the centroid, we must take the limit as the sizes of the parts approach zero. We obtain this limit by replacing Eqs. (7.5) by the integrals

$$\bar{x} = \frac{\int_A x \, dA}{\int_A dA}, \tag{7.6}$$

$$\bar{y} = \frac{\int_A y\, dA}{\int_A dA}, \tag{7.7}$$

where x and y are the coordinates of the differential element of area dA (Fig. 7.2c). The subscript A on the integral signs means the integration is carried out over the entire area. The centroid of the area is shown in Fig. 7.2d.

RESULTS

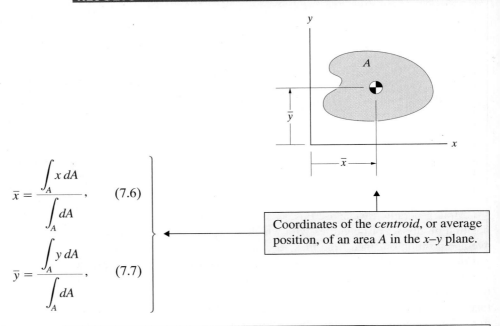

$$\bar{x} = \frac{\int_A x\, dA}{\int_A dA}, \tag{7.6}$$

$$\bar{y} = \frac{\int_A y\, dA}{\int_A dA}, \tag{7.7}$$

Coordinates of the *centroid*, or average position, of an area A in the x–y plane.

Keeping in mind that the centroid of an area is its average position will often help in locating it. If an area has "mirror image" symmetry about an axis, its centroid lies on the axis. If an area is symmetric about two axes, the centroid lies at the intersection of the axes.

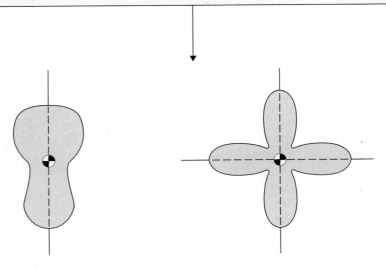

Active Example 7.1   **Centroid of an Area by Integration** (► *Related Problem 7.1*)

Determine the x coordinate of the centroid of the triangular area.

Strategy

We will evaluate Eq. (7.6) using an element of area dA in the form of a vertical "strip" of width dx.

Solution

$$\bar{x} = \frac{\displaystyle\int_A x\,dA}{\displaystyle\int_A dA} = \frac{\displaystyle\int_0^b x\left(\frac{h}{b}x\,dx\right)}{\displaystyle\int_0^b \frac{h}{b}x\,dx} = \frac{\dfrac{h}{b}\left[\dfrac{x^3}{3}\right]_0^b}{\dfrac{h}{b}\left[\dfrac{x^2}{2}\right]_0^b} = \frac{2}{3}b.$$

The height of a strip of width dx at position x is $(h/b)x$, so its area is $dA = (h/b)x\,dx$. Use this expression to evaluate Eq. (7.6).

Practice Problem Determine the y coordinate of the centroid of the triangular area. Evaluate Eq. (7.7) by using an element of area dA in the form of a vertical "strip" of width dx, and let y be the height of the midpoint of the strip.

Answer: $\bar{y} = \dfrac{1}{3}h.$

| **Example 7.2** | **Area Defined by Two Equations** (▶ *Related Problems 7.2, 7.3*) |

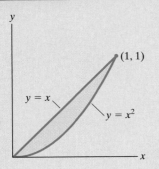

Determine the centroid of the area.

Strategy
We can determine the coordinates of the centroid using an element of area in the form of a vertical strip, just as we did in Active Example 7.1. In this case the strip must be defined so that it extends from the lower curve ($y = x^2$) to the upper curve ($y = x$).

Solution
Let dA be the vertical strip in Fig. a. The height of the strip is $x - x^2$, so $dA = (x - x^2)\,dx$. The x coordinate of the centroid is

$$\bar{x} = \frac{\displaystyle\int_A x\,dA}{\displaystyle\int_A dA} = \frac{\displaystyle\int_0^1 x(x - x^2)\,dx}{\displaystyle\int_0^1 (x - x^2)\,dx} = \frac{\left[\dfrac{x^3}{3} - \dfrac{x^4}{4}\right]_0^1}{\left[\dfrac{x^2}{2} - \dfrac{x^3}{3}\right]_0^1} = \frac{1}{2}.$$

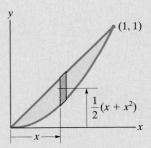

(a) A vertical strip of width dx. The height of the strip is equal to the difference in the two functions.

(b) The y coordinate of the midpoint of the strip.

The y coordinate of the midpoint of the strip is $x^2 + \frac{1}{2}(x - x^2) = \frac{1}{2}(x + x^2)$ (Fig. b). Substituting this expression for y in Eq. (7.7), we obtain the y coordinate of the centroid:

$$\bar{y} = \frac{\displaystyle\int_A y\,dA}{\displaystyle\int_A dA} = \frac{\displaystyle\int_0^1 \left[\frac{1}{2}(x + x^2)\right](x - x^2)\,dx}{\displaystyle\int_0^1 (x - x^2)\,dx} = \frac{\dfrac{1}{2}\left[\dfrac{x^3}{3} - \dfrac{x^5}{5}\right]_0^1}{\left[\dfrac{x^2}{2} - \dfrac{x^3}{3}\right]_0^1} = \frac{2}{5}.$$

Critical Thinking
Notice the generality of the approach we use in this example. It can be used to determine the x and y coordinates of the centroid of any area whose upper and lower boundaries are defined by two functions.

Problems

▶ **7.1** In Active Example 7.1, suppose that the triangular area is oriented as shown. Use integration to determine the x and y coordinates of its centroid. (Notice that you already know the answers based on the results of Active Example 7.1.)

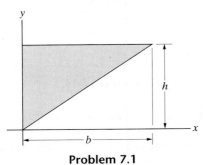

Problem 7.1

▶ **7.2** In Example 7.2, suppose that the area is redefined as shown. Determine the x coordinate of the centroid.

▶ **7.3** In Example 7.2, suppose that the area is redefined as shown. Determine the y coordinate of the centroid.

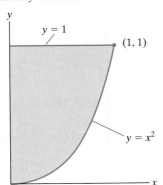

Problems 7.2/7.3

7.4 Determine the centroid of the area.

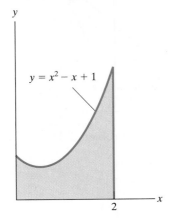

Problem 7.4

7.5 Determine the coordinates of the centroid of the area.

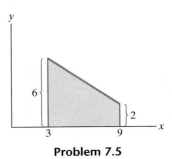

Problem 7.5

7.6 Determine x coordinate of the centroid of the area and compare your answer to the value given in Appendix B.

7.7 Determine the y coordinate of the centroid of the area and compare your answer to the value given in Appendix B.

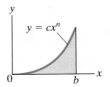

Problems 7.6/7.7

7.8 Suppose that an art student wants to paint a panel of wood as shown, with the horizontal and vertical lines passing through the centroid of the painted area, and asks you to determine the coordinates of the centroid. What are they?

Problem 7.8

7.9 Determine the value of the constant c so that the y coordinate of the centroid of the area is $\bar{y} = 2$. What is the x coordinate of the centroid?

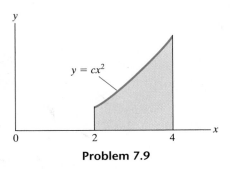

$y = cx^2$

Problem 7.9

7.10 Determine the coordinates of the centroid of the metal plate's cross-sectional area.

$y = 4 - \dfrac{1}{4}x^2$ ft

Problem 7.10

7.11 An architect wants to build a wall with the profile shown. To estimate the effects of wind loads, he must determine the wall's area and the coordinates of its centroid. What are they?

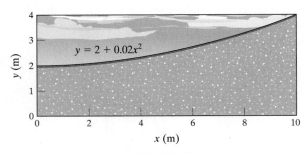

$y = 2 + 0.02x^2$

Problem 7.11

7.12 Determine the coordinates of the centroid of the area.

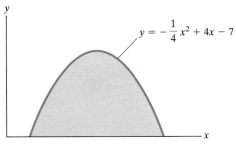

$y = -\dfrac{1}{4}x^2 + 4x - 7$

Problem 7.12

7.13 Determine the coordinates of the centroid of the area.

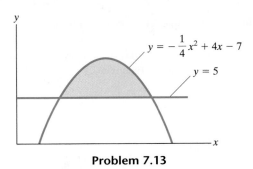

$y = -\dfrac{1}{4}x^2 + 4x - 7$

$y = 5$

Problem 7.13

7.14 Determine the x coordinate of the centroid of the area.

7.15 Determine the y coordinate of the centroid of the area.

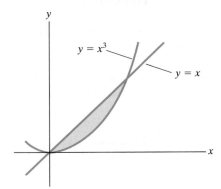

$y = x^3$

$y = x$

Problems 7.14/7.15

7.16 Determine the x component of the centroid of the area.

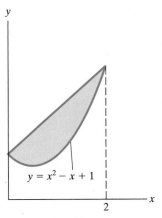

$y = x^2 - x + 1$

Problem 7.16

7.17 Determine the x coordinate of the centroid of the area.

7.18 Determine the y coordinate of the centroid of the area.

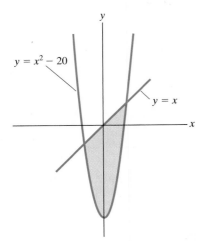

$y = x^2 - 20$

$y = x$

Problems 7.17/7.18

7.19 What is the x coordinate of the centroid of the area?

7.20 What is the y coordinate of the centroid of the area?

$y = -\dfrac{1}{6}x^2 + 2x$

Problems 7.19/7.20

7.21 An agronomist wants to measure the rainfall at the centroid of a plowed field between two roads. What are the coordinates of the point where the rain gauge should be placed?

Problem 7.21

7.22 The cross section of an earth-fill dam is shown. Determine the coefficients a and b so that the y coordinate of the centroid of the cross section is 10 m.

$y = ax - bx^3$

Problem 7.22

7.23 The Supermarine Spitfire used by Great Britain in World War II had a wing with an elliptical profile. Determine the coordinates of its centroid.

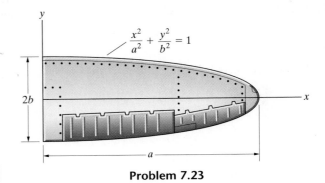

$\dfrac{x^2}{a^2} + \dfrac{y^2}{b^2} = 1$

Problem 7.23

7.24 Determine the coordinates of the centroid of the area.

Strategy: Write the equation for the circular boundary in the form $y = (R^2 - x^2)^{1/2}$ and use a vertical "strip" of width dx as the element of area dA.

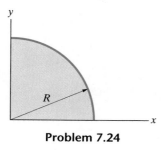

Problem 7.24

7.25* If $R = 6$ and $b = 3$, what is the y coordinate of the centroid of the area?

7.26* What is the x coordinate of the centroid of the area in Problem 7.25?

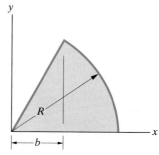

Problems 7.25/7.26

7.2 Composite Areas

BACKGROUND

Although centroids of areas can be determined by integration, the process becomes difficult and tedious for complicated areas. In this section we describe a much easier approach that can be used if an area consists of a combination of simple areas, which we call a *composite area*. We can determine the centroid of a composite area without integration if the centroids of its parts are known.

The area in Fig. 7.3a consists of a triangle, a rectangle, and a semicircle, which we call parts 1, 2, and 3. The x coordinate of the centroid of the composite area is

$$\bar{x} = \frac{\displaystyle\int_A x \, dA}{\displaystyle\int_A dA} = \frac{\displaystyle\int_{A_1} x \, dA + \int_{A_2} x \, dA + \int_{A_3} x \, dA}{\displaystyle\int_{A_1} dA + \int_{A_2} dA + \int_{A_3} dA}. \tag{7.8}$$

The x coordinates of the centroids of the parts are shown in Fig. 7.3b. From the equation for the x coordinate of the centroid of part 1,

$$\bar{x}_1 = \frac{\displaystyle\int_{A_1} x \, dA}{\displaystyle\int_{A_1} dA},$$

we obtain

$$\int_{A_1} x \, dA = \bar{x}_1 A_1.$$

Using this equation and equivalent equations for parts 2 and 3, we can write Eq. (7.8) as

$$\bar{x} = \frac{\bar{x}_1 A_1 + \bar{x}_2 A_2 + \bar{x}_3 A_3}{A_1 + A_2 + A_3}.$$

(a)

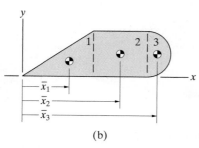

(b)

Figure 7.3
(a) A composite area composed of three simple areas.
(b) The centroids of the parts.

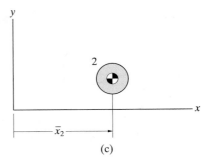

(a)   (b)   (c)

Figure 7.4
(a) An area with a cutout.
(b) The triangular area.
(c) The area of the cutout.

We have obtained an equation for the x coordinate of the composite area in terms of those of its parts. The coordinates of the centroid of a composite area with an arbitrary number of parts are

$$\bar{x} = \frac{\sum\limits_i \bar{x}_i A_i}{\sum\limits_i A_i}, \qquad \bar{y} = \frac{\sum\limits_i \bar{y}_i A_i}{\sum\limits_i A_i}. \tag{7.9}$$

When we can divide an area into parts whose centroids are known, we can use these expressions to determine its centroid. The centroids of some simple areas are tabulated in Appendix B.

We began our discussion of the centroid of an area by dividing an area into finite parts and writing equations for its weighted average position. The results, Eqs. (7.5), are approximate because of the uncertainty in the positions of the parts of the area. The exact Eqs. (7.9) are identical except that the positions of the parts are their centroids.

The area in Fig. 7.4a consists of a triangular area with a circular hole, or cutout. Designating the triangular area (without the cutout) as part 1 of the composite area (Fig. 7.4b) and the area of the cutout as part 2 (Fig. 7.4c), we obtain the x coordinate of the centroid of the composite area:

$$\bar{x} = \frac{\displaystyle\int_{A_1} x\, dA - \int_{A_2} x\, dA}{\displaystyle\int_{A_1} dA - \int_{A_2} dA} = \frac{\bar{x}_1 A_1 - \bar{x}_2 A_2}{A_1 - A_2}.$$

This equation is identical in form to the first of Eqs. (7.9) except that the terms corresponding to the cutout are negative. As this example demonstrates, we can use Eqs. (7.9) to determine the centroids of composite areas containing cutouts by treating the cutouts as negative areas.

We see that determining the centroid of a composite area requires three steps:

1. Choose the parts—Try to divide the composite area into parts whose centroids you know or can easily determine.
2. Determine the values for the parts—Determine the centroid and the area of each part. Watch for instances of symmetry that can simplify your task.
3. Calculate the centroid—Use Eqs. (7.9) to determine the centroid of the composite area.

RESULTS

$$\bar{x} = \frac{\bar{x}_1 A_1 + \bar{x}_2 A_2 + \cdots}{A_1 + A_2 + \cdots} = \frac{\sum\limits_i \bar{x}_i A_i}{\sum\limits_i A_i}$$

$$\bar{y} = \frac{\bar{y}_1 A_1 + \bar{y}_2 A_2 + \cdots}{A_1 + A_2 + \cdots} = \frac{\sum\limits_i \bar{y}_i A_i}{\sum\limits_i A_i}$$

(7.9)

Coordinates of the centroid of a composite area consisting of parts 1, 2, ... The term A_i is the area of the ith part, and x_i, y_i are the coordinates of the centroid of A_i.

If an area contains a hole or *cutout*, the centroid of the area can be determined from Eqs. (7.9) by treating the cutout as a negative area.

A triangular area with a circular cutout.

The triangular area without the cutout. Let its area be A_1 and let $\bar{x}_1$ be the x coordinate of its centroid.

The area of the circular cutout. Let its area be A_2 and let $\bar{x}_2$ be the x coordinate of its centroid.

The x coordinate of the centroid of the triangular area with the cutout is

$$\bar{x} = \frac{\bar{x}_1 A_1 - \bar{x}_2 A_2}{A_1 - A_2}.$$

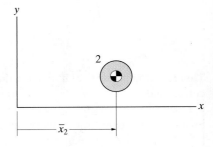

| Active Example 7.3 | Centroid of a Composite Area (▶ *Related Problem 7.27*) |

Determine the x coordinate of the centroid of the composite area.

Strategy
We must divide the area into simple parts (in this example the parts are obvious), determine the areas and centroid locations for the parts, and apply Eq. $(7.9)_1$.

Solution

| **Choose the Parts**
Divide the area into simple parts. The x coordinates of the centroids of the parts are shown. |

| **Determine the Values for the Parts**
Tabulate the terms needed to apply Eq. $(7.9)_1$.
See Appendix B. |

	$\bar{x}_i$	A_i	$\bar{x}_i A_i$
Part 1 (triangle)	$\dfrac{2}{3}b$	$\dfrac{1}{2}b(2R)$	$\left(\dfrac{2}{3}b\right)\left[\dfrac{1}{2}b(2R)\right]$
Part 2 (rectangle)	$b + \dfrac{1}{2}c$	$c(2R)$	$\left(b + \dfrac{1}{2}c\right)[c(2R)]$
Part 3 (semicircle)	$b + c + \dfrac{4R}{3\pi}$	$\dfrac{1}{2}\pi R^2$	$\left(b + c + \dfrac{4R}{3\pi}\right)\left(\dfrac{1}{2}\pi R^2\right)$

$$\bar{x} = \frac{\bar{x}_1 A_1 + \bar{x}_2 A_2 + \bar{x}_3 A_3}{A_1 + A_2 + A_3}$$

$$= \frac{\left(\dfrac{2}{3}b\right)\left[\dfrac{1}{2}b(2R)\right] + \left(b + \dfrac{1}{2}c\right)[c(2R)] + \left(b + c + \dfrac{4R}{3\pi}\right)\left(\dfrac{1}{2}\pi R^2\right)}{\dfrac{1}{2}b(2R) + c(2R) + \dfrac{1}{2}\pi R^2}$$

| **Calculate the Centroid**
Use Eq. $(7.9)_1$ to determine the x component of the centroid. |

Practice Problem Determine the y coordinate of the centroid of the composite area.

Answer: $\bar{y} = \dfrac{\left[\dfrac{1}{3}(2R)\right]\left[\dfrac{1}{2}b(2R)\right] + R[c(2R)] + R\left(\dfrac{1}{2}\pi R^2\right)}{\dfrac{1}{2}b(2R) + c(2R) + \dfrac{1}{2}\pi R^2}$.

Example 7.4	**Centroid of an Area with a Cutout** (▶ *Related Problem 7.28*)

Determine the centroid of the area.

Strategy

Instead of attempting to divide the area into parts, a simpler approach is to treat it as a composite of a rectangular area with a semicircular cutout. Then we can apply Eq. (7.9) by treating the cutout as a negative area.

Solution

Choose the Parts   We call the rectangle without the semicircular cutout and the area of the cutout parts 1 and 2, respectively (Fig. a).

(a) The rectangle and the semicircular cutout.

Determine the Values for the Parts   From Appendix B, the x coordinate of the centroid of the cutout is

$$\bar{x}_2 = \frac{4R}{3\pi} = \frac{4(100)}{3\pi} \text{ mm.}$$

The information for determining the x coordinate of the centroid is summarized in the table. Notice that we treat the cutout as a negative area.

Information for determining $\bar{x}$

	$\bar{x}_i$ (mm)	A_i (mm^2)	$\bar{x}_i A_i$ (mm^3)
Part 1 (rectangle)	100	(200)(280)	(100)[(200)(280)]
Part 2 (cutout)	$\dfrac{4(100)}{3\pi}$	$-\frac{1}{2}\pi(100)^2$	$-\dfrac{4(100)}{3\pi}\left[\frac{1}{2}\pi(100)^2\right]$

Calculate the Centroid   The x coordinate of the centroid is

$$\bar{x} = \frac{\bar{x}_1 A_1 + \bar{x}_2 A_2}{A_1 + A_2} = \frac{(100)[(200)(280)] - \dfrac{4(100)}{3\pi}\left[\frac{1}{2}\pi(100)^2\right]}{(200)(280) - \frac{1}{2}\pi(100)^2} = 122 \text{ mm}$$

Because of the symmetry of the area, $\bar{y} = 0$.

Critical Thinking

If you try to divide the area into simple parts, you will gain appreciation for the approach we used. We were able to determine the centroid by dealing with two simple areas, the rectangular area without the cutout and the semicircular cutout. Determining centroids of areas can often be simplified in this way.

Problems

▶ **7.27** In Active Example 7.3, suppose that the area is placed as shown. Let the dimensions $R = 6$ in, $c = 14$ in, and $b = 18$ in. Use Eq. $(7.9)_1$ to determine the x coordinate of the centroid.

Problem 7.27

▶ **7.28** In Example 7.4, suppose that the area is given a second semicircular cutout as shown. Determine the x coordinate of the centroid.

Problem 7.28

For Problems 7.29–7.36, determine the coordinates of the centroids.

Problem 7.29

Problem 7.30

Problem 7.31

Problem 7.32

Problem 7.33

Problem 7.34

Problem 7.35

Problem 7.36

7.37 The dimensions $b = 42$ mm and $h = 22$ mm. Determine the y coordinate of the centroid of the beam's cross section.

7.38 If the cross-sectional area of the beam is 8400 mm² and the y coordinate of the centroid of the area is $\bar{y} = 90$ mm, what are the dimensions b and h?

Problems 7.37/7.38

7.39 Determine the y coordinate of the centroid of the beam's cross section.

Problem 7.39

7.40 Determine the coordinates of the centroid of the airplane's vertical stabilizer.

Problem 7.40

7.41 The area has elliptical boundaries. If $a = 30$ mm, $b = 15$ mm, and $\varepsilon = 6$ mm, what is the x coordinate of the centroid of the area?

7.42 By determining the x coordinate of the centroid of the area shown in Problem 7.41 in terms of a, b, and ε, and evaluating its limit as $\varepsilon \to 0$, show that the x coordinate of the centroid of a quarter-elliptical line is

$$\bar{x} = \frac{4a(a + 2b)}{3\pi(a + b)}.$$

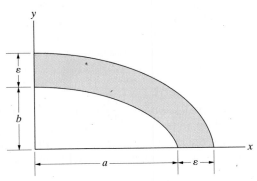

Problems 7.41/7.42

7.43 Three sails of a New York pilot schooner are shown. The coordinates of the points are in feet. Determine the centroid of sail 1.

7.44 Determine the centroid of sail 2.

7.45 Determine the centroid of sail 3.

Problems 7.43–7.45

7.3 Distributed Loads

BACKGROUND

The load exerted on a beam (stringer) supporting a floor of a building is distributed over the beam's length (Fig. 7.5a). The load exerted by wind on a television transmission tower is distributed along the tower's height (Fig. 7.5b). In many engineering applications, loads are continuously distributed along lines. We will show that the concept of the centroid of an area can be useful in the analysis of objects subjected to such loads.

Figure 7.5
Examples of distributed forces:
(a) Uniformly distributed load exerted on a beam of a building's frame by the floor.
(b) Wind load distributed along the height of a tower.

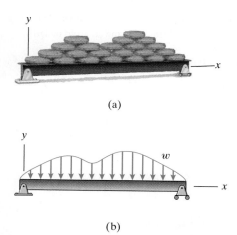

(a)

(b)

Figure 7.6
(a) Loading a beam with bags of sand.
(b) The distributed load w models the load exerted by the bags.

Describing a Distributed Load

We can use a simple example to demonstrate how such loads are expressed analytically. Suppose that we pile bags of sand on a beam, as shown in Fig. 7.6a. It is clear that the load exerted by the bags is distributed over the length of the beam and that its magnitude at a given position x depends on how high the bags are piled at that position. To describe the load, we define a function w such that the *downward* force exerted on an infinitesimal element dx of the beam is $w\,dx$. With this function we can model the varying magnitude of the load exerted by the sand bags (Fig. 7.6b). The arrows in the figure indicate that the load acts in the downward direction. Loads distributed along lines, from simple examples such as a beam's own weight to complicated ones such as the lift distributed along the length of an airplane's wing, are modeled by the function w. Since the product of w and dx is a force, the dimensions of w are (force)/(length). For example, w can be expressed in newtons per meter in SI units or in pounds per foot in U.S. Customary units.

Determining Force and Moment

Let's assume that the function w describing a particular distributed load is known (Fig. 7.7a). The graph of w is called the *loading curve*. Since the force acting on an element dx of the line is $w\,dx$, we can determine the total force F exerted by the distributed load by integrating the loading curve with respect to x:

$$F = \int_L w\,dx. \tag{7.10}$$

We can also integrate to determine the moment about a point exerted by the distributed load. For example, the moment about the origin due to the force exerted on the element dx is $xw\,dx$, so the total moment about the origin due to the distributed load is

$$M = \int_L xw\,dx. \tag{7.11}$$

When you are concerned only with the total force and moment exerted by a distributed load, you can represent it by a single equivalent force F (Fig. 7.7b). For equivalence, the force must act at a position $\bar{x}$ on the x axis such that the moment of F about the origin is equal to the moment of the distributed load about the origin:

$$\bar{x}F = \int_L xw\,dx.$$

Therefore the force F is equivalent to the distributed load if we place it at the position

$$\bar{x} = \frac{\int_L xw\,dx}{\int_L w\,dx}. \tag{7.12}$$

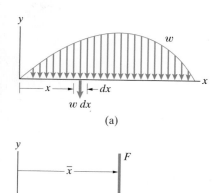

(a)

(b)

Figure 7.7
(a) A distributed load and the force exerted on a differential element dx.
(b) The equivalent force.

The Area Analogy

Notice that the term $w\,dx$ is equal to an element of "area" dA between the loading curve and the x axis (Fig. 7.8a). (We use quotation marks because $w\,dx$ is actually a force and not an area.) Interpreted in this way, Eq. (7.10) states that the total force exerted by the distributed load is equal to the "area" A between the loading curve and the x axis:

$$F = \int_L w\,dx = \int_A dA = A. \tag{7.13}$$

Substituting $w\,dx = dA$ into Eq. (7.12), we obtain

$$\bar{x} = \frac{\displaystyle\int_L xw\,dx}{\displaystyle\int_L w\,dx} = \frac{\displaystyle\int_A x\,dA}{\displaystyle\int_A dA}. \tag{7.14}$$

The force F is equivalent to the distributed load if it acts at the centroid of the "area" between the loading curve and the x axis (Fig. 7.8b). Using this analogy to represent a distributed load by an equivalent force can be very useful when the loading curve is relatively simple.

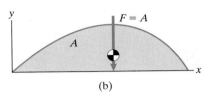

Figure 7.8
(a) Determining the "area" between the function w and the x axis.
(b) The equivalent force is equal to the "area," and the line of action passes through its centroid.

RESULTS

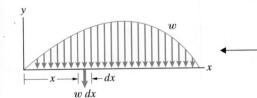

To represent a load that is distributed along the x axis, we define a function w such that the downward force on an element dx of the x axis is $w\,dx$. The graph of w is called the *loading curve*.

$$F = \int_L w\,dx, \qquad (7.10)$$

$$M = \int_L xw\,dx, \qquad (7.11)$$

The total downward force and the total clockwise moment about the origin due to a distributed load w acting on an interval L of the x axis can be determined by integration.

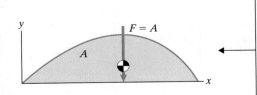

The total downward force F due to a distributed load is equal to the "area" A between the loading curve and the x axis. When this force is represented by a vector, the force vector is *equivalent* to the distributed load if it is placed at the centroid of the "area." (That is, the clockwise moment about the origin due to the force vector is equal to M.) This is called the *area analogy*.

Active Example 7.5 **Beam with a Distributed Load** (▶ *Related Problem 7.46*)

The beam is subjected to a "triangular" distributed load whose value at B is 100 N/m. (That is, the function w increases linearly from $w = 0$ at A to $w = 100$ N/m at B.) Determine the reactions on the beam at A and B.

Strategy
We can use the area analogy to represent the distributed load by an equivalent force. Then we can apply the equilibrium equations to determine the reactions at A and B.

Solution

The "area" of the triangular distributed load is one-half its base times its height, or $\frac{1}{2}(12$ m$) \times (100$ N/m$) = 600$ N. The centroid of the triangular "area" is located at $\bar{x} = \frac{2}{3}(12$ m$) = 8$ m.

$$\Sigma F_x = A_x = 0,$$
$$\Sigma F_y = A_y + B - 600 \text{ N} = 0,$$
$$\Sigma M_{\text{point } A} = (12 \text{ m})B - (8 \text{ m})(600 \text{ N}) = 0.$$

Apply equilibrium.

Solving yields $A_x = 0$, $A_y = 200$ N, and $B = 400$ N.

Practice Problem (a) Determine w as a function of x for the triangular distributed load in this example. (b) Use Eqs. (7.10) and (7.11) to determine the total downward force and the total clockwise moment about the left end of the beam due to the triangular distributed load.

Answer: (a) $w = \dfrac{100}{12}x$ N/m. (b) $F = 600$ N, $M = 4800$ N-m.

| Example 7.6 | Beam Subjected to Distributed Loads (▶ *Related Problem 7.48*) |

The beam is subjected to two distributed loads. Determine the reactions at A and B.

Strategy

We can easily apply the area analogy to the uniformly distributed load between A and B. We will treat the distributed load on the vertical section of the beam as the sum of uniform and triangular distributed loads and use the area analogy to represent each distributed load by an equivalent force.

Solution

We draw the free-body diagram of the beam in Fig. a, expressing the left distributed load as the sum of uniform and triangular loads. In Fig. b, we represent the three distributed loads by equivalent forces. The "area" of the uniform distributed load on the right is $(6\text{ m}) \times (400\text{ N/m}) = 2400\text{ N}$, and its centroid is 3 m from B. The area of the uniform distributed load on the vertical part of the beam is $(6\text{ m}) \times (400\text{ N/m}) = 2400\text{ N}$, and its centroid is located at $y = 3$ m. The area of the triangular distributed load is $\frac{1}{2}(6\text{ m}) \times (400\text{ N/m}) = 1200\text{ N}$, and its centroid is located at $y = \frac{1}{3}(6\text{ m}) = 2$ m.

From the equilibrium equations

$$\Sigma F_x = A_x + 1200\text{ N} + 2400\text{ N} = 0,$$

$$\Sigma F_y = A_y + B - 2400\text{ N} = 0,$$

$$\Sigma M_{\text{point } A} = (6\text{ m})B - (3\text{ m})(2400\text{ N}) - (2\text{ m})(1200\text{ N}) - (3\text{ m})(2400\text{ N}) = 0,$$

we obtain $A_x = -3600$ N, $A_y = -400$ N, and $B = 2800$ N.

(a) Free-body diagram of the beam.

(b) Representing the distributed loads by equivalent forces.

Critical Thinking

When you analyze a problem involving distributed loads, should you always use the area analogy to represent them as we did in this example? The area analogy is useful when a loading curve is sufficiently simple that its area and the location of its centroid are easy to determine. When that is not the case, you can use Eqs. (7.10) and (7.11) to determine the force and moment exerted by a distributed load. We illustrate this approach in Example 7.7.

Example 7.7	**Beam with a Distributed Load** (▶ *Related Problem 7.49*)

The beam is subjected to a distributed load, a force, and a couple. The distributed load is $w = 300x - 50x^2 + 0.3x^4$ lb/ft. Determine the reactions at the fixed support A.

Strategy

Since we know the function w, we can use Eqs. (7.10) and (7.11) to determine the force and moment exerted on the beam by the distributed load. We can then use the equilibrium equations to determine the reactions at A.

Solution

We isolate the beam and show the reactions at the fixed support in Fig. a. The *downward* force exerted by the distributed load is

$$\int_L w \, dx = \int_0^{10} (300x - 50x^2 + 0.3x^4) \, dx = 4330 \text{ lb}.$$

The *clockwise* moment about A exerted by the distributed load is

$$\int_L xw \, dx = \int_0^{10} x(300x - 50x^2 + 0.3x^4) \, dx = 25{,}000 \text{ ft-lb}.$$

From the equilibrium equations

$$\Sigma F_x = A_x = 0,$$

$$\Sigma F_y = A_y - 4330 \text{ lb} + 2000 \text{ lb} = 0,$$

$$\Sigma M_{\text{point } A} = M_A - 25{,}000 \text{ ft-lb} + (20 \text{ ft})(2000 \text{ lb}) + 10{,}000 \text{ ft-lb} = 0,$$

we obtain $A_x = 0$, $A_y = 2330$ lb, and $M_A = -25{,}000$ ft-lb.

(a) Free-body diagram of the beam.

Critical Thinking

When you use Eq. (7.11), it is important to be aware that you are calculating the *clockwise* moment due to the distributed load w *about the origin* $x = 0$.

Problems

▶ **7.46** In Active Example 7.5, suppose that the distributed load is modified as shown. Determine the reactions on the beam at A and B.

Problem 7.46

7.47 Determine the reactions at A and B.

Problem 7.47

▶ **7.48** In Example 7.6, suppose that the distributed loads are modified as shown. Determine the reactions on the beam at A and B.

Problem 7.48

▶ **7.49** In Example 7.7, suppose that the distributed load acting on the beam from $x = 0$ to $x = 10$ ft is given by $w = 350 + 0.3x^3$ lb/ft. (a) Determine the downward force and the clockwise moment about A exerted by the distributed load. (b) Determine the reactions at the fixed support.

7.50 Determine the reactions at the fixed support A.

Problem 7.50

7.51 An engineer measures the forces exerted by the soil on a 10-m section of a building foundation and finds that they are described by the distributed load $w = -10x - x^2 + 0.2x^3$ kN/m.
(a) Determine the magnitude of the total force exerted on the foundation by the distributed load.
(b) Determine the magnitude of the moment about A due to the distributed load.

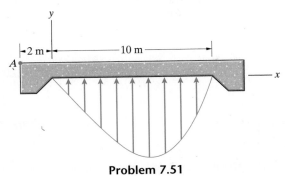

Problem 7.51

7.52 Determine the reactions on the beam at A and B.

Problem 7.52

7.53 The aerodynamic lift of the wing is described by the distributed load $w = -300 \sqrt{1 - 0.04x^2}$ N/m. The mass of the wing is 27 kg, and its center of mass is located 2 m from the wing root R.

(a) Determine the magnitudes of the force and the moment about R exerted by the lift of the wing.

(b) Determine the reactions on the wing at R.

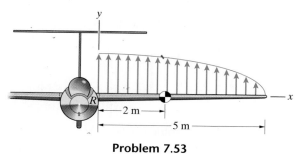

Problem 7.53

7.54 Determine the reactions on the bar at A and B.

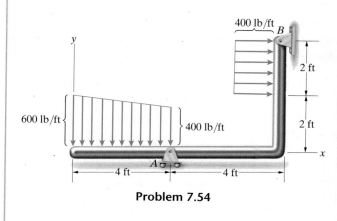

Problem 7.54

7.55 Determine the reactions on member AB at A and B.

Problem 7.55

7.56 Determine the axial forces in members BD, CD, and CE of the truss and indicate whether they are in tension (T) or compression (C).

Problem 7.56

7.57 Determine the reactions on member *ABC* at *A* and *B*.

7.58 Determine the forces on member *ABC* of the frame.

Problem 7.58

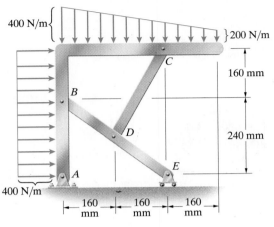

Problem 7.57

7.4 Centroids of Volumes and Lines

BACKGROUND

Here we define the centroids, or average positions, of volumes and lines, and show how to determine the centroids of composite volumes and lines. We will show in Section 7.7 that knowing the centroids of volumes and lines allows you to determine the centers of mass of certain types of objects, which tells you where their weights effectively act.

Volumes Consider a volume *V*, and let *dV* be a differential element of *V* with coordinates *x*, *y*, and *z* (Fig. 7.9). By analogy with Eqs. (7.6) and (7.7), the coordinates of the centroid of *V* are

$$
\bar{x} = \frac{\int_V x\,dV}{\int_V dV}, \qquad \bar{y} = \frac{\int_V y\,dV}{\int_V dV}, \qquad \bar{z} = \frac{\int_V z\,dV}{\int_V dV}. \qquad (7.15)
$$

The subscript *V* on the integral signs means that the integration is carried out over the entire volume.

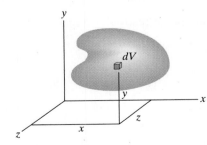

Figure 7.9
A volume *V* and differential element *dV*.

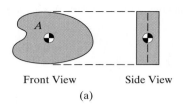

Front View   Side View

(a)

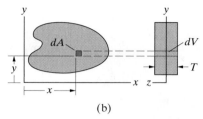

(b)

Figure 7.10
(a) A volume of uniform thickness.
(b) Obtaining dV by projecting dA through the volume.

If a volume has the form of a plate with uniform thickness and cross-sectional area A (Fig. 7.10a), its centroid coincides with the centroid of A and lies at the midpoint between the two faces. To show that this is true, we obtain a volume element dV by projecting an element dA of the cross-sectional area through the thickness T of the volume, so that $dV = T\,dA$ (Fig. 7.10b). Then the x and y coordinates of the centroid of the volume are

$$\bar{x} = \frac{\displaystyle\int_V x\,dV}{\displaystyle\int_V dV} = \frac{\displaystyle\int_A xT\,dA}{\displaystyle\int_A T\,dA} = \frac{\displaystyle\int_A x\,dA}{\displaystyle\int_A dA},$$

$$\bar{y} = \frac{\displaystyle\int_V y\,dV}{\displaystyle\int_V dV} = \frac{\displaystyle\int_A yT\,dA}{\displaystyle\int_A T\,dA} = \frac{\displaystyle\int_A y\,dA}{\displaystyle\int_A dA}.$$

The coordinate $\bar{z} = 0$ by symmetry. Thus you know the centroid of this type of volume if you know (or can determine) the centroid of its cross-sectional area.

Lines   The coordinates of the centroid of a line L are

$$\bar{x} = \frac{\displaystyle\int_L x\,dL}{\displaystyle\int_L dL}, \qquad \bar{y} = \frac{\displaystyle\int_L y\,dL}{\displaystyle\int_L dL}, \qquad \bar{z} = \frac{\displaystyle\int_L z\,dL}{\displaystyle\int_L dL}, \qquad (7.16)$$

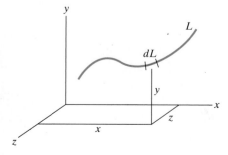

Figure 7.11
A line L and differential element dL.

where dL is a differential length of the line with coordinates x, y, and z. (Fig. 7.11).

RESULTS

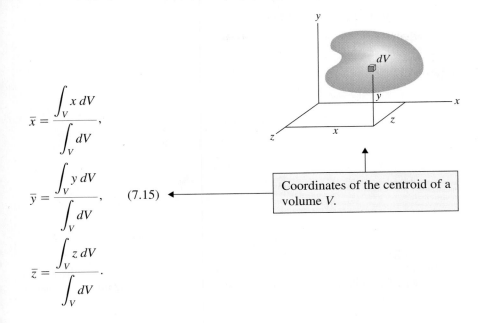

$$\bar{x} = \frac{\int_V x \, dV}{\int_V dV},$$

$$\bar{y} = \frac{\int_V y \, dV}{\int_V dV}, \qquad (7.15)$$

$$\bar{z} = \frac{\int_V z \, dV}{\int_V dV}.$$

Coordinates of the centroid of a volume V.

If a volume has the form of a plate with uniform thickness and cross-sectional area A, its centroid coincides with the centroid of A and lies at the midpoint between the two faces.

Front View   Side View

$$\bar{x} = \frac{\int_L x \, dL}{\int_L dL},$$

$$\bar{y} = \frac{\int_L y \, dL}{\int_L dL}, \qquad (7.16)$$

$$\bar{z} = \frac{\int_L z \, dL}{\int_L dL}.$$

Coordinates of the centroid of a line L.

Active Example 7.8   **Centroid of a Cone by Integration** (▶ *Related Problem 7.59*)

Determine the centroid of the cone.

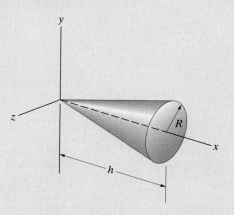

Strategy

Because of the axial symmetry of the cone, the centroid must lie on the x axis. We will determine the x coordinate of the centroid by applying Eq. $(7.15)_1$ using an element of volume dV in the form of a disk of thickness dx.

Solution

An element of volume in the form of a disk.

The radius of the disk at position x is $(R/h)x$. The volume of the disk is the product of the area of the disk and its thickness:

$$dV = \pi\left[\left(\frac{R}{h}\right)x\right]^2 dx.$$

$$\bar{x} = \frac{\displaystyle\int_V x\,dV}{\displaystyle\int_V dV} = \frac{\displaystyle\int_0^h x\pi\left[\left(\frac{R}{h}\right)x\right]^2\,dx}{\displaystyle\int_0^h \pi\left[\left(\frac{R}{h}\right)x\right]^2\,dx} = \frac{3}{4}h.$$

Apply Eq. $(7.15)_1$.

Practice Problem The radius in feet of the circular cross section of the truncated cone is given as a function of x by $r = 1 + \frac{1}{4}x$. Determine the x coordinate of its centroid.

Answer: 2.43 ft.

Example 7.9   **Centroid of a Line by Integration** (▶ *Related Problem 7.66*)

The line L is defined by the function $y = x^2$. Determine the x coordinate of its centroid.

Strategy
We can express a differential element dL of a line (Fig. a) in terms of dx and dy:

$$dL = \sqrt{dx^2 + dy^2} = \sqrt{1 + \left(\frac{dy}{dx}\right)^2}\, dx.$$

From the equation describing the line, the derivative $dy/dx = 2x$, so we obtain an expression for dL in terms of x:

$$dL = \sqrt{1 + 4x^2}\, dx.$$

Solution
To integrate over the entire line, we must integrate from $x = 0$ to $x = 1$. The x coordinate of the centroid is

$$\bar{x} = \frac{\displaystyle\int_L x\, dL}{\displaystyle\int_L dL} = \frac{\displaystyle\int_0^1 x\sqrt{1 + 4x^2}\, dx}{\displaystyle\int_0^1 \sqrt{1 + 4x^2}\, dx} = 0.574.$$

(a) A differential line element dL.

Critical Thinking
Our approach in this example is appropriate to determine the centroid of a line that is described by a function of the form $y = f(x)$. In Example 7.10 we show how to determine the centroid of a line that is described in terms of polar coordinates.

Example 7.10	**Centroid of a Semicircular Line by Integration** (▶ *Related Problem 7.70*)

Determine the centroid of the semicircular line.

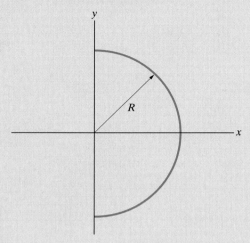

Strategy

Because of the symmetry of the line, the centroid lies on the x axis. To determine $\bar{x}$, we will integrate in terms of polar coordinates. By letting θ change by an amount $d\theta$, we obtain a differential line element of length $dL = R\, d\theta$ (Fig. a). The x coordinate of dL is $x = R \cos \theta$.

Solution

To integrate over the entire line, we must integrate with respect to θ from $\theta = -\pi/2$ to $\theta = +\pi/2$:

$$\bar{x} = \frac{\displaystyle\int_L x\, dL}{\displaystyle\int_L dL} = \frac{\displaystyle\int_{-\pi/2}^{\pi/2} (R \cos \theta) R\, d\theta}{\displaystyle\int_{-\pi/2}^{\pi/2} R\, d\theta} = \frac{R^2 [\sin \theta]_{-\pi/2}^{\pi/2}}{R[\theta]_{-\pi/2}^{\pi/2}} = \frac{2R}{\pi}.$$

(a) A differential line element $dL = R\, d\theta$.

Critical Thinking

Notice that our integration procedure gives the correct length of the line:

$$\int_L dL = \int_{-\pi/2}^{\pi/2} R\, d\theta = R[\theta]_{-\pi/2}^{\pi/2} = \pi R.$$

Problems

▶ **7.59** Use the method described in Active Example 7.8 to determine the centroid of the truncated cone.

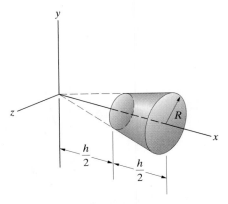

Problem 7.59

7.60 A grain storage tank has the form of a surface of revolution with the profile shown. The height of the tank is 7 m and its diameter at ground level is 10 m. Determine the volume of the tank and the height *above ground level* of the centroid of its volume.

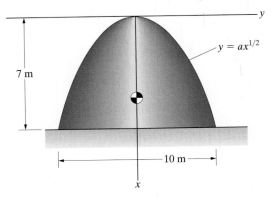

Problem 7.60

7.61 The object shown, designed to serve as a pedestal for a speaker, has a profile obtained by revolving the curve $y = 0.167x^2$ about the x axis. What is the x coordinate of the centroid of the object?

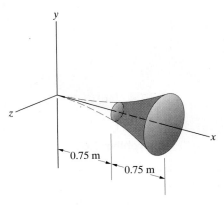

Problem 7.61

7.62 The volume of a nose cone is generated by rotating the function $y = x - 0.2x^2$ about the x axis.

(a) What is the volume of the nose cone?

(b) What is the x coordinate of the centroid of the volume?

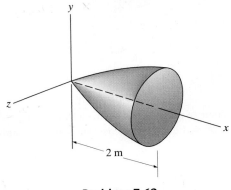

Problem 7.62

7.63 Determine the centroid of the hemispherical volume.

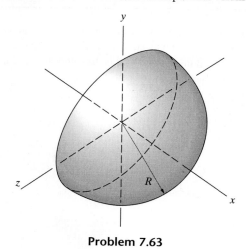

Problem 7.63

7.64 The volume consists of a segment of a sphere of radius R. Determine its centroid.

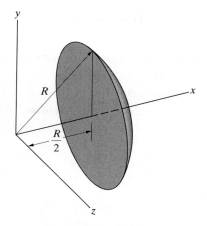

Problem 7.64

7.65 A volume of revolution is obtained by revolving the curve $x^2/a^2 + y^2/b^2 = 1$ about the x axis. Determine its centroid.

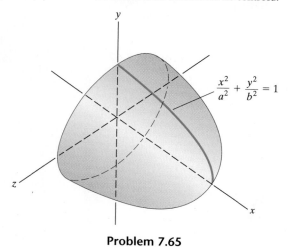

Problem 7.65

▶ **7.66** In Example 7.9, determine the y coordinate of the centroid of the line.

7.67 Determine the coordinates of the centroid of the line.

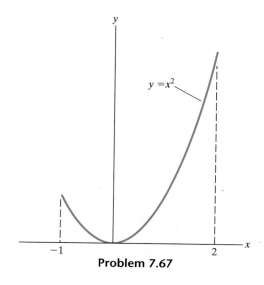

Problem 7.67

7.68 Determine the x coordinate of the centroid of the line.

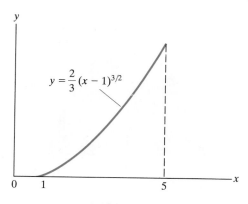

Problem 7.68

7.69 Determine the x coordinate of the centroid of the line.

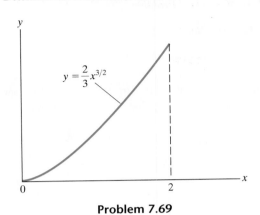

$y = \frac{2}{3}x^{3/2}$

Problem 7.69

▶ **7.70** Use the method described in Example 7.10 to determine the centroid of the circular arc.

Problem 7.70

7.5 Composite Volumes and Lines

BACKGROUND

The centroids of composite volumes and lines can be derived using the same approach we applied to areas. The coordinates of the centroid of a composite volume are

$$\bar{x} = \frac{\sum\limits_i \bar{x}_i V_i}{\sum\limits_i V_i}, \qquad \bar{y} = \frac{\sum\limits_i \bar{y}_i V_i}{\sum\limits_i V_i}, \qquad \bar{z} = \frac{\sum\limits_i \bar{z}_i V_i}{\sum\limits_i V_i}, \qquad (7.17)$$

and the coordinates of the centroid of a composite line are

$$\bar{x} = \frac{\sum\limits_i \bar{x}_i L_i}{\sum\limits_i L_i}, \qquad \bar{y} = \frac{\sum\limits_i \bar{y}_i L_i}{\sum\limits_i L_i}, \qquad \bar{z} = \frac{\sum\limits_i \bar{z}_i L_i}{\sum\limits_i L_i}. \qquad (7.18)$$

The centroids of some simple volumes and lines are tabulated in Appendices B and C.

Determining the centroid of a composite volume or line requires three steps:

1. Choose the parts—Try to divide the composite into parts whose centroids you know or can easily determine.

2. Determine the values for the parts—Determine the centroid and the volume or length of each part. Watch for instances of symmetry that can simplify your task.

3. Calculate the centroid—Use Eqs. (7.17) or (7.18) to determine the centroid of the composite volume or line.

Active Example 7.11 **Centroid of a Composite Volume** (▶ *Related Problem 7.71*)

Determine the x coordinate of the centroid of the composite volume.

Strategy

We must divide the volume into simple parts (in this example the parts are obvious), determine the volumes and centroid locations for the parts, and apply Eq. $(7.17)_1$.

Solution

Choose the Parts
Divide the volume into simple parts. The x coordinates of the centroids of the parts are shown. See Appendix C.

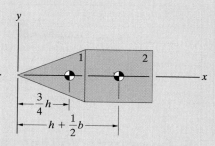

Determine the Values for the Parts
Tabulate the terms needed to apply Eq. $(7.17)_1$.

	$\bar{x}_i$	V_i	$\bar{x}_i V_i$
Part 1 (cone)	$\dfrac{3}{4}h$	$\dfrac{1}{3}\pi R^2 h$	$\left(\dfrac{3}{4}h\right)\left(\dfrac{1}{3}\pi R^2 h\right)$
Part 2 (cylinder)	$h + \dfrac{1}{2}b$	$\pi R^2 b$	$\left(h + \dfrac{1}{2}b\right)(\pi R^2 b)$

$$\bar{x} = \frac{\bar{x}_1 V_1 + \bar{x}_2 V_2}{V_1 + V_2}.$$

$$= \frac{\left(\dfrac{3}{4}h\right)\left(\dfrac{1}{3}\pi R^2 h\right) + \left(h + \dfrac{1}{2}b\right)(\pi R^2 b)}{\dfrac{1}{3}\pi R^2 h + \pi R^2 b}.$$

Calculate the Centroid
Use Eq. $(7.17)_1$ to determine the x component of the centroid.

Practice Problem The composite volume consists of a circular cylinder and a hemisphere. Determine the x coordinate of its centroid.

Answer: $\bar{x} = \dfrac{\left(\frac{1}{2}b\right)(\pi R^2 b) + \left(b + \frac{3}{8}R\right)\left(\frac{2}{3}\pi R^3\right)}{\pi R^2 b + \frac{2}{3}\pi R^3}.$

Example 7.12	Centroid of a Volume Containing a Cutout (▶ *Related Problem 7.72*)

Determine the centroid of the volume.

Strategy

We can divide this volume into the five simple parts shown in Fig. a. Notice that parts 2 and 3 *do not* have the cutout. It is assumed to be "filled in," which simplifies the geometries of those parts. Part 5, which is the volume of the 20-mm-diameter hole, will be treated as a negative volume in Eqs. (7.17).

Solution

Choose the Parts We can divide the volume into the five simple parts shown in Fig. a. Part 5 is the volume of the 20-mm-diameter hole.

Determine the Values for the Parts The centroids of parts 1 and 3 are located at the centroids of their semicircular cross sections (Fig. b). The information for determining the x-coordinate of the centroid is summarized in the table. Part 5 is a negative volume.

Side View

End View

Information for determining $\bar{x}$

	$\bar{x}_i$ (mm)	V_i (mm³)	$\bar{x}_i V_i$ (mm⁴)
Part 1	$-\dfrac{4(25)}{3\pi}$	$\dfrac{\pi(25)^2}{2}(20)$	$\left[-\dfrac{4(25)}{3\pi}\right]\left[\dfrac{\pi(25)^2}{2}(20)\right]$
Part 2	100	$(200)(50)(20)$	$(100)[(200)(50)(20)]$
Part 3	$200+\dfrac{4(25)}{3\pi}$	$\dfrac{\pi(25)^2}{2}(20)$	$\left[200+\dfrac{4(25)}{3\pi}\right]\left[\dfrac{\pi(25)^2}{2}(20)\right]$
Part 4	0	$\pi(25)^2(40)$	0
Part 5	200	$-\pi(10)^2(20)$	$-(200[\pi(10)^2(20)]$

(a) Dividing the line into three parts.

Calculate the Centroid The x coordinate of the centroid of the composite volume is

$$\bar{x} = \frac{\bar{x}_1 V_1 + \bar{x}_2 V_2 + \bar{x}_3 V_3 + \bar{x}_4 V_4 + \bar{x}_5 V_5}{V_1 + V_2 + V_3 + V_4 + V_5}$$

$\dfrac{4(25)}{3\pi}$ mm ⟶ | ⟵ 200 mm ⟶ | ⟵ $\dfrac{4(25)}{3\pi}$ mm

(b) Positions of the centroids of parts 1 and 3.

$$=\frac{\begin{aligned}&\left[-\dfrac{4(25)}{3\pi}\right]\left[\dfrac{\pi(25)^2}{2}(20)\right]+(100)[(200)(50)(20)]\\[2mm]&\quad+\left[200+\dfrac{4(25)}{3\pi}\right]\left[\dfrac{\pi(25)^2}{2}(20)\right]+0-(200)[\pi(10)^2(20)]\end{aligned}}{\dfrac{\pi(25)^2}{2}(20)+(200)(50)(20)+\dfrac{\pi(25)^2}{2}(20)+\pi(25)^2(40)-\pi(10)^2(20)}$$

$$= 72.77 \text{ mm}.$$

The z coordinates of the centroids of the parts are zero except $\bar{z}_4 = 30$ mm. Therefore the z coordinate of the centroid of the composite volume is

$$\bar{z} = \frac{\bar{z}_4 V_4}{V_1 + V_2 + V_3 + V_4 + V_5}$$

$$= \frac{30[\pi(25)^2(40)]}{\dfrac{\pi(25)^2}{2}(20) + (200)(50)(20) + \dfrac{\pi(25)^2}{2}(20) + \pi(25)^2(40) - \pi(10)^2(20)}$$

$$= 7.56 \text{ mm.}$$

Because of symmetry, $\bar{y} = 0$.

Critical Thinking
You can recognize that the volume in this example could be part of a mechanical device. Many manufactured parts have volumes that are composites of simple volumes, and the method used in this example can be used to determine their centroids and, if they are homogeneous, their centers of mass.

Example 7.13 | **Centroid of a Composite Line** (▶ *Related Problem 7.81*)

Determine the centroid of the line. The quarter-circular arc lies in the y–z plane.

Strategy
We must divide the line into parts (in this case the quarter-circular arc and the two straight segments), determine the centroids of the parts, and apply Eqs. (7.18).

Solution
Choose the Parts The line consists of a quarter-circular arc and two straight segments, which we call parts 1, 2, and 3 (Fig. a).

Determine the Values for the Parts From Appendix B, the coordinates of the centroid of the quarter-circular arc are $\bar{x}_1 = 0$, $\bar{y}_1 = \bar{z}_1 = 2(2)/\pi$ m. The centroids of the straight segments lie at their midpoints. For segment 2, $\bar{x}_2 = 2$ m, $\bar{y}_2 = 0$, and $\bar{z}_2 = 2$ m, and for segment 3, $\bar{x}_3 = 2$ m, $\bar{y}_3 = 1$ m, and $\bar{z}_3 = 1$ m. The length of segment 3 is $L_3 = \sqrt{(4)^2 + (2)^2 + (2)^2} = 4.90$ m. This information is summarized in the table.

(a) Dividing the line into three parts.

Information for determining the centroid.

	$\bar{x}_i$ (m)	$\bar{y}_i$ (m)	$\bar{z}_i$ (m)	L_i (m)
Part 1	0	$2(2)/\pi$	$2(2)/\pi$	$\pi(2)/2$
Part 2	2	0	2	4
Part 3	2	1	1	4.90

Calculate the Centroid The coordinates of the centroid of the composite line are

$$\bar{x} = \frac{\bar{x}_1 L_1 + \bar{x}_2 L_2 + \bar{x}_3 L_3}{L_1 + L_2 + L_3} = \frac{0 + (2)(4) + (2)(4.90)}{\pi + 4 + 4.90} = 1.478 \text{ m},$$

$$\bar{y} = \frac{\bar{y}_1 L_1 + \bar{y}_2 L_2 + \bar{y}_3 L_3}{L_1 + L_2 + L_3} = \frac{[2(2)/\pi][\pi(2)/2] + 0 + (1)(4.90)}{\pi + 4 + 4.90} = 0.739 \text{ m},$$

$$\bar{z} = \frac{\bar{z}_1 L_1 + \bar{z}_2 L_2 + \bar{z}_3 L_3}{L_1 + L_2 + L_3} = \frac{[2(2)/\pi][\pi(2)/2] + (2)(4) + (1)(4.90)}{\pi + 4 + 4.90} = 1.404 \text{ m}.$$

Critical Thinking

What possible reason could you have for wanting to know the centroid (average position) of a line? In Section 7.7 we show that the center of mass of a slender homogeneous bar, which is the point at which the weight of the bar can be represented by an equivalent force, lies approximately at the centroid of the bar's axis.

Problems

▶ **7.71** In Active Example 7.11, suppose that the cylinder is hollow with inner radius $R/2$ as shown. If the dimensions $R = 6$ in, $h = 12$ in, and $b = 10$ in, what is the x coordinate of the centroid of the volume?

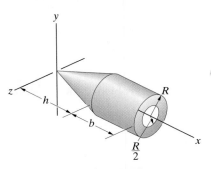

Problem 7.71

▶ **7.72** Use the procedure described in Example 7.12 to determine the x component of the centroid of the volume.

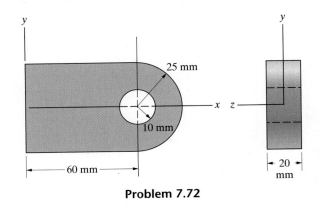

Problem 7.72

For Problems 7.73–7.78, determine the centroids of the volumes.

Problem 7.73

Problem 7.74

Problem 7.75

Problem 7.76

Problem 7.77

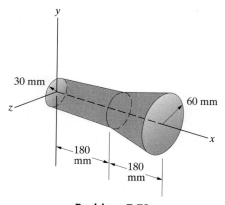

Problem 7.78

7.79 The dimensions of the *Gemini* spacecraft (in meters) were $a = 0.70$, $b = 0.88$, $c = 0.74$, $d = 0.98$, $e = 1.82$, $f = 2.20$, $g = 2.24$, and $h = 2.98$. Determine the centroid of its volume.

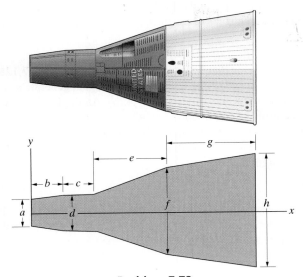

Problem 7.79

7.80 Two views of a machine element are shown. Determine the centroid of its volume.

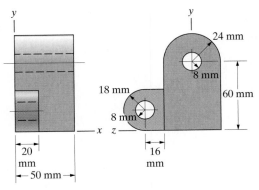

Problem 7.80

▶ **7.81** In Example 7.13, suppose that the circular arc is replaced by a straight line as shown. Determine the centroid of the three-segment line.

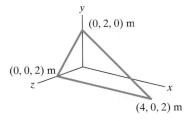

Problem 7.81

For Problems 7.82 and 7.83, determine the centroids of the lines.

Problem 7.82

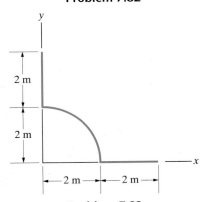

Problem 7.83

7.84 The semicircular part of the line lies in the *x–z* plane. Determine the centroid of the line.

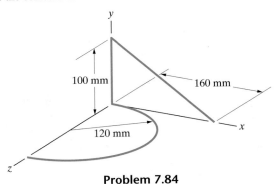

Problem 7.84

7.85 Determine the centroid of the line.

Problem 7.85

7.6 The Pappus–Guldinus Theorems

BACKGROUND

In this section we discuss two simple and useful theorems relating surfaces and volumes of revolution to the centroids of the lines and areas that generate them.

First Theorem

Consider a line L in the x–y plane that does not intersect the x axis (Fig. 7.12a). Let the coordinates of the centroid of the line be $(\bar{x}, \bar{y})$. We can generate a surface by revolving the line about the x axis (Fig. 7.12b). As the line revolves about the x axis, the centroid of the line moves in a circular path of radius $\bar{y}$.

The first Pappus–Guldinus theorem states that the area of the surface of revolution is equal to the product of the distance through which the centroid of the line moves and the length of the line:

$$A = 2\pi\bar{y}\,L. \tag{7.19}$$

To prove this result, we observe that as the line revolves about the x axis, the area dA generated by an element dL of the line is $dA = 2\pi y\,dL$, where y is the y coordinate of the element dL (Fig. 7.12c). Therefore, the total area of the surface of revolution is

$$A = 2\pi \int_L y\,dL. \tag{7.20}$$

From the definition of the y coordinate of the centroid of the line,

$$\bar{y} = \frac{\displaystyle\int_L y\,dL}{\displaystyle\int_L dL},$$

we obtain

$$\int_L y\,dL = \bar{y}L.$$

Substituting this result into Eq. (7.20), we obtain Eq. (7.19).

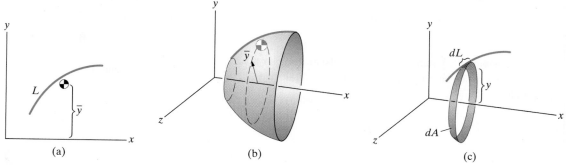

Figure 7.12
(a) A line L and the y coordinate of its centroid.
(b) The surface generated by revolving the line L about the x axis and the path followed by the centroid of the line.
(c) An element dL of the line and the element of area dA it generates.

Second Theorem

Consider an area A in the x–y plane that does not intersect the x axis (Fig. 7.13a). Let the coordinates of the centroid of the area be $(\bar{x}, \bar{y})$. We can generate a volume by revolving the area about the x axis (Fig. 7.13b). As the area revolves about the x axis, the centroid of the area moves in a circular path of length $2\pi\bar{y}$.

The second Pappus–Guldinus theorem states that the volume V of the volume of revolution is equal to the product of the distance through which the centroid of the area moves and the area:

$$V = 2\pi\bar{y}A. \tag{7.21}$$

As the area revolves about the x axis, the volume dV generated by an element dA of the area is $dV = 2\pi y \, dA$, where y is the y coordinate of the element dA (Fig. 7.13c). Therefore, the total volume is

$$V = 2\pi \int_A y \, dA. \tag{7.22}$$

From the definition of the y coordinate of the centroid of the area,

$$\bar{y} = \frac{\displaystyle\int_A y \, dA}{\displaystyle\int_A dA},$$

we obtain

$$\int_A y \, dA = \bar{y}A.$$

Substituting this result into Eq. (7.22), we obtain Eq. (7.21).

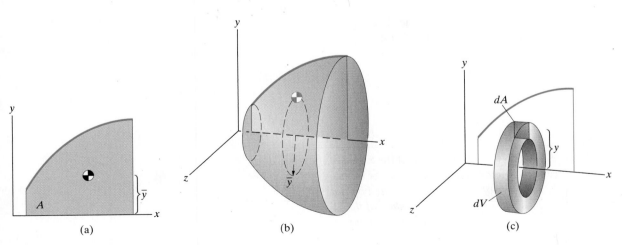

(a)   (b)   (c)

Figure 7.13
(a) An area A and the y coordinate of its centroid.
(b) The volume generated by revolving the area A about the x axis and the path followed by the centroid of the area.
(c) An element dA of the area and the element of volume dV it generates.

RESULTS

The First Pappus–Guldinus Theorem

The line L is in the x–y plane. The y coordinate of the centroid of L is $\bar{y}$.

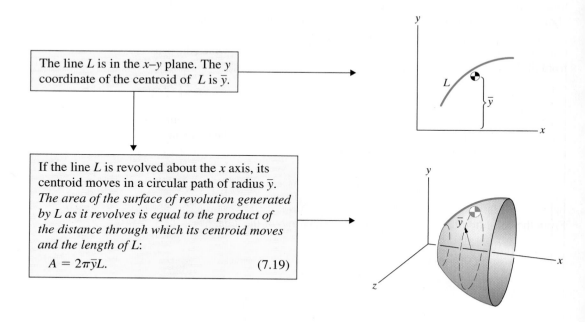

If the line L is revolved about the x axis, its centroid moves in a circular path of radius $\bar{y}$. *The area of the surface of revolution generated by L as it revolves is equal to the product of the distance through which its centroid moves and the length of L:*

$$A = 2\pi\bar{y}L. \tag{7.19}$$

The Second Pappus–Guldinus Theorem

The area A is in the x–y plane. The y coordinate of the centroid of A is $\bar{y}$.

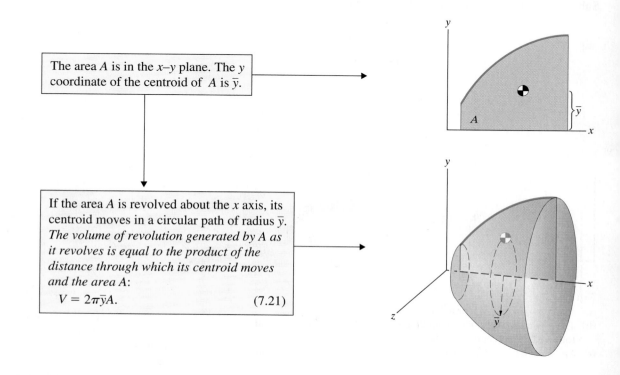

If the area A is revolved about the x axis, its centroid moves in a circular path of radius $\bar{y}$. *The volume of revolution generated by A as it revolves is equal to the product of the distance through which its centroid moves and the area A:*

$$V = 2\pi\bar{y}A. \tag{7.21}$$

Active Example 7.14   The Pappus–Guldinus Theorems (▶ *Related Problem 7.86*)

Use the first Pappus–Guldinus theorem to determine the surface area of the cone.

Strategy
We can generate the curved surface of the cone by revolving a straight line about an axis. Because the location of the centroid of the straight line is known, we can use the first Pappus–Guldinus theorem to determine the area of the curved surface.

Solution

> Revolving this straight line about the x axis generates the curved surface of the cone. The y coordinate of the centroid of the line is shown.
>
> The length of the line is $L = \sqrt{h^2 + R^2}$.
> The area of the curved surface is
>
> $$A = 2\pi \bar{y}_L L = \pi R \sqrt{h^2 + R^2}.$$
>
> Adding the area of the base, the total surface area of the cone is $\pi R \sqrt{h^2 + R^2} + \pi R^2$.

$\bar{y}_L = \dfrac{1}{2}R$

Practice Problem Use the second Pappus–Guldinus theorem to determine the volume of the cone.

Answer: $V = \dfrac{1}{3}\pi h R^2$.

Example 7.15   Determining a Centroid with a Pappus–Guldinus Theorem (▶ *Related Problem 7.88*)

The circumference of a sphere of radius R is $2\pi R$ and its surface area is $4\pi R^2$. Use this information to determine the centroid of a semicircular line.

Strategy
Revolving a semicircular line about an axis generates a spherical area. Knowing the area, we can use the first Pappus–Guldinus theorem to determine the centroid of the generating line.

Solution
The length of the semicircular line is $L = \pi R$, and $\bar{y}_L$ is the y coordinate of its centroid. Rotating the line about the x axis generates the surface of a sphere. The first Pappus–Guldinus theorem states that the surface area of the sphere is

$$(2\pi \bar{y}_L)L = 2\pi^2 R \bar{y}_L.$$

By equating this expression to the given surface area $4\pi R^2$, we obtain $\bar{y}_L$:

$$\bar{y}_L = \frac{2R}{\pi}.$$

Critical Thinking
If you can obtain a result by using the Pappus–Guldinus theorems, you will often save time and effort in comparison with other approaches. Compare this example with Example 7.10, in which we used integration to determine the centroid of a semicircular line.

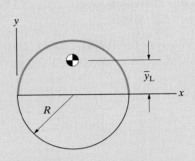

Revolving a semicircular line about the x axis.

Problems

▶ **7.86** Use the method described in Active Example 7.14 to determine the area of the curved part of the surface of the truncated cone.

7.87 Use the second Pappus–Guldinus theorem to determine the volume of the truncated cone.

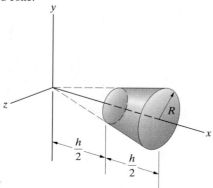

Problems 7.86/7.87

▶ **7.88** The area of the shaded semicircle is $\frac{1}{2}\pi R^2$. The volume of a sphere is $\frac{4}{3}\pi R^3$. Extend the approach described in Example 7.15 to the second Pappus–Guldinus theorem and determine the centroid $\bar{y}_S$ of the semicircular area.

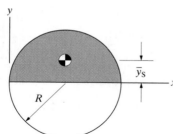

Problem 7.88

7.89 Use the second Pappus–Guldinus theorem to determine the volume generated by revolving the curve about the y axis.

7.90 The length of the curve is $L = 1.479$, and the area generated by rotating it about the x axis is $A = 3.810$. Use the first Pappus–Guldinus theorem to determine the y coordinate of the centroid of the curve.

7.91 Use the first Pappus–Guldinus theorem to determine the area of the surface generated by revolving the curve about the y axis.

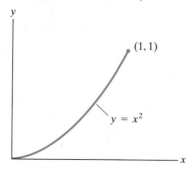

Problems 7.89–7.91

7.92 A nozzle for a large rocket engine is designed by revolving the function $y = \frac{2}{3}(x - 1)^{3/2}$ about the y axis. Use the first Pappus–Guldinus theorem to determine the surface area of the nozzle.

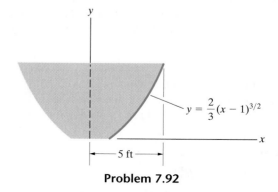

Problem 7.92

7.93 The coordinates of the centroid of the line are $\bar{x} = 332$ mm and $\bar{y} = 118$ mm. Use the first Pappus–Guldinus theorem to determine the area of the surface of revolution obtained by revolving the line about the x axis.

7.94 The coordinates of the centroid of the area between the x axis and the line are $\bar{x} = 355$ mm and $\bar{y} = 78.4$ mm. Use the second Pappus–Guldinus theorem to determine the volume obtained by revolving the area about the x axis.

Problems 7.93/7.94

7.95 The volume of revolution contains a hole of radius R.

(a) Use integration to determine its volume.

(b) Use the second Pappus–Guldinus theorem to determine its volume.

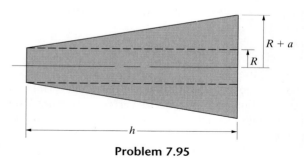

Problem 7.95

7.96 Determine the volume of the volume of revolution.

7.97 Determine the surface area of the volume of revolution.

7.98 The volume of revolution has an elliptical cross section. Determine its volume.

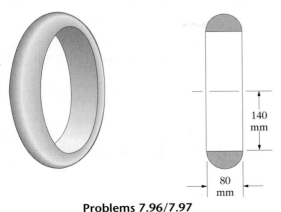

140 mm

80 mm

Problems 7.96/7.97

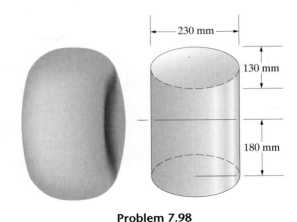

230 mm

130 mm

180 mm

Problem 7.98

7.7 Centers of Mass of Objects

BACKGROUND

The *center of mass* of an object is the centroid, or average position, of its mass. Here we give the analytical definition of the center of mass and demonstrate one of its most important properties: *An object's weight can be represented by a single equivalent force acting at its center of mass.* We then discuss how to locate centers of mass and show that for particular classes of objects, the center of mass coincides with the centroid of a volume, area, or line.

The center of mass of an object is defined by

$$\bar{x} = \frac{\int_m x \, dm}{\int_m dm}, \qquad \bar{y} = \frac{\int_m y \, dm}{\int_m dm}, \qquad \bar{z} = \frac{\int_m z \, dm}{\int_m dm}, \qquad (7.23)$$

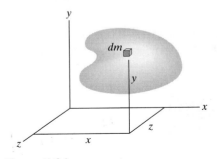

Figure 7.14
An object and differential element of mass *dm*.

where *x*, *y*, and *z* are the coordinates of the differential element of mass *dm* (Fig. 7.14). The subscripts *m* indicate that the integration must be carried out over the entire mass of the object.

Before considering how to determine the center of mass of an object, we will demonstrate that the weight of an object can be represented by a single equivalent force acting at its center of mass. Consider an element of mass *dm* of an object (Fig. 7.15a). If the *y* axis of the coordinate system points upward, the weight of *dm* is $-dm\,g\,\mathbf{j}$. Integrating this expression over the mass *m*, we obtain the total weight of the object,

$$\int_m - g\mathbf{j} \, dm = -mg\mathbf{j} = -W\mathbf{j}.$$

The moment of the weight of the element *dm* about the origin is

$$(x\mathbf{i} + y\mathbf{j} + z\mathbf{k}) \times (-dm\,g\,\mathbf{j}) = gz\mathbf{i} \, dm - gx\mathbf{k} \, dm.$$

Integrating this expression over *m*, we obtain the total moment about the origin due to the weight of the object:

$$\int_m (gz\mathbf{i} \, dm - gx\mathbf{k} \, dm) = mg\bar{z}\mathbf{i} - mg\bar{x}\mathbf{k} = W\bar{z}\mathbf{i} - W\bar{x}\mathbf{k}.$$

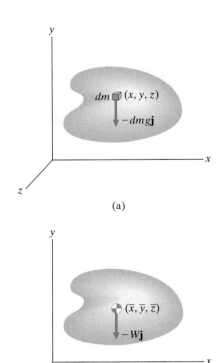

(a)

(b)

Figure 7.15
(a) Weight of the element dm.
(b) Representing the weight by a single
force at the center of mass.

If we represent the weight of the object by the force $-W\mathbf{j}$ acting at the center of mass (Fig. 7.15b), the moment of this force about the origin is equal to the total moment due to the weight:

$$(\bar{x}\mathbf{i} + \bar{y}\mathbf{j} + \bar{z}\mathbf{k}) \times (-W\mathbf{j}) = W\bar{z}\mathbf{i} - W\bar{x}\mathbf{k}.$$

This result shows that when we are concerned only with the total force and total moment exerted by the weight of an object, we can assume that its weight acts at the center of mass.

To apply Eqs. (7.23) to specific objects, we will change the variable of integration from mass to volume by introducing the *density*. The density ρ of an object is defined such that the mass of a differential element dV of the volume of the object is $dm = \rho\, dV$. The dimensions of ρ are therefore (mass/volume). For example, it can be expressed in kg/m^3 in SI units or in $slug/ft^3$ in U.S. Customary units. The total mass of an object is

$$m = \int_m dm = \int_V \rho\, dV. \tag{7.24}$$

An object whose density is uniform throughout its volume is said to be *homogeneous*. In this case, the total mass equals the product of the density and the volume:

$$m = \rho \int_V dV = \rho V. \quad \text{Homogeneous object} \tag{7.25}$$

The *weight density* is defined by $\gamma = g\rho$. It can be expressed in N/m^3 in SI units or in lb/ft^3 in U.S. Customary units. The weight of an element of volume dV of an object is $dW = \gamma\, dV$, and the total weight of a homogeneous object equals γV.

By substituting $dm = \rho\, dV$ into Eq. (7.23), we can express the coordinates of the center of mass in terms of volume integrals:

$$\bar{x} = \frac{\displaystyle\int_V \rho x\, dV}{\displaystyle\int_V \rho\, dV}, \qquad \bar{y} = \frac{\displaystyle\int_V \rho y\, dV}{\displaystyle\int_V \rho\, dV}, \qquad \bar{z} = \frac{\displaystyle\int_V \rho z\, dV}{\displaystyle\int_V \rho\, dV}. \tag{7.26}$$

If ρ is known as a function of position in an object, these expressions determine its center of mass. Furthermore, we can use these expressions to show that the centers of mass of particular classes of objects coincide with centroids of volumes, areas, and lines:

- **The center of mass of a homogeneous object coincides with the centroid of its volume.** If an object is homogeneous, ρ = constant and Eqs. (7.26) become the equations for the centroid of the volume,

$$\bar{x} = \frac{\displaystyle\int_V x\, dV}{\displaystyle\int_V dV}, \qquad \bar{y} = \frac{\displaystyle\int_V y\, dV}{\displaystyle\int_V dV}, \qquad \bar{z} = \frac{\displaystyle\int_V z\, dV}{\displaystyle\int_V dV}.$$

- **The center of mass of a homogeneous plate of uniform thickness coincides with the centroid of its cross-sectional area** (Fig. 7.16). The center of mass of the plate coincides with the centroid of its volume, and we showed in Section 7.4 that the centroid of the volume of a plate of uniform thickness coincides with the centroid of its cross-sectional area.

- **The center of mass of a homogeneous slender bar of uniform cross-sectional area coincides approximately with the centroid of the axis of**

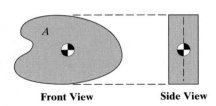

Front View   Side View

Figure 7.16
A plate of uniform thickness.

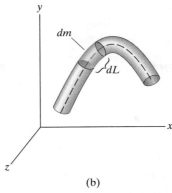

(a)

(b)

Figure 7.17
(a) A slender bar and the centroid of its axis.
(b) The element dm.

the bar (Fig. 7.17a). The axis of the bar is defined to be the line through the centroid of its cross section. Let $dm = \rho A \, dL$, where A is the cross-sectional area of the bar and dL is a differential element of length of its axis (Fig. 7.17b). If we substitute this expression into Eqs. (7.26), they become the equations for the centroid of the axis:

$$\bar{x} = \frac{\int_L x \, dL}{\int_L dL}, \qquad \bar{y} = \frac{\int_L y \, dL}{\int_L dL}, \qquad \bar{z} = \frac{\int_L z \, dL}{\int_L dL}.$$

This result is approximate because the center of mass of the element dm does not coincide with the centroid of the cross section in regions where the bar is curved.

RESULTS

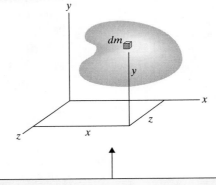

$$\bar{x} = \frac{\int_m x \, dm}{\int_m dm} = \frac{\int_V \rho x \, dV}{\int_i \rho \, dV},$$

$$\bar{y} = \frac{\int_m y \, dm}{\int_m dm} = \frac{\int_V \rho y \, dV}{\int_V \rho \, dV}, \qquad (7.23), (7.26). \ \blacktriangleleft$$

$$\bar{z} = \frac{\int_m z \, dm}{\int_m dm} = \frac{\int_V \rho z \, dV}{\int_V \rho \, dV}.$$

Coordinates of the center of mass of an object, where dm is an infinitesimal element of its mass and ρ is its *density*.

An object is *homogeneous* if its density ρ is constant, or uniform. *The center of mass of a homogeneous object coincides with the centroid of its volume.*

The center of mass of a homogeneous plate of uniform thickness coincides with the centroid of its cross-sectional area.

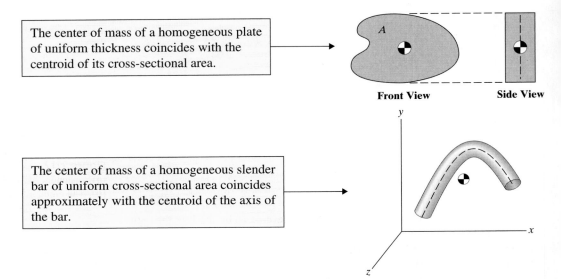

Front View   **Side View**

The center of mass of a homogeneous slender bar of uniform cross-sectional area coincides approximately with the centroid of the axis of the bar.

Active Example 7.16 **Representing the Weight of an L-Shaped Bar** (▶ *Related Problem 7.99*)

The mass of the homogeneous slender bar is 80 kg. What are the reactions at A and B?

Strategy
We can determine the reactions in two ways.

First Method By representing the weight of each straight segment of the bar by a force acting at the center of mass of the segment.

Second Method By determining the center of mass of the entire bar, which is located at the centroid of its axis, and representing the weight of the entire bar by a force acting at its center of mass.

Solution
First Method

$\Sigma F_x = A_x - B = 0,$

$\Sigma F_y = A_y - (40)(9.81)\text{ N} - (40)(9.81)\text{ N} = 0,$

$\Sigma M_{\text{point } A} = (1\text{ m})B - (1\text{ m})[(40)(9.81)\text{ N}] - (0.5\text{ m})[(40)(9.81)\text{ N}] = 0.$

Solving yields $A_x = 589$ N, $A_y = 785$ N, and $B = 589$ N.

Represent the weight of each straight segment by a force acting at the center of mass of the segment, and apply equilibrium.

Second Method

$$\bar{x} = \frac{\bar{x}_1 L_1 + \bar{x}_2 L_2}{L_1 + L_2} = \frac{(0.5)(1) + (1)(1)}{1 + 1} = 0.75 \text{ m,}$$

$$\bar{y} = \frac{\bar{y}_1 L_1 + \bar{y}_2 L_2}{L_1 + L_2} = \frac{(0)(1) + (0.5)(1)}{1 + 1} = 0.25 \text{ m,}$$

Treat the axis of the bar as a composite line with parts 1 and 2 and calculate the coordinates of its centroid.

$$\Sigma F_x = A_x - B = 0,$$

$$\Sigma F_y = A_y - (80)(9.81) \text{ N} = 0,$$

$$\Sigma M_{\text{point } A} = (1 \text{ m})B - (0.75 \text{ m}) [(80)(9.81) \text{ N}] = 0.$$

Solving again yields $A_x = 589$ N, $A_y = 785$ N, and $B = 589$ N.

Place the weight of the entire bar at its center of mass and apply equilibrium.

Practice Problem The mass of the homogeneous circular bar is 80 kg. What are the reactions at A and B?

Answer: $A_x = 500$ N, $A_y = 785$ N, $B = 500$ N.

Example 7.17 | Cylinder with Nonuniform Density (▶ *Related Problem 7.105*)

Determine the mass of the cylinder and the position of its center of mass if (a) it is homogeneous with density ρ_0; (b) its density is given by the equation $\rho = \rho_0(1 + x/L)$.

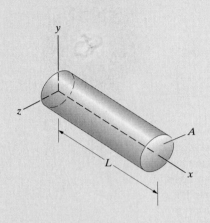

Strategy
In (a), the mass of the cylinder is simply the product of its density and its volume and the center of mass is located at the centroid of its volume. In (b), the cylinder is inhomogeneous and we must use Eqs. (7.24) and (7.26) to determine its mass and center of mass.

Solution
(a) The volume of the cylinder is LA, so its mass is $\rho_0 LA$. Since the center of mass is coincident with the centroid of the volume of the cylinder, the coordinates of the center of mass are $\bar{x} = \frac{1}{2}L, \bar{y} = 0, \bar{z} = 0$.

(b) We can determine the mass of the cylinder by using an element of volume dV in the form of a disk of thickness dx (Fig. a). The volume $dV = A\, dx$. The mass of the cylinder is

$$m = \int_v \rho\, dV = \int_0^L \rho_0\left(1 + \frac{x}{L}\right)A\, dx = \frac{3}{2}\rho_0 AL.$$

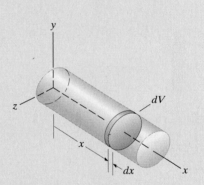

(a) An element of volume dV in the form of a disk.

The x coordinate of the center of mass is

$$\bar{x} = \frac{\int_v x\rho\, dV}{\int_v \rho\, dV} = \frac{\int_0^L \rho_0\left(x + \frac{x^2}{L}\right)A\, dx}{\frac{3}{2}\rho_0 AL} = \frac{5}{9}L.$$

Because the density does not depend on y or z, we know from symmetry that $\bar{y} = 0$ and $\bar{z} = 0$.

Critical Thinking
Notice that the center of mass of the inhomogeneous cylinder is *not* located at the centroid of its volume. Its density increases from left to right, so the center of mass is located to the right of the midpoint of the cylinder. Many of the objects we deal with in engineering are not homogeneous, but it is not common for an object's density to vary continuously through its volume as in this example. More often, objects consist of assemblies of parts (composites) that have different densities because they consist of different materials. Frequently the individual parts are approximately homogeneous. We discuss the determination of the centers of mass of such composite objects in the next section.

Problems

▶ **7.99** Suppose that the bar in Active Example 7.16 is replaced with this 100-kg homogeneous bar. (a) What is the x coordinate of the bar's center of mass? (b) Determine the reactions at A and B.

Problem 7.99

7.100 The mass of the homogeneous flat plate is 50 kg. Determine the reactions at the supports A and B.

Problem 7.100

7.101 The suspended sign is a homogeneous flat plate that has a mass of 130 kg. Determine the axial forces in members AD and CE. (Notice that the y axis is positive downward.)

$y = 1 + 0.0625x^2$

Problem 7.101

7.102 The bar has a mass of 80 kg. What are the reactions at A and B?

Problem 7.102

7.103 The mass of the bar per unit length is 2 kg/m. Choose the dimension b so that part BC of the suspended bar is horizontal. What is the dimension b, and what are the resulting reactions on the bar at A?

Problem 7.103

7.104 The semicircular part of the homogeneous slender bar lies in the x–z plane. Determine the center of mass of the bar.

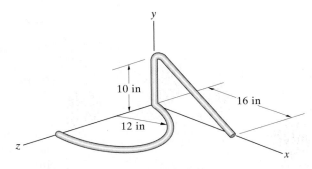

Problem 7.104

▶ **7.105** The density of the cone is given by the equation $\rho = \rho_0(1 + x/h)$, where ρ_0 is a constant. Use the procedure described in Example 7.17 to show that the mass of the cone is given by $m = (7/4)\rho_0 V$, where V is the volume of the cone, and that the x coordinate of the center of mass of the cone is $\bar{x} = (27/35)h$.

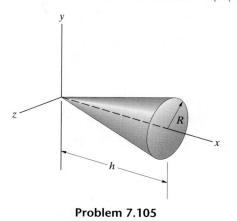

Problem 7.105

7.106 A horizontal cone with 800-mm length and 200-mm radius has a fixed support at A. Its density is $\rho = 6000(1 + 0.4x^2)$ kg/m³, where x is in meters. What are the reactions at A?

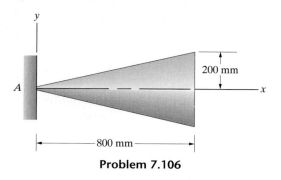

Problem 7.106

7.8 Centers of Mass of Composite Objects

BACKGROUND

The center of mass of an object consisting of a combination of parts can be determined if the centers of mass of its parts are known. The coordinates of the center of mass of a composite object composed of parts with masses $m_1, m_2, \ldots$, are

$$\bar{x} = \frac{\sum_i \bar{x}_i m_i}{\sum_i m_i}, \qquad \bar{y} = \frac{\sum_i \bar{y}_i m_i}{\sum_i m_i}, \qquad \bar{z} = \frac{\sum_i \bar{z}_i m_i}{\sum_i m_i}, \qquad (7.27)$$

where $\bar{x}_i, \bar{y}_i, \bar{z}_i$ are the coordinates of the centers of mass of the parts. Because the weights of the parts are related to their masses by $W_i = gm_i$, Eqs. (7.27) can also be expressed as

$$\bar{x} = \frac{\sum_i \bar{x}_i W_i}{\sum_i W_i}, \qquad \bar{y} = \frac{\sum_i \bar{y}_i W_i}{\sum_i W_i}, \qquad \bar{z} = \frac{\sum_i \bar{z}_i W_i}{\sum_i W_i}. \qquad (7.28)$$

When the masses or weights and the centers of mass of the parts of a composite object are known, these equations determine its center of mass.

Determining the center of mass of a composite object requires three steps:

1. **Choose the parts**—Try to divide the object into parts whose centers of mass you know or can easily determine.

2. **Determine the values for the parts**—Determine the center of mass and the mass or weight of each part. Watch for instances of symmetry that can simplify your task.

3. **Calculate the center of mass**—Use Eqs. (7.27) or (7.28) to determine the center of mass of the composite object.

Active Example 7.18 **Center of Mass of a Composite Object** (▶ *Related Problem 7.107*)

The L-shaped machine part is composed to two homogeneous bars. Bar 1 is tungsten alloy with a density of $14,000 \text{ kg/m}^3$. Bar 2 is steel with a density of 7800 kg/m^3. Determine the x coordinate of the center of mass of the machine part.

Strategy

We can determine the mass and the x coordinate of the center of mass of each homogeneous bar and apply Eq. $(7.27)_1$.

Solution

The volume of bar 1 is
$$V_1 = (80 \text{ mm})(240 \text{ mm})(40 \text{ mm})$$
$$= 7.68 \times 10^5 \text{ mm}^3$$
$$= 7.68 \times 10^{-4} \text{ m}^3,$$
so its mass is
$$m_1 = \rho_1 V_1$$
$$= (14,000 \text{ kg/m}^3)(7.68 \times 10^{-4} \text{ m}^3)$$
$$= 10.8 \text{ kg}.$$

⟩ ← [Mass of bar 1.]

The center of mass coincides with the centroid of the volume of the bar, so
$$\bar{x}_1 = \frac{1}{2}(80 \text{ mm}) = 40 \text{ mm}.$$

⟩ ← [Center of mass of bar 1.]

Bar 2 has the same volume as bar 1, so the mass of bar 2 is
$$m_2 = \rho_2 V_2$$
$$= (7800 \text{ kg/m}^3)(7.68 \times 10^{-4} \text{ m}^3)$$
$$= 5.99 \text{ kg}.$$

⟩ ← [Mass of bar 2.]

The x coordinate of the centroid of the volume of the bar is
$$\bar{x}_2 = 80 \text{ mm} + \frac{1}{2}(240 \text{ mm}) = 200 \text{ mm}.$$

⟩ ← [Center of mass of bar 2.]

$$\bar{x} = \frac{\bar{x}_1 m_1 + \bar{x}_2 m_2}{m_1 + m_2}$$
$$= \frac{(40 \text{ mm})(10.8 \text{ kg}) + (200 \text{ mm})(5.99 \text{ kg})}{10.8 \text{ kg} + 5.99 \text{ kg}}$$
$$= 97.2 \text{ mm}.$$

⟩ ← [Apply Eq. $(7.27)_1$.]

Practice Problem Determine the y coordinate of the center of mass of the L-shaped machine part.

Answer: $\bar{y} = 91.4 \text{ mm}$.

Example 7.19 | **Center of Mass of a Composite Object** (▶ *Related Problem 7.109*)

The composite object consists of a bar welded to a cylinder. The homogeneous bar is aluminum (weight density 168 lb/ft³), and the homogeneous cylinder is bronze (weight density 530 lb/ft³). Determine the center of mass of the object.

Strategy

We can determine the weight of each homogeneous part by multiplying its volume by its weight density. We also know that the center of mass of each part coincides with the centroid of its volume. The centroid of the cylinder is located at its center, but we must determine the location of the centroid of the bar by treating it as a composite volume.

Solution

The volume of the cylinder is

$$V_{\text{cylinder}} = (12 \text{ in})[\pi(4 \text{ in})^2 - \pi(2 \text{ in})^2]$$

$$= 452 \text{ in}^3 = 0.262 \text{ ft}^3,$$

so its weight is

$$W_{\text{cylinder}} = (0.262 \text{ ft}^3)(530 \text{ lb/ft}^3) = 138.8 \text{ lb}.$$

The *x* coordinate of its center of mass is $\bar{x}_{\text{cylinder}} = 10$ in. The volume of the bar is

$$V_{\text{bar}} = (10 \text{ in})(8 \text{ in})(2 \text{ in}) + \tfrac{1}{2}\pi(4 \text{ in})^2(2 \text{ in}) - \tfrac{1}{2}\pi(4 \text{ in})^2(2 \text{ in})$$

$$= 160 \text{ in}^3 = 0.0926 \text{ ft}^3,$$

and its weight is

$$W_{\text{bar}} = (0.0926 \text{ ft}^3)(168 \text{ lb/ft}^3) = 15.6 \text{ lb}.$$

We can determine the centroid of the volume of the bar by treating it as a composite volume consisting of three parts (Fig. a). Part 3 is a semicircular "cutout." The centroids of part 1 and the semicircular cutout 3 are located at the

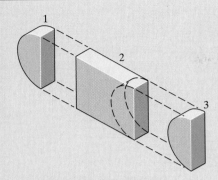

(a) Dividing the bar into three parts.

(b) The centroids of the two semicircular parts.

centroids of their semicircular cross sections (Fig b). Using the information summarized in the table, we have

$$\bar{x}_{bar} = \frac{\bar{x}_1 V_1 + \bar{x}_2 V_2 + \bar{x}_3 V_3}{V_1 + V_2 + V_3}$$

$$= \frac{-\dfrac{4(4)}{3\pi}\left[\frac{1}{2}\pi(4)^2(2)\right] + 5[(10)(8)(2)] - \left[10 - \dfrac{4(4)}{3\pi}\right]\left[\frac{1}{2}\pi(4)^2(2)\right]}{\frac{1}{2}\pi(4)^2(2) + (10)(8)(2) - \frac{1}{2}\pi(4)^2(2)}$$

$$= 1.86 \text{ in.}$$

Information for determining the x coordinate of the centroid of the bar

	$\bar{x}_i$ (in)	V_i (in^3)	$\bar{x}_i V_i$ (in^4)
Part 1	$-\dfrac{4(4)}{3\pi}$	$\frac{1}{2}\pi(4)^2(2)$	$-\dfrac{4(4)}{3\pi}\left[\frac{1}{2}\pi(4)^2(2)\right]$
Part 2	5	$(10)(8)(2)$	$5[(10)(8)(2)]$
Part 3	$10 - \dfrac{4(4)}{3\pi}$	$-\frac{1}{2}\pi(4)^2(2)$	$-\left[10 - \dfrac{4(4)}{3\pi}\right]\left[\frac{1}{2}\pi(4)^2(2)\right]$

Therefore, the x coordinate of the center of mass of the composite object is

$$\bar{x} = \frac{\bar{x}_{bar} W_{bar} + \bar{x}_{cylinder} W_{cylinder}}{W_{bar} + W_{cylinder}}$$

$$= \frac{(1.86 \text{ in})(15.6 \text{ lb}) + (10 \text{ in})(138.8 \text{ lb})}{15.6 \text{ lb} + 138.8 \text{ lb}}$$

$$= 9.18 \text{ in.}$$

Because of the symmetry of the bar, the y and z coordinates of its center of mass are $\bar{y} = 0$ and $\bar{z} = 0$.

Critical Thinking

The composite object in this example is not homogeneous, which means we could not assume that its center of mass coincides with the centroid of its volume. But the bar and the cylinder are each homogeneous, so we *could* determine their individual centers of mass by finding the centroids of their volumes. The primary challenge in this example was determining the centroid of the volume of the bar with its semicircular end and semicircular cutout.

| **Example 7.20** | **Centers of Mass of Vehicles** (▶ *Related Problems 7.115, 7.116*) |

A car is placed on a platform that measures the normal force exerted by each tire independently. Measurements made with the platform horizontal and with the platform tilted at $\alpha = 15°$ are shown in the table. Determine the position of the car's center of mass.

Measurements of the normal forces exerted by the tires

Wheelbase = 2.82 m		
Track = 1.55 m		**Measured Loads (N)**
	$\alpha = 0$	$\alpha = 15°$
Left front wheel, N_{LF}	5104	4463
Right front wheel, N_{RF}	5027	4396
Left rear wheel, N_{LR}	3613	3956
Right rear wheel, N_{RR}	3559	3898

Strategy

The given measurements tell us the normal reactions exerted on the car's tires by the platform. By drawing free-body diagrams of the car in the two positions and applying equilibrium equations, we will obtain equations that can be solved for the unknown coordinates of the car's center of mass.

Solution

We draw the free-body diagram of the car when the platform is in the horizontal position in Figs. a and b. The car's weight is

$$W = N_{\text{LF}} + N_{\text{RF}} + N_{\text{LR}} + N_{\text{RR}}$$

$$= 5104 + 5027 + 3613 + 3559$$

$$= 17{,}303 \text{ N}.$$

From Fig. a, we obtain the equilibrium equation

$$\Sigma M_{z \text{ axis}} = (\text{wheelbase})(N_{\text{LF}} + N_{\text{RF}}) - \bar{x}W = 0,$$

(a) Side view of the free-body diagram with the platform horizontal.

which we can solve for $\bar{x}$:

$$\bar{x} = \frac{(\text{wheelbase})(N_{LF} + N_{RF})}{W}$$

$$= \frac{(2.82 \text{ m})(5104 \text{ N} + 5027 \text{ N})}{17{,}303 \text{ N}}$$

$$= 1.651 \text{ m}.$$

From Fig. b,

$$\Sigma M_{x \text{ axis}} = \bar{z}W - (\text{track})(N_{RF} + N_{RR}) = 0,$$

which we can solve for $\bar{z}$:

$$\bar{z} = \frac{(\text{track})(N_{RF} + N_{RR})}{W}$$

$$= \frac{(1.55 \text{ m})(5027 \text{ N} + 3559 \text{ N})}{17{,}303 \text{ N}}$$

$$= 0.769 \text{ m}.$$

(b) Front view of the free-body diagram with the platform horizontal.

Now that we know $\bar{x}$, we can determine $\bar{y}$ from the free-body diagram of the car when the platform is in the tilted position (Fig. c). From the equilibrium equation

$$\Sigma M_{z \text{ axis}} = (\text{wheelbase})(N_{LF} + N_{RF}) + \bar{y}W \sin 15° - \bar{x}W \cos 15°$$

$$= 0,$$

we obtain

$$\bar{y} = \frac{\bar{x}W \cos 15° - (\text{wheelbase})(N_{LF} + N_{RF})}{W \sin 15°}$$

$$= \frac{(1.651 \text{ m})(17{,}303 \text{ N}) \cos 15° - (2.82 \text{ m})(4463 \text{ N} + 4396 \text{ N})}{(17{,}303 \text{ N}) \sin 15°}$$

$$= 0.584 \text{ m}.$$

Notice that we could not have determined $\bar{y}$ without the measurements made with the car in the tilted position.

(c) Side view of the free-body diagram with the platform tilted.

Problems

▶ **7.107** In Active Example 7.18, suppose that bar 1 is replaced by a bar with the same dimensions that consists of aluminum alloy with a density of 2600 kg/m³. Determine the x coordinate of the center of mass of the machine part.

7.108 The cylindrical tube is made of aluminum with density 2700 kg/m³. The cylindrical plug is made of steel with density 7800 kg/m³. Determine the coordinates of the center of mass of the composite object.

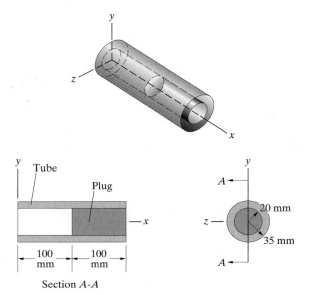

Section A-A

Problem 7.108

▶ **7.109** In Example 7.19, suppose that the object is redesigned so that the radius of the hole in the hollow cylinder is increased from 2 in to 3 in. What is the x coordinate of the center of mass of the object?

7.110 A machine consists of three parts. The masses and the locations of the centers of mass of two of the parts are

Part	Mass (kg)	$\bar{x}$ (mm)	$\bar{y}$ (mm)	$\bar{z}$ (mm)
1	2.0	100	50	−20
2	4.5	150	70	0

The mass of part 3 is 2.5 kg. The design engineer wants to position part 3 so that the center of mass location of the machine is $\bar{x} = 120$ mm, $\bar{y} = 80$ mm, $\bar{z} = 0$. Determine the necessary position of the center of mass of part 3.

7.111 Two views of a machine element are shown. Part 1 is aluminum alloy with density 2800 kg/m³, and part 2 is steel with density 7800 kg/m³. Determine the coordinates of its center of mass.

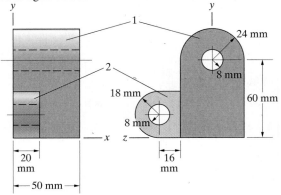

Problem 7.111

7.112 The loads $F_1 = F_2 = 25$ kN. The mass of the truss is 900 kg. The members of the truss are homogeneous bars with the same uniform cross section. (a) What is the x coordinate of the center of mass of the truss? (b) Determine the reactions at A and G.

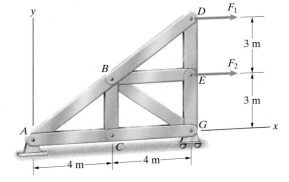

Problem 7.112

7.113 With its engine removed, the mass of the car is 1100 kg and its center of mass is at C. The mass of the engine is 220 kg.

(a) Suppose that you want to place the center of mass E of the engine so that the center of mass of the car is midway between the front wheels A and the rear wheels B. What is the distance b?

(b) If the car is parked on a 15° slope facing up the slope, what total normal force is exerted by the road on the rear wheels B?

Problem 7.113

7.114 The airplane is parked with its landing gear resting on scales. The weights measured at A, B, and C are 30 kN, 140 kN, and 146 kN, respectively. After a crate is loaded onto the plane, the weights measured at A, B, and C are 31 kN, 142 kN, and 147 kN, respectively. Determine the mass and the x and y coordinates of the center of mass of the crate.

Problem 7.114

▶ **7.115** A suitcase with a mass of 90 kg is placed in the trunk of the car described in Example 7.20. The position of the center of mass of the suitcase is $\bar{x}_s = -0.533$ m, $\bar{y}_s = 0.762$ m, $\bar{z}_s = -0.305$ m. If the suitcase is regarded as part of the car, what is the new position of the car's center of mass?

▶ **7.116** A group of engineering students constructs a miniature device of the kind described in Example 7.20 and uses it to determine the center of mass of a miniature vehicle. The data they obtain are shown in the following table:

Wheelbase = 36 in		
Track = 30 in	**Measured Loads (lb)**	
	$\alpha = 0$	$\alpha = 10°$
Left front wheel, N_{LF}	35	32
Right front wheel, N_{RF}	36	33
Left rear wheel, N_{LR}	27	34
Right rear wheel, N_{RR}	29	30

Determine the center of mass of the vehicle. Use the same coordinate system as in Example 7.20.

Review Problems

7.117 Determine the centroid of the area by letting dA be a vertical strip of width dx.

7.118 Determine the centroid of the area by letting dA be a horizontal strip of height dy.

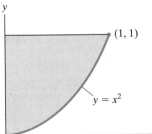

Problems 7.117/7.118

7.119 Determine the centroid of the area.

Problem 7.119

7.120 Determine the centroid of the area.

Problem 7.120

7.121 The cantilever beam is subjected to a triangular distributed load. What are the reactions at A?

200 N/m

10 m

Problem 7.121

7.122 What is the axial load in member BD of the frame?

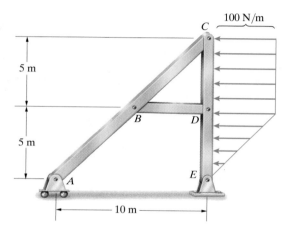

100 N/m

5 m

5 m

10 m

Problem 7.122

7.123 An engineer estimates that the maximum wind load on the 40-m tower in Fig. a is described by the distributed load in Fig. b. The tower is supported by three cables, A, B, and C, from the top of the tower to equally spaced points 15 m from the bottom of the tower (Fig. c). If the wind blows from the west and cables B and C are slack, what is the tension in cable A? (Model the base of the tower as a ball and socket support.)

200 N/m

40 m

15 m

400 N/m

(a)   (b)   (c)

Problem 7.123

7.124 Determine the reactions on member $ABCD$ at A and D.

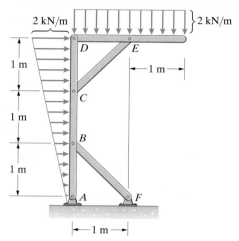

2 kN/m   2 kN/m

1 m

1 m

1 m

1 m

1 m

Problem 7.124

7.125 Estimate the centroid of the volume of the *Apollo* lunar return configuration (not including its rocket nozzle) by treating it as a cone and a cylinder.

12.8 ft

10 ft   14 ft

Nozzle

Problem 7.125

7.126 The shape of the rocket nozzle of the *Apollo* lunar return configuration is approximated by revolving the curve shown around the x axis. In terms of the coordinate system shown, determine the centroid of the volume of the nozzle.

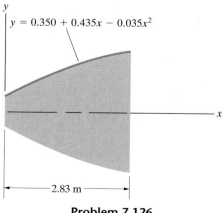

$y = 0.350 + 0.435x - 0.035x^2$

2.83 m

Problem 7.126

7.127 Determine the coordinates of the centroid of the volume.

Problem 7.127

7.128 Determine the surface area of the volume of revolution.

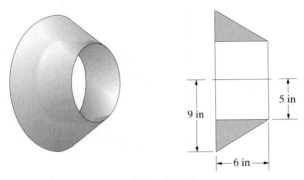

Problem 7.128

7.129 Determine the y coordinate of the center of mass of the homogeneous steel plate.

Problem 7.129

7.130 Determine the x coordinate of the center of mass of the homogeneous steel plate.

Problem 7.130

7.131 The area of the homogeneous plate is 10 ft^2. The vertical reactions on the plate at A and B are 80 lb and 84 lb, respectively. Suppose that you want to equalize the reactions at A and B by drilling a 1-ft-diameter hole in the plate. What horizontal distance from A should the center of the hole be? What are the resulting reactions at A and B?

Problem 7.131

7.132 The plate is of uniform thickness and is made of homogeneous material whose mass per unit area of the plate is 2 kg/m^2. The vertical reactions at A and B are 6 N and 10 N, respectively. What is the x coordinate of the centroid of the hole?

Problem 7.132

7.133 Determine the center of mass of the homogeneous sheet of metal.

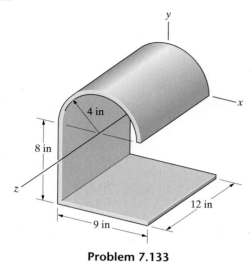

Problem 7.133

7.134 Determine the center of mass of the homogeneous object.

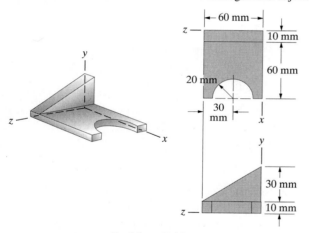

Problem 7.134

7.135 Determine the center of mass of the homogeneous object.

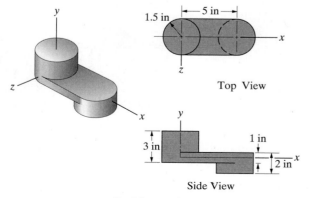

Top View

Side View

Problem 7.135

7.136 The arrangement shown can be used to determine the location of the center of mass of a person. A horizontal board has a pin support at A and rests on a scale that measures weight at B. The distance from A to B is 2.3 m. When the person is not on the board, the scale at B measures 90 N.

(a) When a 63-kg person is in position (1), the scale at B measures 496 N. What is the x coordinate of the person's center of mass?

(b) When the same person is in position (2), the scale measures 523 N. What is the x coordinate of his center of mass?

Problem 7.136

7.137 If a string is tied to the slender bar at A and the bar is allowed to hang freely, what will be the angle between AB and the vertical?

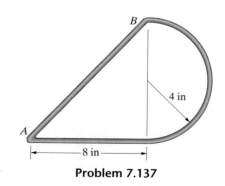

Problem 7.137

7.138 When the truck is unloaded, the total reactions at the front and rear wheels are $A = 54$ kN and $B = 36$ kN. The density of the load of gravel is $\rho = 1600$ kg/m^3. The dimension of the load in the z direction is 3 m, and its surface profile, given by the function shown, does not depend on z. What are the total reactions at the front and rear wheels of the loaded truck?

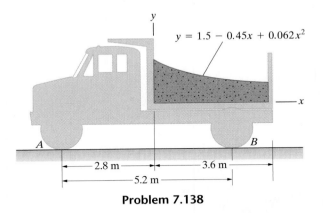

$$y = 1.5 - 0.45x + 0.062x^2$$

Problem 7.138

7.139 The mass of the moon is 0.0123 times the mass of the earth. If the moon's center of mass is 383,000 km from the center of mass of the earth, what is the distance from the center of mass of the earth to the center of mass of the earth–moon system?

Design Project

7.140 Construct a homogeneous thin flat plate with the shape shown in Fig. a. (Use the cardboard back of a pad of paper to construct the plate. Choose your dimensions so that the plate is as large as possible.) Calculate the location of the center of mass of the plate. Measuring as carefully as possible, mark the center of mass clearly on both sides of the plate. Then carry out the following experiments.

(a) Balance the plate on your finger (Fig. b) and observe that it balances at its center of mass. Explain the result of this experiment by drawing a free-body diagram of the plate.

(b) This experiment requires a needle or slender nail, a length of string, and a small weight. Tie the weight to one end of the string and make a small loop at the other end. Stick the needle through the plate at any point other than its center of mass. Hold the needle horizontal so that the plate hangs freely from it (Fig. c). Use the loop to hang the weight from the needle, and let the weight hang freely so that the string lies along the face of the plate. Observe that the string passes through the center of mass of the plate. Repeat this experiment several times, sticking the needle through various points on the plate. Explain the results of this experiment by drawing a free-body diagram of the plate.

(c) Hold the plate so that the plane of the plate is vertical, and throw the plate upward, spinning it like a Frisbee. Observe that the plate spins about its center of mass.

(a)

(b)   (c)

Moments of Inertia

Quantities called moments of inertia arise repeatedly in analyses of engineering problems. Moments of inertia of areas are used in the study of distributed forces and in calculating deflections of beams. The moment exerted by the pressure on a submerged flat plate can be expressed in terms of the moment of inertia of the plate's area. In dynamics, mass moments of inertia are used in calculating the rotational motions of objects. We show how to calculate the moments of inertia of simple areas and objects and then use results called parallel-axis theorems to calculate moments of inertia of more complex areas and objects.

◀ A beam's resistance to bending and ability to support loads depend on a property of its cross section called the moment of inertia. In this chapter we define and show how to calculate moments of inertia of areas.

AREAS

8.1 Definitions

Consider an area A in the x–y plane (Fig. 8.1a). Four moments of inertia of A are defined:

1. **Moment of inertia about the x axis:**

$$I_x = \int_A y^2 \, dA,$$
(8.1)

where y is the y coordinate of the differential element of area dA (Fig. 8.1b). This moment of inertia is sometimes expressed in terms of the *radius of gyration* about the x axis, k_x, which is defined by

$$I_x = k_x^2 A.$$
(8.2)

2. **Moment of inertia about the y axis:**

$$I_y = \int_A x^2 \, dA,$$
(8.3)

where x is the x coordinate of the element dA (Fig. 8.1b). The radius of gyration about the y axis, k_y, is defined by

$$I_y = k_y^2 A.$$
(8.4)

(a)

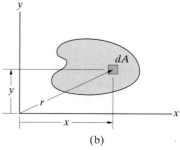

(b)

Figure 8.1
(a) An area A in the x–y plane.
(b) A differential element of A.

3. Product of inertia:

$$I_{xy} = \int_A xy \, dA. \tag{8.5}$$

4. Polar moment of inertia:

$$J_O = \int_A r^2 \, dA, \tag{8.6}$$

where r is the radial distance from the origin of the coordinate system to dA (Fig. 8.1b). The radius of gyration about the origin, k_O, is defined by

$$J_O = k_O^2 A. \tag{8.7}$$

The polar moment of inertia is equal to the sum of the moments of inertia about the x and y axes:

$$J_O = \int_A r^2 \, dA = \int_A (y^2 + x^2) \, dA = I_x + I_y.$$

Substituting the expressions for the moments of inertia in terms of the radii of gyration into this equation, we obtain

$$k_O^2 = k_x^2 + k_y^2.$$

The dimensions of the moments of inertia of an area are $(\text{length})^4$, and the radii of gyration have dimensions of length. Notice that the definitions of the moments of inertia I_x, I_y, and J_O and the radii of gyration imply that they have positive values for any area. They cannot be negative or zero.

If an area A is symmetric about the x axis, for each element dA with coordinates (x, y), there is a corresponding element dA with coordinates $(x, -y)$, as shown in Fig. 8.2. The contributions of these two elements to the product of inertia I_{xy} of the area cancel: $xy \, dA + (-xy) \, dA = 0$. This means that the product of inertia of the area is zero. The same kind of argument can be used for an area that is symmetric about the y axis. *If an area is symmetric about either the x axis or the y axis, its product of inertia is zero.*

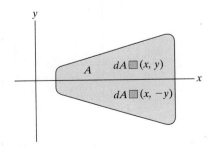

Figure 8.2

Active Example 8.1 **Moments of Inertia of a Triangular Area** (▶ *Related Problems 8.1–8.3*)

Determine I_x and I_y for the triangular area.

Strategy

Equation (8.3) for the moment of inertia about the y axis is very similar in form to the equation for the x coordinate of the centroid of an area, and we can evaluate it in exactly the same way, by using a differential element of area dA in the form of a vertical strip of width dx. We will then show that I_x can be evaluated using the same element of area.

Solution

$$I_y = \int_A x^2 \, dA$$

$$= \int_0^b x^2 f(x) \, dx$$

$$= \int_0^b x^2 \left(\frac{h}{b}\right) x \, dx$$

$$= \frac{1}{4} h b^3.$$

The height of a strip of width dx at position x is $f(x) = (h/b)x$, so its area is $dA = f(x)\,dx$. Use this expression to evaluate Eq. (8.3).

To evaluate I_x, we will first determine the moment of inertia about the x axis of the vertical strip dA.

$$(I_x)_{\text{strip}} = \int_{\text{strip}} y^2 \, dA_s$$

$$= \int_0^{f(x)} (y^2 dx) \, dy$$

$$= \frac{1}{3} [f(x)]^3 \, dx.$$

Let dA_s be an element of the vertical strip dA and apply Eq. (8.1).

$$I_x = \int_0^b \frac{1}{3} [f(x)]^3 \, dx$$

$$= \int_0^b \frac{1}{3} \left(\frac{h}{b} x\right)^3 dx$$

$$= \frac{1}{12} bh^3.$$

> Integrate the expression for $(I_x)_{\text{strip}}$ with respect to x from $x = 0$ to $x = b$ to determine I_x for the triangle.

Practice Problem Determine I_{xy} for the triangular area. Do so by determining the product of inertia for the vertical strip dA and then integrating the resulting expression with respect to x from $x = 0$ to $x = b$.

Answer: $I_{xy} = \dfrac{1}{8} b^2 h^2$.

Example 8.2   Moments of Inertia of a Circular Area (▶ Related Problem 8.21)

Determine the moments of inertia and radii of gyration of the circular area.

Strategy
We will first determine the polar moment of inertia J_O by integrating in terms of polar coordinates. We know from the symmetry of the area that $I_x = I_y$, and since $I_x + I_y = J_O$, the moments of inertia I_x and I_y are each equal to $\frac{1}{2} J_O$. We also know from the symmetry of the area that $I_{xy} = 0$.

Solution
By letting r change by an amount dr, we obtain an annular element of area $dA = 2\pi r \, dr$ (Fig. a). The polar moment of inertia is

$$J_O = \int_A r^2 \, dA = \int_0^R 2\pi r^3 \, dr = 2\pi \left[\frac{r^4}{4}\right]_0^R = \frac{1}{2}\pi R^4,$$

and the radius of gyration about O is

$$k_O = \sqrt{\frac{J_O}{A}} = \sqrt{\frac{(1/2)\pi R^4}{\pi R^2}} = \frac{1}{\sqrt{2}} R.$$

The moments of inertia about the x and y axes are

$$I_x = I_y = \frac{1}{2} J_O = \frac{1}{4}\pi R^4,$$

and the radii of gyration about the x and y axes are

$$k_x = k_y = \sqrt{\frac{I_x}{A}} = \sqrt{\frac{(1/4)\pi R^4}{\pi R^2}} = \frac{1}{2} R.$$

The product of inertia is zero:

$$I_{xy} = 0.$$

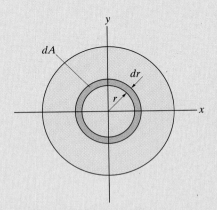

(a) An annular element dA.

Critical Thinking
The symmetry of this example saved us from having to integrate to determine I_x, I_y, and I_{xy}. Be alert for symmetry that can shorten your work. In particular, remember that $I_{xy} = 0$ if the area is symmetric about either the x or the y axis.

Problems

▶ **8.1** Use the method described in Active Example 8.1 to determine I_y and k_y for the rectangular area.

▶ **8.2** Use the method described in Active Example 8.1 to determine I_x and k_x for the rectangular area.

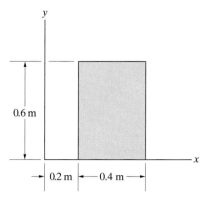

Problems 8.1/8.2

▶ **8.3** In Active Example 8.1, suppose that the triangular area is reoriented as shown. Use integration to determine I_y and k_y.

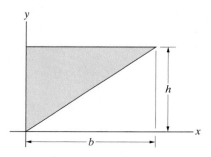

Problem 8.3

8.4 (a) Determine the moment of inertia I_y of the beam's rectangular cross section about the y axis.

(b) Determine the moment of inertia $I_{y'}$ of the beam's cross section about the y' axis. Using your numerical values, show that $I_y = I_{y'} + d_x^2 A$, where A is the area of the cross section.

8.5 (a) Determine the polar moment of inertia J_O of the beam's rectangular cross section about the origin O.

(b) Determine the polar moment of inertia $J_{O'}$ of the beam's cross section about the origin O'. Using your numerical values, show that $J_O = J_{O'} + (d_x^2 + d_y^2)A$, where A is the area of the cross section.

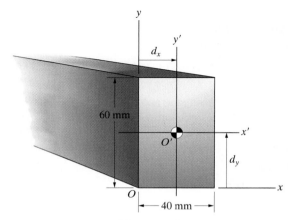

Problems 8.4/8.5

8.6 Determine I_y and k_y.

8.7 Determine J_O and k_O.

8.8 Determine I_{xy}.

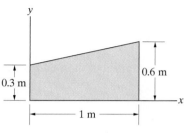

Problems 8.6–8.8

8.9 Determine I_y.

8.10 Determine I_x.

8.11 Determine J_O.

8.12 Determine I_{xy}.

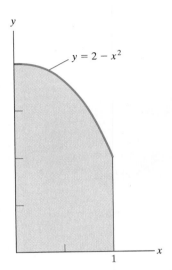

$y = 2 - x^2$

Problems 8.9–8.12

8.13 Determine I_y and k_y.

8.14 Determine I_x and k_x.

8.15 Determine J_O and k_O.

8.16 Determine I_{xy}.

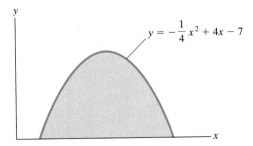

$y = -\dfrac{1}{4}x^2 + 4x - 7$

Problems 8.13–8.16

8.17 Determine I_y and k_y.

8.18 Determine I_x and k_x.

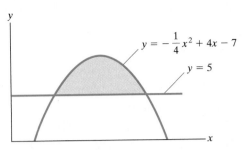

$y = -\dfrac{1}{4}x^2 + 4x - 7$

$y = 5$

Problems 8.17/8.18

8.19 (a) Determine I_y and k_y by letting dA be a vertical strip of width dx.

(b) The polar moment of inertia of a circular area with its center at the origin is $J_O = \frac{1}{2}\pi R^4$. Explain how you can use this information to confirm your answer to (a).

8.20 (a) Determine I_x and k_x by letting dA be a horizontal strip of height dy.

(b) The polar moment of inertia of a circular area with its center at the origin is $J_O = \frac{1}{2}\pi R^4$. Explain how you can use this information to confirm your answer to (a).

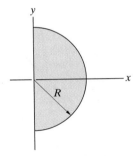

R

Problems 8.19/8.20

▶ **8.21** Use the procedure described in Example 8.2 to determine the moments of inertia I_x and I_y for the annular ring.

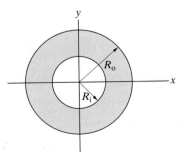

R_o

R_i

Problem 8.21

8.22 What are the values of I_y and k_y for the elliptical area of the airplane's wing?

8.23 What are the values of I_x and k_x for the elliptical area of the airplane's wing?

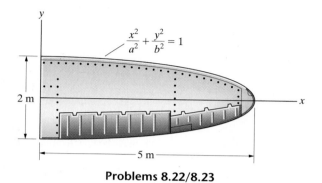

Problems 8.22/8.23

8.24 Determine I_y and k_y.

8.25 Determine I_x and k_x.

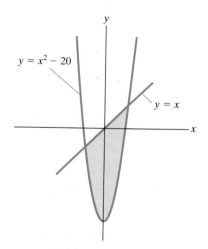

Problems 8.24/8.25

8.26 A vertical plate of area A is beneath the surface of a stationary body of water. The pressure of the water subjects each element dA of the surface of the plate to a force $(p_O + \gamma y)\, dA$, where p_O is the pressure at the surface of the water and γ is the weight density of the water. Show that the magnitude of the moment about the x axis due to the pressure on the front face of the plate is

$$M_{x\,\text{axis}} = p_O \bar{y} A + \gamma I_x,$$

where $\bar{y}$ is the y coordinate of the centroid of A and I_x is the moment of inertia of A about the x axis.

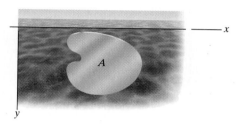

Problem 8.26

8.2 Parallel-Axis Theorems

BACKGROUND

The values of the moments of inertia of an area depend on the position of the coordinate system relative to the area.

In some situations the moments of inertia of an area are known in terms of a particular coordinate system but we need their values in terms of a different coordinate system. When the coordinate systems are parallel, the desired moments of inertia can be obtained by using the theorems we describe in this section. Furthermore, these theorems make it possible for us to determine the moments of inertia of a composite area when the moments of inertia of its parts are known.

Suppose that we know the moments of inertia of an area A in terms of a coordinate system $x'y'$ with its origin at the centroid of the area, and we wish to determine the moments of inertia in terms of a parallel coordinate system xy (Fig. 8.3a). We denote the coordinates of the centroid of A in the xy coordinate system by (d_x, d_y), and $d = \sqrt{d_x^2 + d_y^2}$ is the distance from the origin of the xy coordinate system to the centroid (Fig. 8.3b).

We need two preliminary results before deriving the parallel-axis theorems. In terms of the $x'y'$ coordinate system, the coordinates of the centroid of A are

$$\bar{x}' = \frac{\displaystyle\int_A x'\,dA}{\displaystyle\int_A dA}, \qquad \bar{y}' = \frac{\displaystyle\int_A y'\,dA}{\displaystyle\int_A dA}.$$

But the origin of the $x'y'$ coordinate system is located at the centroid of A, so $\bar{x}' = 0$ and $\bar{y}' = 0$. Therefore,

$$\int_A x'\,dA = 0, \qquad \int_A y'\,dA = 0. \tag{8.8}$$

Moment of Inertia About the x Axis   In terms of the xy coordinate system, the moment of inertia of A about the x axis is

$$I_x = \int_A y^2\,dA, \tag{8.9}$$

where y is the coordinate of the element of area dA relative to the xy coordinate system. From Fig. 8.3b, we see that $y = y' + d_y$, where y' is the coordinate of dA relative to the $x'y'$ coordinate system. Substituting this expression into Eq. (8.9), we obtain

$$I_x = \int_A (y' + d_y)^2\,dA = \int_A (y')^2\,dA + 2d_y\int_A y'\,dA + d_y^2\int_A dA.$$

The first integral on the right is the moment of inertia of A about the x' axis. From Eq. (8.8), the second integral on the right equals zero. Therefore, we obtain

$$I_x = I_{x'} + d_y^2 A. \tag{8.10}$$

(a)

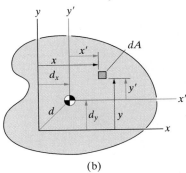

(b)

Figure 8.3
(a) The area A and the coordinate systems $x'y'$ and xy.
(b) The differential element dA.

Figure 8.4
The parallel-axis theorem for the moment of inertia about the x axis.

This is a *parallel-axis theorem*. It relates the moment of inertia of A about the x' axis through the centroid to the moment of inertia about the parallel axis x (Fig. 8.4).

Moment of Inertia About the y Axis   In terms of the xy coordinate system, the moment of inertia of A about the y axis is

$$
I_y = \int_A x^2 \, dA = \int_A (x' + d_x)^2 \, dA
$$

$$
= \int_A (x')^2 \, dA + 2d_x \int_A x' dA + d_x^2 \int_A dA.
$$

From Eq. (8.8), the second integral on the right equals zero. Therefore, the parallel-axis theorem that relates the moment of inertia of A about the y' axis through the centroid to the moment of inertia about the parallel axis y is

$$
I_y = I_{y'} + d_x^2 A. \tag{8.11}
$$

Product of Inertia   In terms of the xy coordinate system, the product of inertia is

$$
I_{xy} = \int_A xy \, dA = \int_A (x' + d_x)(y' + d_y) \, dA
$$

$$
= \int_A x'y' dA + d_y \int_A x' dA + d_x \int_A y' dA + d_x d_y \int_A dA.
$$

The second and third integrals equal zero from Eq. (8.8). We see that the parallel-axis theorem for the product of inertia is

$$
I_{xy} = I_{x'y'} + d_x d_y A. \tag{8.12}
$$

Polar Moment of Inertia   The polar moment of inertia $J_O = I_x + I_y$. Summing Eqs. (8.10) and (8.11), the parallel axis theorem for the polar moment of inertia is

$$
J_O = J'_O + (d_x^2 + d_y^2)A = J'_O + d^2 A, \tag{8.13}
$$

where d is the distance from the origin of the $x'y'$ coordinate system to the origin of the xy coordinate system.

How can the parallel-axis theorems be used to determine the moments of inertia of a composite area? Suppose that we want to determine the moment of inertia about the y axis of the area in Fig. 8.5a. We can divide it into a triangle, a semicircle, and a circular cutout, denoted as parts 1, 2, and 3 (Fig. 8.5b). By using the parallel-axis theorem for I_y, we can determine the moment of inertia of each part about the y axis. For example, the moment of inertia of part 2 (the semicircle) about the y axis is (Fig. 8.5c)

$$(I_y)_2 = (I_{y'})_2 + (d_x)_2^2 A_2.$$

We must determine the values of $(I_{y'})_2$ and $(d_x)_2$. Moments of inertia and centroid locations for some simple areas are tabulated in Appendix B. Once this procedure is carried out for each part, the moment of inertia of the composite area is

$$I_y = (I_y)_1 + (I_y)_2 - (I_y)_3.$$

Notice that the moment of inertia of the circular cutout is subtracted.

We see that determining a moment of inertia of a composite area in terms of a given coordinate system involves three steps:

1. **Choose the parts**—Try to divide the composite area into parts whose moments of inertia you know or can easily determine.

2. **Determine the moments of inertia of the parts**—Determine the moment of inertia of each part in terms of a parallel coordinate system with its origin at the centroid of the part, and then use the parallel-axis theorem to determine the moment of inertia in terms of the given coordinate system.

3. **Sum the results**—Sum the moments of inertia of the parts (or subtract in the case of a cutout) to obtain the moment of inertia of the composite area.

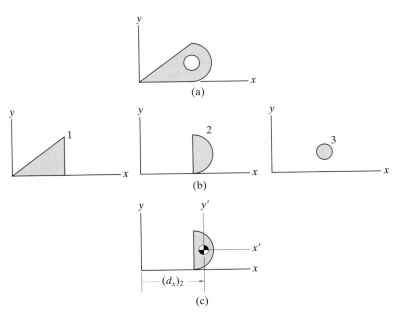

(a)

(b)

(c)

Figure 8.5
(a) A composite area.
(b) The three parts of the area.
(c) Determining $(I_y)_2$.

RESULTS

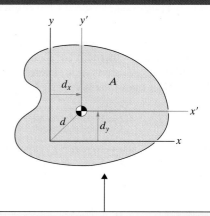

$$I_x = I_{x'} + d_y^2 A, \qquad (8.10)$$

$$I_y = I_{y'} + d_x^2 A, \qquad (8.11)$$

$$I_{xy} = I_{x'y'} + d_x d_y A, \qquad (8.12)$$

$$J_O = J'_O + d^2 A. \qquad (8.13)$$

The parallel-axis theorems are relationships between the moments and product of inertia of an area expressed in terms of a coordinate system $x'y'z'$—with its origin at the centroid of the area—and a parallel coordinate system xyz.

The parallel-axis theorems make it possible to determine the moments and product of inertia of a composite area in terms of a given coordinate system xyz when the moments and products of inertia of each part of the composite area are known in terms of a parallel coordinate system with its origin at the centroid of the part. The values of the moments and product of inertia of the parts in terms of the xyz coordinate system can be summed (or subtracted in the case of a cutout) to obtain the values for the composite area.

| Active Example 8.3 | Moments of Inertia of a Composite Area (▶ *Related Problem 8.27*) |

Determine I_x for the composite area.

Strategy
We can divide this area into two rectangles. We must use the parallel-axis theorems to determine I_x for each rectangle in terms of the xy coordinate system. The values can be summed to determine I_x for the composite area.

Solution

Divide the composite area into two rectangles.

From Appendix B, the moment of inertia of area 1 about the x' axis is

$$(I_{x'})_1 = \frac{1}{12}(1\ \text{m})(4\ \text{m})^3 = 5.33\ \text{m}^4.$$

Applying the parallel-axis theorem, the moment of inertia of area 1 about the x axis is

$$(I_x)_1 = 5.33\ \text{m}^4 + (2\ \text{m})^2(1\ \text{m})(4\ \text{m}) = 21.3\ \text{m}^4.$$

Apply Eq. (8.10) to area 12.

The moment of inertia of area 2 about the x' axis is

$$(I_{x'})_2 = \frac{1}{12}(2\ \text{m})(1\ \text{m})^3 = 0.167\ \text{m}^4.$$

Applying the parallel-axis theorem, the moment of inertia of area 2 about the x axis is

$$(I_x)_2 = 0.167\ \text{m}^4 + (0.5\ \text{m})^2(2\ \text{m})(1\ \text{m}) = 0.667\ \text{m}^4.$$

Apply Eq. (8.10) to area 2.

The moment of inertia of the composite area about the x axis is

$$I_x = (I_x)_1 + (I_x)_2$$
$$= 21.3\ \text{m}^4 + 0.667\ \text{m}^4$$
$$= 22.0\ \text{m}^4.$$

Sum the values for the parts.

Practice Problem Determine I_{xy} for the composite area.

Answer: $I_{xy} = 6\ \text{m}^4.$

Example 8.4　Moments of Inertia of a Composite Area (▶ *Related Problem 8.30*)

Determine I_y and k_y for the composite area.

Strategy

We can divide this area into a rectangle *without the semicircular cutout*, a semicircle *without the semicircular cutout*, and a circular cutout. We can use a parallel-axis theorem to determine I_y for each part in terms of the xy coordinate system. Then, by adding the values for the rectangle and semicircle and subtracting the value for the circular cutout, we can determine I_y for the composite area. Then we can use Eq. (8.4) to determine the radius of gyration k_y for the composite area.

Solution

Choose the Parts　We divide the area into a rectangle, a semicircle, and the circular cutout, calling them parts 1, 2, and 3, respectively (Fig. a).

Determine the Moments of Inertia of the Parts　The moments of inertia of the parts in terms of the $x'y'$ coordinate systems and the location of the centroid of the semicircular part are given in Appendix B. In the table we use the parallel-axis theorem to determine the moment of inertia of each part about the y axis.

Determining the moments of inertia of the parts

	d_x (mm)	A (mm^2)	$I_{y'}$ (mm^4)	$I_y = I_{y'} + d_x^2 A$ (mm^4)
Part 1	60	$(120)(80)$	$\frac{1}{12}(80)(120)^3$	4.608×10^7
Part 2	$120 + \dfrac{4(40)}{3\pi}$	$\frac{1}{2}\pi(40)^2$	$\left(\dfrac{\pi}{8} - \dfrac{8}{9\pi}\right)(40)^4$	4.744×10^7
Part 3	120	$\pi(20)^2$	$\frac{1}{4}\pi(20)^4$	1.822×10^7

Sum the Results　The moment of inertia of the composite area about the y axis is

$$I_y = (I_y)_1 + (I_y)_2 - (I_y)_3 = (4.608 + 4.744 - 1.822) \times 10^7 \text{ mm}^4$$
$$= 7.530 \times 10^7 \text{ mm}^4.$$

The total area is

$$A = A_1 + A_2 - A_3 = (120 \text{ mm})(80 \text{ mm}) + \frac{1}{2}\pi(40 \text{ mm})^2 - \pi(20 \text{ mm})^2$$
$$= 1.086 \times 10^4 \text{ mm}^2,$$

so the radius of gyration about the y axis is

$$k_y = \sqrt{\frac{I_y}{A}} = \sqrt{\frac{7.530 \times 10^7 \text{ mm}^4}{1.086 \times 10^4 \text{ mm}^2}} = 83.3 \text{ mm}.$$

Critical Thinking

Integration is an additive process, which is why the moments of inertia of composite areas can be determined by adding (or, in the case of a cutout, subtracting) the moments of inertia of the parts. But you can't determine the radii of gyration of composite areas by adding or subtracting the radii of gyration of the parts. This can be seen from the equations relating the moments of inertia, radii of gyration, and area. For this example, we can demonstrate it numerically. The operation

$$(k_y)_1 + (k_y)_2 - (k_y)_3 = \sqrt{\frac{(I_y)_1}{A_1}} + \sqrt{\frac{(I_y)_2}{A_2}} - \sqrt{\frac{(I_y)_3}{A_3}} = 86.3 \text{ mm}$$

does not yield the correct radius of gyration of the composite area.

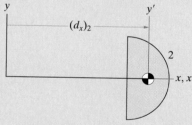

(a) Parts 1, 2, and 3.

Example 8.5   Beam Cross Sections (▶ *Related Problems 8.81–8.84*)

The equal areas are candidates for the cross section of a beam. (A beam with the second cross section shown is called an I-beam.) Compare their moments of inertia about the *x* axis.

Strategy
We can obtain the moment of inertia of the square cross section from Appendix B. We will divide the I-beam cross section into three rectangles and use the parallel-axis theorem to determine its moment of inertia.

Solution
Square Cross Section   From Appendix B, the moment of inertia of the square cross section about the *x* axis is

$$I_x = \frac{1}{12}(144.2 \text{ mm})(144.2 \text{ mm})^3 = 3.60 \times 10^7 \text{ mm}^4.$$

I-Beam Cross Section   We can divide the area into the rectangular parts shown in Fig. a. Introducing coordinate systems $x'y'$ with their origins at the centroids of the parts (Fig. b), we use the parallel-axis theorem to determine the moments of inertia about the *x* axis (see table). Their sum is

$$I_x = (I_x)_1 + (I_x)_2 + (I_x)_3 = (5.23 + 0.58 + 5.23) \times 10^7 \text{ mm}^4$$

$$= 11.03 \times 10^7 \text{ mm}^4.$$

(a) Dividing the I-beam cross section into parts.

(b) Parallel coordinate systems $x'y'$ with origins at the centroids of the parts.

Determining the moments of inertia of the parts about the *x* axis.

	d_y (mm)	A (mm^2)	$I_{x'}$ (mm^4)	$I_x = I_{x'} + d_y^2 A$ (mm^4)
Part 1	80	$(200)(40)$	$\frac{1}{12}(200)(40)^3$	5.23×10^7
Part 2	0	$(40)(120)$	$\frac{1}{12}(40)(120)^3$	0.58×10^7
Part 3	−80	$(200)(40)$	$\frac{1}{12}(200)(40)^3$	5.23×10^7

Critical Thinking
The moment of inertia of the I-beam is 3.06 times that of the square cross section of equal area. Generally a beam with a larger moment of inertia has greater resistance to bending and greater ability to support lateral loads. The cross sections of I-beams are designed to obtain large moments of inertia.

Problems

▶ **8.27** Using the procedure described in Active Example 8.3, determine I_x and k_x for the composite area by dividing it into rectangles 1 and 2 as shown.

8.28 Determine I_y and k_y for the composite area by dividing it into rectangles 1 and 2 as shown.

Problems 8.27/8.28

8.29 Determine I_x and k_x.

Problem 8.29

▶ **8.30** In Example 8.4, determine I_x and k_x for the composite area.

8.31 Determine I_x and k_x.

8.32 Determine I_y and k_y.

8.33 Determine J_O and k_O.

Problems 8.31–8.33

8.34 If you design the beam cross section so that $I_x = 6.4 \times 10^5 \text{ mm}^4$, what are the resulting values of I_y and J_O?

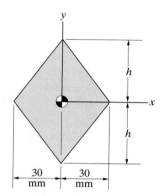

Problem 8.34

8.35 Determine I_y and k_y.

8.36 Determine I_x and k_x.

8.37 Determine I_{xy}.

Problems 8.35–8.37

8.38 Determine I_x and k_x.

8.39 Determine I_y and k_y.

8.40 Determine I_{xy}.

Problems 8.38–8.40

8.41 Determine I_x and k_x.

8.42 Determine J_O and k_O.

8.43 Determine I_{xy}.

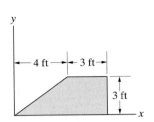

Problems 8.41–8.43

8.44 Determine I_x and k_x.

8.45 Determine J_O and k_O.

8.46 Determine I_{xy}.

Problems 8.44–8.46

8.47 Determine I_x and k_x.

8.48 Determine J_O and k_O.

8.49 Determine I_{xy}.

Problems 8.47–8.49

8.50 Determine I_x and k_x.

8.51 Determine I_y and k_y.

8.52 Determine J_O and k_O.

Problems 8.50–8.52

8.53 Determine I_y and k_y.

8.54 Determine J_O and k_O.

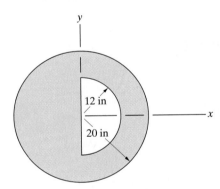

Problems 8.53/8.54

8.55 Determine I_y and k_y if $h = 3$ m.

8.56 Determine I_x and k_x if $h = 3$ m.

8.57 If $I_y = 5$ m^4, what is the dimension h?

Problems 8.55–8.57

8.58 Determine I_y and k_y.

8.59 Determine I_x and k_x.

8.60 Determine I_{xy}.

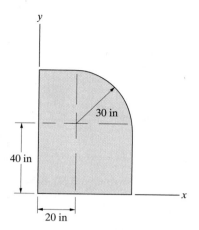

Problems 8.58–8.60

8.61 Determine I_y and k_y.

8.62 Determine I_x and k_x.

8.63 Determine I_{xy}.

Problems 8.61–8.63

8.64 Determine I_y and k_y.

8.65 Determine I_x and k_x.

8.66 Determine I_{xy}.

Problems 8.64–8.66

8.67 Determine I_y and k_y.

8.68 Determine J_O and k_O.

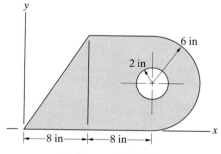

Problems 8.67/8.68

8.69 Determine I_y and k_y.

8.70 Determine I_x and k_x.

8.71 Determine I_{xy}.

Problems 8.69–8.71

8.72 Determine I_y and k_y.

8.73 Determine I_x and k_x.

8.74 Determine I_{xy}.

Problems 8.72–8.74

8.75 Determine I_y and k_y.

8.76 Determine J_O and k_O.

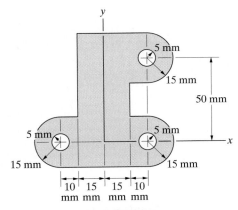

Problems 8.75/8.76

8.77 Determine I_x and I_y for the beam's cross section.

Problem 8.77

8.78 Determine I_x and I_y for the beam's cross section.

Problem 8.78

8.79 The area $A = 2 \times 10^4$ mm². Its moment of inertia about the y axis is $I_y = 3.2 \times 10^8$ mm⁴. Determine its moment of inertia about the $\hat{y}$ axis.

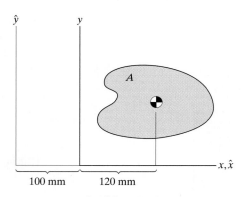

Problem 8.79

8.80 The area $A = 100$ in² and it is *symmetric* about the x' axis. The moments of inertia $I_{x'} = 420$ in⁴, $I_{y'} = 580$ in⁴, $J_O = 11,000$ in⁴, and $I_{xy} = 4800$ in⁴. What are I_x and I_y?

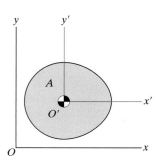

Problem 8.80

▶ **8.81** Determine the moment of inertia of the beam cross section about the x axis. Compare your result with the moment of inertia of a solid square cross section of equal area. (See Example 8.5.)

Problem 8.81

▶ **8.82** The area of the beam cross section is 5200 mm². Determine the moment of inertia of the beam cross section about the x axis. Compare your result with the moment of inertia of a solid square cross section of equal area. (See Example 8.5.)

Problem 8.82

▶ **8.83** If the beam in Fig. a is subjected to couples of magnitude M about the x axis (Fig. b), the beam's longitudinal axis bends into a circular arc whose radius R is given by

$$R = \frac{EI_x}{M},$$

where I_x is the moment of inertia of the beam's cross section about the x axis. The value of the term E, which is called the *modulus of elasticity*, depends on the material of which the beam is constructed. Suppose that a beam with the cross section shown in Fig. c is subjected to couples of magnitude $M = 180$ N-m. As a result, the beam's axis bends into a circular arc with radius $R = 3$ m. What is the modulus of elasticity of the beam's material? (See Example 8.5.)

(a) Unloaded.

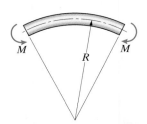

(b) Subjected to couples at the ends.

(c) Beam cross section.

Problem 8.83

▶ **8.84** Suppose that you want to design a beam made of material whose density is 8000 kg/m³. The beam is to be 4 m in length and have a mass of 320 kg. Design a cross section for the beam so that $I_x = 3 \times 10^{-5}$ m⁴. (See Example 8.5.)

8.85 The area in Fig. a is a **C230×30** American Standard Channel beam cross section. Its cross-sectional area is $A = 3790$ mm² and its moments of inertia about the x and y axes are $I_x = 25.3 \times 10^6$ mm⁴ and $I_y = 1 \times 10^6$ mm⁴. Suppose that two beams with **C230×30** cross sections are riveted together to obtain a composite beam with the cross section shown in Fig. b. What are the moments of inertia about the x and y axes of the composite beam?

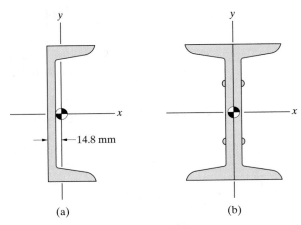

(a)   (b)

Problem 8.85

8.86 The area in Fig. a is an **L152×102×12.7** Angle beam cross section. Its cross-sectional area is $A = 3060$ mm² and its moments of inertia about the x and y axes are $I_x = 7.24 \times 10^6$ mm⁴ and $I_y = 2.61 \times 10^6$ mm⁴. Suppose that four beams with **L152×102×12.7** cross sections are riveted together to obtain a composite beam with the cross section shown in Fig. b. What are the moments of inertia about the x and y axes of the composite beam?

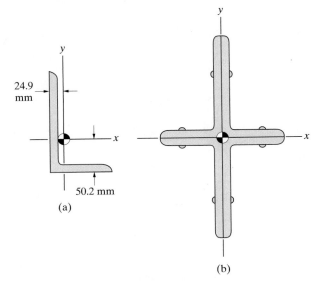

(a)

(b)

Problem 8.86

8.3 Rotated and Principal Axes

BACKGROUND

Suppose that Fig. 8.6a is the cross section of a cantilever beam. If a vertical force is applied to the end of the beam, a larger vertical deflection results if the cross section is oriented as shown in Fig. 8.6b than if it is oriented as shown in Fig. 8.6c. The minimum vertical deflection results when the beam's cross section is oriented so that the moment of inertia I_x is a maximum (Fig. 8.6d).

In many engineering applications we must determine moments of inertia of areas with various angular orientations relative to a coordinate system and also determine the orientation for which the value of a moment of inertia is a maximum or minimum. We discuss these procedures in this section.

Rotated Axes

Consider an area A, a coordinate system xy, and a second coordinate system $x'y'$ that is rotated through an angle θ relative to the xy coordinate system (Fig. 8.7a). Suppose that we know the moments of inertia of A in terms of the xy coordinate system. Our objective is to determine the moments of inertia in terms of the $x'y'$ coordinate system.

In terms of the radial distance r to a differential element of area dA and the angle α in Fig. 8.7b, the coordinates of dA in the xy coordinate system are

$$x = r \cos \alpha, \tag{8.14}$$

$$y = r \sin \alpha. \tag{8.15}$$

Figure 8.6
(a) A beam cross section.
(b)–(d) Applying a lateral load with different orientations of the cross section.

(a)

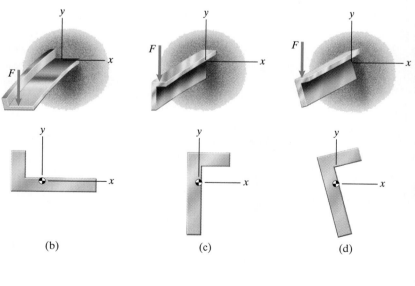

(b)   (c)   (d)

Figure 8.7
(a) The $x'y'$ coordinate system is rotated through an angle θ relative to the xy coordinate system.
(b) A differential element of area dA.

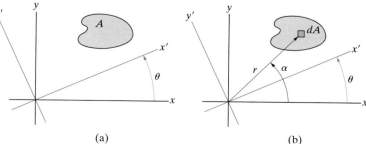

(a)   (b)

The coordinates of dA in the $x'y'$ coordinate system are

$$x' = r\cos(\alpha - \theta) = r(\cos\alpha\cos\theta + \sin\alpha\sin\theta), \qquad (8.16)$$

$$y' = r\sin(\alpha - \theta) = r(\sin\alpha\cos\theta - \cos\alpha\sin\theta). \qquad (8.17)$$

In Eqs. (8.16) and (8.17), we use identities for the cosine and sine of the difference of two angles (Appendix A). By substituting Eqs. (8.14) and (8.15) into Eqs. (8.16) and (8.17), we obtain equations relating the coordinates of dA in the two coordinate systems:

$$x' = x\cos\theta + y\sin\theta, \qquad (8.18)$$

$$y' = -x\sin\theta + y\cos\theta. \qquad (8.19)$$

We can use these expressions to derive relations between the moments of inertia of A in terms of the xy and $x'y'$ coordinate systems.

Moment of Inertia About the x' Axis

$$I_{x'} = \int_A (y')^2\, dA = \int_A (-x\sin\theta + y\cos\theta)^2\, dA$$

$$= \cos^2\theta \int_A y^2\, dA - 2\sin\theta\cos\theta \int_A xy\, dA + \sin^2\theta \int_A x^2\, dA.$$

From this equation we obtain

$$I_{x'} = I_x\cos^2\theta - 2I_{xy}\sin\theta\cos\theta + I_y\sin^2\theta. \qquad (8.20)$$

Moment of Inertia About the y' Axis

$$I_{y'} = \int_A (x')^2\, dA = \int_A (x\cos\theta + y\sin\theta)^2\, dA$$

$$= \sin^2\theta \int_A y^2\, dA + 2\sin\theta\cos\theta \int_A xy\, dA + \cos^2\theta \int_A x^2\, dA.$$

This equation gives us the result

$$I_{y'} = I_x\sin^2\theta + 2I_{xy}\sin\theta\cos\theta + I_y\cos^2\theta. \qquad (8.21)$$

Product of Inertia   In terms of the $x'y'$ coordinate system, the product of inertia of A is

$$I_{x'y'} = (I_x - I_y)\sin\theta\cos\theta + (\cos^2\theta - \sin^2\theta)I_{xy}. \qquad (8.22)$$

Polar Moment of Inertia   From Eqs. (8.20) and (8.21), the polar moment of inertia in terms of the $x'y'$ coordinate system is

$$J_O' = I_{x'} + I_{y'} = I_x + I_y = J_O.$$

Thus the value of the polar moment of inertia is unchanged by a rotation of the coordinate system.

Principal Axes

We have seen that the moments of inertia of A in terms of the $x'y'$ coordinate system depend on the angle θ in Fig. 8.7a. Consider the following question: For what values of θ is the moment of inertia $I_{x'}$ a maximum or minimum?

To answer this question, it is convenient to use the identities

$$\sin 2\theta = 2 \sin \theta \cos \theta,$$

$$\cos 2\theta = \cos^2 \theta - \sin^2 \theta = 1 - 2 \sin^2 \theta = 2 \cos^2 \theta - 1.$$

With these expressions, we can write Eqs. (8.20)–(8.22) in the forms

$$I_{x'} = \frac{I_x + I_y}{2} + \frac{I_x - I_y}{2} \cos 2\theta - I_{xy} \sin 2\theta, \tag{8.23}$$

$$I_{y'} = \frac{I_x + I_y}{2} - \frac{I_x - I_y}{2} \cos 2\theta + I_{xy} \sin 2\theta, \tag{8.24}$$

$$I_{x'y'} = \frac{I_x - I_y}{2} \sin 2\theta + I_{xy} \cos 2\theta. \tag{8.25}$$

We will denote a value of θ at which $I_{x'}$ is a maximum or minimum by θ_p. To determine θ_p, we evaluate the derivative of Eq. (8.23) with respect to 2θ and equate it to zero, obtaining

$$\tan 2\theta_p = \frac{2I_{xy}}{I_y - I_x}. \tag{8.26}$$

If we set the derivative of Eq. (8.24) with respect to 2θ equal to zero to determine a value of θ for which $I_{y'}$ is a maximum or minimum, we again obtain Eq. (8.26). The second derivatives of $I_{x'}$ and $I_{y'}$ with respect to 2θ are opposite in sign; that is,

$$\frac{d^2 I_{x'}}{d(2\theta)^2} = -\frac{d^2 I_{y'}}{d(2\theta)^2},$$

which means that at an angle θ_p for which $I_{x'}$ is a maximum, $I_{y'}$ is a minimum; and at an angle θ_p for which $I_{x'}$ is a minimum, $I_{y'}$ is a maximum.

A rotated coordinate system $x'y'$ that is oriented so that $I_{x'}$ and $I_{y'}$ have maximum or minimum values is called a set of *principal axes* of the area A. The corresponding moments of inertia $I_{x'}$ and $I_{y'}$ are called the *principal moments of inertia*. In the next section we show that the product of inertia $I_{x'y'}$ corresponding to a set of principal axes equals zero.

Because the tangent is a periodic function, Eq. (8.26) does not yield a unique solution for the angle θ_p. We show, however, that it does determine the orientation of the principal axes within an arbitrary multiple of 90°. Observe in Fig. 8.8 that if $2\theta_0$ is a solution of Eq. (8.26), then $2\theta_0 + n(180°)$ is also a solution for any integer n. The resulting orientations of the $x'y'$ coordinate system are shown in Fig. 8.9.

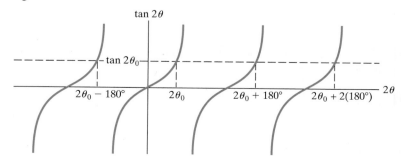

Figure 8.8
For a given value of $\tan 2\theta_0$, there are multiple roots $2\theta_0 + n(180°)$.

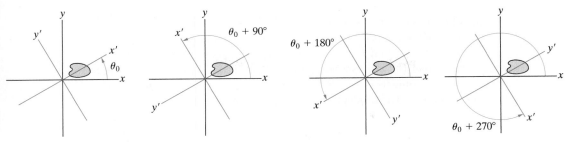

Figure 8.9
The orientation of the $x'y'$ coordinate system
is determined within a multiple of $90°$.

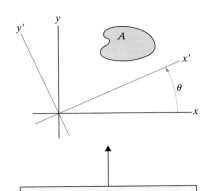

$$I_{x'} = I_x \cos^2 \theta - 2I_{xy} \sin \theta \cos \theta + I_y \sin^2 \theta, \qquad (8.20)$$

$$I_{y'} = I_x \sin^2 \theta + 2I_{xy} \sin \theta \cos \theta + I_y \cos^2 \theta, \qquad (8.21)$$

$$I_{x'y'} = (I_x - I_y) \sin \theta \cos \theta + (\cos^2 \theta - \sin^2 \theta) I_{xy}. \qquad (8.22)$$

The moments and product of inertia of an area in terms of a rotated coordinate system $x'y'$ can be expressed in terms of the moments and products of inertia in terms of the xy coordinate system and the angle θ.

Equations (8.20) to (8.22) can be expressed in useful alternative forms:

$$I_{x'} = \frac{I_x + I_y}{2} + \frac{I_x - I_y}{2} \cos 2\theta - I_{xy} \sin 2\theta, \qquad (8.23)$$

$$I_{y'} = \frac{I_x + I_y}{2} - \frac{I_x - I_y}{2} \cos 2\theta + I_{xy} \sin 2\theta, \qquad (8.24)$$

$$I_{x'y'} = \frac{I_x - I_y}{2} \sin 2\theta + I_{xy} \cos 2\theta, \qquad (8.24)$$

A value of θ for which the moment of inertia $I_{x'}$ obtained from Eq. (8.23) is a maximum or minimum is denoted by θ_p. If $I_{x'}$ is a maximum at $\theta = \theta_p$, $I_{y'}$ is a minimum at $\theta = \theta_p$, and if $I_{x'}$ is a minimum, $I_{y'}$ is a maximum. The rotated coordinate system $x'y'$ corresponding to $\theta = \theta_p$ defines the *principal axes* of the area A, and the moments of inertia about the principal axes are the *principal moments of inertia*. The product of inertia $I_{x'y'}$ corresponding to $\theta = \theta_p$ equals zero.

For given values of I_x, I_y, and I_{xy}, the angle θ_p can be determined from the equation

$$\tan 2\theta_p = \frac{2I_{xy}}{I_y - I_x}. \tag{8.26}$$

This equation uniquely defines the principal axes, but defines the orientation of the $x'y'$ coordinate system only within a multiple of 90°. For example, if θ_0 is a solution of Eq. (8.26), then $\theta_0 + 90°$, $\theta_0 + 180°$, and $\theta_0 + 270°$ are also solutions, resulting in four valid orientations of the $x'y'$ coordinate system.

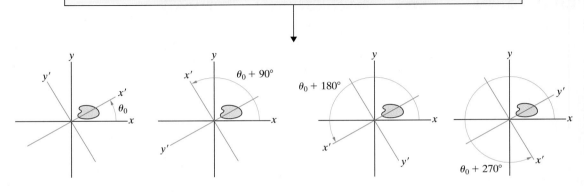

Determining principal axes and principal moments of inertia for a given area A and coordinate system xy involves three steps:

1. Determine I_x, I_y, and I_{xy}.
2. Use Eq. (8.26) to determine θ_p to within a multiple of 90°.
3. Choose the orientation of the $x'y'$ coordinate system and use Eqs.(8.23) and (8.24) to determine the principal moments of inertia.

Active Example 8.6   **Principal Axes and Moments of Inertia** (▶ *Related Problem 8.87*)

Determine a set of principal axes and the corresponding principal moments of inertia for the triangular area.

Strategy

We can obtain the moments and product of inertia of the triangular area in terms of the xy coordinate system from Appendix B. Then we can use Eq. (8.26) to determine the orientation of the principal axes and evaluate the principal moments of inertia with Eqs. (8.23) and (8.24).

Solution

$$I_x = \frac{1}{12}(4 \text{ m})(3 \text{ m})^3 = 9 \text{ m}^4,$$

$$I_y = \frac{1}{4}(4 \text{ m})^3(3 \text{ m}) = 48 \text{ m}^4,$$

$$I_{xy} = \frac{1}{8}(4 \text{ m})^2(3 \text{ m})^2 = 18 \text{ m}^4.$$

Determine the moments and products of inertia from Appendix B.

$$\tan 2\theta_p = \frac{2I_{xy}}{I_y - I_x} = \frac{2(18)}{48 - 9} = 0.923.$$

This yields $\theta_p = 21.4°$.

Determine θ_p from Eq. (8.26).

$$I_{x'} = \frac{I_x + I_y}{2} + \frac{I_x - I_y}{2}\cos 2\theta - I_{xy}\sin 2\theta,$$

$$= \left(\frac{9 + 48}{2}\right) + \left(\frac{9 - 48}{2}\right)\cos[2(21.4°)] - (18)\sin[2(21.4°)]$$

$$= 1.96 \text{ m}^4,$$

Calculate the principal moments of inertia from Eqs. (8.23) and (8.24).

$$I_{y'} = \frac{I_x + I_y}{2} - \frac{I_x - I_y}{2}\cos 2\theta + I_{xy}\sin 2\theta,$$

$$= \left(\frac{9 + 48}{2}\right) - \left(\frac{9 - 48}{2}\right)\cos[2(21.4°)] + (18)\sin[2(21.4°)]$$

$$= 55.0 \text{ m}^4,$$

Practice Problem The moments and product of inertia of the triangular area are $I_x = 9 \text{ m}^4$, $I_y = 16 \text{ m}^4$, and $I_{xy} = 6 \text{ m}^4$. Determine a set of principal axes and the corresponding principal moments of inertia.

Answer: $\theta_p = 29.9°$, $I_{x'} = 5.55 \text{ m}^4$, $I_{y'} = 19.4 \text{ m}^4$.

Example 8.7	**Rotated and Principal Axes** (▶ *Related Problems 8.88, 8.89*)

The moments of inertia of the area in terms of the xy coordinate system shown are $I_x = 22$ ft^4, $I_y = 10$ ft^4, and $I_{xy} = 6$ ft^4. (a) Determine $I_{x'}$, $I_{y'}$, and $I_{x'y'}$ for $\theta = 30°$. (b) Determine a set of principal axes and the corresponding principal moments of inertia.

Strategy

(a) We can determine the moments of inertia in terms of the $x'y'$ coordinate system by substituting $\theta = 30°$ into Eqs. (8.23)–(8.25).

(b) The orientation of the principal axes is determined by solving Eq. (8.26) for θ_p. Once θ_p has been determined, the moments of inertia about the principal axes can be determined from Eqs. (8.23) and (8.24).

Solution

(a) Determine $I_{x'}$, $I_{y'}$, and $I_{x'y'}$ By setting $\theta = 30°$ in Eqs. (8.23)–(8.25), we obtain (with moments of inertia in ft^4)

$$I_{x'} = \frac{I_x + I_y}{2} + \frac{I_x - I_y}{2}\cos 2\theta - I_{xy}\sin 2\theta$$

$$= \left(\frac{22 + 10}{2}\right) + \left(\frac{22 - 10}{2}\right)\cos[2(30°)] - (6)\sin[2(30°)] = 13.8 \text{ ft}^4,$$

$$I_{y'} = \frac{I_x + I_y}{2} - \frac{I_x - I_y}{2}\cos 2\theta + I_{xy}\sin 2\theta$$

$$= \left(\frac{22 + 10}{2}\right) - \left(\frac{22 - 10}{2}\right)\cos[2(30°)] + (6)\sin[2(30°)] = 18.2 \text{ ft}^4,$$

$$I_{x'y'} = \frac{I_x - I_y}{2}\sin 2\theta + I_{xy}\cos 2\theta$$

$$= \left(\frac{22 - 10}{2}\right)\sin[2(30°)] + (6)\cos[2(30°)] = 8.2 \text{ ft}^4.$$

(b) Determine θ_p   We substitute the moments of inertia in terms of the xy coordinate system into Eq. (8.26), yielding

$$\tan 2\theta_p = \frac{2I_{xy}}{I_y - I_x} = \frac{2(6)}{10 - 22} = -1.$$

Thus, $\theta_p = -22.5°$. The principal axes corresponding to this value of θ_p are shown in Fig. a.

Calculate $I_{x'}$ and $I_{y'}$ We substitute $\theta_p = -22.5°$ into Eqs. (8.23) and (8.24), obtaining the principal moments of inertia:

$$I_{x'} = 24.5 \text{ ft}^4, \quad I_{y'} = 7.5 \text{ ft}^4.$$

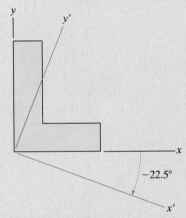

(a) The set of principal axes corresponding to $\theta_p = -22.5°$.

Critical Thinking
Remember that the orientation of the principal axes is only determined within an arbitrary multiple of 90°. In this example we chose to designate the axes in Fig. a as the positive x' and y' axes, but any of these four choices is equally valid.

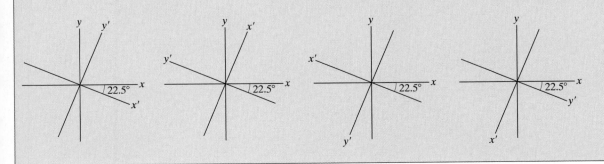

Problems

▶ **8.87** In Active Example 8.6, suppose that the vertical 3-m dimension of the triangular area is increased to 4 m. Determine a set of principal axes and the corresponding principal moments of inertia.

▶ **8.88** In Example 8.7, suppose that the area is reoriented as shown. Determine the moments of inertia $I_{x'}$, $I_{y'}$, and $I_{x'y'}$ if $\theta = 30°$.

▶ **8.89** In Example 8.7, suppose that the area is reoriented as shown. Determine a set of principal axes and the corresponding principal moments of inertia. Based on the results of Example 8.7, can you predict a value of θ_p without using Eq. (8.26)?

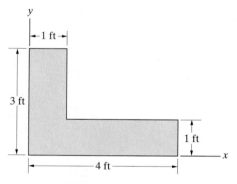

Problems 8.88/8.89

8.90 The moments of inertia of the area are $I_x = 1.26 \times 10^6$ in^4, $I_y = 6.55 \times 10^5$ in^4, and $I_{xy} = -1.02 \times 10^5$ in^4. Determine the moments of inertia of the area $I_{x'}$, $I_{y'}$, and $I_{x'y'}$ if $\theta = 30°$.

8.91 The moments of inertia of the area are $I_x = 1.26 \times 10^6$ in^4, $I_y = 6.55 \times 10^5$ in^4, and $I_{xy} = -1.02 \times 10^5$ in^4. Determine a set of principal axes and the corresponding principal moments of inertia.

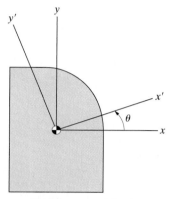

Problems 8.90/8.91

8.92* Determine a set of principal axes and the corresponding principal moments of inertia.

Problem 8.92

8.4 Mohr's Circle

Given the moments of inertia of an area in terms of a particular coordinate system, we have presented equations that determine the moments of inertia in terms of a rotated coordinate system, the orientation of the principal axes, and the principal moments of inertia. We can also obtain this information by using a graphical method called *Mohr's circle*, which is very useful for visualizing the solutions of Eqs. (8.23)–(8.25).

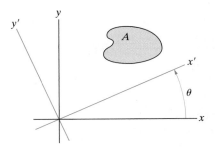

Figure 8.10
The *xy* coordinate system and the rotated *x'y'* coordinate system.

Determining $I_{x'}$, $I_{y'}$, and $I_{x'y'}$

We first describe how to construct Mohr's circle and then explain why it works. Suppose we know the moments of inertia I_x, I_y, and I_{xy} of an area in terms of a coordinate system *xy* and we want to determine the moments of inertia for a rotated coordinate system $x'y'$ (Fig. 8.10). Constructing Mohr's circle involves three steps:

1. Establish a set of horizontal and vertical axes and plot two points: point 1 with coordinates (I_x, I_{xy}) and point 2 with coordinates $(I_y, -I_{xy})$ as shown in Fig. 8.11a.
2. Draw a straight line connecting points 1 and 2. Using the intersection of the straight line with the horizontal axis as the center, draw a circle that passes through the two points (Fig. 8.11b).
3. Draw a straight line through the center of the circle at an angle 2θ measured counterclockwise from point 1. This line intersects the circle at point $1'$ with coordinates $(I_{x'}, I_{x'y'})$ and point $2'$ with coordinates $(I_{y'}, -I_{x'y'})$, as shown in Fig. 8.11c.

Thus, for a given angle θ, the coordinates of points $1'$ and $2'$ determine the moments of inertia of the rotated coordinate system. Why does this graphical construction work? In Fig. 8.12, we show the points 1 and 2 and Mohr's circle. Notice that the horizontal coordinate of the center of the circle is $(I_x + I_y)/2$. The sine and cosine of the angle β are

$$\sin\beta = \frac{I_{xy}}{R}, \qquad \cos\beta = \frac{I_x - I_y}{2R},$$

where R, the radius of the circle, is given by

$$R = \sqrt{\left(\frac{I_x - I_y}{2}\right)^2 + (I_{xy})^2}.$$

Figure 8.13 shows the construction of the points $1'$ and $2'$. The horizontal coordinate of point $1'$ is

$$\frac{I_x + I_y}{2} + R\cos(\beta + 2\theta)$$

$$= \frac{I_x + I_y}{2} + R(\cos\beta\cos 2\theta - \sin\beta\sin 2\theta)$$

$$= \frac{I_x + I_y}{2} + \frac{I_x - I_y}{2}\cos 2\theta - I_{xy}\sin 2\theta = I_{x'},$$

(a)

(b)

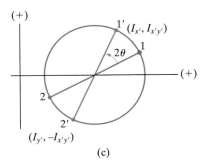

(c)

Figure 8.11
(a) Plotting the points 1 and 2.
(b) Drawing Mohr's circle. The center of the circle is the intersection of the line from 1 to 2 with the horizontal axis.
(c) Finding the points $1'$ and $2'$.

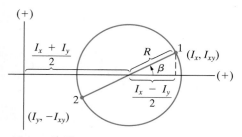

Figure 8.12
The points 1 and 2 and Mohr's circle.

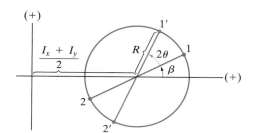

Figure 8.13
The points 1′ and 2′.

and the horizontal coordinate of point 2′ is

$$\frac{I_x + I_y}{2} - R\cos(\beta + 2\theta)$$

$$= \frac{I_x + I_y}{2} - R(\cos\beta\cos 2\theta - \sin\beta\sin 2\theta)$$

$$= \frac{I_x + I_y}{2} - \frac{I_x - I_y}{2}\cos 2\theta + I_{xy}\sin 2\theta = I_{y'}.$$

The vertical coordinate of point 1′ is

$$R\sin(\beta + 2\theta) = R(\sin\beta\cos 2\theta + \cos\beta\sin 2\theta)$$

$$= I_{xy}\cos 2\theta + \frac{I_x - I_y}{2}\sin 2\theta = I_{x'y'},$$

and the vertical coordinate of point 2′ is

$$-R\sin(\beta + 2\theta) = -I_{x'y'}.$$

We have shown that the coordinates of point 1′ are $(I_{x'}, I_{x'y'})$ and the coordinates of point 2′ are $(I_{y'}, -I_{x'y'})$.

Determining Principal Axes and Principal Moments of Inertia

Because the moments of inertia $I_{x'}$ and $I_{y'}$ are the horizontal coordinates of points 1′ and 2′ of Mohr's circle, their maximum and minimum values occur when points 1′ and 2′ coincide with the intersections of the circle with the horizontal axis (Fig. 8.14). (Which intersection you designate as 1′ is arbitrary. In Fig. 8.14, we have designated the minimum moment of inertia as point 1′.) You can determine the orientation of the principal axes by measuring the angle $2\theta_p$ from point 1 to point 1′, and the coordinates of points 1′ and 2′ are the principal moments of inertia.

Notice that Mohr's circle demonstrates that the product of inertia $I_{x'y'}$ corresponding to a set of principal axes (the vertical coordinate of point 1′ in Fig. 8.14) is always zero. Furthermore, we can use Fig. 8.12 to obtain an analytical expression for the horizontal coordinates of the points where the circle intersects the horizontal axis, which are the principal moments of inertia:

$$\text{Principal moments of inertia} = \frac{I_x + I_y}{2} \pm R$$

$$= \frac{I_x + I_y}{2} \pm \sqrt{\left(\frac{I_x - I_y}{2}\right)^2 + (I_{xy})^2}.$$

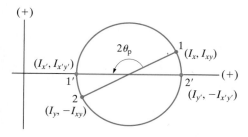

Figure 8.14
To determine the orientation of a set of principal axes, let points 1′ and 2′ be the points where the circle intersects the horizontal axis.

RESULTS

When the values of I_x, I_y, and I_{xy} are known for an area A, Mohr's circle can be used to determine the values of $I_{x'}$, $I_{y'}$, and $I_{x'y'}$ for a given angle θ:

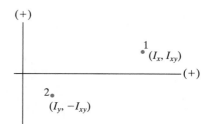

| Establish a set of horizontal and vertical axes and plot two points: point 1 with coordinates (I_x, I_{xy}) and point 2 with coordinates $(I_y, -I_{xy})$. |

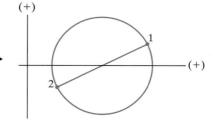

| Draw a straight line connecting points 1 and 2. Using the intersection of the straight line with the horizontal axis as the center, draw a circle that passes through the two points. |

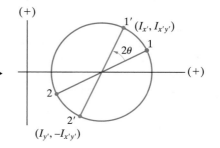

| Draw a straight line through the center of the circle at an angle 2θ measured counterclockwise from point 1. This line intersects the circle at point $1'$ with coordinates $(I_{x'}, I_{x'y'})$ and point $2'$ with coordinates $(I_{y'}, -I_{x'y'})$. |

Mohr's circle can also be used to determine the orientation of the principal axes and the principal moments of interia:

| Place the point $1'$ at one of the points where Mohr's circle intersects the horizontal axis. Then the values of $I_{x'}$ and $I_{y'}$ obtained from points $1'$ and $2'$ are the principal moments of inertia. The angle measured counterclockwise from point 1 to point $1'$ is $2\theta_p$, so the orientation of the principal axes can be determined. |

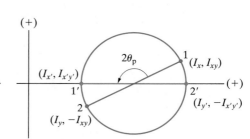

Active Example 8.8   *Mohr's Circle* (▶ *Related Problems 8.94, 8.95*)

The moments and product of inertia of the area in terms of the *xy* coordinate system are $I_x = 22 \text{ ft}^4$, $I_y = 10 \text{ ft}^4$, and $I_{xy} = 6 \text{ ft}^4$. Use Mohr's circle to determine the moments of inertia $I_{x'}$, $I_{y'}$, and $I_{x'y'}$ for $\theta = 30°$.

Strategy

By using the given values of I_x, I_y, and I_{xy} to construct Mohr's circle, we can determine $I_{x'}$, $I_{y'}$, and $I_{x'y'}$ for $\theta = 30°$.

Solution

Plot point 1 with coordinates $(I_x, I_{xy}) = (22, 6) \text{ ft}^4$ and point 2 with coordinates $(I_y, -I_{xy}) = (10, -6) \text{ ft}^4$.

Draw a straight line connecting points 1 and 2. Using the intersection of the straight line with the horizontal axis as the center, draw a circle that passes through the two points.

Draw a straight line through the center of the circle at an angle $2\theta = 60°$ measured counterclockwise from point 1. From the coordinates of points 1' and 2', $I_{x'} = 14 \text{ ft}^4$, $I_{y'} = 18 \text{ ft}^4$, and $I_{x'y'} = 8 \text{ ft}^4$.

Practice Problem Use Mohr's circle to determine the orientation of the principal axes of the area and the corresponding principal moments of inertia.

Answer: $\theta_p = 67.5°$, $I_{x'} = 7.5 \text{ ft}^4$, $I_{y'} = 24.5 \text{ ft}^4$.

Problems

8.93 Solve Problem 8.87 by using Mohr's circle.

▶ **8.94** Solve Problem 8.88 by using Mohr's circle.

▶ **8.95** Solve Problem 8.89 by using Mohr's circle.

8.96 Solve Problem 8.90 by using Mohr's circle.

8.97 Solve Problem 8.91 by using Mohr's circle.

8.98* Solve Problem 8.92 by using Mohr's circle.

8.99 Derive Eq. (8.22) for the product of inertia by using the same procedure we used to derive Eqs. (8.20) and (8.21).

MASSES

8.5 Simple Objects

BACKGROUND

The acceleration of an object that results from the forces acting on it depends on its mass. The angular acceleration, or rotational acceleration, that results from the forces and couples acting on an object depends on quantities called the mass moments of inertia of the object. In this section we discuss methods for determining mass moments of inertia of particular objects. We show that for special classes of objects, their mass moments of inertia can be expressed in terms of moments of inertia of areas, which explains how the names of those area integrals originated.

An object and a line or "axis" L_O are shown in Fig. 8.15a. The *moment of inertia* of the object about the axis L_O is defined by

$$I_O = \int_m r^2 \, dm, \tag{8.27}$$

where r is the perpendicular distance from the axis to the differential element of mass dm (Fig. 8.15b). Often L_O is an axis about which the object rotates, and the value of I_O is required to determine the angular acceleration, or the rate of change of the rate of rotation, caused by a given couple about L_O. The dimensions of the moment of inertia of an object are (mass) $\times$ (length)^2. Notice that the definition implies that its value must be positive.

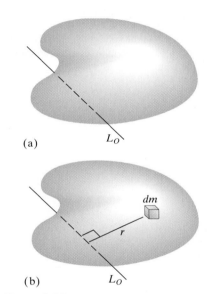

(a)

(b)

Figure 8.15
(a) An object and axis L_O.
(b) A differential element of mass dm.

Slender Bars

Let us determine the moment of inertia of a straight, slender bar about a perpendicular axis L through the center of mass of the bar (Fig. 8.16a). "Slender" means that we assume that the bar's length is much greater than its width. Let the bar have length l, cross-sectional area A, and mass m. We assume that A is uniform along the length of the bar and that the material is homogeneous.

Consider a differential element of the bar of length dr at a distance r from the center of mass (Fig. 8.16b). The element's mass is equal to the product of its volume and the density: $dm = \rho A \, dr$. Substituting this expression into Eq. (8.27), we obtain the moment of inertia of the bar about a perpendicular axis through its center of mass:

$$I = \int_m r^2 \, dm = \int_{-l/2}^{l/2} \rho A r^2 \, dr = \frac{1}{12} \rho A l^3.$$

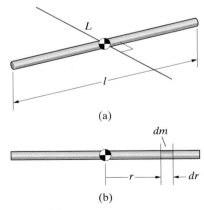

Figure 8.16
(a) A slender bar.
(b) A differential element of length dr.

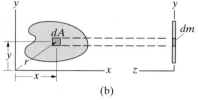

Figure 8.17

(a) A plate of arbitrary shape and uniform thickness T.

(b) An element of volume obtained by projecting an element of area dA through the plate.

The mass of the bar equals the product of the mass density and the volume of the bar, $m = \rho A l$, so we can express the moment of inertia as

$$I = \frac{1}{12}ml^2. \tag{8.28}$$

We have neglected the lateral dimensions of the bar in obtaining this result. That is, we treated the differential element of mass dm as if it were concentrated on the axis of the bar. As a consequence, Eq. (8.28) is an approximation for the moment of inertia of a bar. In the next section we determine the moments of inertia for a bar of finite lateral dimension and show that Eq. (8.28) is a good approximation when the width of the bar is small in comparison to its length.

Thin Plates

Consider a homogeneous flat plate that has mass m and uniform thickness T. We will leave the shape of the cross-sectional area of the plate unspecified. Let a cartesian coordinate system be oriented so that the plate lies in the x–y plane (Fig. 8.17a). Our objective is to determine the moments of inertia of the plate about the x, y, and z axes.

We can obtain a differential element of volume of the plate by projecting an element of area dA through the thickness T of the plate (Fig. 8.17b). The resulting volume is $T\,dA$. The mass of this element of volume is equal to the product of the density and the volume: $dm = \rho T\,dA$. Substituting this expression into Eq. (8.27), we obtain the moment of inertia of the plate about the z axis in the form

$$I_{z\,\text{axis}} = \int_m r^2\,dm = \rho T \int_A r^2\,dA,$$

where r is the distance from the z axis to dA. Since the mass of the plate is $m = \rho T A$, where A is the cross-sectional area of the plate, $\rho T = m/A$. The integral on the right is the polar moment of inertia J_O of the cross-sectional area of the plate. We can therefore write the moment of inertia of the plate about the z axis as

$$I_{z\,\text{axis}} = \frac{m}{A}J_O. \tag{8.29}$$

From Fig. 8.17b, we see that the perpendicular distance from the x axis to the element of area dA is the y coordinate of dA. Therefore, the moment of inertia of the plate about the x axis is

$$I_{x\,\text{axis}} = \int_m y^2\,dm = \rho T \int_A y^2\,dA = \frac{m}{A}I_x, \tag{8.30}$$

where I_x is the moment of inertia of the cross-sectional area of the plate about the x axis. The moment of inertia of the plate about the y axis is

$$I_{y\,\text{axis}} = \int_m x^2\,dm = \rho T \int_A x^2\,dA = \frac{m}{A}I_y, \tag{8.31}$$

where I_y is the moment of inertia of the cross-sectional area of the plate about the y axis.

Because the sum of the area moments of inertia I_x and I_y is equal to the polar moment of inertia J_O, the mass moment of inertia of the thin plate about the z axis is equal to the sum of its moments of inertia about the x and y axes:

$$I_{z\,\text{axis}} = I_{x\,\text{axis}} + I_{y\,\text{axis}}. \qquad \text{Thin plate} \tag{8.32}$$

We have expressed the moments of inertia of a thin homogeneous plate of uniform thickness in terms of the moments of inertia of the cross-sectional area of the plate. In fact, these results explain why the area integrals I_x, I_y, and J_O are called moments of inertia. The use of the same terminology and similar symbols for moments of inertia of areas and moments of inertia of objects can be confusing, but is entrenched in engineering practice. The type of moment of inertia being referred to can be determined either from the context or from the units: $(\text{length})^4$ for moments of inertia of areas and $(\text{mass}) \times (\text{length})^2$ for moments of inertia of masses.

RESULTS

Moment of Inertia of an Object
The moment of inertia of an object about an axis L_O is defined by

$$I_O = \int_m r^2 \, dm, \tag{8.27}$$

where r is the perpendicular distance from L_O to the differential element of mass dm.

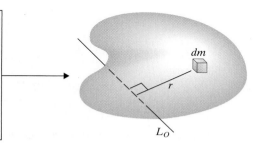

Slender Bars
The differential element of mass $dm = \rho A \, dr$, where ρ is the density of the homogeneous bar and A is its uniform cross-sectional area. The moment of inertia of the slender bar of length l about the perpendicular axis L through its center of mass is

$$I = \int_m r^2 \, dm = \int_{-l/2}^{l/2} \rho A r^2 \, dr = \frac{1}{12} \rho A l^3.$$

In terms of the mass of the bar $m = \rho A l$,

$$I = \frac{1}{12} m l^2. \tag{8.28}$$

Thin Plates
The moments of inertia of a thin homogeneous plate of uniform thickness and mass m that lies in the x–y plane can be expressed in terms of the moments of inertia of the cross-sectional area A of the plate:

$$I_{x\text{ axis}} = \frac{m}{A} I_x, \tag{8.30}$$

$$I_{y\text{ axis}} = \frac{m}{A} I_y, \tag{8.31}$$

$$I_{z\text{ axis}} = \frac{m}{A} J_O = I_{x\text{ axis}} + I_{y\text{ axis}}. \tag{8.29}$$

Here I_x is the moment of inertia of A about the x axis, I_y is the moment of inertia of A about the y axis, and J_O is the polar moment of inertia of A about the origin.

Active Example 8.9   **Moment of Inertia of a Triangular Plate** (▶ *Related Problem 8.104*)

The thin homogeneous plate is of uniform thickness and mass m. Determine its moment of inertia about the x axis.

Strategy

The moment of inertia of the plate about the x axis is given by Eq. (8.30) in terms of the moment of inertia of the area of the plate about the x axis. We can obtain the moment of inertia of the area from Appendix B.

Solution

From Appendix B,

$$I_x = \frac{1}{12} bh^3.$$

Determine the moment of inertia of the area of the plate about the x axis.

The moment of inertia of the plate about the x axis is

$$I_{x\,\text{axis}} = \frac{m}{A} I_x$$

$$= \frac{m}{\frac{1}{2} bh} \left(\frac{1}{12} bh^3 \right)$$

$$= \frac{1}{6} mh^2.$$

Apply Eq. (8.30).

Practice Problem Determine the moment of inertia of the plate about the y axis.

Answer: $I_{y\,\text{axis}} = \frac{1}{2} mb^2.$

Example 8.10   **Moments of Inertia of a Slender Bar** (▶ *Related Problem 8.100*)

Two homogeneous slender bars, each of length l, mass m, and cross-sectional area A, are welded together to form the L-shaped object. Determine the moment of inertia of the object about the axis L_O through point O. (The axis L_O is perpendicular to the two bars.)

Strategy
Using the same integration procedure we used for a single bar, we will determine the moment of inertia of each bar about L_O and sum the results.

Solution
Our first step is to introduce a coordinate system with the z axis along L_O and the x axis collinear with bar 1 (Fig. a). The mass of the differential element of bar 1 of length dx is $dm = \rho A\, dx$. The moment of inertia of bar 1 about L_O is

$$(I_O)_1 = \int_m r^2\, dm = \int_0^l \rho A x^2\, dx = \frac{1}{3}\rho A l^3.$$

In terms of the mass of the bar, $m = \rho A l$, we can write this result as

$$(I_O)_1 = \frac{1}{3}ml^2.$$

The mass of an element of bar 2 of length dy, shown in Fig. b, is $dm = \rho A\, dy$. From the figure we see that the perpendicular distance from L_O to the element is $r = \sqrt{l^2 + y^2}$. Therefore, the moment of inertia of bar 2 about L_O is

(a) Differential element of bar 1.

$$(I_O)_2 = \int_m r^2\, dm = \int_0^l \rho A (l^2 + y^2)\, dy = \frac{4}{3}\rho A l^3.$$

In terms of the mass of the bar, we obtain

$$(I_O)_2 = \frac{4}{3}ml^2.$$

The moment of inertia of the L-shaped object about L_O is

$$I_O = (I_O)_1 + (I_O)_2 = \frac{1}{3}ml^2 + \frac{4}{3}ml^2 = \frac{5}{3}ml^2.$$

(b) Differential element of bar 2.

Critical Thinking
In this example we used integration to determine a moment of inertia of an object consisting of two straight bars. The same procedure could be applied to more complicated objects made of such bars, but it would obviously be cumbersome. Once we have used integration to determine a moment of inertia of a single bar, such as Eq. (8.28), it would be very convenient to use that result to determine moments of inertia of composite objects made of bars without having to resort to integration. We show how this can be done in the next section.

Problems

▶ **8.100** The axis L_O is perpendicular to both segments of the L-shaped slender bar. The mass of the bar is 6 kg and the material is homogeneous. Use the method described in Example 8.10 to determine the moment of inertia of the bar about L_O.

Problem 8.100

8.101 Two homogeneous slender bars, each of mass m and length l, are welded together to form the T-shaped object. Use integration to determine the moment of inertia of the object about the axis through point O that is perpendicular to the bars.

Problem 8.101

8.102 The slender bar lies in the x–y plane. Its mass is 6 kg and the material is homogeneous. Use integration to determine its moment of inertia about the z axis.

8.103 Use integration to determine the moment of inertia of the slender 6-kg bar about the y axis.

Problems 8.102/8.103

▶ **8.104** The homogeneous thin plate has mass $m = 12$ kg and dimensions $b = 2$ m and $h = 1$ m. Use the procedure described in Active Example 8.9 to determine the moments of inertia of the plate about the x and y axes.

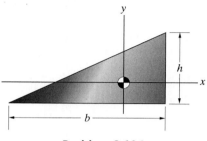

Problem 8.104

8.105 The homogeneous thin plate is of uniform thickness and mass m.

(a) Determine its moments of inertia about the x and z axes.

(b) Let $R_i = 0$ and compare your results with the values given in Appendix C for a thin circular plate.

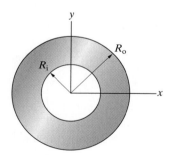

Problem 8.105

8.106 The homogeneous thin plate is of uniform thickness and weighs 20 lb. Determine its moment of inertia about the y axis.

8.107 Determine the moment of inertia of the plate about the x axis.

$y = 4 - \dfrac{1}{4}x^2$ ft

Problems 8.106/8.107

8.6 Parallel-Axis Theorem

BACKGROUND

The parallel-axis theorem allows us to determine the moment of inertia of an object about any axis when the moment of inertia about a parallel axis through the center of mass is known. This theorem can be used to calculate the moment of inertia of a composite object about an axis given the moments of inertia of each of its parts about parallel axes.

Suppose that we know the moment of inertia I about an axis L through the center of mass of an object, and we wish to determine its moment of inertia I_O about a parallel axis L_O (Fig. 8.18a). To determine I_O, we introduce parallel coordinate systems xyz and $x'y'z'$ with the z axis along L_O and the z' axis along L, as shown in Fig. 8.18b. (In this figure the axes L_O and L are perpendicular to the page.) The origin O of the xyz coordinate system is contained in the $x'-y'$ plane. The terms d_x and d_y are the coordinates of the center of mass relative to the xyz coordinate system.

The moment of inertia of the object about L_O is

$$I_O = \int_m r^2 \, dm = \int_m (x^2 + y^2) \, dm, \tag{8.33}$$

where r is the perpendicular distance from L_O to the differential element of mass dm, and x, y are the coordinates of dm in the x–y plane. The coordinates of dm in the two coordinate systems are related by

$$x = x' + d_x, \qquad y = y' + d_y.$$

By substituting these expressions into Eq. (8.33), we can write it as

$$I_O = \int_m [(x')^2 + (y')^2] dm + 2d_x \int_m x' dm + 2d_y \int_m y' dm$$

$$+ \int_m (d_x^2 + d_y^2) \, dm. \tag{8.34}$$

Since $(x')^2 + (y')^2 = (r')^2$, where r' is the perpendicular distance from L to dm, the first integral on the right side of this equation is the moment of inertia I of the object about L. Recall that the x' and y' coordinates of the center of mass of the object relative to the $x'y'z'$ coordinate system are defined by

$$\bar{x}' = \frac{\int_m x' \, dm}{\int_m dm}, \qquad \bar{y}' = \frac{\int_m y' \, dm}{\int_m dm}.$$

Because the center of mass of the object is at the origin of the $x'y'z'$ system, $\bar{x}' = 0$ and $\bar{y}' = 0$. Therefore the integrals in the second and third terms on the right side of Eq. (8.34) are equal to zero. From Fig. 8.18b, we see that $d_x^2 + d_y^2 = d^2$, where d is the perpendicular distance between the axes L and L_O. Therefore, we obtain

$$I_O = I + d^2 m. \tag{8.35}$$

This is the *parallel-axis theorem* for moments of inertia of objects. Equation (8.35) relates the moment of inertia I of an object about an axis *through the center of mass* to its moment of inertia I_O about any parallel axis, where d is the perpendicular distance between the two axes and m is the mass of the object.

(a)

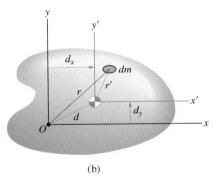

(b)

Figure 8.18
(a) An axis L through the center of mass of an object and a parallel axis L_O.
(b) The xyz and $x'y'z'$ coordinate systems.

Determining the moment of inertia of an object about a given axis L_O typically requires three steps:

1. Choose the parts—Try to divide the object into parts whose mass moments of inertia you know or can easily determine.
2. Determine the moments of inertia of the parts—You must first determine the moment of inertia of each part about the axis through its center of mass parallel to L_O. Then you can use the parallel-axis theorem to determine its moment of inertia about L_O.
3. Sum the results—Sum the moments of inertia of the parts (or subtract in the case of a hole or cutout) to obtain the moment of inertia of the composite object.

RESULTS

Parallel-Axis Theorem
The moment of inertia of an object of mass m about an axis L_O is given by

$$I_O = I + d^2 m, \tag{8.35}$$

where I is the moment of inertia of the object about a parallel axis through the center of mass and d is the perpendicular distance between the two axes.

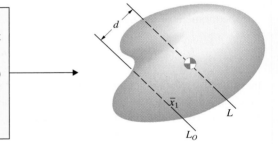

Composite Objects
The parallel-axis theorem makes it possible to determine the moment of inertia of a composite object about a given axis L_O. The moment of inertia of each part must be determined about an axis through the center of mass of the part that is parallel to L_O. Then the parallel-axis theorem can be applied to each part to determine its moment of inertia about L_O. Summing the results gives the moment of inertia of the composite object about L_O.

Active Example 8.11 **Parallel-Axis Theorem** (▶ *Related Problem 8.111*)

The homogeneous slender bar has mass m and length l. The axis L_O is perpendicular to the bar.
(a) Use integration to determine the moment of inertia of the bar about L_O.
(b) The moment of inertia of the bar about an axis through the center of mass of the bar that is perpendicular to the bar is $I = (1/12)ml^2$. Use this result and the parallel-axis theorem to determine the moment of inertia of the bar about L_O.

Solution

The differential element of mass $dm = \rho A\, dr$, where ρ is the density of the homogeneous bar and A is its uniform cross-sectional area. The moment of inertia is

$$I_O = \int_m r^2\, dm = \int_O^l \rho A r^2\, dr = \frac{1}{3}\rho A l^3.$$

In terms of the mass of the bar $m = \rho A l$,

$$I_O = \frac{1}{3}ml^2.$$

(a) Integrate to determine the moment of inertia about L_O.

$$\begin{aligned} I_O &= I + d^2 m \\ &= \frac{1}{12}ml^2 + \left(\frac{1}{2}l\right)^2 m \\ &= \frac{1}{3}\,ml^2. \end{aligned}$$

(b) Apply the parallel-axis theorem.

Practice Problem Two homogeneous slender bars, each of length l and mass m, are welded together to form an L-shaped object. Use the parallel-axis theorem to determine the moment of inertia of the object about the axis L_O. (The axis L_O is perpendicular to both bars.)

Answer: $I_O = \frac{5}{3}ml^2.$

Example 8.12 **Moment of Inertia of a Composite Object** (▶ *Related Problem 8.127*)

The object consists of a slender, 3-kg bar welded to a thin, circular 2-kg disk. Determine its moment of inertia about the axis L through its center of mass. (The axis L is perpendicular to the bar and disk.)

Strategy

We must first locate the center of mass of the composite object and then apply the parallel-axis theorem to the parts separately and sum the results.

Solution

Choose the Parts The parts are the bar and the disk. Introducing the coordinate system in Fig. a, the x coordinate of the center of mass of the composite object is

$$\bar{x} = \frac{\bar{x}_{\text{bar}} m_{\text{bar}} + \bar{x}_{\text{disk}} m_{\text{disk}}}{m_{\text{bar}} + m_{\text{disk}}}$$

$$= \frac{(0.3 \text{ m})(3 \text{ kg}) + (0.6 \text{ m} + 0.2 \text{ m})(2 \text{ kg})}{(3 \text{ kg}) + (2 \text{ kg})} = 0.5 \text{ m}.$$

(a) The coordinate $\bar{x}$ of the center of mass of the object.

Determine the Moments of Inertia of the Parts The distance from the center of mass of the bar to the center of mass of the composite object is 0.2 m (Fig. b). Therefore, the moment of inertia of the bar about L is

$$I_{\text{bar}} = \frac{1}{12}(3 \text{ kg})(0.6 \text{ m})^2 + (3 \text{ kg})(0.2 \text{ m})^2 = 0.210 \text{ kg-m}^2.$$

(b) Distance from L to the center of mass of the bar.

The distance from the center of mass of the disk to the center of mass of the composite object is 0.3 m (Fig. c). The moment of inertia of the disk about L is

$$I_{\text{disk}} = \frac{1}{2}(2\text{ kg})(0.2\text{ m})^2 + (2\text{ kg})(0.3\text{ m})^2 = 0.220\text{ kg-m}^2.$$

(c) Distance from L to the center of mass of the disk.

Sum the Results   The moment of inertia of the composite object about L is

$$I = I_{\text{bar}} + I_{\text{disk}} = 0.430\text{ kg-m}^2.$$

Critical Thinking
This example demonstrates the most common procedure for determining moments of inertia of objects in engineering applications. Objects usually consist of assemblies of parts. The center of mass of each part and its moment of inertia about the axis through its center of mass must be determined. (It may be necessary to determine this information experimentally, or it is sometimes supplied by manufacturers of subassemblies.) Then the center of mass of the composite object is determined and the parallel-axis theorem is used to determine the moment of inertia of each part about the axis through the center of mass of the composite object. Finally, the individual moments of inertia are summed to obtain the moment of inertia of the composite object.

Example 8.13 | **Moments of Inertia of a Cylinder** (▶ *Related Problems 8.122, 8.123, 8.125, 8.126*)

The homogeneous cylinder has mass m, length l, and radius R. Determine its moments of inertia about the x, y, and z axes.

Strategy

We first determine the moments of inertia about the x, y, and z axes of an infinitesimal element of the cylinder consisting of a disk of thickness dz. We then integrate the results with respect to z to obtain the moments of inertia of the cylinder. We must apply the parallel-axis theorem to determine the moments of inertia of the disk about the x and y axes.

Solution

Consider an element of the cylinder of thickness dz at a distance z from the center of the cylinder (Fig. a). (You can imagine obtaining this element by "slicing" the cylinder perpendicular to its axis.) The mass of the element is equal to the product of the mass density and the volume of the element, $dm = \rho(\pi R^2 \, dz)$. We obtain the moments of inertia of the element by using the values for a thin circular plate given in Appendix C. The moment of inertia about the z axis is

$$dI_{z \text{ axis}} = \frac{1}{2} dm R^2 = \frac{1}{2}(\rho \pi R^2 \, dz) R^2.$$

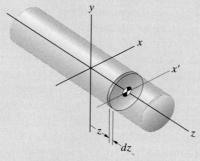

(a) A differential element of the cylinder in the form of a disk.

By integrating this result with respect to z from $-l/2$ to $l/2$, we sum the mass moments of inertia of the infinitesimal disk elements that make up the cylinder. The result is the moment of inertia of the cylinder about the z axis:

$$I_{z \text{ axis}} = \int_{-l/2}^{l/2} \frac{1}{2} \rho \pi R^4 \, dz = \frac{1}{2} \rho \pi R^4 l.$$

We can write this result in terms of the mass of the cylinder, $m = \rho(\pi R^2 l)$, as

$$I_{z \text{ axis}} = \frac{1}{2} m R^2.$$

The moment of inertia of the disk element about the x' axis is

$$dI_{x' \text{ axis}} = \frac{1}{4} dm \, R^2 = \frac{1}{4} (\rho \pi R^2 \, dz) R^2.$$

We can use this result and the parallel-axis theorem to determine the moment of inertia of the element about the x axis:

$$dI_{x \text{ axis}} = dI_{x' \text{ axis}} + z^2 \, dm = \frac{1}{4} (\rho \pi R^2 \, dz) R^2 + z^2 (\rho \pi R^2 \, dz).$$

Integrating this expression with respect to z from $-l/2$ to $l/2$, we obtain the moment of inertia of the cylinder about the x axis:

$$I_{x \text{ axis}} = \int_{-l/2}^{l/2} \left(\frac{1}{4} \rho \pi R^4 + \rho \pi R^2 z^2 \right) dz = \frac{1}{4} \rho \pi R^4 l + \frac{1}{12} \rho \pi R^2 l^3.$$

In terms of the mass of the cylinder,

$$I_{x \text{ axis}} = \frac{1}{4} m R^2 + \frac{1}{12} m l^2.$$

Due to the symmetry of the cylinder,

$$I_{y \text{ axis}} = I_{x \text{ axis}}.$$

Critical Thinking

When the cylinder is very long in comparison to its width, $l \gg R$, the first term in the equation for $I_{x \text{ axis}}$ can be neglected, and we obtain the moment of inertia of a slender bar about a perpendicular axis, Eq. (8.28). Conversely, when the radius of the cylinder is much greater than its length, $R \gg l$, the second term in the equation for $I_{x \text{ axis}}$ can be neglected, and we obtain the moment of inertia for a thin circular disk about an axis parallel to the disk. This indicates the sizes of the terms you neglect when you use the approximate expressions for the moments of inertia of a "slender" bar and a "thin" disk.

Problems

8.108 The mass of the object is 10 kg. Its moment of inertia about L_1 is 10 kg-m². What is its moment of inertia about L_2? (The three axes lie in the same plane.)

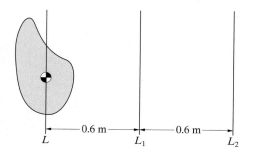

— 0.6 m — | — 0.6 m —

L L_1 L_2

Problem 8.108

8.109 An engineer gathering data for the design of a maneuvering unit determines that the astronaut's center of mass is at $x = 1.01$ m, $y = 0.16$ m and that her moment of inertia about the z axis is 105.6 kg-m². Her mass is 81.6 kg. What is her moment of inertia about the z' axis through her center of mass?

Problem 8.109

8.110 Two homogeneous slender bars, each of mass m and length l, are welded together to form the T-shaped object. Use the parallel-axis theorem to determine the moment of inertia of the object about the axis through point O that is perpendicular to the bars.

▶ **8.111** Use the parallel-axis theorem to determine the moment of inertia of the T-shaped object about the axis through the center of mass of the object that is perpendicular to the two bars. (See Active Example 8.11.)

Problems 8.110/8.111

8.112 The mass of the homogeneous slender bar is 20 kg. Determine its moment of inertia about the z axis.

8.113 Determine the moment of inertia of the 20-kg bar about the z' axis through its center of mass.

Problems 8.112/8.113

8.114 The homogeneous slender bar weighs 5 lb. Determine its moment of inertia about the z axis.

8.115 Determine the moment of inertia of the 5-lb bar about the z' axis through its center of mass.

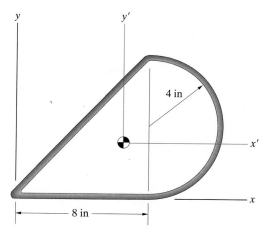

Problems 8.114/8.115

8.116 The rocket is used for atmospheric research. Its weight and its moment of inertia about the z axis through its center of mass (including its fuel) are 10 kip and 10,200 slug-ft^2, respectively. The rocket's fuel weighs 6000 lb, its center of mass is located at $x = -3$ ft, $y = 0$, $z = 0$, and the moment of inertia of the fuel about the axis through the fuel's center of mass parallel to z is 2200 slug-ft^2. When the fuel is exhausted, what is the rocket's moment of inertia about the axis through its new center of mass parallel to z?

Problem 8.116

8.117 The mass of the homogeneous thin plate is 36 kg. Determine its moment of inertia about the x axis.

8.118 Determine the moment of inertia of the 36-kg plate about the z axis.

Problems 8.117/8.118

8.119 The homogeneous thin plate weighs 10 lb. Determine its moment of inertia about the x axis.

8.120 Determine the moment of inertia of the 10-lb plate about the y axis.

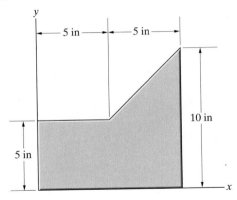

Problems 8.119/8.120

8.121 The thermal radiator (used to eliminate excess heat from a satellite) can be modeled as a homogeneous thin rectangular plate. Its mass is 5 slugs. Determine its moments of inertia about the x, y, and z axes.

Problem 8.121

▶ **8.122** The homogeneous cylinder has mass m, length l, and radius R. Use integration as described in Example 8.13 to determine its moment of inertia about the x axis.

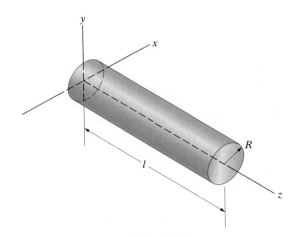

Problem 8.122

▶ **8.123** The homogeneous cone is of mass m. Determine its moment of inertia about the z axis, and compare your result with the value given in Appendix C. (See Example 8.13.)

8.124 Determine the moments of inertia of the homogeneous cone of mass m about the x and y axes, and compare your results with the values given in Appendix C.

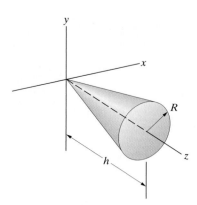

Problems 8.123/8.124

▶ **8.125** The mass of the homogeneous wedge is m. Use integration as described in Example 8.13 to determine its moment of inertia about the z axis. (Your answer should be in terms of m, a, b, and h.)

▶ **8.126** The mass of the homogeneous wedge is m. Use integration as described in Example 8.13 to determine its moment of inertia about the x axis. (Your answer should be in terms of m, a, b, and h.)

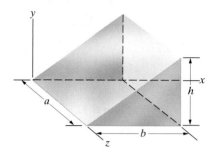

Problems 8.125/8.126

▶ **8.127** In Example 8.12, suppose that part of the 3-kg bar is sawed off so that the bar is 0.4 m long and its mass is 2 kg. Determine the moment of inertia of the composite object about the perpendicular axis L through the center of mass of the modified object.

8.128 The L-shaped machine part is composed of two homogeneous bars. Bar 1 is tungsten alloy with density 14,000 kg/m³, and bar 2 is steel with density 7800 kg/m³. Determine its moment of inertia about the x axis.

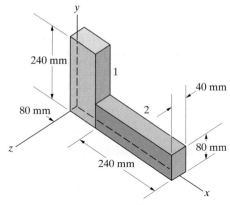

Problem 8.128

8.129 The homogeneous object is a cone with a conical hole. The dimensions $R_1 = 2$ in, $R_2 = 1$ in, $h_1 = 6$ in, and $h_2 = 3$ in. It consists of aluminum alloy with a density of 5 slug/ft³. Determine its moment of inertia about the x axis.

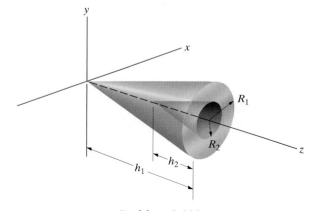

Problem 8.129

8.130 The circular cylinder is made of aluminum (Al) with density 2700 kg/m³ and iron (Fe) with density 7860 kg/m³. Determine its moments of inertia about the x' and y' axes.

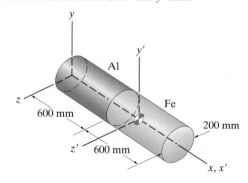

Problem 8.130

8.131 The homogeneous half-cylinder is of mass m. Determine its moment of inertia about the axis L through its center of mass.

Problem 8.131

8.132 The homogeneous machine part is made of aluminum alloy with density $\rho = 2800$ kg/m^3. Determine its moment of inertia about the z axis.

8.133 Determine the moment of inertia of the machine part described in Problem 8.132 about the x axis.

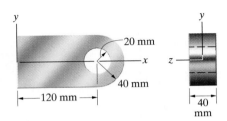

Problems 8.132/8.133

8.134 The object consists of steel of density $\rho = 7800$ kg/m^3. Determine its moment of inertia about the axis L_O.

8.135 Determine the moment of inertia of the object in Problem 8.134 about the axis through the center of mass of the object parallel to L_O.

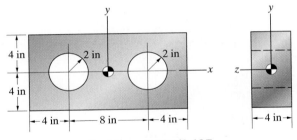

Problems 8.134/8.135

8.136 The thick plate consists of steel of density $\rho = 15$ slug/ft^3. Determine its moment of inertia about the z axis.

8.137 Determine the moment of inertia of the plate in Problem 8.136 about the x axis.

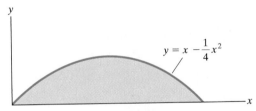

Problems 8.136/8.137

Review Problems

8.138 Determine I_y and k_y.

8.139 Determine I_x and k_x.

8.140 Determine J_O and k_O.

8.141 Determine I_{xy}.

8.142 Determine I_y and k_y.

8.143 Determine I_x and k_x.

8.144 Determine I_{xy}.

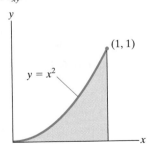

Problems 8.138–8.141

$y = x - \frac{1}{4}x^2$

Problems 8.142–8.144

8.145 Determine $I_{y'}$ and $k_{y'}$.

8.146 Determine $I_{x'}$ and $k_{x'}$.

8.147 Determine $I_{x'y'}$.

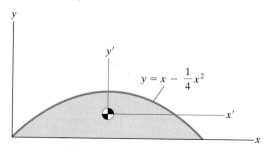

$$y = x - \frac{1}{4}x^2$$

Problems 8.145–8.147

8.148 Determine I_y and k_y.

8.149 Determine I_x and k_x.

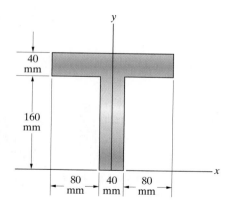

Problems 8.148/8.149

8.150 Determine I_x and k_x.

8.151 Determine J_O and k_O.

Problems 8.150/8.151

8.152 Determine I_y and k_y.

8.153 Determine J_O and k_O.

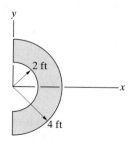

Problems 8.152/8.153

8.154 Determine I_x and k_x.

8.155 Determine I_y and k_y.

Problems 8.154/8.155

8.156 The moments of inertia of the area are $I_x = 36$ m^4, $I_y = 145$ m^4, and $I_{xy} = 44.25$ m^4. Determine a set of principal axes and the principal moments of inertia.

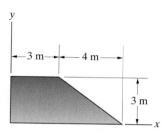

Problem 8.156

8.157 The moment of inertia of the 31-oz bat about a perpendicular axis through point B is 0.093 slug-ft². What is the bat's moment of inertia about a perpendicular axis through point A? (Point A is the bat's "instantaneous center," or center of rotation, at the instant shown.)

Problem 8.157

8.158 The mass of the thin homogeneous plate is 4 kg. Determine its moment of inertia about the y axis.

8.159 Determine the moment of inertia of the 4-kg plate about the z axis.

Problems 8.158/8.159

8.160 The homogeneous pyramid is of mass m. Determine its moment of inertia about the z axis.

8.161 Determine the moments of inertia of the homogeneous pyramid of mass m about the x and y axes.

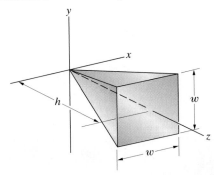

Problems 8.160/8.161

8.162 The homogeneous object weighs 400 lb. Determine its moment of inertia about the x axis.

8.163 Determine the moments of inertia of the 400-lb object about the y and z axes.

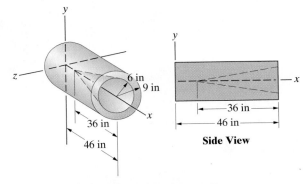

Side View

Problems 8.162/8.163

8.164 Determine the moment of inertia of the 14-kg flywheel about the axis L.

Problem 8.164

Friction

Friction forces have many important effects, both desirable and undesirable, in engineering applications. The Coulomb theory of friction allows us to estimate the maximum friction forces that can be exerted by contacting surfaces and the friction forces exerted by sliding surfaces. This opens the path to the analysis of important new classes of supports and machines, including wedges (shims), threaded connections, bearings, and belts.

◀ The workpiece exerts normal and friction forces on the grinding wheel. In this chapter we analyze friction forces between contacting surfaces.

9.1 Theory of Dry Friction

BACKGROUND

Suppose that a person climbs a ladder that leans against a smooth wall. Figure 9.1a shows the free-body diagram of the person and ladder. If the person is stationary on the ladder, we can use the equilibrium equations to determine the friction force. But there is another question that we cannot answer using the equilibrium equations alone: Will the ladder remain in place, or will it slip on the floor? If a truck is parked on an incline, the total friction force exerted on its tires by the road prevents it from sliding down the incline (Fig. 9.1b). We can use the equilibrium equations to determine the total friction force. But here too there is another question that we cannot answer: What is the steepest incline on which the truck could be parked without slipping?

To answer these questions, we must examine the nature of friction forces in more detail. Place a book on a table and push it with a small horizontal force, as shown in Fig. 9.2a. If the force you exert is sufficiently small, the book does not move. The free-body diagram of the book is shown in Fig. 9.2b. The force W is the book's weight, and N is the total normal force exerted by the table on the surface of the book that is in contact with the table. The force F is the horizontal force you apply, and f is the total friction force exerted by the table. Because the book is in equilibrium, $f = F$.

Now slowly increase the force you apply to the book. As long as the book remains in equilibrium, the friction force must increase correspondingly, since it equals the force you apply. When the force you apply becomes too large, the book moves. It slips on the table. After reaching some maximum value, the friction force can no longer maintain the book in equilibrium. Also, notice that the force you must apply to keep the book moving on the table is smaller than the force required to cause it to slip. (You are familiar with this phenomenon if you've ever pushed a piece of furniture across a floor.)

Friction force

(a)

Friction force

(b)

Figure 9.1
Objects supported by friction forces.

(a)

(b)

Figure 9.2
(a) Exerting a horizontal force on a book.
(b) The free-body diagram of the book.

How does the table exert a friction force on the book? Why does the book slip? Why is less force required to slide the book across the table than is required to start it moving? If the surfaces of the table and the book are magnified sufficiently, they will appear rough (Fig. 9.3). Friction forces arise in part from the interactions of the roughnesses, or *asperities,* of the contacting surfaces. We can gain insight into this mechanism of friction by considering a simple two-dimensional model of the rough surfaces of the book and table.

Suppose that we idealize the asperities of the book and table as the mating two-dimensional "saw-tooth" profiles in Fig. 9.4a. As the horizontal force F increases, the book will remain stationary until the force is sufficiently large to cause the book to slide upward as shown in Fig. 9.4b. What horizontal force is necessary for this to occur? To find out, we must determine the value of F necessary for the book to be in equilibrium in the "slipped" position in Fig. 9.4b. The normal force C_i exerted on the ith saw-tooth asperity of the book is shown in Fig. 9.4c. (Notice that in this simple model we assume the contacting surfaces of the asperities to be smooth.) Denoting the sum of the normal forces exerted on the asperities of the book by the table by $C = \sum_i C_i$, we obtain the equilibrium equations

$$\Sigma F_x = F - C \sin \alpha = 0,$$
$$\Sigma F_y = C \cos \alpha - W = 0.$$

Eliminating C from these equations, we obtain the force necessary to cause the book to slip on the table:

$$F = (\tan \alpha)W.$$

We see that *the force necessary to cause the book to slip is proportional to the force pressing the saw-tooth surfaces together* (the book's weight). Think about stacking increasing numbers of books and applying a horizontal force to them. A progressively larger force is required to cause them to slip as the number of books increases. Also, in our two-dimensional thought experiment, *the angle α is a measure of the roughness of the saw-tooth surfaces.* As $\alpha \rightarrow 0$, the surfaces become smooth and the force necessary to cause the book to slip approaches zero. As α increases, the roughness increases and the force necessary to cause the book to slip increases.

Figure 9.3
The roughnesses of the surfaces can be seen in a magnified view.

Figure 9.4
(a) Two-dimensional model of rough surfaces in contact.
(b) Slip of the book relative to the table.
(c) Normal force on one of the book's asperities.

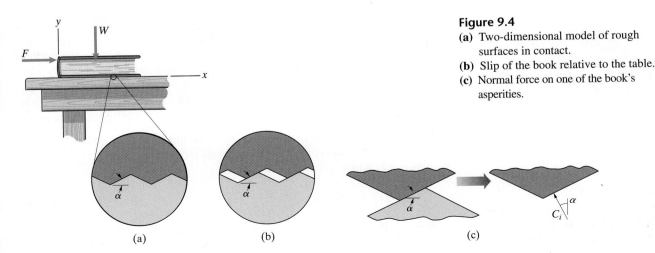

(a)　　　　　(b)　　　　　(c)

In the sections that follow, we present a theory that incorporates the basic phenomena we have just described and that has been found useful for determining friction forces between dry surfaces. (Friction between lubricated surfaces is a hydrodynamic phenomenon and must be analyzed in the context of fluid mechanics.)

Coefficients of Friction

The theory of dry friction, or *Coulomb friction,* predicts the maximum friction forces that can be exerted by dry, contacting surfaces that are stationary relative to each other. It also predicts the friction forces exerted by the surfaces when they are in relative motion, or sliding. We first consider surfaces that are not in relative motion.

The Static Coefficient   The magnitude of the *maximum* friction force that can be exerted between two plane, dry surfaces in contact that are not in motion relative to one another is

$$f = \mu_s N, \tag{9.1}$$

where N is the normal component of the contact force between the surfaces and μ_s is a constant called the *coefficient of static friction*. The value of μ_s is assumed to depend only on the materials of the contacting surfaces and the conditions (smoothness and degree of contamination by other materials) of the surfaces. Typical values of μ_s for various materials are shown in Table 9.1. The relatively large range of values for each pair of materials reflects the sensitivity of μ_s to the conditions of the surfaces. In engineering applications it is usually necessary to measure the value of μ_s for the actual surfaces used.

Let us return to the example of the book on the table (Fig. 9.2). If a *specified* horizontal force F is applied to the book, and the book remains in equilibrium, what friction force is exerted on the book by the table? We can see from the free-body diagram in Fig. 9.2b that $f = F$. Notice that we do not use Eq. (9.1) to answer this question. But suppose that we want to know the *largest* force F that can be applied to the book without causing it to slip. If we know the coefficient of static friction μ_s between the book and the table, Eq. (9.1) tells us the largest friction force that the table can exert on the book. Therefore, the largest force F that can be applied without causing the book to slip is $F = f = \mu_s N$. We also know from the free-body diagram in Fig. 9.2b that $N = W$, so the largest force that will not cause the book to slip is $F = \mu_s W$.

Equation (9.1) determines the magnitude of the maximum friction force but not its direction. The friction force is a maximum, and Eq. (9.1) is applicable, when two surfaces are on the verge of slipping relative to each other. We say that slip is *impending,* and the friction forces resist the impending motion. In Fig. 9.5a, suppose that the lower surface is fixed and slip of the upper surface toward the right is impending. The friction force on the upper surface resists its impending motion (Fig. 9.5b). The friction force on the lower surface is in the opposite direction.

The Kinetic Coefficient   According to the theory of dry friction, the magnitude of the friction force between two plane dry contacting surfaces that are in motion (sliding) relative to each other is

$$f = \mu_k N, \tag{9.2}$$

where N is the normal force between the surfaces and μ_k is the *coefficient of kinetic friction*. The value of μ_k is assumed to depend only on the compositions

TABLE 9.1   Typical values of the coefficient of static friction.

Materials	Coefficient of Static Friction μ_s
Metal on metal	0.15–0.20
Masonry on masonry	0.60–0.70
Wood on wood	0.25–0.50
Metal on masonry	0.30–0.70
Metal on wood	0.20–0.60
Rubber on concrete	0.50–0.90

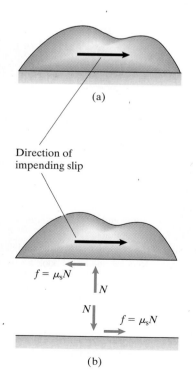

Direction of impending slip

Figure 9.5
(a) The upper surface is on the verge of slipping to the right.
(b) Directions of the friction forces.

Direction of relative motion

$f = \mu_k N$

N

N

$f = \mu_k N$

(a)

(b)

Figure 9.6
(a) The upper surface is moving to the right relative to the lower surface.
(b) Directions of the friction forces.

of the surfaces and their conditions. For a given pair of surfaces, its value is generally smaller than that of μ_s.

Once you have caused the book in Fig. 9.2 to begin sliding on the table, the friction force $f = \mu_k N = \mu_k W$. Therefore, the force you must exert to keep the book in uniform motion is $F = f = \mu_k W$.

When two surfaces are sliding relative to each other, the friction forces resist the relative motion. In Fig. 9.6a, suppose that the lower surface is fixed and the upper surface is moving to the right. The friction force on the upper surface acts in the direction opposite to its motion (Fig. 9.6b). The friction force on the lower surface is in the opposite direction.

Angles of Friction

We have expressed the reaction exerted on a surface due to its contact with another surface in terms of its components parallel and perpendicular to the surface, the friction force f and normal force N (Fig. 9.7a). In some situations it is more convenient to express the reaction in terms of its magnitude R and the *angle of friction* θ between the reaction and the normal to the surface (Fig. 9.7b). The forces f and N are related to R and θ by

$$f = R \sin\theta, \tag{9.3}$$

$$N = R \cos\theta. \tag{9.4}$$

The value of θ when slip is impending is called the *angle of static friction* θ_s, and its value when the surfaces are sliding relative to each other is called the *angle of kinetic friction* θ_k. By using Eqs. (9.1)–(9.4), we can express the angles of static and kinetic friction in terms of the coefficients of friction:

$$\tan\theta_s = \mu_s, \tag{9.5}$$

$$\tan\theta_k = \mu_k. \tag{9.6}$$

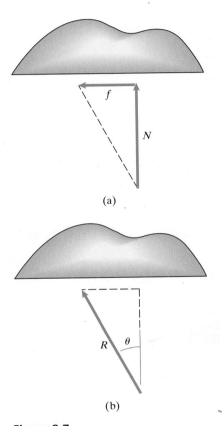

f

N

(a)

R

θ

(b)

Figure 9.7
(a) The friction force f and the normal force N.
(b) The magnitude R and the angle of friction θ.

RESULTS

The forces resulting from the contact of plane surfaces can be expressed in two alternative ways:

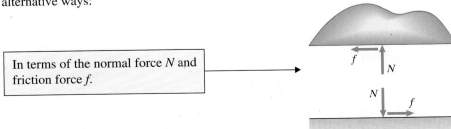

In terms of the normal force N and friction force f.

f

N

N

f

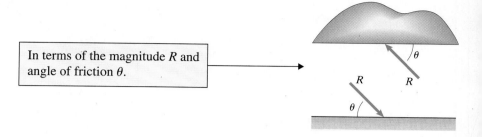

In terms of the magnitude R and angle of friction θ.

Friction Coefficients

The magnitude of the maximum friction force that can be exerted by dry surfaces that are stationary relative to each other (that is, when slip is impending) is

$$f = \mu_s N, \tag{9.1}$$

where μ_s is the *coefficient of static friction*. The angle of friction when slip is impending is related to the coefficient of static friction by

$$\tan \theta_s = \mu_s. \tag{9.5}$$

The magnitude of the friction force exerted by dry surfaces that are in motion (sliding) relative to each other is

$$f = \mu_k N, \tag{9.2}$$

where μ_k is the *coefficient of kinetic friction*. The angle of friction when the surfaces are sliding is related to the coefficient of kinetic friction by

$$\tan \theta_k = \mu_k \tag{9.6}$$

Evaluating the friction force and angle of friction requires a sequence of decisions:

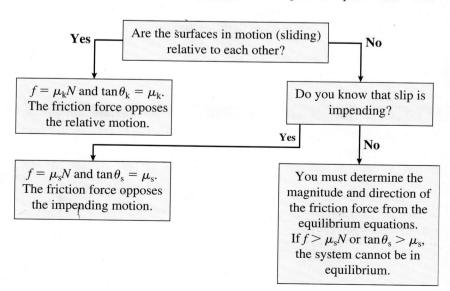

Active Example 9.1   Determining a Friction Force (▶ *Related Problem 9.1*)

The rope exerts a horizontal force on the stationary 180-lb crate. The coefficient of static friction between the crate and the ramp is $\mu_s = 0.4$. If the rope exerts a 90-lb force on the crate, what friction force is exerted on the crate by the ramp?

Strategy
The crate is not sliding on the ramp, and we don't know whether slip is impending, so we must determine the friction force by applying the equilibrium equations.

Solution

Draw the free-body diagram of the crate. The direction of the friction force f is not known, so choose it arbitrarily. The sign of the answer for f will indicate its direction.

Apply equilibrium. The negative value of the friction force indicates that its direction is down the ramp.

$$\Sigma F_x = f + T \cos 20° - W \sin 20° = 0,$$
$$\Sigma F_y = N - T \sin 20° - W \cos 20° = 0.$$

Setting $W = 180$ lb and $T = 90$ lb and solving yields $N = 200$ lb and $f = -23.0$ lb.

Calculate the maximum friction force the surfaces will support to confirm that it is not exceeded by the magnitude of the friction force necessary for equilibrium.

$$\mu_s N = (0.4)(200 \text{ lb}) = 80 \text{ lb}.$$

Practice Problem What is the largest horizontal force the rope can exert on the crate without causing it to start sliding up the ramp?

Answer: 161 lb.

| Example 9.2 | **Analyzing a Friction Brake** (▶ *Related Problem 9.22*) |

The motion of the disk is controlled by the friction force exerted at C by the brake ABC. The hydraulic actuator BE exerts a horizontal force of magnitude F on the brake at B. The coefficients of friction between the disk and the brake are μ_s and μ_k. What couple M is necessary to rotate the disk at a constant rate in the counterclockwise direction?

Strategy
We can use the free-body diagram of the disk to obtain a relation between M and the reaction exerted on the disk by the brake, then use the free-body diagram of the brake to determine the reaction in terms of F.

Solution
We draw the free-body diagram of the disk in Fig. a, representing the force exerted by the brake by a single force R. The force R opposes the counterclockwise rotation of the disk, and the friction angle is the angle of kinetic friction $\theta_k = \arctan \mu_k$. Summing moments about D, we obtain

$$\Sigma M_{\text{point } D} = M - (R \sin \theta_k)r = 0.$$

(a) The free-body diagram of the disk.

Then, from the free-body diagram of the brake (Fig. b), we get

$$\Sigma M_{\text{point } A} = -F\left(\frac{1}{2}h\right) + (R \cos \theta_k)h - (R \sin \theta_k)b = 0.$$

We can solve these two equations for M and R. The solution for the couple M is

$$M = \frac{(1/2)hr\, F \sin \theta_k}{h \cos \theta_k - b \sin \theta_k} = \frac{(1/2)hr\, F \mu_k}{h - b \mu_k}.$$

(b) The free-body diagram of the brake.

Critical Thinking
If the friction coefficient μ_k is sufficiently small, the denominator in our solution for the couple M, the term $h \cos \theta_k - b \sin \theta_k$, is positive. As μ_k increases, the denominator becomes smaller, because $\cos \theta_k$ decreases and $\sin \theta_k$ increases. As the denominator approaches zero, the couple required to rotate the disk approaches infinity. To understand this result, notice that the denominator equals zero when $\tan \theta_k = h/b$, which means that the line of action of the force R passes through point A (Fig. c). As μ_k increases and the line of action of R approaches point A, the magnitude of R necessary to balance the moment due to F about A approaches infinity. As a result, the *analytical prediction* for M approaches infinity. Of course, at some value of M, the forces F and R would exceed the values the brake could support.

(c) The line of action of R passing through point A.

Example 9.3   Determining Whether an Object Will Tip Over (▶ *Related Problem 9.45*)

Suppose that we want to push the tool chest across the floor by applying the horizontal force F. If we apply the force at too great a height h, the chest will tip over before it slips. If the coefficient of static friction between the floor and the chest is μ_s, what is the largest value of h for which the chest will slip before it tips over?

Strategy
When the chest is on the verge of tipping over, it is in equilibrium with no reaction at B. We can use this condition to determine F in terms of h. Then, by determining the value of F that will cause the chest to slip, we will obtain the value of h that causes the chest to be on the verge of tipping over *and* on the verge of slipping.

Solution
We draw the free-body diagram of the chest when it is on the verge of tipping over in Fig. a. Summing moments about A, we obtain

$$\Sigma M_{\text{point } A} = Fh - W\left(\frac{1}{2}b\right) = 0.$$

Equilibrium also requires that $f = F$ and $N = W$.
    When the chest is on the verge of slipping,

$$f = \mu_s N,$$

so

$$F = f = \mu_s N = \mu_s W.$$

Substituting this expression into the moment equation, we obtain

$$\mu_s Wh - W\left(\frac{1}{2}b\right) = 0.$$

(a) The free-body diagram when the chest is on the verge of tipping over.

Solving this equation for h, we find that when the chest is on the verge of slipping, it is also on the verge of tipping over if it is pushed at the height

$$h = \frac{b}{2\mu_s}.$$

If h is smaller than this value, the chest will begin sliding before it tips over.

Critical Thinking
Notice that the largest value of h for which the chest will slip before it tips over is independent of F. Whether the chest will tip over depends only on where the force is applied, not how large it is. What is the motivation for the solution in this example? The possibility of heavy objects falling over is an obvious safety hazard, and analyses of this kind can influence their design. Once they are in use, safety engineers can establish guidelines (for example, by marking a horizontal line on a vertical cabinet or machine above which it should not be pushed) to prevent tipping.

Problems

▶ **9.1** In Active Example 9.1, suppose that the coefficient of static friction between the 180-lb crate and the ramp is $\mu_s = 0.3$. What is the magnitude of the smallest horizontal force the rope must exert on the crate to prevent it from sliding down the ramp?

9.2 A person places a 2-lb book on a table that is tilted at 15° relative to the horizontal. She finds that if she exerts a very small force on the book as shown, the book remains in equilibrium, but if she removes the force, the book slides down the table. What force would she need to exert on the book (in the direction parallel to the table) to cause it to slide up the table?

15°

Problem 9.2

9.3 A student pushes a 200-lb box of books across the floor. The coefficient of kinetic friction between the carpet and the box is $\mu_k = 0.15$.

(a) If he exerts the force F at angle $\alpha = 25°$, what is the magnitude of the force he must exert to slide the box across the floor?

(b) If he bends his knees more and exerts the force F at angle $\alpha = 10°$, what is the magnitude of the force he must exert to slide the box?

α

F

Problem 9.3

9.4 The 2975-lb car is parked on a sloped street. The brakes are applied to both its front and rear wheels.

(a) If the coefficient of static friction between the car's tires and the road is $\mu_s = 0.8$, what is the steepest slope (in degrees relative to the horizontal) on which the car could remain in equilibrium?

(b) If the street were icy and the coefficient of static friction between the car's tires and the road was $\mu_s = 0.2$, what is the steepest slope on which the car could remain in equilibrium?

Problem 9.4

9.5 The truck's winch exerts a horizontal force on the 200-kg crate in an effort to pull it down the ramp. The coefficient of static friction between the crate and the ramp is $\mu_s = 0.6$.

(a) If the winch exerts a 200-N horizontal force on the crate, what is the magnitude of the friction force exerted on the crate by the ramp?

(b) What is the magnitude of the horizontal force the winch must exert on the crate to cause it to start moving down the ramp?

20°

Problem 9.5

9.6 The device shown is designed to position pieces of luggage on a ramp. It exerts a force parallel to the ramp. The suitcase weighs 40 lb. The coefficients of friction between the suitcase and the ramp are $\mu_s = 0.20$ and $\mu_k = 0.18$.

(a) Will the suitcase remain stationary on the ramp when the device exerts no force on it?

(b) What force must the device exert to push the suitcase up the ramp at a constant speed?

20°

Problem 9.6

9.7 The coefficient of static friction between the 50-kg crate and the ramp is $\mu_s = 0.35$. The unstretched length of the spring is 800 mm, and the spring constant is $k = 660$ N/m. What is the minimum value of x at which the crate can remain stationary on the ramp?

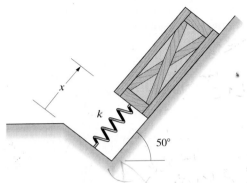

Problem 9.7

9.8 The coefficient of kinetic friction between the 40-kg crate and the slanting floor is $\mu_k = 0.3$. If the angle $\alpha = 20°$, what tension must the person exert on the rope to move the crate at constant speed?

9.9 In Problem 9.8, for what angle α is the tension necessary to move the crate at constant speed a minimum? What is the necessary tension?

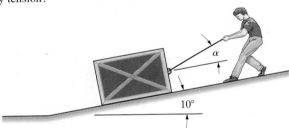

Problems 9.8/9.9

9.10 Box A weighs 100 lb and box B weighs 30 lb. The coefficients of friction between box A and the ramp are $\mu_s = 0.30$ and $\mu_k = 0.28$. What is the magnitude of the friction force exerted on box A by the ramp?

9.11 Box A weighs 100 lb, and the coefficients of friction between box A and the ramp are $\mu_s = 0.30$ and $\mu_k = 0.28$. For what range of weights of the box B will the system remain stationary?

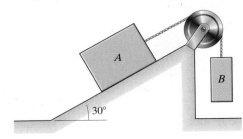

Problems 9.10/9.11

9.12 The mass of the box on the left is 30 kg, and the mass of the box on the right is 40 kg. The coefficient of static friction between each box and the inclined surface is $\mu_s = 0.2$. Determine the minimum angle α for which the boxes will remain stationary.

Problem 9.12

9.13 The coefficient of kinetic friction between the 100-kg box and the inclined surface is 0.35. Determine the tension T necessary to pull the box up the surface at a constant rate.

Problem 9.13

9.14 The box is stationary on the inclined surface. The coefficient of static friction between the box and the surface is μ_s.

(a) If the mass of the box is 10 kg, $\alpha = 20°$, $\beta = 30°$, and $\mu_s = 0.24$, what force T is necessary to start the box sliding up the surface?

(b) Show that the force T necessary to start the box sliding up the surface is a minimum when $\tan \beta = \mu_s$.

9.15 In explaining observations of ship launchings at the port of Rochefort in 1779, Coulomb analyzed the system shown to determine the minimum force T necessary to hold the box stationary on the inclined surface. Show that the result is

$$T = \frac{(\sin \alpha - \mu_s \cos \alpha)mg}{\cos \beta - \mu_s \sin \beta}.$$

Problems 9.14/9.15

9.16 Two sheets of plywood A and B lie on the bed of the truck. They have the same weight W, and the coefficient of static friction between the two sheets of wood and between sheet B and the truck bed is μ_s.

(a) If you apply a horizontal force to sheet A and apply no force to sheet B, can you slide sheet A off the truck without causing sheet B to move? What force is necessary to cause sheet A to start moving?

(b) If you prevent sheet A from moving by exerting a horizontal force on it, what horizontal force on sheet B is necessary to start it moving?

Problem 9.16

9.17 The weights of the two boxes are $W_1 = 100$ lb and $W_2 = 50$ lb. The coefficients of friction between the left box and the inclined surface are $\mu_s = 0.12$ and $\mu_k = 0.10$. Determine the tension the man must exert on the rope to pull the boxes upward at a constant rate.

9.18 In Problem 9.17, for what range of tensions exerted on the rope by the man will the boxes remain stationary?

Problems 9.17/9.18

9.19 Each box weighs 10 lb. The coefficient of static friction between box A and box B is 0.24, and the coefficient of static friction between box B and the inclined surface is 0.3. What is the largest angle α for which box B will not slip?

Strategy: Draw individual free-body diagrams of the two boxes and write their equilibrium equations assuming that slip of box B is impending.

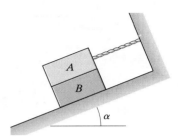

Problem 9.19

9.20 The masses of the boxes are $m_A = 15$ kg and $m_B = 60$ kg. The coefficient of static friction between boxes A and B and between box B and the inclined surface is 0.12. What is the largest force F for which the boxes will not slip?

9.21 In Problem 9.20, what is the smallest force F for which the boxes will not slip?

Problems 9.20/9.21

▶ **9.22** In Example 9.2, what clockwise couple M would need to be applied to the disk to cause it to rotate at a constant rate in the clockwise direction?

9.23 The homogeneous horizontal bar *AB* weighs 20 lb. The homogeneous disk weighs 30 lb. The coefficient of kinetic friction between the disk and the sloping surface is $\mu_k = 0.24$. What is the magnitude of the couple that would need to be applied to the disk to cause it to rotate at a constant rate in the clockwise direction?

9.24 The homogeneous horizontal bar *AB* weighs 20 lb. The homogeneous disk weighs 30 lb. The coefficient of kinetic friction between the disk and the sloping surface is $\mu_k = 0.24$. What is the magnitude of the couple that would need to be applied to the disk to cause it to rotate at a constant rate in the counterclockwise direction?

Problems 9.23/9.24

9.25 The mass of the bar is 4 kg. The coefficient of static friction between the bar and the floor is 0.3. Neglect friction between the bar and the wall.

(a) If $\alpha = 20°$, what is the magnitude of the friction force exerted on the bar by the floor?

(b) What is the maximum angle α for which the bar will not slip?

9.26 The coefficient of static friction between the bar and the floor and between the 4-kg bar and the wall is 0.3. What is the maximum angle α for which the bar will not slip?

Problems 9.25/9.26

9.27 The ladder and the person weigh 30 lb and 180 lb, respectively. The center of mass of the 12-ft ladder is at its midpoint. The angle $\alpha = 30°$. Assume that the wall exerts a negligible friction force on the ladder.

(a) If $x = 4$ ft, what is the magnitude of the friction force exerted on the ladder by the floor?

(b) What minimum coefficient of static friction between the ladder and the floor is necessary for the person to be able to climb to the top of the ladder without slipping?

9.28 The ladder and the person weigh 30 lb and 180 lb, respectively. The center of mass of the 12-ft ladder is at its midpoint. The coefficient of static friction between the ladder and the floor is $\mu_s = 0.5$. What is the largest value of the angle α for which the person could climb to the top of the ladder without it slipping?

9.29 The ladder and the person weigh 30 lb and 180 lb, respectively. The center of mass of the 12-ft ladder is at its midpoint. The coefficient of static friction between the ladder and the floor is 0.5 and the coefficient of friction between the ladder and the wall is 0.3. What is the largest value of the angle α for which the person could climb to the top of the ladder without it slipping? Compare your answer to the answer to Problem 9.28.

Problems 9.27–9.29

9.30 The disk weighs 50 lb and the bar weighs 25 lb. The coefficients of friction between the disk and the inclined surface are $\mu_s = 0.6$ and $\mu_k = 0.5$.

(a) What is the largest couple M that can be applied to the stationary disk without causing it to start rotating?

(b) What couple M is necessary to rotate the disk at a constant rate?

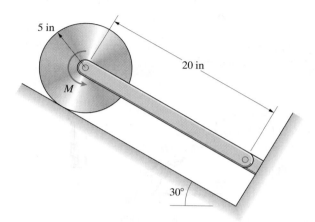

Problem 9.30

9.31 The radius of the 40-kg homogeneous cylinder is $R = 0.15$ m. The slanted wall is smooth and the angle $\alpha = 30°$. The coefficient of static friction between the cylinder and the floor is $\mu_s = 0.2$. What is the largest couple M that can be applied to the cylinder without causing it to slip?

9.32 The homogeneous cylinder has weight W. The coefficient of static friction between the cylinder and both surfaces is μ_s. What is the largest couple M that can be applied to the cylinder without causing it to slip? (Assume that the cylinder slips before rolling up the inclined surface.)

9.33 The homogeneous cylinder has weight W. The coefficient of static friction between the cylinder and both surfaces is μ_s. What is the minimum value of μ_s for which the couple M will cause the cylinder to roll up the inclined surface without slipping?

Problems 9.31–9.33

9.34 The coefficient of static friction between the blades of the shears and the object they are gripping is 0.36. What is the largest value of the angle α for which the object will not slip out? Neglect the object's weight.

Strategy: Draw the free-body diagram of the object and assume that slip is impending.

Problem 9.34

9.35 A stationary disk of 300-mm radius is attached to a pin support at D. The disk is held in place by the brake ABC in contact with the disk at C. The hydraulic actuator BE exerts a horizontal 400-N force on the brake at B. The coefficients of friction between the disk and the brake are $\mu_s = 0.6$ and $\mu_k = 0.5$. What couple must be applied to the stationary disk to cause it to slip in the counterclockwise direction?

Problem 9.35

9.36 The figure shows a preliminary conceptual idea for a device to exert a braking force on a rope when the rope is pulled downward by the force T. The coefficient of kinetic friction between the rope and the two bars is $\mu_k = 0.28$. Determine the force T necessary to pull the rope downward at a constant rate if $F = 10$ lb and (a) $\alpha = 30°$; (b) $\alpha = 20°$.

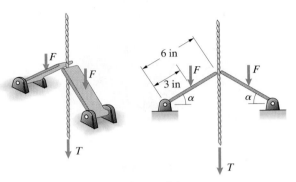

Problem 9.36

9.37 The mass of block B is 8 kg. The coefficient of static friction between the surfaces of the clamp and the block is $\mu_s = 0.2$. When the clamp is aligned as shown, what minimum force must the spring exert to prevent the block from slipping out?

9.38 By altering its dimensions, redesign the clamp in Problem 9.37 so that the minimum force the spring must exert to prevent the block from slipping out is 180 N. Draw a sketch of your new design.

Problems 9.37/9.38

9.39 The horizontal bar is attached to a collar that slides on the smooth vertical bar. The collar at P slides on the smooth horizontal bar. The total mass of the horizontal bar and the two collars is 12 kg. The system is held in place by the pin in the circular slot. The pin contacts only the lower surface of the slot, and the coefficient of static friction between the pin and the slot is 0.8. If the system is in equilibrium and $y = 260$ mm, what is the magnitude of the friction force exerted on the pin by the slot?

9.40 In Problem 9.39, what is the minimum height y at which the system can be in equilibrium?

Problems 9.39/9.40

9.41 The rectangular 100-lb plate is supported by the pins A and B. If friction can be neglected at A and the coefficient of static friction between the pin at B and the slot is $\mu_s = 0.4$, what is the largest angle α for which the plate will not slip?

9.42 If you can neglect friction at B and the coefficient of static friction between the pin at A and the slot is $\mu_s = 0.4$, what is the largest angle α for which the 100-lb plate will not slip?

Problems 9.41/9.42

9.43 The airplane's weight is W = 2400 lb. Its brakes keep the rear wheels locked, and the coefficient of static friction between the wheels and the runway is μ_s = 0.6. The front (nose) wheel can turn freely and so exerts a negligible friction force on the runway. Determine the largest horizontal thrust force T the plane's propeller can generate without causing the rear wheels to slip.

Problem 9.43

9.44 The refrigerator weighs 220 lb. It is supported at A and B. The coefficient of static friction between the supports and the floor is μ_s = 0.2. If you assume that the refrigerator does not tip over before it slips, what force F is necessary for impending slip?

▶ **9.45** The refrigerator weighs 220 lb. It is supported at A and B. The coefficient of static friction between the supports and the floor is μ_s = 0.2. The distance h = 60 in and the dimension b = 30 in. When the force F is applied to push the refrigerator across the floor, will it tip over before it slips? (See Example 9.3.)

Problems 9.44/9.45

9.46 To obtain a preliminary evaluation of the stability of a turning car, imagine subjecting the stationary car to an increasing lateral force F at the height of its center of mass, and determine whether the car will slip (skid) laterally before it tips over. Show that this will be the case if $b/h > 2\mu_s$. (Notice the importance of the height of the center of mass relative to the width of the car. This reflects on recent discussions of the stability of sport utility vehicles and vans that have relatively high centers of mass.)

Problem 9.46

9.47 The man exerts a force P on the car at an angle α = 20°. The 1760-kg car has front wheel drive. The driver spins the front wheels, and the coefficient of kinetic friction is μ_k = 0.02. Snow behind the rear tires exerts a horizontal resisting force S. Getting the car to move requires overcoming a resisting force S = 420 N. What force P must the man exert?

9.48 In Problem 9.47, what value of the angle α minimizes the magnitude of the force P the man must exert to overcome the resisting force S = 420 N exerted on the rear tires by the snow? What force must he exert?

Problems 9.47/9.48

9.49 The coefficient of static friction between the 3000-lb car's tires and the road is $\mu_s = 0.5$. Determine the steepest grade (the largest value of the angle α) the car can drive up at constant speed if the car has (a) rear-wheel drive; (b) front-wheel drive; (c) four-wheel drive.

Problem 9.49

9.50 The stationary cabinet has weight W. Determine the force F that must be exerted to cause it to move if (a) the coefficient of static friction at A and at B is μ_s; (b) the coefficient of static friction at A is μ_{sA} and the coefficient of static friction at B is μ_{sB}.

Problem 9.50

9.51 The table weighs 50 lb and the coefficient of static friction between its legs and the inclined surface is 0.7.

(a) If you apply a force at A parallel to the inclined surface to push the table up the inclined surface, will the table tip over before it slips? If not, what force is required to start the table moving up the surface?

(b) If you apply a force at B parallel to the inclined surface to push the table down the inclined surface, will the table tip over before it slips? If not, what force is required to start the table moving down the surface?

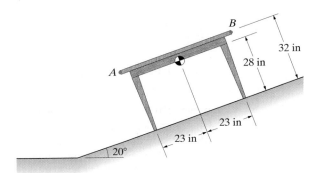

Problem 9.51

9.52 The coefficient of static friction between the right bar and the surface at A is $\mu_s = 0.6$. Neglect the weights of the bars. If $\alpha = 20°$, what is the magnitude of the friction force exerted at A?

9.53 The coefficient of static friction between the right bar and the surface at A is $\mu_s = 0.6$. Neglect the weights of the bars. What is the largest angle α at which the truss will remain stationary without slipping?

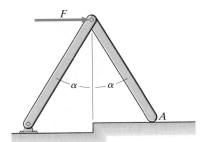

Problems 9.52/9.53

9.54 The bar BC is supported by a rough floor at C. If $F = 2$ kN and bar BC does not slip at C, what is the magnitude of the friction force exerted on the bar at C?

9.55 The bar BC is supported by a rough floor at C. If $F = 2$ kN, what is the minimum coefficient of static friction for which bar BC will not slip at C?

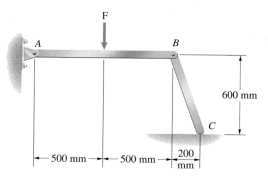

Problems 9.54/9.55

9.56 The weight of the box is 20 lb and the coefficient of static friction between the box and the floor is $\mu_s = 0.65$. Neglect the weights of the bars. What is the largest value of the force F that will not cause the box to slip?

Problem 9.56

9.57 The mass of the suspended object is 6 kg. The structure is supported at B by the normal and friction forces exerted on the plate by the wall. Neglect the weights of the bars.

(a) What is the magnitude of the friction force exerted on the plate at B?

(b) What is the minimum coefficient of static friction at B necessary for the structure to remain in equilibrium?

9.58 Suppose that the lengths of the bars in Problem 9.57 are $L_{AB} = 1.2$ m and $L_{AC} = 1.0$ m and their masses are $m_{AB} = 3.6$ kg and $m_{AC} = 3.0$ kg.

(a) What is the magnitude of the friction force exerted on the plate at B?

(b) What is the minimum coefficient of static friction at B necessary for the structure to remain in equilibrium?

Problems 9.57/9.58

9.59 The frame is supported by the normal and friction forces exerted on the plates at A and G by the fixed surfaces. The coefficient of static friction at A is $\mu_s = 0.6$. Will the frame slip at A when it is subjected to the loads shown?

Problem 9.59

9.60 The frame is supported by the normal and friction forces exerted on the plate at *A* by the wall.

(a) What is the magnitude of the friction force exerted on the plate at *A*?

(b) What is the minimum coefficient of static friction at *A* necessary for the structure to remain in equilibrium?

Problem 9.60

9.61 The direction cosines of the crane's cable are $\cos \theta_x = 0.588$, $\cos \theta_y = 0.766$, $\cos \theta_z = 0.260$. The *y* axis is vertical. The stationary caisson to which the cable is attached weighs 2000 lb and rests on horizontal ground. If the coefficient of static friction between the caisson and the ground is $\mu_s = 0.4$, what tension in the cable is necessary to cause the caisson to slip?

Problem 9.61

9.62* The 10-lb metal disk *A* is at the center of the inclined surface. The tension in the string *AB* is 5 lb. What minimum coefficient of static friction between the disk and the surface is necessary to keep the disk from slipping?

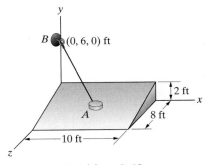

Problem 9.62

9.63* The 5-kg box is at rest on the sloping surface. The *y* axis points upward. The unit vector $0.557\mathbf{i} + 0.743\mathbf{j} + 0.371\mathbf{k}$ is perpendicular to the sloping surface. What is the magnitude of the friction force exerted on the box by the surface?

9.64* In Problem 9.63, what is the minimum coefficient of static friction necessary for the box to remain at rest on the sloping surface?

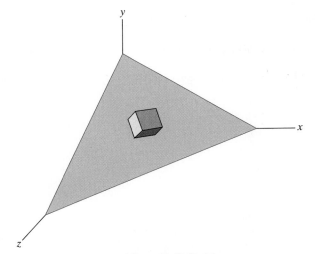

Problems 9.63/9.64

9.2 Wedges

A *wedge* is a bifacial tool with the faces set at a small acute angle (Figs. 9.8a and b). When a wedge is pushed forward, the faces exert large normal forces as a result of the small angle between them (Fig. 9.8c). In various forms, wedges are used in many engineering applications.

The large lateral force generated by a wedge can be used to lift a load (Fig. 9.9a). Let W_L be the weight of the load and W_W the weight of the wedge. To determine the force F necessary to start raising the load, we assume that slip of the load and wedge are impending (Fig. 9.9b). From the free-body diagram of the load, we obtain the equilibrium equations

$$\Sigma F_x = Q - N \sin \alpha - \mu_s N \cos \alpha = 0$$

and

$$\Sigma F_y = N \cos \alpha - \mu_s N \sin \alpha - \mu_s Q - W_L = 0.$$

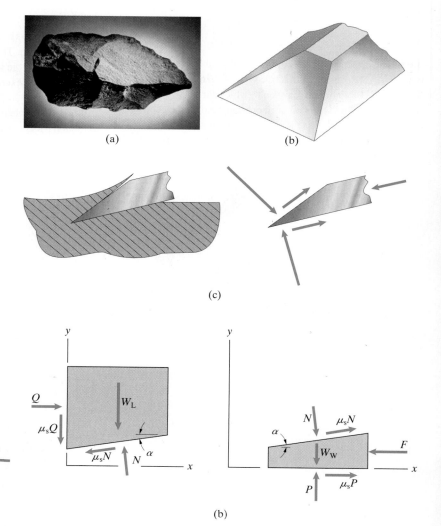

Figure 9.8

(a) An early wedge tool—a bifacial "hand axe" from Olduvai Gorge, East Africa.

(b) A modern chisel blade.

(c) The faces of a wedge can exert large lateral forces.

(a)

(b)

Figure 9.9

(a) Raising a load with a wedge.

(b) Free-body diagrams of the load and the wedge when slip is impending.

From the free-body diagram of the wedge, we obtain the equations

$$\Sigma F_x = N \sin \alpha + \mu_s N \cos \alpha + \mu_s P - F = 0$$

and

$$\Sigma F_y = P - N \cos \alpha + \mu_s N \sin \alpha - W_W = 0.$$

These four equations determine the three normal forces Q, N, and P, and the force F. The solution for F is

$$F = \mu_s W_W + \left[\frac{(1 - \mu_s^2) \tan \alpha + 2\mu_s}{(1 - \mu_s^2) - 2\mu_s \tan \alpha}\right] W_L.$$

Suppose that $W_W = 0.2W_L$ and $\alpha = 10°$. If $\mu_s = 0$, the force necessary to lift the load is only $0.176W_L$. But if $\mu_s = 0.2$, the force becomes $0.680W_L$, and if $\mu_s = 0.4$, it becomes $1.44W_L$. From this standpoint, friction is undesirable. But if there were no friction, the wedge would not remain in place when the force F is removed.

Active Example 9.4   Forces on a Wedge (▶ Related Problems 9.65, 9.66, 9.67)

A wedge is used to split a log. The angle $\alpha = 10°$. The coefficients of friction between the wedge and the log are $\mu_s = 0.22$ and $\mu_k = 0.20$. If the wedge is driven into the log at a constant speed by a vertical force F, what are the magnitudes of the normal forces exerted on the log by the wedge (that is, what are the magnitudes of the forces causing the log to split)?

Strategy
The friction forces exerted on the wedge by the log resist the motion of the wedge into the log and are of magnitude $\mu_k N$. We can apply equilibrium to the wedge to determine N in terms of F.

Solution

Draw the free-body diagram of the wedge.

The sum of the forces in the vertical direction is

$$2N \sin\left(\frac{\alpha}{2}\right) + 2\mu_k N \cos\left(\frac{\alpha}{2}\right) - F = 0.$$

Solving for N yields

$$N = \frac{F}{2[\sin(\alpha/2) + \mu_k \cos(\alpha/2)]}$$

$$= \frac{F}{2[\sin(10°/2) + (0.20)\cos(10°/2)]}$$

$$= 1.75F.$$

Apply equilibrium.

Practice Problem If the force F is removed, will the wedge remain in place in the log?

Answer: Yes.

Problems

▶ **9.65** In Active Example 9.4, the coefficients of friction between the wedge and the log are $\mu_s = 0.22$ and $\mu_k = 0.20$. What is the largest value of the wedge angle α for which the wedge would remain in place in the log when the force F is removed?

▶ **9.66** The wedge shown is being used to split the log. The wedge weighs 20 lb and the angle α equals 30°. The coefficient of kinetic friction between the faces of the wedge and the log is 0.28. If the normal force exerted by each face of the wedge must equal 150 lb to split the log, what vertical force F is necessary to drive the wedge into the log at a constant rate? (See Active Example 9.4.)

▶ **9.67** The coefficient of static friction between the faces of the wedge and the log in Problem 9.66 is 0.30. Will the wedge remain in place in the log when the vertical force F is removed? (See Active Example 9.4.)

Problems 9.66/9.67

9.68 The weights of the blocks are $W_A = 100$ lb and $W_B = 25$ lb. Between all of the contacting surfaces, $\mu_s = 0.32$ and $\mu_k = 0.30$. What force F is necessary to move B to the left at a constant rate?

Problem 9.68

9.69 The masses of the blocks are $m_A = 30$ kg and $m_B = 70$ kg. Between all of the contacting surfaces, $\mu_s = 0.1$. What is the largest force F that can be applied without causing the blocks to slip?

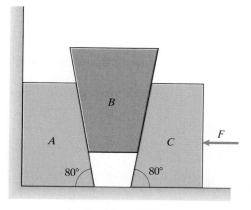

Problem 9.69

9.70 Each block weighs 200 lb. Between all of the contacting surfaces, $\mu_s = 0.1$. What is the largest force F that can be applied without causing block B to slip upward?

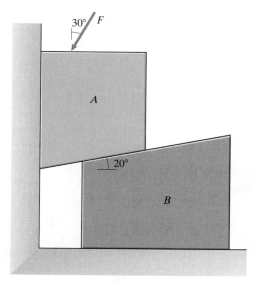

Problem 9.70

9.71 Small wedges called *shims* can be used to hold an object in place. The coefficient of kinetic friction between the contacting surfaces is 0.4. What force F is needed to push the shim downward until the horizontal force exerted on the object A is 200 N?

9.72 The coefficient of static friction between the contacting surfaces is 0.44. If the shims are in place and exert a 200-N horizontal force on the object A, what upward force must be exerted on the left shim to loosen it?

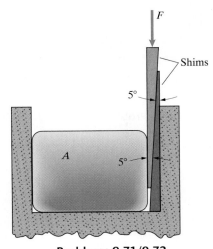

Problems 9.71/9.72

9.73 The crate A weighs 600 lb. Between all contacting surfaces, $\mu_s = 0.32$ and $\mu_k = 0.30$. Neglect the weights of the wedges. What force F is required to move A to the right at a constant rate?

9.74 Suppose that between all contacting surfaces, $\mu_s = 0.32$ and $\mu_k = 0.30$. Neglect the weights of the 5° wedges. If a force $F = 800$ N is required to move A to the right at a constant rate, what is the mass of A?

Problems 9.73/9.74

9.75 The box A has a mass of 80 kg, and the wedge B has a mass of 40 kg. Between all contacting surfaces, $\mu_s = 0.15$ and $\mu_k = 0.12$. What force F is required to raise A at a constant rate?

9.76 Suppose that A weighs 800 lb and B weighs 400 lb. The coefficients of friction between all of the contacting surfaces are $\mu_s = 0.15$ and $\mu_k = 0.12$. Will B remain in place if the force F is removed?

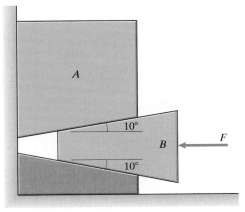

Problems 9.75/9.76

9.77 Between A and B, $\mu_s = 0.20$, and between B and C, $\mu_s = 0.18$. Between C and the wall, $\mu_s = 0.30$. The weights $W_B = 20$ lb and $W_C = 80$ lb. What force F is required to start C moving upward?

9.78 The masses of A, B, and C are 8 kg, 12 kg, and 80 kg, respectively. Between all contacting surfaces, $\mu_s = 0.4$. What force F is required to start C moving upward?

Problem 9.78

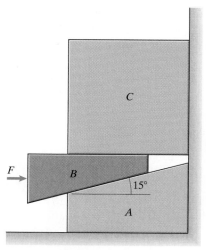

Problem 9.77

9.3 Threads

BACKGROUND

Threads are familiar from their use on wood screws, machine screws, and other machine elements. We show a shaft with square threads in Fig. 9.10a. The axial distance p from one thread to the next is called the *pitch* of the thread, and the angle α is its *slope*. We will consider only the case in which the shaft has a single continuous thread, so the relation between the pitch and slope is

$$\tan \alpha = \frac{p}{2\pi r}, \tag{9.7}$$

where r is the mean radius of the thread.

Figure 9.10
(a) A shaft with a square thread.
(b) The shaft within a sleeve with a mating groove and the direction of M that can cause the shaft to start moving in the axial direction opposite to F.
(c) A differential element of the thread when slip is impending.

Suppose that the threaded shaft is enclosed in a fixed sleeve with a mating groove and is subjected to an axial load F (Fig. 9.10b). Applying a couple M in the direction shown will tend to cause the shaft to start rotating and moving in the axial direction opposite to F. Our objective is to determine the couple M necessary to cause the shaft to start rotating.

We draw the free-body diagram of a differential element of the thread of length dL in Fig. 9.10c, representing the reaction exerted by the mating groove by the force dR. If the shaft is on the verge of rotating, dR resists the impending motion and the friction angle is the angle of static friction θ_s. The vertical component of the reaction on the element is $dR \cos(\theta_s + \alpha)$. To determine the total vertical force on the thread, we must integrate this expression over the length L of the thread. For equilibrium, the result must equal the axial force F acting on the shaft:

$$\cos(\theta_s + \alpha) \int_L dR = F. \qquad (9.8)$$

The moment about the center of the shaft due to the reaction on the element is $r\, dR \sin(\theta_s + \alpha)$. The total moment must equal the couple M exerted on the shaft:

$$r \sin(\theta_s + \alpha) \int_L dR = M.$$

Dividing this equation by Eq. (9.8), we obtain the couple M necessary for the shaft to be on the verge of rotating and moving in the axial direction opposite to F:

$$M = rF \tan(\theta_s + \alpha). \qquad (9.9)$$

Replacing the angle of static friction θ_s in this expression with the angle of kinetic friction θ_k gives the couple required to cause the shaft to rotate at a constant rate.

If the couple M is applied to the shaft in the opposite direction (Fig. 9.11a), the shaft tends to start rotating and moving in the axial direction of the load F. Figure 9.11b shows the reaction on a differential element of the thread of length dL when slip is impending. The direction of the reaction opposes the rotation of the shaft. In this case, the vertical component of the reaction on the element is $dR \cos(\theta_s - \alpha)$. Equilibrium requires that

$$\cos(\theta_s - \alpha) \int_L dR = F. \qquad (9.10)$$

The moment about the center of the shaft due to the reaction is $r\, dR \sin(\theta_s - \alpha)$, so

$$r \sin(\theta_s - \alpha) \int_L dR = M.$$

Dividing this equation by Eq. (9.10), we obtain the couple M necessary for the shaft to be on the verge of rotating and moving in the direction of the force F:

$$M = rF \tan(\theta_s - \alpha). \qquad (9.11)$$

Replacing θ_s with θ_k in this expression gives the couple necessary to rotate the shaft at a constant rate.

Notice in Eq. (9.11) that the couple required for impending motion is zero when $\theta_s = \alpha$. When the angle of static friction is less than this value, the shaft will rotate and move in the direction of the force F with no couple applied.

Figure 9.11
(a) The direction of M that can cause the shaft to move in the axial direction of F.
(b) A differential element of the thread when slip is impending.

The slope α of the thread is related to its pitch p and the radius r by

$$\tan \alpha = \frac{p}{2\pi r} . \qquad (9.7)$$

$$M = rF \tan(\theta_s + \alpha). \qquad (9.9)$$

The couple M required for impending rotation and axial motion of the shaft opposite to the direction of F, where

$$\theta_s = \arctan \mu_s.$$

$$M = rF \tan(\theta_s - \alpha). \qquad (9.11)$$

The couple M (opposite to the direction shown) required for impending rotation and axial motion of the shaft in the direction of F. If $\theta_s < \alpha$, the shaft will rotate and move in the direction of F with no couple applied.

Active Example 9.5 **Rotating a Threaded Collar** (▶ *Related Problem 9.79*)

The right end of bar AB is pinned to an unthreaded collar B that rests on a threaded collar C. The mean radius of the threaded vertical shaft is $r = 1.6$ in and its pitch is $p = 0.2$ in. The coefficients of friction between the threads of the collar C and the vertical shaft are $\mu_s = 0.25$ and $\mu_k = 0.22$. The 400-lb suspended object can be raised or lowered by rotating the collar C. When the system is in the position shown, with bar AB horizontal, what is the magnitude of the couple that must be applied to the collar C to cause it to turn at a constant rate and move the suspended object upward?

Strategy

By drawing the free-body diagram of bar AB and the collar B, we can determine the axial force exerted on the collar C. Then we can use Eq. (9.9), with θ_s replaced by θ_k, to determine the required couple.

Solution

Draw the free-body diagram of the bar and collar B.

A_x

A_y

B

400 lb

F

From the sum of the moments about point A,

$\Sigma M_{\text{point } A} = (40 \text{ in})F - (20 \text{ in})(400 \text{ lb}) = 0,$

the force $F = 200$ lb. This is the axial force exerted on collar C.

Apply equilibrium.

F

C

From Eq. (9.7),

$\tan \alpha = \dfrac{p}{2\pi r} = \dfrac{0.2 \text{ in}}{2\pi(1.6 \text{ in})} = 0.0199,$

the slope of the thread is $\alpha = 1.14°$. The kinetic angle of friction is

$\theta_k = \arctan \mu_k = \arctan(0.22) = 12.4°.$

Substituting these values into Eq. (9.9),

$M = rF \tan(\theta_k + \alpha)$

$= (1.6 \text{ in})(200 \text{ lb}) \tan(12.4° + 1.14°)$

$= 77.1 \text{ in-lb}.$

Apply Eq. (9.9).

Practice Problem When the system is in the position shown, with bar AB horizontal, what is the magnitude of the couple that must be applied to the collar C to cause it to turn at a constant rate and move the suspended object downward?

Answer: 63.8 in-lb.

Problems

▶ **9.79** In Active Example 9.5, suppose that the pitch of the thread is changed from $p = 0.2$ in to $p = 0.24$ in. What is the slope of the thread? What is the magnitude of the couple that must be applied to the collar C to cause it to turn at a constant rate and move the suspended object upward?

9.80 The pitch of the threaded shaft is $p = 2$ mm and the mean radius of the thread is $r = 20$ mm. The coefficients of friction between the thread and the mating groove are $\mu_s = 0.22$ and $\mu_k = 0.20$. The weight $W = 500$ N. Neglect the weight of the threaded shaft. What couple must be applied to the threaded shaft to lower the weight at a constant rate?

Problem 9.80

9.81 The position of the horizontal beam can be adjusted by turning the machine screw A. Neglect the weight of the beam. The pitch of the screw is $p = 1$ mm, and the mean radius of the thread is $r = 4$ mm. The coefficients of friction between the thread and the mating groove are $\mu_s = 0.20$ and $\mu_k = 0.18$. If the system is initially stationary, determine the couple that must be applied to the screw to cause the beam to start moving (a) upward; (b) downward.

400 N

A

100 mm   300 mm

Problem 9.81

9.82 The pitch of the threaded shaft of the C clamp is $p = 0.05$ in, and the mean radius of the thread is $r = 0.15$ in. The coefficients of friction between the threaded shaft and the mating collar are $\mu_s = 0.18$ and $\mu_k = 0.16$.

(a) What maximum couple must be applied to the shaft to exert a 30-lb force on the clamped object?

(b) If a 30-lb force is exerted on the clamped object, what couple must be applied to the shaft to begin loosening the clamp?

Problem 9.82

9.83 The mass of block A is 60 kg. Neglect the weight of the 5° wedge. The coefficient of kinetic friction between the contacting surfaces of the block A, the wedge, the table, and the wall is $\mu_k = 0.4$. The pitch of the threaded shaft is 5 mm, the mean radius of the thread is 15 mm, and the coefficient of kinetic friction between the thread and the mating groove is 0.2. What couple must be exerted on the threaded shaft to raise the block A at a constant rate?

A

5°

Problem 9.83

9.84 The vise exerts 80-lb forces on A. The threaded shafts are subjected only to axial loads by the jaws of the vise. The pitch of their threads is $p = 1/8$ in, the mean radius of the threads is $r = 1$ in, and the coefficient of static friction between the threads and the mating grooves is 0.2. Suppose that you want to loosen the vise by turning one of the shafts. Determine the couple you must apply (a) to shaft B; (b) to shaft C.

9.85 Suppose that you want to tighten the vise in Problem 9.84 by turning one of the shafts. Determine the couple you must apply
(a) to shaft B;
(b) to shaft C.

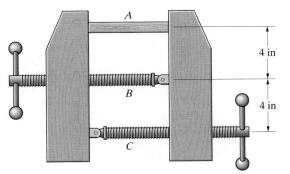

Problems 9.84/9.85

9.86 The threaded shaft has a ball and socket support at B. The 400-lb load A can be raised or lowered by rotating the threaded shaft, causing the threaded collar at C to move relative to the shaft. Neglect the weights of the members. The pitch of the shaft is $p = \frac{1}{4}$ in, the mean radius of the thread is $r = 1$ in, and the coefficient of static friction between the thread and the mating groove is 0.24. If the system is stationary in the position shown, what couple is necessary to start the shaft rotating to raise the load?

9.87 In Problem 9.86, if the system is stationary in the position shown, what couple is necessary to start the shaft rotating to lower the load?

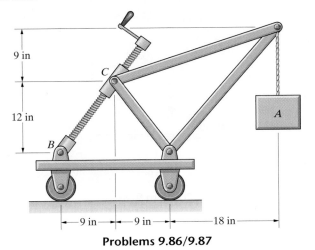

Problems 9.86/9.87

9.88 The car jack is operated by turning the horizontal threaded shaft at A. The threaded shaft fits into a mating threading collar at B. As the shaft turns, points A and B move closer together or farther apart, thereby raising or lowering the jack. The pitch of the threaded shaft is $p = 0.1$ in, the mean radius of the thread is $r = 0.2$ in, and the coefficient of kinetic friction between the threaded shaft and the mating collar at B is 0.15. What couple must be applied at A to rotate the shaft at a constant rate and raise the jack when it is in the position shown if the load $L = 1400$ lb?

9.89 The car jack is operated by turning the horizontal threaded shaft at A. The threaded shaft fits into a mating threading collar at B. As the shaft turns, points A and B move closer together or farther apart, thereby raising or lowering the jack. The pitch of the threaded shaft is $p = 0.1$ in, the mean radius of the thread is $r = 0.2$ in, and the coefficient of kinetic friction between the threaded shaft and the mating collar at B is 0.15. What couple must be applied at A to rotate the shaft at a constant rate and lower the jack when it is in the position shown if the load $L = 1400$ lb?

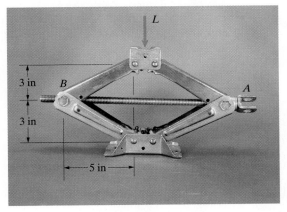

Problems 9.88/9.89

9.90 A *turnbuckle,* used to adjust the length or tension of a bar or cable, is threaded at both ends. Rotating it draws threaded ends of the bar or cable together or moves them apart. Suppose that the pitch of the threads is $p = 0.05$ in, their mean radius is $r = 0.25$ in, and the coefficient of static friction between the threads and the mating grooves is 0.24. If $T = 200$ lb, what couple must be exerted on the turnbuckle to start tightening it?

9.91 Suppose that the pitch of the threads of the turnbuckle is $p = 0.05$ in, their mean radius is $r = 0.25$ in, and the coefficient of static friction between the threads and the mating grooves is 0.24. If $T = 200$ lb, what couple must be exerted on the turnbuckle to start loosening it?

Problems 9.90/9.91

9.92 Member BE of the frame has a turnbuckle. (See Problem 9.90.) The threads have pitch $p = 1$ mm, their mean radius is $r = 6$ mm, and the coefficient of static friction between the threads and the mating grooves is 0.2. What couple must be exerted on the turnbuckle to start loosening it?

9.93 In Problem 9.92, what couple must be exerted on the turnbuckle to start tightening it?

Problems 9.92/9.93

9.94 Members CD and DG of the truss have turnbuckles. (See Problem 9.90.) The pitch of the threads is $p = 4$ mm, their mean radius is $r = 10$ mm, and the coefficient of static friction between the threads and the mating grooves is 0.18. What couple must be exerted on the turnbuckle of member CD to start loosening it?

9.95 In Problem 9.94, what couple must be exerted on the turnbuckle of member DG to start loosening it?

Problems 9.94/9.95

9.96* The load $W = 800$ N can be raised or lowered by rotating the threaded shaft. The distances are $b = 75$ mm and $h = 200$ mm. The pinned bars are each 300 mm in length. The pitch of the threaded shaft is $p = 5$ mm, the mean radius of the thread is $r = 15$ mm, and the coefficient of kinetic friction between the thread and the mating groove is 0.2. When the system is in the position shown, what couple must be exerted to turn the threaded shaft at a constant rate, raising the load?

Problem 9.96

| (a) | (b) | (c) | (d) | (e) | (f) |

Figure 9.12
(a) A shaft supported by journal bearings.
(b) A pulley supported by the shaft.
(c) The shaft and bearing when no couple is applied to the shaft.
(d) A couple causes the shaft to roll within the bearing.
(e) Free-body diagram of the shaft.
(f) The two forces on the shaft must be equal and opposite.

9.4 Journal Bearings

BACKGROUND

A *bearing* is a support. This term usually refers to supports designed to allow the supported object to move. For example, in Fig. 9.12a, a horizontal shaft is supported by two *journal bearings*, which allow the shaft to rotate. The shaft can then be used to support a load perpendicular to its axis, such as that subjected by a pulley (Fig. 9.12b).

Here we analyze journal bearings consisting of brackets with holes through which the shaft passes. The radius of the shaft is slightly smaller than the radius of the holes in the bearings. Our objective is to determine the couple that must be applied to the shaft to cause it to rotate in the bearings. Let F be the total load supported by the shaft including the weight of the shaft itself. When no couple is exerted on the shaft, the force F presses it against the bearings as shown in Fig. 9.12c. When a couple M is exerted on the shaft, it rolls up the surfaces of the bearings (Fig. 9.12d). The term α is the angle from the original point of contact of the shaft to its point of contact when M is applied.

In Fig. 9.12e, we draw the free-body diagram of the shaft when M is sufficiently large that slip is impending. The force R is the total reaction exerted on the shaft by the two bearings. Since R and F are the only forces acting on the shaft, equilibrium requires that $\alpha = \theta_s$ and $R = F$ (Fig. 9.12f). The reaction exerted on the shaft by the bearings is displaced a distance $r \sin \theta_s$ from the vertical line through the center of the shaft. By summing moments about the center of the shaft, we obtain the couple M that causes the shaft to be on the verge of slipping:

$$M = rF \sin \theta_s. \tag{9.12}$$

This is the largest couple that can be exerted on the shaft without causing it to start rotating. Replacing θ_s in this expression by the angle of kinetic friction θ_k gives the couple necessary to rotate the shaft at a constant rate.

The simple type of journal bearing we have described is too primitive for most applications. The surfaces where the shaft and bearing are in contact would quickly become worn. Designers usually incorporate "ball" or "roller" bearings in journal bearings to minimize friction (Fig. 9.13).

(a)

Wheel

(b)

Figure 9.13
(a) A journal bearing with one row of balls.
(b) Journal bearing assembly of the wheel of a car. There are two rows of balls between the rotating wheel and the fixed inner cylinder.

RESULTS

A journal bearing has a circular hole slightly larger than the circular shaft it supports.

The couple M that must be applied to the circular shaft in order for slip to impend relative to a journal bearing is

$$M = rF \sin \theta_s, \qquad (9.12)$$

where r is the radius of the shaft, F is the lateral load supported by the shaft, and $\theta_s = \arctan \mu_s$.

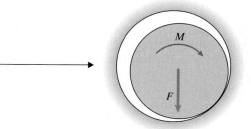

Active Example 9.6 **Pulley Supported by Journal Bearings (▶ Related Problem 9.97)**

The weight of the suspended load is $W = 1000$ lb. The pulley P has a 6-in radius and is rigidly attached to a horizontal circular shaft that is supported by journal bearings. The radius of the shaft is 0.5 in, and the coefficient of kinetic friction between the shaft and the bearings is $\mu_k = 0.2$. The weights of the pulley and shaft are negligible. What tension must the winch A exert on the rope to raise the load at a constant rate?

Strategy

Eq. (9.12) with θ_s replaced by θ_k relates the couple M required to turn the pulley at a constant rate to the lateral force F supported by the shaft. By expressing M and F in terms of the forces exerted on the pulley by the rope and applying Eq. (9.12), we can obtain an equation for the tension the winch must exert.

Solution

Forces exerted on the pulley by the rope. The force T is the tension exerted by the winch.

$F = \sqrt{(W + T \sin 45°)^2 + (T \cos 45°)^2}.$

The vector sum of the forces exerted on the pulley by the rope is the lateral force F the shaft of the pulley must support. The magnitude of F can be expressed in terms of W and T.

$M = (6 \text{ in})(T - W).$

The pulley moves in the clockwise direction. Express the clockwise couple on the pulley in terms of T and W.

The angle of kinetic friction is

$\theta_k = \arctan \mu_k = \arctan(0.2) = 11.3°.$

Equation (9.12) is

$M = rF \sin \theta_k:$

Apply Eq. (9.12).

$(6 \text{ in})(T - W) = (0.5 \text{ in}) \sqrt{(W + T \sin 45°)^2 + (T \cos 45°)^2} \sin 11.3°.$

Setting $W = 1000$ lb and solving yields

$T = 1030$ lb.

Practice Problem What tension must the winch A exert on the rope to lower the load at a constant rate?

Answer: $T = 970$ lb.

Problems

▶ **9.97** In Active Example 9.6, suppose that the placement of the winch at A is changed so that the angle between the rope from A to P and the horizontal increases from 45° to 60°. If the suspended load weighs 1500 lb, what tension must the winch exert on the rope to raise the load at a constant rate?

9.98 The radius of the pulley is 4 in. The pulley is rigidly attached to the horizontal shaft, which is supported by two journal bearings. The radius of the shaft is 1 in, and the combined weight of the pulley and shaft is 20 lb. The coefficients of friction between the shaft and the bearings are $\mu_s = 0.30$ and $\mu_k = 0.28$. Determine the largest weight W that can be suspended as shown without causing the stationary shaft to slip in the bearings.

9.99 In Problem 9.98, suppose that the weight $W = 4$ lb. What couple would have to be applied to the horizontal shaft to raise the weight at a constant rate?

Problems 9.98/9.99

9.100 The pulley is mounted on a horizontal shaft supported by journal bearings. The coefficient of kinetic friction between the shaft and the bearings is $\mu_k = 0.3$. The radius of the shaft is 20 mm, and the radius of the pulley is 150 mm. The mass $m = 10$ kg. Neglect the masses of the pulley and shaft. What force T must be applied to the cable to move the mass upward at a constant rate?

9.101 In Problem 9.100, what force T must be applied to the cable to lower the mass at a constant rate?

Problems 9.100/9.101

9.102 The pulley of 8-in radius is mounted on a shaft of 1-in radius. The shaft is supported by two journal bearings. The coefficient of static friction between the bearings and the shaft is $\mu_s = 0.15$. Neglect the weights of the pulley and shaft. The 50-lb block A rests on the floor. If sand is slowly added to the bucket B, what do the bucket and sand weigh when the shaft slips in the bearings?

Problem 9.102

9.103 The pulley of 50-mm radius is mounted on a shaft of 10-mm radius. The shaft is supported by two journal bearings. The mass of the block A is 8 kg. Neglect the weights of the pulley and shaft. If a force $T = 84$ N is necessary to raise block A at a constant rate, what is the coefficient of kinetic friction between the shaft and the bearings?

Problem 9.103

9.104 The mass of the suspended object is 4 kg. The pulley has a 100-mm radius and is rigidly attached to a horizontal shaft supported by journal bearings. The radius of the horizontal shaft is 10 mm and the coefficient of kinetic friction between the shaft and the bearings is 0.26. What tension must the person exert on the rope to raise the load at a constant rate?

9.105 In Problem 9.104, what tension must the person exert to lower the load at a constant rate?

Problems 9.104/9.105

9.106 The radius of the pulley is 200 mm, and it is mounted on a shaft of 20-mm radius. The coefficient of static friction between the pulley and shaft is $\mu_s = 0.18$. If $F_A = 200$ N, what is the largest force F_B that can be applied without causing the pulley to turn? Neglect the weight of the pulley.

Problem 9.106

9.107 The masses of the boxes are $m_A = 15$ kg and $m_B = 60$ kg. The coefficient of static friction between boxes A and B and between box B and the inclined surface is 0.12. The pulley has a radius of 60 mm and is mounted on a shaft of 10-mm radius. The coefficient of static friction between the pulley and shaft is 0.16. What is the largest force F for which the boxes will not slip?

Problem 9.107

9.108 The two pulleys have a radius of 4 in and are mounted on shafts of 1-in radius supported by journal bearings. Neglect the weights of the pulleys and shafts. The tension in the spring is 40 lb. The coefficient of kinetic friction between the shafts and the bearings is $\mu_k = 0.3$. What couple M is required to turn the left pulley at a constant rate?

Problem 9.108

9.109 The weights of the boxes are $W_A = 65$ lb and $W_B = 130$ lb. The coefficient of static friction between boxes A and B and between box B and the floor is 0.12. The pulley has a radius of 4 in and is mounted on a shaft of 0.8-in radius. The coefficient of static friction between the pulley and shaft is 0.16. What is the largest force F for which the boxes will not slip?

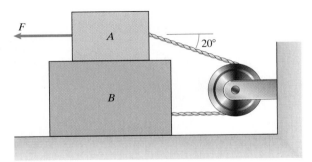

Problem 9.109

9.110 The coefficient of kinetic friction between the 100-kg box and the inclined surface is 0.35. Each pulley has a radius of 100 mm and is mounted on a shaft of 5-mm radius supported by journal bearings. The coefficient of kinetic friction between the shafts and the journal bearings is 0.18. Determine the tension T necessary to pull the box up the surface at a constant rate.

Problem 9.110

9.5 Thrust Bearings and Clutches

BACKGROUND

A *thrust bearing* supports a rotating shaft that is subjected to an axial load. In the type shown in Figs. 9.14a and b, the conical end of the shaft is pressed against the mating conical cavity by an axial load F. Let us determine the couple M necessary to rotate the shaft.

The differential element of area dA in Fig. 9.14c is

$$dA = 2\pi r \, ds = 2\pi r \left(\frac{dr}{\cos \alpha} \right).$$

Integrating this expression from $r = r_i$ to $r = r_o$, we obtain the area of contact:

$$A = \frac{\pi(r_o^2 - r_i^2)}{\cos \alpha}.$$

If we assume that the mating surface exerts a uniform pressure p, the axial component of the total force due to p must equal F: $pA \cos \alpha = F$. Therefore, the pressure is

$$p = \frac{F}{A \cos \alpha} = \frac{F}{\pi(r_o^2 - r_i^2)}.$$

As the shaft rotates about its axis, the moment about the axis due to the friction force on the element dA is $r\mu_k(p \, dA)$. The total moment is

$$M = \int_A \mu_k r p \, dA = \int_{r_i}^{r_o} \mu_k r \left[\frac{F}{\pi(r_o^2 - r_i^2)} \right] \left(\frac{2\pi r \, dr}{\cos \alpha} \right).$$

Integrating, we obtain the couple M necessary to rotate the shaft at a constant rate:

$$M = \frac{2\mu_k F}{3 \cos \alpha} \left(\frac{r_o^3 - r_i^3}{r_o^2 - r_i^2} \right). \tag{9.13}$$

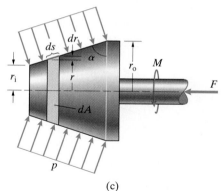

Figure 9.14

(a), (b) A thrust bearing supports a shaft subjected to an axial load.

(c) The differential element dA and the uniform pressure p exerted by the cavity.

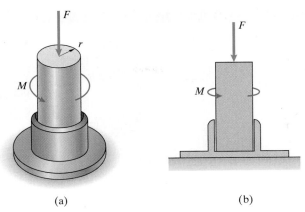

(a)   (b)

Figure 9.15
A thrust bearing that supports a flat-ended shaft.

A simpler thrust bearing is shown in Figs. 9.15a and b. The bracket supports the flat end of a shaft of radius r that is subjected to an axial load F. We can obtain the couple necessary to rotate the shaft at a constant rate from Eq. (9.13) by setting $\alpha = 0$, $r_i = 0$, and $r_o = r$:

$$M = \frac{2}{3}\mu_k Fr. \qquad (9.14)$$

Although they are good examples of the analysis of friction forces, the thrust bearings we have described would become worn too quickly to be used in most applications. The designer of the thrust bearing in Fig. 9.16 minimizes friction by incorporating "roller" bearings.

A *clutch* is a device used to connect and disconnect two coaxial rotating shafts. The type shown in Figs. 9.17a and b consists of disks of radius r attached to the ends of the shafts. When the disks are separated (Fig. 9.17a), the clutch is *disengaged,* and the shafts can rotate freely relative to each other. When the clutch is engaged by pressing the disks together with axial forces F (Fig. 9.17b), the shafts can support a couple M due to the friction forces between the disks. If the couple M becomes too large, the clutch slips.

The friction forces exerted on one face of the clutch by the other face are identical to the friction forces exerted on the flat-ended shaft by the bracket in Fig. 9.15. We can therefore determine the largest couple the clutch can support without slipping by replacing μ_k with μ_s in Eq. (9.14):

$$M = \frac{2}{3}\mu_s Fr. \qquad (9.15)$$

Figure 9.16
A thrust bearing with two rows of cylindrical rollers between the shaft and the fixed support.

(a)   (b)

Figure 9.17
A clutch.
(a) Disengaged position.
(b) Engaged position.

$$M = \frac{2\mu_k F}{3 \cos \alpha} \left(\frac{r_o^3 - r_i^3}{r_o^2 - r_i^2} \right). \qquad (9.13)$$

The couple required to rotate a shaft supported by a *thrust bearing* in terms of the axial force supported by the shaft.

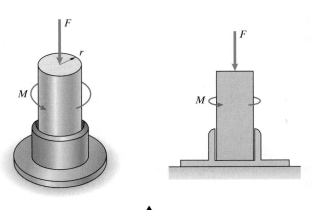

$$M = \frac{2}{3} \mu_k Fr. \qquad (9.14)$$

The couple required to rotate a flat-ended shaft supported by a thrust bearing in terms of the axial force supported by the shaft.

$$M = \frac{2}{3} \mu_s Fr. \qquad (9.15)$$

The couple required to cause impending slip of a *clutch* in terms of the axial force applied to the clutch.

Active Example 9.7   Thrust Bearing (▶ *Related Problem 9.111*)

The axial force on the thrust bearing is $F = 200$ lb. The diameters $D_o = 3\frac{1}{2}$ in and $D_i = 1$ in, and the angle $\alpha = 72°$. The coefficient of kinetic friction is $\mu_k = 0.18$. What couple is required to turn the shaft at a constant rate?

Strategy
The couple is given by Eq. (9.13).

Solution

The radii $r_o = 1.75$ in and $r_i = 0.5$ in.

$$M = \frac{2\mu_k F}{3 \cos \alpha} \left(\frac{r_o^3 - r_i^3}{r_o^2 - r_i^2} \right)$$

$$= \frac{2(0.18)(200 \text{ lb})}{3 \cos 72°} \left[\frac{(1.75 \text{ in})^3 - (0.5 \text{ in})^3}{(1.75 \text{ in})^2 - (0.5 \text{ in})^2} \right]$$

$$= 145 \text{ in-lb.}$$

> Apply Eq. (9.13).

Practice Problem The axial force on the thrust bearing is $F = 200$ lb. The diameters $D_o = 3\frac{1}{2}$ in and $D_i = 1$ in, and the dimension $b = 5$ in. The coefficient of kinetic friction is $\mu_k = 0.18$. What couple is required to turn the shaft at a constant rate?

Answer: $M = 184$ in-lb.

| Example 9.8 | Friction on a Disk Sander (▶ *Related Problem 9.118*) |

D

The handheld sander has a rotating disk D of 4-in radius with sandpaper bonded to it. The total downward force exerted by the operator and the weight of the sander is 15 lb. The coefficient of kinetic friction between the sandpaper and the surface is $\mu_k = 0.6$. What couple (torque) M must the motor exert to turn the sander at a constant rate?

Strategy
As the disk D rotates, it is subjected to friction forces analogous to the friction forces exerted on a flat-ended shaft supported by a thrust bearing. We can determine the couple required to turn the disk D at a constant rate from Eq. (9.14).

Solution
The couple required to turn the disk at a constant rate is

$$M = \frac{2}{3}\mu_k rF = \frac{2}{3}(0.6)(4 \text{ in})(15 \text{ lb}) = 24 \text{ in-lb}.$$

Critical Thinking
Equations (9.13)–(9.15) were derived under the assumption that the normal force (and consequently the friction force) is uniformly distributed over the contacting surfaces. Evaluating and improving upon this assumption would require analysis of the deformations of the contacting surfaces in specific applications such as the disk sander in this example.

Problems

▶ **9.111** In Active Example 9.7, suppose that the diameters $D_o = 3\frac{1}{2}$ in and $D_i = 1\frac{1}{2}$ in and the angle $\alpha = 72°$. What couple is required to turn the shaft at a constant rate?

9.112 The circular flat-ended shaft is pressed into the thrust bearing by an axial load of 600 lb. The weight of the shaft is negligible. The coefficients of friction between the end of the shaft and the bearing are $\mu_s = 0.20$ and $\mu_k = 0.15$. What is the largest couple M that can be applied to the stationary shaft without causing it to rotate in the bearing?

9.113 The circular flat-ended shaft is pressed into the thrust bearing by an axial load of 600 lb. The weight of the shaft is negligible. The coefficients of friction between the end of the shaft and the bearing are $\mu_s = 0.20$ and $\mu_k = 0.15$. What couple M is required to rotate the shaft at a constant rate?

600 lb

2 in

M

Problems 9.112/9.113

9.114 The disk D is rigidly attached to the vertical shaft. The shaft has flat ends supported by thrust bearings. The disk and the shaft together have a mass of 220 kg and the diameter of the shaft is 50 mm. The vertical force exerted on the end of the shaft by the upper thrust bearing is 440 N. The coefficient of kinetic friction between the ends of the shaft and the bearings is 0.25. What couple M is required to rotate the shaft at a constant rate?

9.115 Suppose that the ends of the shaft in Problem 9.114 are supported by thrust bearings of the type shown in Fig. 9.14, where $r_o = 25$ mm, $r_i = 6$ mm, $\alpha = 45°$, and $\mu_k = 0.25$. What couple M is required to rotate the shaft at a constant rate?

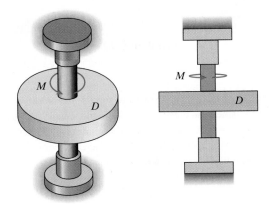

Problems 9.114/9.115

9.116 The shaft is supported by thrust bearings that subject it to an axial load of 800 N. The coefficients of kinetic friction between the shaft and the left and right bearings are 0.20 and 0.26, respectively. What couple is required to rotate the shaft at a constant rate?

15 mm

38 mm

38 mm

Problem 9.116

9.117 A motor is used to rotate a paddle for mixing chemicals. The shaft of the motor is coupled to the paddle using a friction clutch of the type shown in Fig. 9.17. The radius of the disks of the clutch is 120 mm, and the coefficient of static friction between the disks is 0.6. If the motor transmits a maximum torque of 15 N-m to the paddle, what minimum normal force between the plates of the clutch is necessary to prevent slipping?

Clutch

Paddle

Problem 9.117

▶ **9.118** The thrust bearing is supported by contact of the collar C with a fixed plate. The area of contact is an annulus with an inside diameter $D_1 = 40$ mm and an outside diameter $D_2 = 120$ mm. The coefficient of kinetic friction between the collar and the plate is $\mu_k = 0.3$. The force $F = 400$ N. What couple M is required to rotate the shaft at a constant rate? (See Example 9.8.)

F

C

M

F

M

C

D_1

D_2

Problem 9.118

9.119 An experimental automobile brake design works by pressing the fixed red annular plate against the rotating wheel. If $\mu_k = 0.6$, what force F pressing the plate against the wheel is necessary to exert a couple of 200 N-m on the wheel?

9.120 An experimental automobile brake design works by pressing the fixed red annular plate against the rotating wheel. Suppose that $\mu_k = 0.65$ and the force pressing the plate against the wheel is $F = 2$ kN.

(a) What couple is exerted on the wheel?

(b) What percentage increase in the couple exerted on the wheel is obtained if the outer radius of the brake is increased from 90 mm to 100 mm?

Problems 9.119/9.120

9.121 The coefficient of static friction between the plates of the car's clutch is 0.8. If the plates are pressed together with a force $F = 2.60$ kN, what is the maximum torque the clutch will support without slipping?

Problem 9.121

9.122* The "Morse taper" is used to support the workpiece on a machinist's lathe. The taper is driven into the spindle and is held in place by friction. If the spindle exerts a uniform pressure $p = 15$ psi on the taper and $\mu_s = 0.2$, what couple must be exerted about the axis of the taper to loosen it?

Problem 9.122

9.6 Belt Friction

If a rope is wrapped around a fixed post as shown in Fig. 9.18, a large force T_2 exerted on one end can be supported by a relatively small force T_1 applied to the other end. In this section we analyze this familiar phenomenon. It is referred to as *belt friction* because a similar approach can be used to analyze belts used in machines, such as the belts that drive alternators and other devices in a car.

Let us consider a rope wrapped through an angle β around a fixed cylinder (Fig. 9.19a). We will assume that the tension T_1 is known. Our objective is to determine the largest force T_2 that can be applied to the other end of the rope without causing the rope to slip.

We begin by drawing the free-body diagram of an element of the rope whose boundaries are at angles α and $\alpha + \Delta\alpha$ from the point where the rope comes into contact with the cylinder (Figs. 9.19b and c). The force T is the tension in the rope at the position defined by the angle α. We know that the tension in the rope varies with position, because it increases from T_1 at $\alpha = 0$ to T_2 at $\alpha = \beta$. We therefore write the tension in the rope at the position $\alpha + \Delta\alpha$ as $T + \Delta T$. The force ΔN is the normal force exerted on the element by the cylinder. Because we want to determine the largest value of T_2 that will not cause the rope to slip, we assume that the friction force is equal to its maximum possible value $\mu_s\Delta N$, where μ_s is the coefficient of static friction between the rope and the cylinder.

The equilibrium equations in the directions tangential to and normal to the centerline of the rope are

$$\Sigma F_{\text{tangential}} = \mu_s\Delta N + T\cos\left(\frac{\Delta\alpha}{2}\right) - (T + \Delta T)\cos\left(\frac{\Delta\alpha}{2}\right) = 0,$$
(9.16)

$$\Sigma F_{\text{normal}} = \Delta N - (T + \Delta T)\sin\left(\frac{\Delta\alpha}{2}\right) - T\sin\left(\frac{\Delta\alpha}{2}\right) = 0.$$

Eliminating ΔN, we can write the resulting equation as

$$\left[\cos\left(\frac{\Delta\alpha}{2}\right) - \mu_s\sin\left(\frac{\Delta\alpha}{2}\right)\right]\frac{\Delta T}{\Delta\alpha} - \mu_s T\frac{\sin(\Delta\alpha/2)}{\Delta\alpha/2} = 0.$$

Figure 9.18
A rope wrapped around a post.

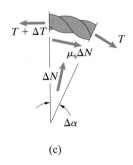

(a) (b) (c)

Figure 9.19
(a) A rope wrapped around a fixed cylinder.
(b) A differential element with boundaries at angles α and $\alpha + \Delta\alpha$.
(c) Free-body diagram of the element.

Evaluating the limit of this equation as $\Delta\alpha \rightarrow 0$ and observing that

$$\frac{\sin(\Delta\alpha/2)}{\Delta\alpha/2} \rightarrow 1,$$

we obtain

$$\frac{dT}{d\alpha} - \mu_s T = 0.$$

This differential equation governs the variation of the tension in the rope. Separating variables yields

$$\frac{dT}{T} = \mu_s \, d\alpha.$$

We can now integrate to determine the tension T_2 in terms of the tension T_1 and the angle β:

$$\int_{T_1}^{T_2} \frac{dT}{T} = \int_0^\beta \mu_s \, d\alpha.$$

Thus, we obtain the largest force T_2 that can be applied without causing the rope to slip when the force on the other end is T_1:

$$T_2 = T_1 e^{\mu_s \beta}. \tag{9.17}$$

The angle β in this equation must be expressed in radians. Replacing μ_s by the coefficient of kinetic friction μ_k gives the force T_2 required to cause the rope to slide at a constant rate.

Equation (9.17) explains why a large force can be supported by a relatively small force when a rope is wrapped around a fixed support. The force required to cause the rope to slip increases exponentially as a function of the angle through which the rope is wrapped. Suppose that $\mu_s = 0.3$. When the rope is wrapped one complete turn around the post ($\beta = 2\pi$), the ratio $T_2/T_1 = 6.59$. When the rope is wrapped four complete turns around the post ($\beta = 8\pi$), the ratio $T_2/T_1 = 1880$.

RESULTS

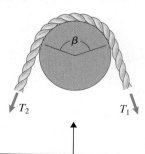

$$T_2 = T_1 e^{\mu_s \beta}. \tag{9.17}$$

The force T_2 necessary for impending slip of the rope relative to the fixed support in the direction of T_2, where the angle β is in radians and μ_s is the coefficient of static friction between the rope and the support.

Active Example 9.9 Rope Wrapped Around Fixed Cylinders (▶ *Related Problem 9.123*)

The 100-lb box is suspended from a rope that passes over two *fixed* cylinders. The coefficient of static friction is 0.2 between the rope and the left cylinder and 0.4 between the rope and the right cylinder. What is the smallest force the woman needs to exert on the rope to support the stationary box?

Strategy

She exerts the smallest necessary force when slip of the rope is impending on both cylinders. If we assume that slip is impending and apply Eq. (9.17) to each cylinder, we can determine the force she must apply.

Solution

Let T be the tension in the rope between the two cylinders. The weight $W = 100$ lb and F is the force the woman exerts. The rope is wrapped around each cylinder through an angle (in radians) $\beta = \pi/2$.

$\beta = \dfrac{\pi}{2}$ T T $\beta = \dfrac{\pi}{2}$

W F

$W = Te^{\mu_s \beta} = Te^{(0.2)(\pi/2)}.$
Solving for T yields
$T = We^{-(0.2)(\pi/2)} = (100 \text{ lb})e^{-(0.2)(\pi/2)} = 73.0 \text{ lb.}$

> Apply Eq. (9.17) to the left cylinder. Assume that slip of the rope in the direction of the force W is impending.

$T = Fe^{\mu_s \beta} = Fe^{(0.4)(\pi/2)}.$
Solving for F yields
$F = Te^{-(0.4)(\pi/2)} = (73.0 \text{ lb})e^{-(0.4)(\pi/2)} = 39.0 \text{ lb.}$

> Apply Eq. (9.17) to the right cylinder. Assume that slip of the rope in the direction of the force T is impending.

Practice Problem What force would the woman need to exert on the rope for slip to be impending in the direction she is pulling? That is, how hard would she have to pull for the box to be on the verge of moving upward? Would she need help?

Answer: 257 lb. Yes.

Example 9.10 **Belts and Pulleys** (▶ *Related Problem 9.134*)

The pulleys turn at a constant rate. The large pulley is attached to a fixed support. The small pulley is supported by a smooth horizontal slot and is pulled to the right by the force $F = 200$ N. The coefficient of static friction between the pulleys and the belt is $\mu_s = 0.8$, the dimension $b = 500$ mm, and the radii of the pulleys are $R_A = 200$ mm and $R_B = 100$ mm. What are the largest values of the couples M_A and M_B for which the belt will not slip?

(a) Free-body diagram of the large pulley.

(b) Free-body diagram of the small pulley.

Strategy
By drawing free-body diagrams of the pulleys, we can use the equilibrium equations to relate the tensions in the belt to M_A and M_B and obtain a relation between the tensions in the belt and the force F. When slip is impending, the tensions are also related by Eq. (9.17). From these equations we can determine M_A and M_B.

Solution
From the free-body diagram of the large pulley (Fig. a), we obtain the equilibrium equation

$$M_A = R_A(T_2 - T_1), \tag{1}$$

and from the free-body diagram of the small pulley (Fig. b), we obtain

$$F = (T_1 + T_2) \cos \alpha, \tag{2}$$

$$M_B = R_B(T_2 - T_1). \tag{3}$$

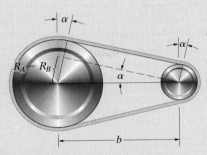

(c) Determining the angle α.

The belt is in contact with the small pulley through the angle $\pi - 2\alpha$ (Fig. c). From the dashed line parallel to the belt, we see that the angle α satisfies the relation

$$\sin \alpha = \frac{R_A - R_B}{b} = \frac{200 \text{ mm} - 100 \text{ mm}}{500 \text{ mm}} = 0.2.$$

Therefore, $\alpha = 11.5° = 0.201$ rad. If we assume that slip impends between the small pulley and the belt, Eq. (9.17) states that

$$T_2 = T_1 e^{\mu_s \beta} = T_1 e^{0.8(\pi - 2\alpha)} = 8.95 T_1.$$

We solve this equation together with Eq. (2) for the two tensions, obtaining $T_1 = 20.5$ N and $T_2 = 183.6$ N. Then from Eqs. (1) and (3), the couples are $M_A = 32.6$ N-m and $M_B = 16.3$ N-m.

  If we assume that slip impends between the large pulley and the belt, we obtain $M_A = 36.3$ N-m and $M_B = 18.1$ N-m, so the belt slips on the small pulley at smaller values of the couples.

Problems

▶ **9.123** In Active Example 9.9, suppose that the left fixed cylinder is replaced by a pulley. Assume that the tensions in the rope on each side of the pulley are approximately equal. What is the smallest force the woman needs to exert on the rope to support the stationary box?

9.124 Suppose that you want to lift a 50-lb crate off the ground by using a rope looped over a tree limb as shown. The coefficient of static friction between the rope and the limb is 0.2, and the rope is wound 135° around the limb. What force must you exert to begin lifting the crate?

Problem 9.124

9.125 *Winches* are used on sailboats to help support the forces exerted by the sails on the ropes (*sheets*) holding them in position. The winch shown is a post that will rotate in the clockwise direction (seen from above), but will not rotate in the counterclockwise direction. The sail exerts a tension $T_S = 800$ N on the sheet, which is wrapped two complete turns around the winch. The coefficient of static friction between the sheet and the winch is $\mu_s = 0.2$. What tension T_C must the crew member exert on the sheet to prevent it from slipping on the winch?

9.126 The coefficient of kinetic friction between the sheet and the winch in Problem 9.125 is $\mu_k = 0.16$. If the crew member wants to let the sheet slip at a constant rate, releasing the sail, what initial tension T_C must he exert on the sheet as it begins slipping?

Problems 9.125/9.126

9.127 The box *A* weighs 20 lb. The rope is wrapped one and one-fourth turns around the fixed wooden post. The coefficients of friction between the rope and post are $\mu_s = 0.15$ and $\mu_k = 0.12$.

(a) What minimum force does the man need to exert to support the stationary box?

(b) What force would the man have to exert to raise the box at a constant rate?

Problem 9.127

9.128 The weight of block *A* is *W*. The disk is supported by a smooth bearing. The coefficient of kinetic friction between the disk and the belt is μ_k. What couple *M* is necessary to turn the disk at a constant rate?

Problem 9.128

9.129 The couple required to turn the wheel of the exercise bicycle is adjusted by changing the weight W. The coefficient of kinetic friction between the wheel and the belt is μ_k. Assume the wheel turns clockwise.

(a) Show that the couple M required to turn the wheel is $M = WR\,(1 - e^{-3.4\mu_k})$.

(b) If $W = 40$ lb and $\mu_k = 0.2$, what force will the scale S indicate when the bicycle is in use?

9.131 The coefficient of static friction between the 50-lb box and the inclined surface is 0.10. The coefficient of static friction between the rope and the fixed cylinder is 0.05. Determine the force the woman must exert on the rope to cause the box to start moving up the inclined surface.

9.132 In Problem 9.131, what is the minimum force the woman must exert on the rope to hold the box in equilibrium on the inclined surface?

Problem 9.129

9.130 The box B weighs 50 lb. The coefficients of friction between the cable and the fixed round supports are $\mu_s = 0.4$ and $\mu_k = 0.3$.

(a) What is the minimum force F required to support the box?

(b) What force F is required to move the box upward at a constant rate?

Problems 9.131/9.132

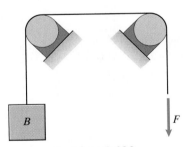

Problem 9.130

9.133 Blocks *B* and *C* each have a mass of 20 kg. The coefficient of static friction at the contacting surfaces is 0.2. Block *A* is suspended by a rope that passes over a fixed cylinder and is attached to block *B*. The coefficient of static friction between the rope and the cylinder is 0.3. What is the largest mass block *A* can have without causing block *B* to slip to the left?

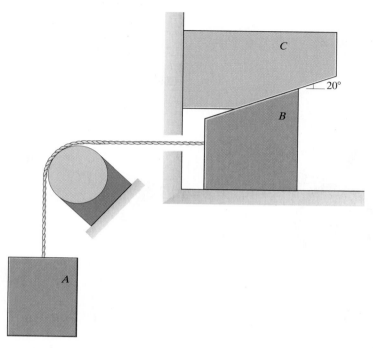

Problem 9.133

▶ **9.134** If the force *F* in Example 9.10 is increased to 400 N, what are the largest values of the couples M_A and M_B for which the belt will not slip?

9.135 The spring exerts a 320-N force on the left pulley. The coefficient of static friction between the flat belt and the pulleys is $\mu_s = 0.5$. The right pulley cannot rotate. What is the largest couple *M* that can be exerted on the left pulley without causing the belt to slip?

Problem 9.135

Review Problems

9.136 The weight of the box is $W = 30$ lb, and the force F is perpendicular to the inclined surface. The coefficient of static friction between the box and the inclined surface is $\mu_s = 0.2$.

(a) If $F = 30$ lb, what is the magnitude of the friction force exerted on the stationary box?

(b) If $F = 10$ lb, show that the box cannot remain at rest on the inclined surface.

9.137 In Problem 9.136, what is the smallest force F necessary to hold the box stationary on the inclined surface?

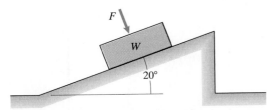

Problems 9.136/9.137

9.138 Blocks A and B are connected by a horizontal bar. The coefficient of static friction between the inclined surface and the 400-lb block A is 0.3. The coefficient of static friction between the surface and the 300-lb block B is 0.5. What is the smallest force F that will prevent the blocks from slipping down the surface?

9.139 What force F is necessary to cause the blocks in Problem 9.138 to start sliding up the plane?

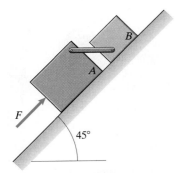

Problems 9.138/9.139

9.140 The masses of crates A and B are 25 kg and 30 kg, respectively. The coefficient of static friction between the contacting surfaces is $\mu_s = 0.34$. What is the largest value of α for which the crates will remain in equilibrium?

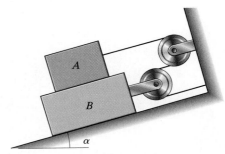

Problem 9.140

9.141 The side of a soil embankment has a 45° slope (Fig. a). If the coefficient of static friction of soil on soil is $\mu_s = 0.6$, will the embankment be stable or will it collapse? If it will collapse, what is the smallest slope that can be stable?

Strategy: Draw a free-body diagram by isolating part of the embankment as shown in Fig. b.

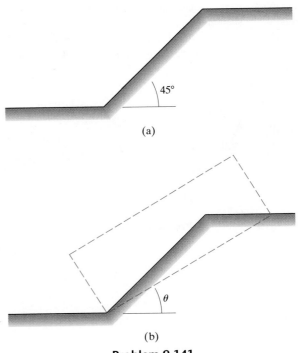

Problem 9.141

9.142 The mass of the van is 2250 kg, and the coefficient of static friction between its tires and the road is 0.6. If its front wheels are locked and its rear wheels can turn freely, what is the largest value of α for which it can remain in equilibrium?

9.143 In Problem 9.142, what is the largest value of α for which the van can remain in equilibrium if it points up the slope?

Problems 9.142/9.143

9.144 The shelf is designed so that it can be placed at any height on the vertical beam. The shelf is supported by friction between the two horizontal cylinders and the vertical beam. The combined weight of the shelf and camera is W. If the coefficient of static friction between the vertical beam and the horizontal cylinders is μ_s, what is the minimum distance b necessary for the shelf to stay in place?

Problem 9.144

9.145 The 20-lb homogeneous object is supported at A and B. The distance $h = 4$ in, friction can be neglected at B, and the coefficient of static friction at A is 0.4. Determine the largest force F that can be exerted without causing the object to slip.

9.146 In Problem 9.145, suppose that the coefficient of static friction at B is 0.36. What is the largest value of h for which the object will slip before it tips over?

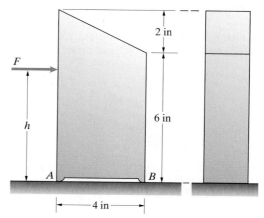

Problems 9.145/9.146

9.147 The 180-lb climber is supported in the "chimney" by the normal and friction forces exerted on his shoes and back. The static coefficients of friction between his shoes and the wall and between his back and the wall are 0.8 and 0.6, respectively. What is the minimum normal force his shoes must exert?

Problem 9.147

9.148 The sides of the 200-lb door fit loosely into grooves in the walls. Cables at A and B raise the door at a constant rate. The coefficient of kinetic friction between the door and the grooves is $\mu_k = 0.3$. What force must the cable at A exert to continue raising the door at a constant rate if the cable at B breaks?

Problem 9.148

9.149 The coefficients of static friction between the tires of the 1000-kg tractor and the ground and between the 450-kg crate and the ground are 0.8 and 0.3, respectively. Starting from rest, what torque must the tractor's engine exert on the rear wheels to cause the crate to move? (The front wheels can turn freely.)

9.150 In Problem 9.149, what is the most massive crate the tractor can cause to move from rest if its engine can exert sufficient torque? What torque is necessary?

Problems 9.149/9.150

9.151 The mass of the vehicle is 900 kg, it has rear-wheel drive, and the coefficient of static friction between its tires and the surface is 0.65. The coefficient of static friction between the crate and the surface is 0.4. If the vehicle attempts to pull the crate up the incline, what is the largest value of the mass of the crate for which it will slip up the incline before the vehicle's tires slip?

Problem 9.151

9.152 Each 1-m bar has a mass of 4 kg. The coefficient of static friction between the bar and the surface at B is 0.2. If the system is in equilibrium, what is the magnitude of the friction force exerted on the bar at B?

9.153 Each 1-m bar has a mass of 4 kg. What is the minimum coefficient of static friction between the bar and the surface at B necessary for the system to be in equilibrium?

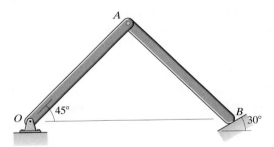

Problems 9.152/9.153

9.154 The collars A and B each have a mass of 2 kg. If friction between collar B and the bar can be neglected, what minimum coefficient of static friction between collar A and the bar is necessary for the collars to remain in equilibrium in the position shown?

9.155 If the coefficient of static friction has the same value μ_s between the 2-kg collars A and B and the bars, what minimum value of μ_s is necessary for the collars to remain in equilibrium in the position shown? (Assume that slip impends at A and B.)

Problems 9.154/9.155

9.156 The clamp presses two pieces of wood together. The pitch of the threads is $p = 2$ mm, the mean radius of the thread is $r = 8$ mm, and the coefficient of kinetic friction between the thread and the mating groove is 0.24. What couple must be exerted on the threaded shaft to press the pieces of wood together with a force of 200 N?

9.157 In Problem 9.156, the coefficient of static friction between the thread and the mating groove is 0.28. After the threaded shaft is rotated sufficiently to press the pieces of wood together with a force of 200 N, what couple must be exerted on the shaft to loosen it?

Problems 9.156/9.157

9.158 The axles of the tram are supported by journal bearings. The radius of the wheels is 75 mm, the radius of the axles is 15 mm, and the coefficient of kinetic friction between the axles and the bearings is $\mu_k = 0.14$. The mass of the tram and its load is 160 kg. If the weight of the tram and its load is evenly divided between the axles, what force P is necessary to push the tram at a constant speed?

Problem 9.158

9.159 The two pulleys have a radius of 6 in and are mounted on shafts of 1-in radius supported by journal bearings. Neglect the weights of the pulleys and shafts. The coefficient of kinetic friction between the shafts and the bearings is $\mu_k = 0.2$. If a force $T = 200$ lb is required to raise the man at a constant rate, what is his weight?

9.160 If the man in Problem 9.159 weighs 160 lb, what force T is necessary to lower him at a constant rate?

Problems 9.159/9.160

9.161 If the two cylinders are held fixed, what is the range of W for which the two weights will remain stationary?

9.162 If the system is initially stationary and the left cylinder is slowly rotated, determine the largest weight W that can be

(a) raised;

(b) lowered.

$\mu_s = 0.34$
$\mu_k = 0.32$

$\mu_s = 0.30$
$\mu_k = 0.28$

100 lb

W

Problems 9.161/9.162

Design Project 1

The wedge is used to split firewood by hammering it into a log as shown (see Active Example 9.4). Suppose that you want to design such a wedge to be marketed at hardware stores. Experiments indicate that the coefficient of static friction between the steel wedge and various types of wood varies from 0.2 to 0.4.

(a) Based on the given range of static friction coefficients, determine the maximum wedge angle α for which the wedge would remain in place in a log with no external force acting on it.

(b) Using the wedge angle determined in part (a), and assuming that the coefficient of kinetic friction is 0.9 times the coefficient of static friction, determine the range of vertical forces necessary to drive the wedge into a log at a constant rate.

(c) Write a brief report describing your analysis and recommending a wedge angle for the manufactured product. Consider whether a margin of safety in the chosen wedge angle might be appropriate.

Design Project 2

Design and build a device to measure the coefficient of static friction μ_s between two materials. Use it to measure μ_s for several of the materials listed in Table 9.1 and compare your results with the values in the table. Discuss possible sources of error in your device and determine how closely your values agree when you perform repeated experiments with the same two materials.

Internal Forces and Moments

We began our study of equilibrium by drawing free-body diagrams of individual objects to determine unknown forces and moments acting on them. In this chapter we carry this process one step further and draw free-body diagrams of parts of individual objects to determine internal forces and moments. In doing so, we arrive at the central concern of the structural design engineer: It is the forces within an object that determine whether it will support the external loads to which it is subjected.

◀ The force exerted by the water on the glass window is distributed over the area of the window. In this chapter we analyze distributed forces in beams, in suspended cables, and in stationary liquids.

BEAMS

10.1 Axial Force, Shear Force, and Bending Moment

To ensure that a structural member will not fail (break or collapse) due to the forces and moments acting on it, the design engineer must know not only the external loads and reactions acting on the member, but also the forces and moments acting *within* it.

Consider a beam subjected to an external load and reactions (Fig. 10.1a). How can we determine the forces and moments within the beam? In Fig. 10.1b, we "cut" the beam by a plane at an arbitrary cross section and isolate the part of the beam to the left of the plane. It is clear that the isolated part cannot be in equilibrium unless it is subjected to some system of forces and moments at the plane where it joins the other part of the beam. These are the internal forces and moments we seek.

In Chapter 4 we demonstrated that *any* system of forces and moments can be represented by an equivalent system consisting of a force and a couple. Since the system of external loads and reactions on the beam is two-dimensional, we can represent the internal forces and moments by an equivalent system consisting of two components of force and a couple (Fig. 10.1c). The component *P* parallel to the beam's axis is called the *axial force*. The component *V* normal to the beam's axis is called the *shear force*, and the couple *M* is called the

(a)

(b)

(c)

(d)

Figure 10.1
(a) A beam subjected to a load and reactions.
(b) Isolating a part of the beam.
(c), (d) The axial force, shear force, and bending moment.

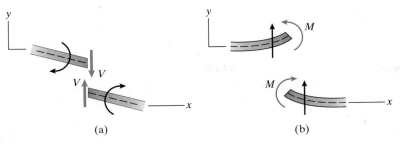

Figure 10.2
(a) Positive shear forces tend to rotate the axis of the beam clockwise.
(b) Positive bending moments tend to bend the axis of the beam upward.

bending moment. The axial force, shear force, and bending moment on the part of the beam to the right of the cutting plane are shown in Fig. 10.1d. Notice that they are equal in magnitude but opposite in direction to the internal forces and moment on the free-body diagram in Fig. 10.1c.

The directions of the axial force, shear force, and bending moment in Figs. 10.1c and 10.1d are the established definitions of the positive directions of these quantities. A positive axial force P subjects the beam to tension. A positive shear force V tends to rotate the axis of the beam clockwise (Fig. 10.2a). A positive bending moment M tends to cause upward curvature of the beam's axis (Fig. 10.2b). Notice that a positive bending moment subjects the upper part of the beam to compression, shortening the beam in the direction parallel to its axis, and subjects the lower part of the beam to tension, lengthening the beam in the direction parallel to its axis.

RESULTS

The axial force P, shear force V, and bending moment M are an equivalent system representing the internal forces and moment at a cross section of a beam. *These are their defined positive directions.*

Determining the values of P, V, and M at a particular cross section of a beam typically involves three steps:
1. Draw the free-body diagram of the beam and determine the reactions at its supports.
2. Pass a plane through the beam at the cross section where the internal forces and moment are to be determined. Draw the free-body diagram of one of the resulting parts of the beam, showing P, V, and M in their defined positive directions.
3. Apply the equilibrium equations to determine P, V, and M.

Active Example 10.1 **Determining the Internal Forces and Moment** (▶ *Related Problem 10.1*)

Determine the internal forces and moment at C.

Strategy

We must first draw the free-body diagram of the entire beam and determine the reactions at A and B. Then we will cut the beam by a plane at C and draw the free-body diagram of the part of the beam to the left of the plane. By applying the equilibrium equations, we will obtain the values of the internal forces and moment at C.

Solution

$$\Sigma F_x = A_x = 0,$$

$$\Sigma F_y = A_y + B - F = 0,$$

$$\Sigma M_{\text{point } A} = LB - \left(\frac{3}{4}L\right)F = 0.$$

Solving yields $A_x = 0$, $A_y = \frac{1}{4}F$, and $B = \frac{3}{4}F$.

> Draw the free-body diagram of the entire beam and apply equilibrium to determine the reactions at the supports.

$$\Sigma F_x = P_C = 0,$$

$$\Sigma F_y = \frac{1}{4}F - V_C = 0,$$

$$\Sigma M_{\text{point } C} = M_C - \left(\frac{1}{4}L\right)\left(\frac{1}{4}F\right) = 0.$$

Solving yields $P_C = 0$, $V_C = \frac{1}{4}F$, and $M_C = \frac{1}{16}LF$.

> Pass a plane through the beam at C. Draw the free-body diagram of the part of the beam to the left of C. Apply equilibrium to determine the internal forces and moment.

Practice Problem Determine the internal forces and moment at C by passing a plane through the beam at C and drawing the free-body diagram of the part of the beam to the right of C.

Answer: $P_C = 0$, $V_C = \frac{1}{4}F$, $M_C = \frac{1}{16}LF$.

| **Example 10.2** | **Determining the Internal Forces and Moment** (▶ *Related Problem 10.8*) |

Determine the internal forces and moment at B.

Strategy
To determine the reactions at the supports, we will represent the triangular distributed load by an equivalent force. Then we will determine the internal forces and moment at B by cutting the beam by a plane at B and drawing the free-body diagram of the part of the beam to the left of the plane, *including the part of the distributed load to the left of the plane.*

Solution
Determine the External Forces and Moments   We draw the free-body diagram of the beam and represent the distributed load by an equivalent force in Fig. a. The equilibrium equations are

$$\Sigma F_x = A_x = 0,$$

$$\Sigma F_y = A_y + C - 180 \text{ N} = 0$$

$$\Sigma M_{\text{point } A} = (12 \text{ m})C - (4 \text{ m})(180 \text{ N}) = 0$$

Solving them, we obtain $A_x = 0$, $A_y = 120$ N, and $C = 60$ N.

Draw the Free-Body Diagram of Part of the Beam   We cut the beam at B, obtaining the free-body diagram in Fig. b. Because point B is at the midpoint of the triangular distributed load, the value of the distributed load at B is 30 N/m. By representing the distributed load in Fig. b by an equivalent force, we obtain the free-body diagram in Fig. c. From the equilibrium equations

$$\Sigma F_x = P_B = 0,$$

$$\Sigma F_y = 120 \text{ N} - 45 \text{ N} - V_B = 0,$$

$$\Sigma M_{\text{point } B} = M_B + (1 \text{ m})(45 \text{ N}) - (3 \text{ m})(120 \text{ N}) = 0,$$

we obtain $P_B = 0$, $V_B = 75$ N, and $M_B = 315$ N-m.

Critical Thinking
If you attempt to determine the internal forces and moment at B by cutting the free-body diagram in Fig. a at B, you do *not* obtain correct results. (You can confirm that the resulting free-body diagram of the part of the beam to the left of B gives $P_B = 0$, $V_B = 120$ N, and $M_B = 360$ N-m.) The reason is that you do not properly account for the effect of the distributed load on your free-body diagram. You must wait until *after* you have isolated part of the beam before representing distributed loads acting on that part by equivalent forces.

(a) Free-body diagram of the entire beam with the distributed load represented by an equivalent force.

(b), (c) Free-body diagram of the part of the beam to the left of B.

Problems

▶ **10.1** In Active Example 10.1, suppose that the distance from point A to point C is increased from $\frac{1}{4}L$ to $\frac{1}{2}L$. Draw a sketch of the beam with C in its new position. Determine the internal forces and moment at C.

10.2 The magnitude of the triangular distributed load is $w_0 = 2$ kN/m. Determine the internal forces and moment at A.

Problem 10.2

10.3 The C clamp exerts 30-lb forces on the clamped object. Determine the internal forces and moment in the clamp at A.

Problem 10.3

10.4 Determine the internal forces and moment at A.

Problem 10.4

10.5 The pipe has a fixed support at the left end. Determine the internal forces and moment at A.

Problem 10.5

10.6 Determine the internal forces and moment at A for each loading.

(a)

(b)

Problem 10.6

10.7 Model the ladder rung as a simply supported (pin supported) beam and assume that the 750-N load exerted by the person's shoe is uniformly distributed. Determine the internal forces and moment at A.

250 mm A

200 mm   100 mm

375 mm

Problem 10.7

▶ **10.8** In Example 10.2, suppose that the distance from point A to point B is increased from 3 m to 4 m. Draw a sketch of the beam with B in its new position. Determine the internal forces and moment at B.

10.9 If $x = 3$ ft, what are the internal forces and moment at A?

10.10 If $x = 4$ ft, what are the internal forces and moment at A?

600 lb/ft

600 lb/ft

3 ft   3 ft

Problems 10.9/10.10

10.11 Determine the internal forces and moment at A for the loadings (a) and (b).

10.12 Determine the internal forces and moment at B for the loadings (a) and (b).

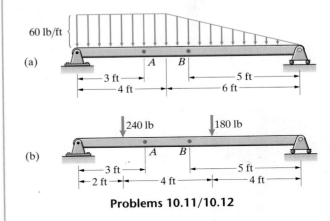

60 lb/ft

(a) A B

3 ft   5 ft

4 ft   6 ft

240 lb   180 lb

(b) A B

3 ft   5 ft

2 ft   4 ft   4 ft

Problems 10.11/10.12

10.13 Determine the internal forces and moment at A.

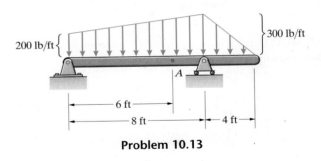

200 lb/ft   300 lb/ft

A

6 ft

8 ft   4 ft

Problem 10.13

10.14 Determine the internal forces and moment at A.

10.15 Determine the internal forces and moment at B.

10 kN

A

B

1 m   1 m   1 m   1 m   1 m

Problems 10.14/10.15

10.16 Determine the internal forces and moment at *A*.

10.17 Determine the internal forces and moment at *B*.

Problems 10.16/10.17

10.18 The tension in the rope is 10 kN. Determine the internal forces and moment at point *A*.

Problem 10.18

10.19 Determine the internal forces and moment at point *A* of the frame.

Problem 10.19

10.20 Determine the internal forces and moment at *A*.

10.21 Determine the internal forces and moment at *B*.

Problems 10.20/10.21

(a)   (b)   (c)

Figure 10.3
(a) A beam loaded by a force F and its free-body diagram.
(b) Cutting the beam at an arbitrary position x to the left of F.
(c) Cutting the beam at an arbitrary position x to the right of F.

10.2 Shear Force and Bending Moment Diagrams

BACKGROUND

To design a beam, an engineer must know the internal forces and moments throughout its length. Of special concern are the maximum and minimum values of the shear force and bending moment and where they occur. In this section we show how the values of P, V, and M can be determined as functions of x and introduce shear force and bending moment diagrams.

Consider a simply supported beam loaded by a force (Fig. 10.3a). Instead of cutting the beam at a specific cross section to determine the internal forces and moment, we cut it at an arbitrary position x between the left end of the beam and the load F (Fig. 10.3b). Applying the equilibrium equations to this free-body diagram, we obtain

$$\left.\begin{aligned} P &= 0 \\ V &= \frac{1}{3}F \\ M &= \frac{1}{3}Fx \end{aligned}\right\} \quad 0 < x < \frac{2}{3}L.$$

To determine the internal forces and moment for values of x greater than $\frac{2}{3}L$, we obtain a free-body diagram by cutting the beam at an arbitrary position x between the load F and the right end of the beam (Fig. 10.3c). The results are

$$\left.\begin{aligned} P &= 0 \\ V &= -\frac{2}{3}F \\ M &= \frac{2}{3}F(L - x) \end{aligned}\right\} \quad \frac{2}{3}L < x < L.$$

The *shear force and bending moment diagrams* are simply the graphs of V and M, respectively, as functions of x (Fig. 10.4). They permit you to see the changes in the shear force and bending moment that occur along the beam's length as well as their maximum and minimum values. (By *maximum* we mean the least upper bound of the shear force or bending moment, and by *minimum* we mean the greatest lower bound.)

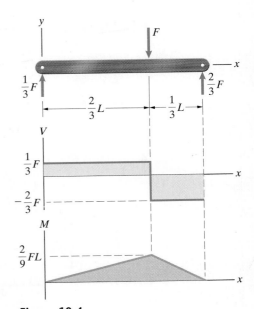

Figure 10.4
The shear force and bending moment diagrams indicating the maximum and minimum values of V and M.

Thus we can determine the distributions of the internal forces and moment in a beam by considering a plane at an arbitrary distance x from the end of the beam and solving for P, V, and M as functions of x. Depending on the complexity of the loading, it may be necessary to draw several free-body diagrams to determine the distributions over the entire length of the beam. The resulting equations for V and M allow us to draw the shear force and bending moment diagrams.

RESULTS

By passing a plane through a beam at an arbitrary position x, the values of P, V, and M can be determined as functions of x. Depending on the loading and supports, it may be necessary to draw several free-body diagrams to determine the distributions for the entire beam.

Shear force and bending moment diagrams for a beam are simply the graphs of V and M as functions of x.

Active Example 10.3   Shear Force and Bending Moment Diagrams (▶ *Related Problem 10.27*)

Determine the shear force V and bending moment M for the beam as functions of x for $0 < x < 2$ m.

Strategy
We must first draw the free-body diagram of the entire beam and determine the reactions at A and B. Then we will cut the beam by a plane at an arbitrary position x between A and B to obtain functions for V and M that are valid in the range $0 < x < 2$ m.

Solution

Draw the free-body diagram of the entire beam. The distributed load is represented by an equivalent force.

$$\Sigma F_x = B_x = 0,$$
$$\Sigma F_y = A + B_y - 80 \text{ kN} + 60 \text{ kN} = 0,$$
$$\Sigma M_{\text{point } A} = (2 \text{ m})B_y - (1 \text{ m})(80 \text{ kN}) + (4 \text{ m})(60 \text{ kN}) = 0.$$
Solving yields $A = 100$ kN, $B_x = 0$, and $B_y = -80$ kN.

> Apply equilibrium to determine the reactions at A and B.

> Pass a plane through the beam at an arbitrary position x between A and B. The distributed load *must not* be represented by an equivalent force before isolating part of the beam.

$$\Sigma F_x = P = 0,$$
$$\Sigma F_y = 100 - 40x - V = 0,$$
$$\Sigma M_{\text{right end}} = M - 100x + \left(\frac{1}{2}x\right)(40x) = 0.$$

Solving yields
$$V = 100 - 40x \text{ kN}$$
$$M = 100x - 20x^2 \text{ kN-m}$$
$$\left. \right\} \quad 0 < x < 2 \text{ m.}$$

> Apply equilibrium to determine V and M.

Practice Problem (a) Determine the shear force V and bending moment M for the beam as functions of x for $2 < x < 4$ m. (b) Draw the shear force and bending moment diagrams for the entire beam.

Answer: $V = -60$ kN, $M = 60(4 - x)$ kN-m.

Problems

10.22 Determine the shear force and bending moment as functions of x.

Strategy: Cut the beam at an arbitrary position x and draw the free-body diagram of the part of the beam to the left of the plane.

Problem 10.22

10.23 (a) Determine the shear force and bending moment as functions of x.

(b) Draw the shear force and bending moment diagrams.

Problem 10.23

10.24 (a) Determine the shear force and bending moment as functions of x.

(b) Show that the equations for V and M as functions of x satisfy the equation $V = dM/dx$.

Strategy: For part (a), cut the beam at an arbitrary position x and draw the free-body diagram of the part of the beam to the right of the plane.

10.25 Draw the shear force and bending moment diagrams.

Problems 10.24/10.25

10.26 Determine the shear force and bending moment as functions of x for $0 < x < 2$ m.

Problem 10.26

▶ **10.27** In Active Example 10.3, suppose that the 40 kN/m distributed load extends all the way across the beam from A to C. Draw a sketch of the beam with its new loading. Determine the shear force V and bending moment M for the beam as functions of x for $2 < x < 4$ m.

10.28 (a) Determine the internal forces and moment as functions of x.

(b) Draw the shear force and bending moment diagrams.

Problem 10.28

10.29 The loads $F = 200$ N and $C = 800$ N-m.

(a) Determine the internal forces and moment as functions of x.

(b) Draw the shear force and bending moment diagrams.

10.30 The beam will safely support shear forces and bending moments of magnitudes 2 kN and 6.5 kN-m, respectively. On the basis of this criterion, can it safely be subjected to the loads $F = 1$ kN, $C = 1.6$ kN-m?

Problems 10.29/10.30

10.31 Model the ladder rung as a simply supported (pin-supported) beam and assume that the 750-N load exerted by the person's shoe is uniformly distributed. Draw the shear force and bending moment diagrams.

10.32 What is the maximum bending moment in the ladder rung in Problem 10.31 and where does it occur?

Problems 10.31/10.32

10.33 Assume that the surface the beam rests on exerts a uniformly distributed load. Draw the shear force and bending moment diagrams.

Problem 10.33

10.34 The homogeneous beams AB and CD weigh 600 lb and 500 lb, respectively. Draw the shear force and bending moment diagrams for beam AB.

10.35 The homogeneous beams AB and CD weigh 600 lb and 500 lb, respectively. Draw the shear force and bending moment diagrams for beam CD.

Problems 10.34/10.35

10.36 Determine the shear force V and bending moment M for the beam as functions of x for $0 < x < 3$ ft.

10.37 Draw the shear force and bending moment diagrams for the beam.

Problems 10.36/10.37

10.38 In preliminary design studies, the vertical forces on an airplane's wing are modeled as shown. The distributed load models aerodynamic forces and the force exerted by the wing's weight. The 80-kN force at $x = 4.4$ m models the force exerted by the weight of the engine. Draw the shear force and bending moment diagrams for the wing for $0 < x < 4.4$ m.

10.39 Draw the shear force and bending moment diagrams for the entire wing.

Problems 10.38/10.39

10.40* Draw the shear force and bending moment diagrams.

Problem 10.40

10.3 Relations Between Distributed Load, Shear Force, and Bending Moment

BACKGROUND

The shear force and bending moment in a beam subjected to a distributed load are governed by simple differential equations. In this section we derive these equations and show that they provide an interesting and enlightening way to obtain shear force and bending moment diagrams. These equations are also useful for determining deformations of beams.

Suppose that a portion of a beam is subjected to a distributed load w (Fig. 10.5a). In Fig. 10.5b we obtain a free-body diagram by cutting the beam at positions x and $x + \Delta x$. The terms ΔP, ΔV, and ΔM are the changes in the axial force, shear force, and bending moment, respectively, from x to $x + \Delta x$. The sum of the forces in the x direction is

$$\Sigma F_x = P + \Delta P - P = 0.$$

Dividing this equation by Δx and taking the limit as $\Delta x \to 0$, we obtain

$$\frac{dP}{dx} = 0,$$

which simply states that the axial force does not depend on x in a portion of a beam subjected only to a lateral distributed load. To sum the forces on the free-body diagram in the y direction, we must determine the force exerted by the distributed load. In Fig. 10.5b we introduce a coordinate $\hat{x}$ that measures distance from the left edge of the free-body diagram. In terms of this coordinate, the downward force exerted on the free-body diagram by the distributed load is

$$\int_0^{\Delta x} w(x + \hat{x}) \, d\hat{x},$$

where $w(x + \hat{x})$ denotes the value of w at $x + \hat{x}$. To evaluate this integral, we express $w(x + \hat{x})$ as a Taylor series in terms of $\hat{x}$:

$$w(x + \hat{x}) = w(x) + \frac{dw(x)}{dx}\hat{x} + \frac{1}{2}\frac{d^2w(x)}{dx^2}\hat{x}^2 + \cdots. \quad (10.1)$$

Substituting this equation into the integral expression for the downward force and integrating term by term, we obtain

$$\int_0^{\Delta x} w(x + \hat{x}) \, d\hat{x} = w(x)\Delta x + \frac{1}{2}\frac{dw(x)}{dx}(\Delta x)^2 + \cdots.$$

The sum of the forces on the free-body diagram in the y direction is therefore

$$\Sigma F_y = V - V - \Delta V - w(x)\Delta x - \frac{1}{2}\frac{dw(x)}{dx}(\Delta x)^2 + \cdots = 0.$$

Dividing by Δx and taking the limit as $\Delta x \to 0$, we obtain

$$\frac{dV}{dx} = -w, \quad (10.2)$$

where $w = w(x)$.

We now want to sum the moments about point Q on the free-body diagram in Fig. 10.5b. The clockwise moment about Q due to the distributed load is

$$\int_0^{\Delta x} \hat{x}w(x + \hat{x}) \, d\hat{x}.$$

Substituting Eq. (10.1) and integrating term by term, the clockwise moment about Q is

$$\int_0^{\Delta x} \hat{x}w(x + \hat{x}) \, d\hat{x} = \frac{1}{2}w(x)(\Delta x)^2 + \frac{1}{3}\frac{dw(x)}{dx}(\Delta x)^3 + \cdots.$$

The sum of the moments about Q is therefore

$$\Sigma M_{\text{point } Q} = M + \Delta M - M - (V + \Delta V)\,\Delta x$$

$$-\frac{1}{2}w(x)(\Delta x)^2 - \frac{1}{3}\frac{dw(x)}{dx}(\Delta x)^3 + \cdots = 0.$$

(a)

(b)

Figure 10.5
(a) A portion of a beam subjected to a distributed force w.
(b) Obtaining the free-body diagram of an element of the beam.

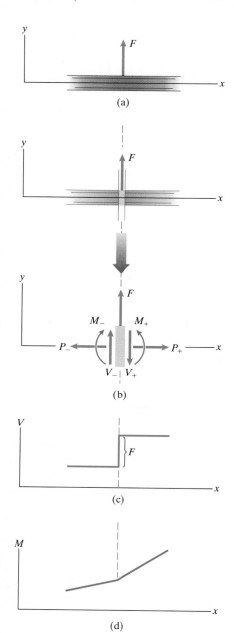

Figure 10.6
(a) A portion of a beam subjected to a distributed force F in the positive y direction.
(b) Obtaining a free-body diagram by cutting the beam to the left and right of F.
(c) The shear force diagram undergoes a positive jump of magnitude F.
(d) The bending moment diagram is continuous.

Dividing by Δx and taking the limit as $\Delta x \rightarrow 0$ gives

$$\frac{dM}{dx} = V. \tag{10.3}$$

In principle, we can use Eqs. (10.2) and (10.3) to determine the shear force and bending moment diagrams for a beam. Equation (10.2) can be integrated to determine V as a function of x, then Eq. (10.3) can be integrated to determine M as a function of x. However, we derived these equations for a segment of beam subjected only to a distributed load. To apply them for a more general loading, we must account for the effects of any point forces and couples acting on the beam.

Let us determine what happens to the shear force and bending moment diagrams where a beam is subjected to a force F in the positive y direction (Fig. 10.6a). By cutting the beam just to the left and just to the right of the force, we obtain the free-body diagram in Fig. 10.6b, where the subscripts $-$ and $+$ denote values to the left and right of the force, respectively. Equilibrium requires that

$$V_+ - V_- = F,$$
$$M_+ - M_- = 0.$$

The shear force diagram undergoes a jump discontinuity of magnitude F (Fig. 10.6c), but the bending moment diagram is continuous (Fig. 10.6d). The jump in the shear force is positive if the force is in the positive y direction.

Now we consider what happens to the shear force and bending moment diagrams when a beam is subjected to a counterclockwise couple C (Fig. 10.7a). Cutting the beam just to the left and just to the right of the couple (Fig. 10.7b), we determine that

$$V_+ - V_- = 0,$$
$$M_+ - M_- = -C.$$

The shear force diagram is continuous (Fig. 10.7c), but the bending moment diagram undergoes a jump discontinuity of magnitude C (Fig. 10.7d), where a beam is subjected to a couple. The jump in the bending moment is *negative* if the couple is in the counterclockwise direction.

We now have the results needed to construct shear force and bending moment diagrams.

Construction of the Shear Force Diagram

In a segment of a beam that is subjected only to a distributed load, we have shown that the shear force is related to the distributed load by

$$\frac{dV}{dx} = -w. \tag{10.4}$$

This equation states that the derivative, or slope, of the shear force with respect to x is equal to the negative of the distributed load. Notice that if there is no distributed load ($w = 0$) throughout the segment, the slope is zero and the shear force is constant. If w is a constant throughout the segment, the slope of the shear force is constant, which means that the shear force diagram for the segment is a straight line. Integrating Eq. (10.4) with respect to x from a position x_A to a position x_B,

$$\int_{x_A}^{x_B} \frac{dV}{dx} dx = -\int_{x_A}^{x_B} w \, dx,$$

yields

$$V_B - V_A = -\int_{x_A}^{x_B} w \, dx.$$

The change in the shear force between two positions is equal to the negative of the area defined by the loading curve between those positions (Fig. 10.8):

$$V_B - V_A = -(\text{area defined by the distributed load from } x_A \text{ to } x_B). \quad (10.5)$$

Where a beam is subjected to a point force of magnitude F in the positive y direction, we have shown that the shear force diagram undergoes an increase of magnitude F. Where a beam is subjected to a couple, the shear force diagram is unchanged (continuous).

Let us demonstrate these results by determining the shear force diagram for the beam in Fig. 10.9. The beam is subjected to a downward force F that results in upward reactions at A and C. Notice that there is no distributed load. Our procedure is to begin at the left end of the beam and construct the diagram from left to right. Figure 10.10a shows the increase in the value of V due to the upward reaction at A. Because there is no distributed load, the value of V remains constant between A and B (Fig. 10.10b). At B, the value of V decreases due to the downward force (Fig. 10.10c). The value of V remains constant between B and C, which completes the shear force diagram (Fig. 10.10d). Compare Fig. 10.10d with the shear force diagram we obtained in Fig. 10.4 by drawing free-body diagrams and applying the equilibrium equations.

Construction of the Bending Moment Diagram

In a segment of a beam subjected only to a distributed load, the bending moment is related to the shear force by

$$\frac{dM}{dx} = V, \quad (10.6)$$

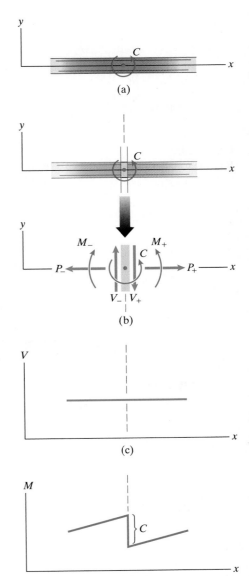

Figure 10.7
(a) A portion of a beam subjected to a counterclockwise couple C.
(b) Obtaining a free-body diagram by cutting the beam to the left and right of C.
(c) The shear force diagram is continuous.
(d) The bending moment diagram undergoes a *negative* jump of magnitude C.

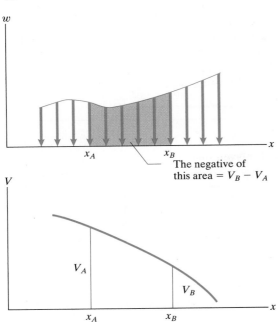

Figure 10.8
The change in the shear force is equal to the negative of the area defined by the loading curve.

Figure 10.9
Beam loaded by a force F and its free-body diagram.

which states that the slope of the bending moment with respect to x is equal to the shear force. If V is constant throughout the segment, the bending moment diagram for the segment is a straight line. Integrating Eq. (10.6) with respect to x from a position x_A to a position x_B yields

$$M_B - M_A = \int_{x_A}^{x_B} V \, dx.$$

The change in the bending moment between two positions is equal to the area defined by the shear force diagram between those positions (Fig. 10.11):

$$M_B - M_A = \text{area defined by the shear force from } x_A \text{ to } x_B. \qquad (10.7)$$

Where a beam is subjected to a counterclockwise couple of magnitude C, the bending moment diagram undergoes a decrease of magnitude C. Where a beam is subjected to a point force, the bending moment diagram is unchanged.

As an example, we will determine the bending moment diagram for the beam in Fig. 10.9. We begin with the shear force diagram we have already determined (Fig. 10.12a) and proceed to construct the bending moment diagram from left to right. The beam is not subjected to a couple at A, so $M_A = 0$. Between A and B, the slope of the bending moment is constant

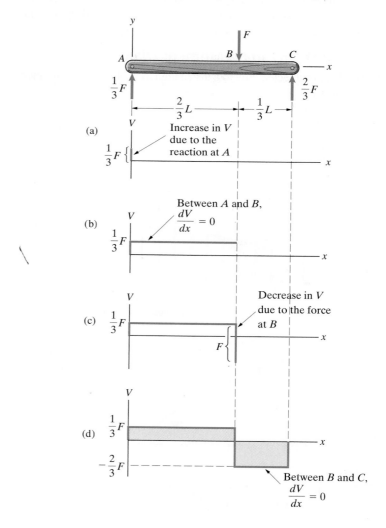

Figure 10.10
Constructing the shear force diagram for the beam in Fig. 10.9.

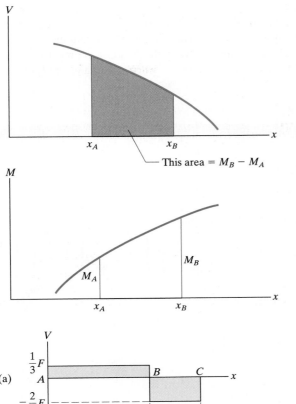

Figure 10.11
The change in the bending moment is equal to the area defined by the shear force diagram.

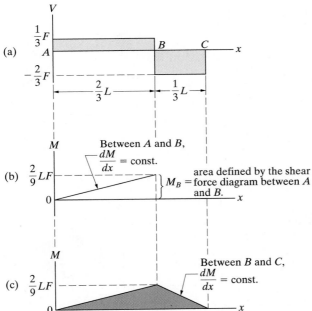

Figure 10.12
Constructing the bending moment diagram for the beam in Fig. 10.9.

$(dM/dx = V = F/3)$, which tells us that the bending moment diagram between A and B is a straight line (10.12b). The change in the bending moment from A to B is equal to the area defined by the shear force from A to B:

$$M_B - M_A = (\tfrac{2}{3}L)(\tfrac{1}{3}F) = \tfrac{2}{9}LF.$$

Therefore, $M_B = 2LF/9$. The slope of the bending moment is also constant between B and C $(dM/dx = V = -2F/3)$, so the bending moment diagram between B and C is a straight line. The change in the bending moment from B to C is equal to the area defined by the shear force from B to C, or

$$M_C - M_B = (\tfrac{1}{3}L)(-\tfrac{2}{3}F) \doteq -\tfrac{2}{9}LF,$$

from which we obtain $M_C = M_B - 2LF/9 = 0$. (Notice that we did not actually need this calculation to conclude that $M_C = 0$, because the beam is not subjected to a couple at C.) The completed bending moment diagram is shown in Fig. 10.12c. Compare it with the bending moment diagram we obtained in Fig. 10.4 by drawing free-body diagrams and applying the equilibrium equations.

RESULTS

In a segment of a beam that is subjected *only* to a distributed load, the shear force is related to the distributed load by

$$\frac{dV}{dx} = -w. \qquad (10.4)$$

The change in the shear force between two positions is equal to the negative of the area defined by the loading curve between those positions.

$$V_B - V_A = -\left(\begin{array}{c}\text{area defined by } w \\ \text{from } x_A \text{ to } x_B\end{array}\right) \qquad (10.5)$$

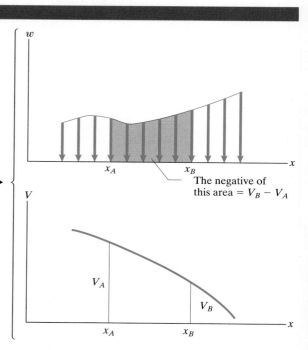

In a segment of a beam that is subjected *only* to a distributed load, the bending moment is related to the shear force by

$$\frac{dM}{dx} = V. \qquad (10.6)$$

The change in the bending moment between two positions is equal to the area defined by the shear force between those positions.

$$M_B - M_A = \left(\begin{array}{c}\text{area defined by } V \\ \text{from } x_A \text{ to } x_B\end{array}\right) \qquad (10.7)$$

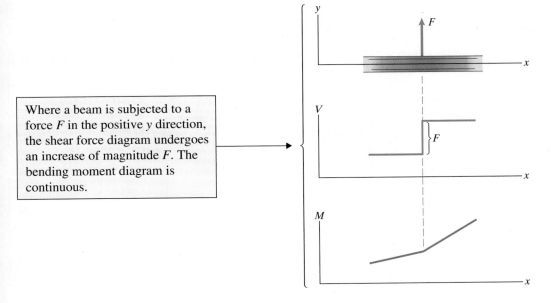

Where a beam is subjected to a force F in the positive y direction, the shear force diagram undergoes an increase of magnitude F. The bending moment diagram is continuous.

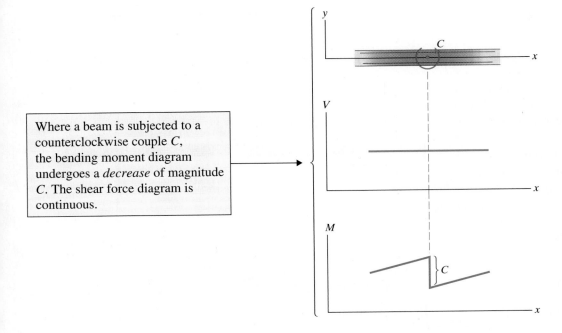

Where a beam is subjected to a counterclockwise couple C, the bending moment diagram undergoes a *decrease* of magnitude C. The shear force diagram is continuous.

Active Example 10.4

Shear Force and Bending Moment Diagrams Using Eqs. (10.4)–(10.7) (▶ *Related Problem 10.45*)

Use Eqs. (10.4) and (10.5) to determine the shear force diagram for the beam.

Strategy

The first step is to draw the free-body diagram of the beam and determine the reactions at A and B. This was done in Active Example 10.3. We will begin at the left end of the beam and progress toward the right constructing the shear force diagram.

Solution

It is helpful to think of beginning just to the *left* of the left end of the beam, with the initial value of the shear force equal to zero. The upward 100-kN reaction at A causes an increase in the shear force of 100-kN magnitude.

Between A and B, the distributed load on the beam is constant. From Eq. (10.4), that means the slope of the shear diagram between A and B is constant—the diagram is a straight line. The change in V between A and B can be determined from Eq. (10.5).

$$V_B - V_A = -(2\text{ m})(40\text{ kN/m})$$
$$= -80\text{ kN.}$$

Therefore V decreases linearly from 100 kN at A to 100 kN − 80 kN = 20 kN at B. This result can also be obtained by integrating Eq. (10.4):

$$\int_{100}^{V} dV = \int_{0}^{x} -40\,dx$$
$$V = 100 - 40x \text{ kN.}$$

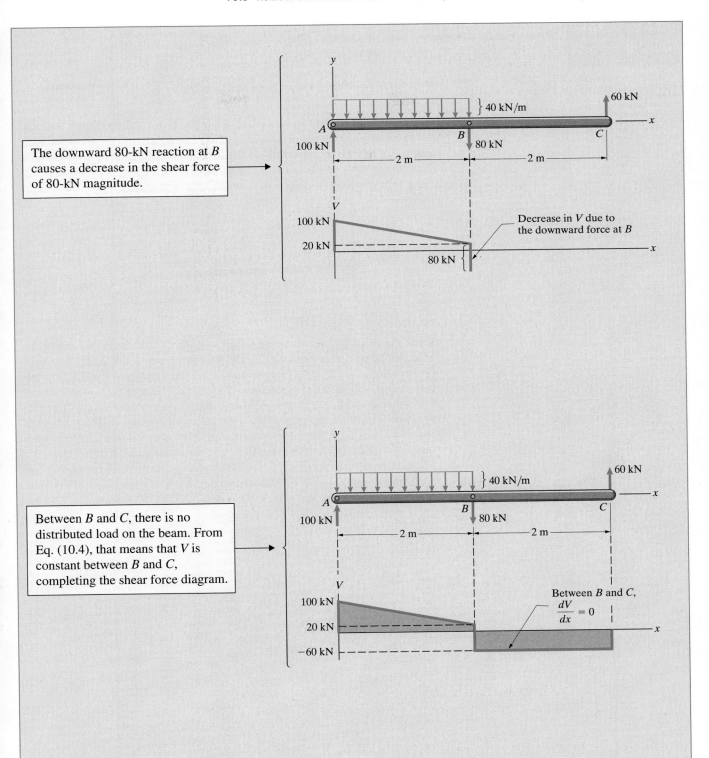

The downward 80-kN reaction at B causes a decrease in the shear force of 80-kN magnitude.

Decrease in V due to the downward force at B

Between B and C, there is no distributed load on the beam. From Eq. (10.4), that means that V is constant between B and C, completing the shear force diagram.

Between B and C,
$$\frac{dV}{dx} = 0$$

Practice Problem Use Eqs. (10.6) and (10.7) to determine the bending moment diagram for the beam.

Example 10.5	Shear Force and Bending Moment Diagrams Using Eqs. (10.4)–(10.7) (▶ *Related Problem 10.44*)

Determine the shear force and bending moment diagrams for the beam.

Strategy

We can begin with the free-body diagram of the beam and use Eqs. (10.4) and (10.5) to construct the shear force diagram. Then we can use the shear force diagram and Eqs. (10.6) and (10.7) to construct the bending moment diagram. In determining both the shear force and bending moment diagrams, we must account for the effects of point forces and couples acting on the beam.

Solution

Shear Force Diagram The first step is to draw the free-body diagram of the beam and determine the reactions at the built-in support A. Using the results of this step, shown in Fig. a, we proceed to construct the shear force diagram from left to right. Figure b shows the increase in the value of V due to the upward force at A. Between A and B, the distributed load on the beam increases linearly from 0 to 300 N/m. Therefore, the slope of the shear force diagram decreases linearly from 0 to −300 N/m. At B, the shear force must be 0, because no force acts there. With this information, we can sketch the shear force diagram qualitatively (Fig. c).

We can also obtain an explicit equation for the shear force between A and B by integrating Eq. (10.4). The distributed load as a function of x is $w = (x/6)300 = 50x$ N/m. We write Eq. (10.4) as

$$dV = -w \, dx = -50x \, dx$$

and integrate to determine V at an arbitrary position x:

$$\int_{V_A}^{V} dV = \int_{0}^{x} - 50x \, dx$$

$$V - V_A = -25x^2.$$

Due to the 900-N upward reaction at A, $V_A = 900$ N, so we obtain

$$V = 900 - 25x^2 \text{ N}. \tag{1}$$

Bending Moment Diagram   We construct the bending moment diagram from left to right. Figure d shows the initial decrease in the value of M due to the counterclockwise couple at A. Between A and B, the slope of the bending moment diagram is equal to the shear force V. We see from the shear force diagram (Fig. c) that at A, the slope of the bending moment diagram has a positive value (900 N). As x increases, the slope begins to decrease, and its rate of decrease grows until the value of the slope reaches zero at B. At B, we know that the value of the bending moment is zero, because no couple acts on the beam at B. Using this information, we can sketch the bending moment diagram qualitatively (Fig. e). Notice that its slope decreases from a positive value at A to zero at B, and the rate at which it decreases grows as x increases.

We can obtain an equation for the bending moment between A and B by integrating Eq. (10.6). The shear force as a function of x is given by Eq. (1). We write Eq. (10.6) as

$$dM = V \, dx = (900 - 25x^2) \, dx$$

and integrate:

$$\int_{M_A}^{M} dM = \int_{0}^{x} (900 - 25x^2) \, dx$$

$$M - M_A = 900x - \frac{25}{3}x^3.$$

As a result of the 3600 N-m counterclockwise couple at A, $M_A = -3600$ N-m, yielding the bending moment distribution

$$M = -3600 + 900x - \frac{25}{3}x^3 \text{ N-m}.$$

(d)   3600 N-m   Decrease in M due to the couple at A

(e)

-3600 N-m

$M_B = 0$

6 m

Between A and B,

$$\frac{dM}{dx} = V$$

Critical Thinking
As demonstrated in this example, Eqs. (10.4)–(10.7) can be applied in two ways. They provide a basis for rapidly obtaining qualitative sketches of shear force and bending moment diagrams. In addition, explicit equations for the diagrams can be obtained by integrating Eqs. (10.4) and (10.6).

Problems

The following problems are to be solved using Eqs. (10.4)–(10.7).

10.41 Draw the shear force and bending moment diagrams.

Problem 10.41

10.42 Draw the shear force and bending moment diagrams.

Problem 10.42

10.43 This arrangement is used to subject a segment of a beam to a uniform bending moment. Draw the shear force and bending moment diagrams.

Problem 10.43

▶ **10.44** Use the procedure described in Example 10.5 to draw the shear force and bending moment diagrams for the beam.

Problem 10.44

▶ **10.45** In Active Example 10.4, suppose that the 40 kN/m distributed load extends all the way across the beam from A to C. Draw a sketch of the beam with its new loading. Draw the shear force diagram for the beam.

10.46 Draw the shear force and bending moment diagrams.

Problem 10.46

10.47 Determine the shear force V and bending moment M for the beam as functions of x.

Problem 10.47

10.48* Draw the shear force and bending moment diagrams.

Problem 10.48

10.49 Draw the shear force and bending moment diagrams for the beam AB.

Problem 10.49

CABLES

Because of their unique combination of strength, lightness, and flexibility, ropes and cables are often used to support loads and transmit forces in structures, machines, and vehicles. The great suspension bridges are supported by enormous steel cables. Architectural engineers use cables to create aesthetic structures with open interior spaces (Fig. 10.13). In the following sections we determine the tensions in ropes and cables subjected to distributed and discrete loads.

Figure 10.13
The use of cables to suspend the roof of this sports stadium provides spectators with a view unencumbered by supporting columns.

10.4 Loads Distributed Uniformly Along Straight Lines

BACKGROUND

The main cable of a suspension bridge is the classic example of a cable subjected to a load uniformly distributed along a straight line (Fig. 10.14). The weight of the bridge is (approximately) uniformly distributed horizontally. The load, transmitted to the main cable by the large number of vertical cables, can be modeled as a distributed load. In this section we determine the shape and the variation in the tension of a cable loaded in this way.

Consider a suspended cable subjected to a load distributed uniformly along a horizontal line (Fig. 10.15a). We neglect the weight of the cable. The origin of the coordinate system is located at the cable's lowest point. Let the function $y(x)$ be the curve described by the cable in the x–y plane. Our objective is to determine the curve $y(x)$ and the tension in the cable.

Shape of the Cable

We obtain a free-body diagram by cutting the cable at its lowest point and at an arbitrary position x (Fig. 10.15b). The term T_0 is the tension in the cable at its lowest point, and T is the tension at x. The downward force exerted by the distributed load is wx. From this free-body diagram, we obtain the equilibrium equations

$$T \cos \theta = T_0,$$

$$T \sin \theta = wx. \tag{10.8}$$

Figure 10.14
(a) Main cable of a suspension bridge.
(b) The load is distributed horizontally.

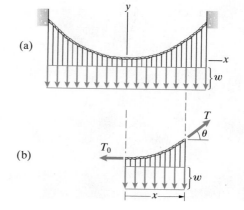

Figure 10.15
(a) A cable subjected to a load uniformly distributed along a horizontal line.
(b) Free-body diagram of the cable between $x = 0$ and an arbitrary position x.

We eliminate the tension T by dividing the second equation by the first one, obtaining

$$\tan \theta = \frac{w}{T_0} x = ax,$$

where

$$a = \frac{w}{T_0}.$$

The slope of the cable at x is $dy/dx = \tan \theta$, so we obtain a differential equation governing the curve described by the cable:

$$\frac{dy}{dx} = ax. \tag{10.9}$$

We have chosen the coordinate system so that $y = 0$ at $x = 0$. Integrating Eq. (10.9),

$$\int_0^y dy = \int_0^x ax \, dx,$$

we find that the curve described by the cable is the parabola

$$y = \frac{1}{2} ax^2. \tag{10.10}$$

Tension of the Cable

To determine the distribution of the tension in the cable, we square both sides of Eqs. (10.8) and then sum them, obtaining

$$T = T_0 \sqrt{1 + a^2 x^2}. \tag{10.11}$$

The tension is a minimum at the lowest point of the cable and increases monotonically with distance from the lowest point.

Length of the Cable

In some applications it is useful to have an expression for the length of the cable in terms of x. We can write the relation $ds^2 = dx^2 + dy^2$, where ds is an element of length of the cable (Fig. 10.16), in the form

$$ds = \sqrt{1 + \left(\frac{dy}{dx}\right)^2} \, dx.$$

Substituting Eq. (10.9) into this expression and integrating, we obtain an equation for the length s of the cable in the horizontal interval from 0 to x:

$$s = \frac{1}{2} \left\{ x \sqrt{1 + a^2 x^2} + \frac{1}{a} \ln \left[ax + \sqrt{1 + a^2 x^2} \right] \right\}. \tag{10.12}$$

Figure 10.16
The length s of the cable in the horizontal interval from 0 to x.

RESULTS

A suspended cable is subjected to a vertical load uniformly distributed *along a horizontal line*. The origin of the coordinate system is at the cable's lowest point. The curve described by the cable is the parabola

$$y = \frac{1}{2}ax^2. \qquad (10.10)$$

The parameter $a = w/T_0$, where w is the magnitude of the distributed load and T_0 is the tension in the cable at its lowest point.

Tension
The tension in the cable in terms of the tension at the lowest point and the horizontal coordinate x relative to the cable's lowest point.

$$T = T_0\sqrt{1 + a^2x^2}. \qquad (10.11)$$

Length
The length of the cable measured from the lowest point $x = 0$ to the point with horizontal coordinate x.

$$s = \frac{1}{2}\left\{x\sqrt{1 + a^2x^2} + \frac{1}{a}\ln\left[ax + \sqrt{1 + a^2x^2}\right]\right\}. \qquad (10.12)$$

Active Example 10.6   Cable with a Horizontally Distributed Load (▶ *Related Problem 10.50*)

The cable supports a distributed load of 100 lb/ft. What is the tension at its lowest point?

Strategy
The horizontal position of the cable's lowest point is not given. However, the coordinates of each attachment point relative to a coordinate system with its origin at the lowest point must satisfy Eq. (10.10). With those conditions we can determine the horizontal coordinates of the attachment points. Equation (10.10) can be used to determine $a = w/T_0$, which tells us the tension at the lowest point.

Solution

Equation (10.10) must be satisfied for both attachment points:

$$y_L = 40 \text{ ft} = \frac{1}{2} ax_L^2, \qquad y_R = 20 \text{ ft} = \frac{1}{2} ax_R^2.$$

Dividing the first equation by the second yields

$$\frac{x_L^2}{x_R^2} = 2.$$

The horizontal span of the cable is

$$x_R - x_L = 40 \text{ ft}.$$

Solving these two equations yields $x_L = -23.4$ ft and $x_R = 16.6$ ft.

> Use Eq. (10.10) to determine the horizontal coordinates of the attachment points.

Substituting the coordinates of the right attachment point into Eq. (10.10),

$$y_R = \frac{1}{2} ax_R^2:$$

$$20 \text{ ft} = \frac{1}{2} a(16.6 \text{ ft})^2,$$

and solving yields $a = 0.146 \text{ ft}^{-1}$. Therefore the tension at the lowest point is

$$T_0 = \frac{w}{a}$$

$$= \frac{100 \text{ lb/ft}}{0.146 \text{ ft}^{-1}}$$

$$= 686 \text{ lb}.$$

> Use Eq. (10.10) to determine the tension at the lowest point.

Practice Problem Determine the maximum tension in the cable.

Answer: 2440 lb.

Example 10.7 | **Cable with a Horizontally Distributed Load** (▶ *Related Problem 10.51*)

The horizontal distance between the supporting towers of the Manhattan Bridge in New York is 1470 ft. The tops of the towers are 145 ft above the lowest point of the main supporting cables. Obtain the equation for the curve described by the cables.

Strategy

We know the coordinates of the cable's attachment points relative to their lowest points. By substituting the coordinates into Eq. (10.10), we can determine the parameter a. Once a is known, Eq. (10.10) describes the shape of the cables.

Solution

The coordinates of the top of the right supporting tower relative to the lowest point of the support cables are $x_R = 735$ ft, $y_R = 145$ ft (Fig. a). By substituting these values into Eq. (10.10),

$$y = \frac{1}{2}ax^2:$$

$$145 \text{ ft} = \frac{1}{2}a(735 \text{ ft})^2,$$

(a) The theoretical curve superimposed on a photograph of the supporting cable.

we obtain $a = 5.37 \times 10^{-4} \text{ ft}^{-1}$. The curve described by the supporting cables is

$$y = \frac{1}{2}ax^2 = (2.68 \times 10^{-4})x^2.$$

Figure a compares this parabola with a photograph of the supporting cables.

Critical Thinking

Knowing the relative locations of the cable's highest and lowest points allowed us to determine the value of a. This parameter not only determines the equation describing the cable's shape, as we demonstrated in this example, but is also the ratio of the distributed load w acting on the cable to the tension in the cable at its lowest point. If the value of w was also known, the tension throughout the cable would be determined by Eq. (10.11).

Problems

▶ **10.50** The cable supports a distributed load $w = 12{,}000$ lb/ft. Using the approach described in Active Example 10.6, determine the maximum tension in the cable.

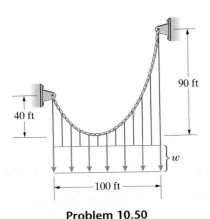

90 ft

40 ft

w

100 ft

Problem 10.50

▶ **10.51** In Example 10.7, suppose that the tension at the lowest point of one of the main supporting cables of the bridge is two million pounds. What is the maximum tension in the cable?

10.52 A cable is used to suspend a pipeline above a river. The towers supporting the cable are 36 m apart. The lowest point of the cable is 1.4 m below the tops of the towers. The mass of the suspended pipe is 2700 kg.

(a) What is the maximum tension in the cable?

(b) What is the suspending cable's length?

10.53 In Problem 10.52, let the lowest point of the cable be a distance h below the tops of the towers supporting the cable.

(a) If the cable will safely support a tension of 70 kN, what is the minimum safe value of h?

(b) If h has the value determined in part (a), what is the suspending cable's length?

Problems 10.52/10.53

10.54 The cable supports a uniformly distributed load $w = 750$ N/m. The lowest point of the cable is 0.18 m below the attachment points C and D. Determine the axial loads in the truss members AC and BC.

Problem 10.54

10.55 The cable supports a railway bridge between two tunnels. The distributed load is $w = 1$ MN/m, and $h = 40$ m.

(a) What is the maximum tension in the cable?

(b) What is the length of the cable?

10.56 The cable in Problem 10.55 will safely support a tension of 40 MN. What is the shortest cable that can be used, and what is the corresponding value of h?

10.57 An oceanographic research ship tows an instrument package from a cable. Hydrodynamic drag subjects the cable to a uniformly distributed force $w = 2$ lb/ft. The tensions in the cable at 1 and 2 are 800 lb and 1300 lb, respectively. Determine the distance h.

10.58 Draw a graph of the shape of the cable in Problem 10.57.

Problems 10.57/10.58

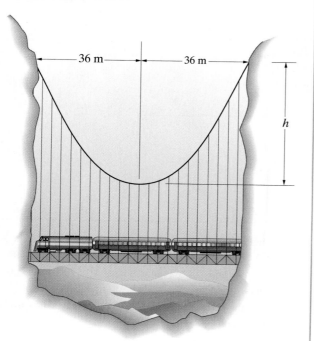

Problems 10.55/10.56

10.5 Loads Distributed Uniformly Along Cables

BACKGROUND

A cable's own weight subjects it to a load that is distributed uniformly along its length. If a cable is subjected to equal, parallel forces spaced uniformly along its length, the load on the cable can often be modeled as a load distributed uniformly along its length. In this section we show how to determine both the cable's resulting shape and the variation in its tension.

Suppose that a cable is acted on by a distributed load that subjects each element ds of its length to a force $w\,ds$, where w is constant. In Fig. 10.17 we show the free-body diagram obtained by cutting the cable at its lowest point and at a point a distance s along its length. The terms T_0 and T are the tensions at the lowest point and at s, respectively. The distributed load exerts a downward force ws. The origin of the coordinate system is located at the lowest point of the cable. Let the function $y(x)$ be the curve described by the cable in the x–y plane. Our objective is to determine $y(x)$ and the tension T.

Shape of the Cable

From the free-body diagram in Fig. 10.17, we obtain the equilibrium equations

$$T \sin \theta = ws, \tag{10.13}$$

$$T \cos \theta = T_0. \tag{10.14}$$

Dividing Eq. (10.13) by Eq. (10.14), we obtain

$$\tan \theta = \frac{w}{T_0} s = as, \tag{10.15}$$

where

$$a = \frac{w}{T_0}. \tag{10.16}$$

Figure 10.17
A cable subjected to a load distributed uniformly along its length.

The slope of the cable $dy/dx = \tan \theta$, so Eq. (10.15) can be written

$$\frac{dy}{dx} = as.$$

The derivative of this equation with respect to x is

$$\frac{d}{dx}\left(\frac{dy}{dx}\right) = a\frac{ds}{dx}. \tag{10.17}$$

By using the relation

$$ds^2 = dx^2 + dy^2,$$

we can write the derivative of s with respect to x as

$$\frac{ds}{dx} = \sqrt{1 + \left(\frac{dy}{dx}\right)^2} = \sqrt{1 + \sigma^2}, \tag{10.18}$$

where

$$\sigma = \frac{dy}{dx} = \tan \theta$$

is the slope. Now, with Eq. (10.18), we can write Eq. (10.17) as

$$\frac{d\sigma}{\sqrt{1 + \sigma^2}} = a\,dx.$$

The slope $\sigma = 0$ at $x = 0$. Integrating this equation yields

$$\int_0^\sigma \frac{d\sigma}{\sqrt{1 + \sigma^2}} = \int_0^x a\,dx,$$

and we obtain the slope as a function of x:

$$\sigma = \frac{dy}{dx} = \frac{1}{2}(e^{ax} - e^{-ax}) = \sinh ax. \tag{10.19}$$

Then, integrating this equation with respect to x yields the curve described by the cable, which is called a *catenary*:

$$y = \frac{1}{2a}(e^{ax} + e^{-ax} - 2) = \frac{1}{a}(\cosh ax - 1). \tag{10.20}$$

Tension of the Cable

Using Eq. (10.14) and the relation $dx = \cos\theta\,ds$, we obtain

$$T = \frac{T_0}{\cos\theta} = T_0\frac{ds}{dx}.$$

Substituting Eq. (10.18) into this expression and using Eq. (10.19) yields the tension in the cable as a function of x:

$$T = T_0\sqrt{1 + \frac{1}{4}(e^{ax} - e^{-ax})^2} = T_0\cosh ax. \qquad (10.21)$$

Length of the Cable

From Eq. (10.15), the length s of the cable from the origin to the point at which the angle between the cable and the x axis equals θ is

$$s = \frac{1}{a}\tan\theta = \frac{\sigma}{a}.$$

Substituting Eq. (10.19) into this equation, we obtain an expression for the length s of the cable in the horizontal interval from its lowest point to x:

$$s = \frac{1}{2a}(e^{ax} - e^{-ax}) = \frac{\sinh ax}{a}. \qquad (10.22)$$

RESULTS

A suspended cable is subjected to a vertical load uniformly distributed *along the length of the cable*. The origin of the coordinate system is at the cable's lowest point. The curve described by the cable is the *catenary*

$$y = \frac{1}{2a}(e^{ax} + e^{-ax} - 2) = \frac{1}{a}(\cosh ax - 1). \quad (10.20)$$

The parameter $a = w/T_0$, where w is the magnitude of the distributed load and T_0 is the tension in the cable at its lowest point.

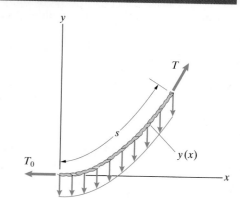

$$T = T_0\sqrt{1 + \frac{1}{4}(e^{ax} - e^{-ax})^2} = T_0\cosh ax. \qquad (10.21)$$

Tension
The tension in the cable in terms of the tension at the lowest point and the horizontal coordinate x relative to the cable's lowest point.

$$s = \frac{1}{2a}(e^{ax} - e^{-ax}) = \frac{\sinh ax}{a}. \qquad (10.22)$$

Length
The length of the cable measured from the lowest point $x = 0$ to the point with horizontal coordinate x.

Active Example 10.8   Cable Loaded by Its Own Weight (▶ *Related Problem 10.59*)

The mass per unit length of the cable is 1 kg/m. The tension at its lowest point is 50 N. Determine the height h of its attachment points relative to the lowest point.

Strategy

The cable is subjected to a load $w = (9.81 \text{ m/s}^2)(1 \text{ kg/m}) = 9.81 \text{ N/m}$ distributed uniformly along its length. Because w and T_0 are known, we can determine $a = w/T_0$. Then we can use Eq. (10.20) to determine h.

Solution

$$a = \frac{w}{T_0} = \frac{9.81 \text{ N/m}}{50 \text{ N}} = 0.196 \text{ m}^{-1}.$$ ◀───── Determine the parameter a.

$$y = \frac{1}{a}(\cosh ax - 1):$$

$$h = \frac{1}{0.196 \text{ m}^{-1}}\{\cosh[(0.196 \text{ m}^{-1})(10 \text{ m})] - 1\}$$

$$= 13.4 \text{ m}.$$

Apply Eq. (10.20).

Practice Problem Determine the maximum tension in the cable.

Answer: 181 N.

Problems

▶ **10.59** The mass of the rope per unit length is 0.1 kg/m. The tension at its lowest point is 4.6 N. Using the approach described in Active Example 10.8, determine (a) the maximum tension in the rope and (b) the rope's length.

Problem 10.59

10.60 The stationary balloon's tether is horizontal at point O where it is attached to the truck. The mass per unit length of the tether is 0.45 kg/m. The tether exerts a 50-N horizontal force on the truck. The horizontal distance from point O to point A where the tether is attached to the balloon is 20 m. What is the height of point A relative to point O?

10.61 In Problem 10.60, determine the magnitudes of the horizontal and vertical components of the force exerted on the balloon at A by the tether.

Problems 10.60/10.61

10.62 The mass per unit length of lines AB and BC is 2 kg/m. The tension at the lowest point of cable AB is 1.8 kN. The two lines exert equal horizontal forces at B.

(a) Determine the sags h_1 and h_2.

(b) Determine the maximum tensions in the two lines.

Problem 10.62

10.63 The rope is loaded by 2-kg masses suspended at 1-m intervals along its length. The mass of the rope itself is negligible. The tension in the rope at its lowest point is 100 N. Determine h and the maximum tension in the rope.

Strategy: Obtain an approximate answer by modeling the discrete loads on the rope as a load uniformly distributed along its length.

Problem 10.63

10.6 Discrete Loads

BACKGROUND

Our first applications of equilibrium in Chapter 3 involved determining the tensions in cables supporting suspended objects. In this section we consider the case of an arbitrary number N of objects suspended from a cable (Fig. 10.18a). We assume that the weight of the cable can be neglected in comparison to the suspended weights and that the cable is sufficiently flexible that we can approximate its shape by a series of straight segments.

Determining the Configuration and Tensions

Suppose that the horizontal distances $b_1, b_2, \ldots, b_{N+1}$ are known and that the vertical distance h_{N+1} specifying the position of the cable's right attachment point is known. We have two objectives: (1) to determine the configuration (shape) of the cable by solving for the vertical distances $h_1, h_2, \ldots, h_N$ specifying the positions of the attachment points of the weights and (2) to determine the tensions in the segments $1, 2, \ldots, N + 1$ of the cable.

We begin by drawing a free-body diagram, cutting the cable at its left attachment point and just to the right of the weight W_1 (Fig. 10.18b). We resolve the tension in the cable at the left attachment point into its horizontal and vertical components T_h and T_v. Summing moments about the attachment point A_1, we obtain the equation

$$\Sigma M_{\text{point } A_1} = h_1 T_h - b_1 T_v = 0.$$

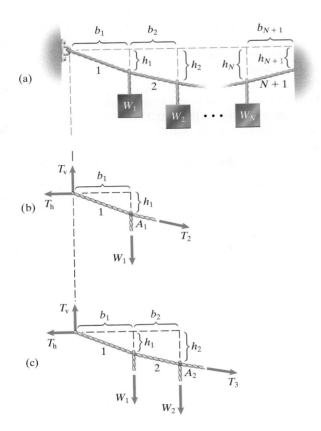

Figure 10.18
(a) N weights suspended from a cable.
(b) The first free-body diagram.
(c) The second free-body diagram.

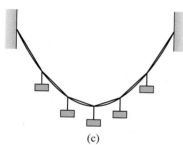

Figure 10.19
(a) Cable subjected to a continuous load.
(b) Cable with three discrete loads.
(c) Cable with five discrete loads.

Our next step is to obtain a free-body diagram by cutting the cable at its left attachment point and just to the right of the weight W_2 (Fig. 10.18c). Summing moments about A_2, we obtain

$$\Sigma M_{\text{point } A_2} = h_2 T_{\text{h}} - (b_1 + b_2)T_{\text{v}} + b_2 W_1 = 0.$$

Proceeding in this way, cutting the cable just to the right of each of the N weights, we obtain N equations. We can also draw a free-body diagram by cutting the cable at its left and right attachment points and sum moments about the right attachment point. In this way, we obtain $N + 1$ equations in terms of $N + 2$ unknowns: the two components of the tension T_{h} and T_{v} and the vertical positions of the attachment points $h_1, h_2, \ldots, h_N$. If the vertical position of just one attachment point is also specified, we can solve the system of equations for the vertical positions of the other attachment points, determining the configuration of the cable.

Once we know the configuration of the cable and the force T_{h}, the tension in any segment can be determined by cutting the cable at the left attachment point and within the segment and summing forces in the horizontal direction.

Comments on Continuous and Discrete Models

By comparing cables subjected to distributed and discrete loads, we can make some observations about how continuous and discrete systems are modeled in engineering. Consider a cable subjected to a horizontally distributed load w (Fig. 10.19a). The total force exerted on it is wL. Since the cable passes through the point $x = L/2$, $y = L/2$, we find from Eq. (10.10) that $a = 4/L$, so the equation for the curve described by the cable is $y = (2/L)x^2$.

In Fig. 10.19b, we compare the shape of the cable with the distributed load to that of a cable of negligible weight subjected to three discrete loads $W = wL/3$ with equal horizontal spacing. (We chose the dimensions of the cable with discrete loads so that the heights of the two cables would be equal at their midpoints.) In Fig. 10.19c, we compare the shape of the cable with the distributed load to that of a cable subjected to five discrete loads $W = wL/5$ with equal horizontal spacing. In Figs. 10.20a and 10.20b, we compare the tension in the cable subjected to the distributed load to those in the cables subjected to three and five discrete loads.

The shape and the tension in the cable with a distributed load are approximated by the shapes and tensions in the cables with discrete loads. Although the approximation of the tension is less impressive than the approximation of the

Figure 10.20
(a) The tension in a cable with a continuous load compared to the cable with three discrete loads.
(b) The tension in a cable with a continuous load compared to the cable with five discrete loads.

shape, it is clear that the former can be improved by increasing the number of discrete loads.

This approach, approximating a continuous distribution by a discrete model, is very important in engineering. It is the starting point of the finite difference and finite element methods. The opposite approach, modeling discrete systems by continuous models, is also widely used, for example when the forces exerted on a bridge by traffic are modeled as a distributed load.

RESULTS

The weights, the horizontal distances $b_1, b_2, ..., b_{N+1}$, and the vertical distance h_{N+1} are known. The objective is to determine the vertical distances $h_1, h_2, ..., h_N$ and the tensions in the segments of the cable.

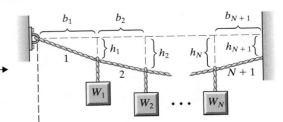

Cut the cable at its left attachment point and just to the right of the weight W_1. Sum the moments about A_1:

$$\Sigma M_{\text{point } A_1} = h_1 T_{\text{h}} - b_1 T_{\text{v}} = 0.$$

Cut the cable at its left attachment point and just to the right of the weight W_2. Sum the moments about A_2:

$$\Sigma M_{\text{point } A_2} = h_2 T_{\text{h}} - (b_1 + b_2) T_{\text{v}} + b_2 W_1 = 0.$$

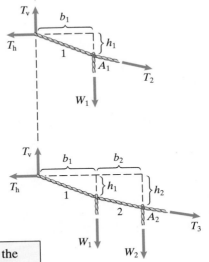

Continuing in this way results in N equations. In addition, cut the cable at its left and right attachment points and sum moments about the right attachment point. This results in $N+1$ equations in $N+2$ unknowns: the two components T_{h} and T_{v} of the tension at the right attachment point and the vertical positions $h_1, h_2, ..., h_N$. *If the vertical position of one weight is specified*, the geometrical configuration of the cable and the components T_{h} and T_{v} can be determined. Once this has been done, the tension in any segment of the cable can be obtained by cutting the cable at the left attachment point and within the segment and summing forces in the horizontal direction.

Active Example 10.9   Cable Subjected to Discrete Loads (▶ *Related Problem 10.64*)

The cable supports two masses $m_1 = 10$ kg and $m_2 = 20$ kg. Determine the vertical distance h_2.

Strategy

By following the procedure described in Results, we can obtain three equations in terms of the horizontal and vertical components of the tension at the right attachment point and the vertical distance h_2.

Solution

$$\Sigma M_{\text{point } A_1} = (1\text{ m})T_\text{h} - (1\text{ m})T_\text{v} = 0.$$

Cut the cable at the left attachment point and just to the right of the mass m_1 and sum moments about A_1.

$$\Sigma M_{\text{point } A_2} = h_2 T_\text{h} - (2\text{ m})T_\text{v} + (1\text{ m})m_1 g = 0.$$

Cut the cable at the left attachment point and just to the right of the mass m_2 and sum moments about A_2.

$$\Sigma M_{\text{point } A_3} = - (3 \text{ m}) T_v + (2 \text{ m}) m_1 g + (1 \text{ m}) m_2 g = 0.$$

Cut the cable at the left and right attachment points and sum moments about A_3.

There are three equations in terms of the unknowns T_h, T_v, and h_2. Solving them yields $T_h = T_v = 131$ N and $h_2 = 1.25$ m.

Practice Problem Determine the tension in segment 2 of the cable.

Answer: 135 N.

Problems

▶ **10.64** In Active Example 10.9, what are the tensions in cable segments 1 and 3?

10.65 Each lamp weighs 12 lb.

(a) What is the length of the wire $ABCD$ needed to suspend the lamps as shown?

(b) What is the maximum tension in the wire?

Problem 10.65

10.66 Two weights, $W_1 = W_2 = 50$ lb, are suspended from a cable. The vertical distance $h_1 = 4$ ft.

(a) Determine the vertical distance h_2.

(b) What is the maximum tension in the cable?

10.67 The weights are $W_1 = 50$ lb and $W_2 = 100$ lb, and the vertical distance $h_1 = 4$ ft.

(a) Determine the vertical distance h_2.

(b) What is the maximum tension in the cable?

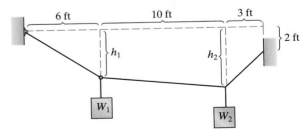

Problems 10.66/10.67

10.68 Three identical masses $m = 10$ kg are suspended from the cable. Determine the vertical distances h_1 and h_3 and draw a sketch of the configuration of the cable.

10.69 In Problem 10.68, what are the tensions in cable segments 1 and 2?

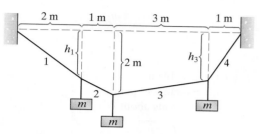

Problem 10.68/10.69

10.70 Three masses are suspended from the cable, where $m = 30$ kg, and the vertical distance $h_1 = 400$ mm. Determine the vertical distances h_2 and h_3.

10.71 In Problem 10.70, what is the maximum tension in the cable, and where does it occur?

Problems 10.70/10.71

10.72 Each suspended object has the same weight W. Determine the vertical distances h_2 and h_3.

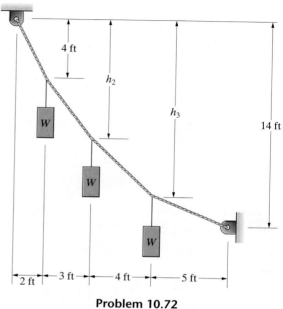

Problem 10.72

LIQUIDS AND GASES

10.7 Pressure and the Center of Pressure

BACKGROUND

Wind forces on buildings and aerodynamic forces on cars and airplanes are examples of forces that are distributed over areas. The downward force exerted on the bed of a dump truck by a load of gravel is distributed over the area of the bed. The upward force that supports a building is distributed over the area of its foundation. Loads distributed over the roofs of buildings by accumulated snow can be hazardous. Many forces of concern in engineering are distributed over areas. In this section we analyze the most familiar example, the force exerted by the pressure of a gas or liquid.

A surface immersed in a gas or liquid is subjected to forces exerted by molecular impacts. If the gas or liquid is stationary, the load can be described by a function p, the *pressure*, defined such that the normal force exerted on a differential element dA of the surface is $p\,dA$ (Figs. 10.21a and b). (Notice the parallel between the pressure and a load w distributed along a line, which is defined such that the force on a differential element dx of the line is $w\,dx$.)

The dimensions of p are (force)/(area). In U.S. Customary units, pressure can be expressed in pounds per square foot or pounds per square inch (psi). In SI units, pressure can be expressed in newtons per square meter, which are called pascals (Pa).

In some applications, it is convenient to use the *gage pressure*

$$p_g = p - p_{\text{atm}}, \tag{10.23}$$

where p_{atm} is the pressure of the atmosphere. Atmospheric pressure varies with location and climatic conditions. Its value at sea level is approximately 1×10^5 Pa in SI units and 14.7 psi or 2120 lb/ft^2 in U.S. Customary units.

Center of Pressure

If the distributed force due to pressure on a surface is represented by an equivalent force, the point at which the line of action of the force intersects the surface

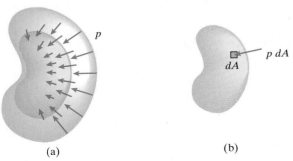

(a)

(b)

Figure 10.21
(a) The pressure on an area.
(b) The force on an element dA is $p\,dA$.

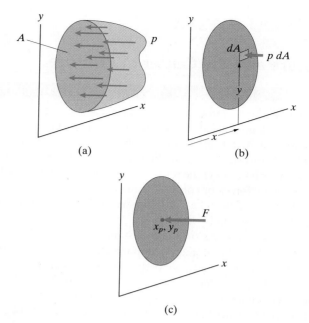

Figure 10.22
(a) A plane area subjected to pressure.
(b) The force on a differential element dA.
(c) The total force acting at the center of pressure.

is called the *center of pressure*. Consider a plane area A subjected to a pressure p and introduce a coordinate system such that the area lies in the x–y plane (Fig. 10.22a). The normal force on each differential element of area dA is $p\,dA$ (Fig. 10.22b), so the total normal force on A is

$$F = \int_A p\,dA. \tag{10.24}$$

Now we will determine the coordinates (x_p, y_p) of the center of pressure (Fig. 10.22c). Equating the moment of F about the origin to the total moment due to the pressure about the origin gives

$$(x_p\mathbf{i} + y_p\mathbf{j}) \times (-F\mathbf{k}) = \int_A (x\mathbf{i} + y\mathbf{j}) \times (-p\,dA\,\mathbf{k}),$$

and using Eq. (10.24), we obtain

$$x_p = \frac{\displaystyle\int_A xp\,dA}{\displaystyle\int_A p\,dA}, \quad y_p = \frac{\displaystyle\int_A yp\,dA}{\displaystyle\int_A p\,dA}. \tag{10.25}$$

These equations determine the position of the center of pressure when the pressure p is known. If the pressure p is uniform, the total normal force

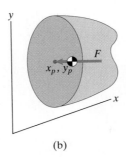

(a)   (b)

Figure 10.23
(a) The differential element $dV = p\, dA$.
(b) The line of action of F passes through the centroid of V.

is $F = pA$ and Eqs. (10.25) indicate that the center of pressure is the centroid of A.

In Chapter 7 it was shown that if we calculate the "area" defined by a load distributed along a line and place the resulting force at its centroid, the force is equivalent to the distributed load. A similar result holds for a pressure distributed on a plane area. The term $p\, dA$ in Eq. (10.24) is equal to a differential element dV of the "volume" between the surface defined by the pressure distribution and the area A (Fig. 10.23a). The total force exerted by the pressure is therefore equal to this "volume":

$$F = \int_V dV = V.$$

Substituting $p\, dA = dV$ into Eqs. (10.25), we obtain

$$x_p = \frac{\displaystyle\int_V x\, dV}{\displaystyle\int_V dV}, \quad y_p = \frac{\displaystyle\int_V y\, dV}{\displaystyle\int_V dV}.$$

The center of pressure coincides with the x and y coordinates of the centroid of the "volume" (Fig. 10.23).

Pressure in a Stationary Liquid

Designers of pressure vessels and piping, ships, dams, and other submerged structures must be concerned with forces and moments exerted by water pressure. The pressure in a liquid at rest increases with depth, which you can confirm by descending to the bottom of a swimming pool and noting the effect of the pressure on your ears. If we restrict ourselves to changes in depth for which changes in the density of the liquid can be neglected, we can determine the dependence of the pressure on depth by using a simple free-body diagram.

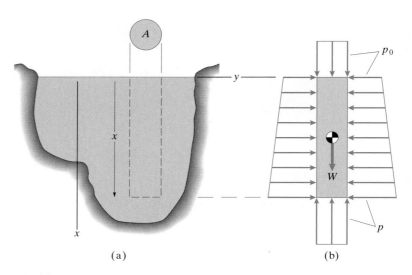

Figure 10.24
(a) A cylindrical volume that extends to a depth x
in a body of stationary liquid.
(b) Free-body diagram of the cylinder.

Introducing a coordinate system with its origin at the surface of the liquid and the positive x axis downward (Fig. 10.24a), we draw a free-body diagram of a cylinder of liquid that extends from the surface to a depth x (Fig. 10.24b). The top of the cylinder is subjected to the pressure at the surface, which we call p_0. The sides and bottom of the cylinder are subjected to pressure by the surrounding liquid, which increases from p_0 at the surface to a value p at the depth x. The volume of the cylinder is Ax, where A is its cross-sectional area. Therefore, its weight is $W = \gamma Ax$, where γ is the weight density of the liquid. (Recall that the weight and mass densities are related by $\gamma = \rho g$.) Since the liquid is stationary, the cylinder is in equilibrium. From the equilibrium equation

$$\Sigma F_x = p_0 A - pA + \gamma Ax = 0,$$

we obtain a simple expression for the pressure p of the liquid at depth x:

$$p = p_0 + \gamma x. \tag{10.26}$$

Thus, the pressure increases linearly with depth, and the derivation we have used illustrates why: The pressure at a given depth literally holds up the liquid above that depth. If the surface of the liquid is open to the atmosphere, $p_0 = p_{\text{atm}}$, and we can write Eq. (10.26) in terms of the gage pressure $p_g = p - p_{\text{atm}}$ as

$$p_g = \gamma x. \tag{10.27}$$

In SI units, the density of water at sea level conditions is $\rho = 1000 \text{ kg/m}^3$, so its weight density is approximately $\gamma = \rho g = 9.81 \text{ kN/m}^3$. In U.S. Customary units, the weight density of water is approximately 62.4 lb/ft^3.

RESULTS

Definition of the Pressure

The pressure p of a liquid or gas is defined such that the normal force exerted on an element of area dA of a surface is $p\,dA$. The force and moment due to a distribution of pressure on a surface can be determined by integration.

The *gage pressure* is defined by

$$p_g = p - p_{atm}, \qquad (10.23)$$

where p_{atm} is atmospheric pressure. Atmospheric pressure at sea level is approximately 1×10^5 Pa in SI units and 14.7 psi or 2120 lb/ft^2 in U.S. Customary units.

Center of Pressure

If the force exerted by a distribution of pressure on an area A is represented by an equivalent force vector, the point where the line of action of the force vector intersects A is called the *center of pressure*.

Volume Analogy

The total force F exerted on a plane area A by a distribution of pressure p is equal to the "volume" between A and the function p. If F is represented by a force vector acting at the centroid of the "volume," the force vector is equivalent to the distribution of pressure—its line of action intersects A at the center of pressure.

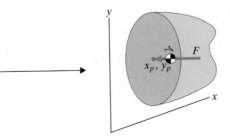

Pressure in a Stationary Liquid

The pressure at a depth x in a stationary liquid is

$$p = p_0 + \gamma x, \qquad (10.26)$$

where p_0 is the pressure at the surface and $\gamma = \rho g$ is the weight density of the liquid. The weight density of water is 9.81 kN/m^3 in SI units and 62.4 lb/ft^3 in U.S. Customary units. If $p_0 = p_{atm}$, Eq. (10.26) can be expressed in terms of the gage pressure as

$$p_g = \gamma x. \qquad (10.27)$$

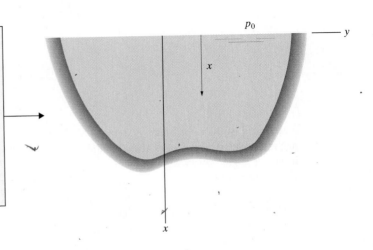

Active Example 10.10 Gate Loaded by a Pressure Distribution (▶ *Related Problem 10.78*)

B

3 ft

2 ft

A

A stationary body of water exerts pressure on the right side of the gate *AB*. The width of the gate (its dimension into the page) is 3 ft, and the gate weighs 100 lb. The weight density of the water is 62.4 lb/ft³. Determine the reactions on the gate at the supports *A* and *B*.

Strategy

We will use integration to determine the force and moment exerted on the gate by the pressure of the water. We can then apply equilibrium to the free-body diagram of the gate to determine the reactions at *A* and *B*.

Solution

The left face of the gate and the right face above the surface of the water are exposed to atmospheric pressure. From Eqs. (10.23) and (10.26), the pressure in the water is the sum of atmospheric pressure and the gage pressure $p_g = \gamma x$, where x is measured downward from the surface of the water. The effects of atmospheric pressure on the gate cancel, so *only the forces and moments exerted on the gate by the gage pressure must be considered.*

Integrate to determine the total force exerted on the gate by the gage pressure.

The origin of the coordinate system is at the surface of the water. The element of area of the gate $dA = (3 \text{ ft}) dx$. The total force exerted by the gage pressure is

$$F = \int_A p_g \, dA = \int_0^2 (\gamma x)(3 \text{ ft}) dx = 374 \text{ lb}.$$

The total moment about the y axis exerted
by the gage pressure is

$$M = \int_A x p_g \, dA = \int_0^2 x(\gamma x)(3\text{ ft}) \, dx = 499\text{ ft-lb.}$$

Integrate to determine the
total moment exerted on the
gate by the gage pressure.

$$x_p = \frac{M}{F} = \frac{499\text{ ft-lb}}{374\text{ lb}} = 1.33\text{ ft.}$$

Determine the position of the
center of pressure from the
condition for equivalence
$M = x_p F$.

$$\Sigma F_x = A_x + 100\text{ lb} = 0,$$
$$\Sigma F_z = A_z + B - 374\text{ lb} = 0,$$
$$\Sigma M_{y\text{ axis}} = (1\text{ ft})B - (2\text{ ft})A_z + (1.33\text{ ft})(374\text{ lb}) = 0.$$
Solving yields $A_x = -100\text{ lb}$, $A_z = 291\text{ lb}$, and $B = 83.2\text{ lb.}$

Draw the free-body diagram
of the gate, placing the total
force exerted by gage pressure
at the center of pressure. Apply
equilibrium to determine the
reactions at A and B.

Practice Problem Determine the reactions on the gate at the supports A and B. To do
so, use the volume analogy to determine the total force exerted by gage pressure on the
gate and the location of the center of pressure.

Answer: $A_x = -100\text{ lb}$, $A_z = 291\text{ lb}$, $B = 83.2\text{ lb.}$

| **Example 10.11** | **Pressure Force and Center of Pressure** (▶ *Related Problems 10.79, 10.80*) |

An engineer making preliminary design studies for a canal lock needs to determine the total pressure force on a submerged rectangular plate and the location of the center of pressure. The top of the plate is 6 m below the surface. Atmospheric pressure is $p_{atm} = 1 \times 10^5$ Pa, and the weight density of the water is $\gamma = 9.81$ kN/m^3.

Strategy

We will determine the pressure force on a differential element of area of the plate in the form of a horizontal strip and integrate to determine the total force and moment exerted by the pressure.

Solution

In terms of a coordinate system with its origin at the surface and the positive x axis downward (Fig. a), the pressure of the water is $p = p_{atm} + \gamma x$. The horizontal strip $dA = (8$ m$)$ dx. Therefore, the total force exerted on the face of the plate by the pressure is

$$F = \int_A p\, dA = \int_6^{18} (p_{atm} + \gamma x)(8\text{ m})\, dx$$

$$= p_{atm}(8\text{ m}) \int_6^{18} dx + \gamma(8\text{ m}) \int_6^{18} x\, dx$$

$$= \left(1 \times 10^5 \frac{\text{N}}{\text{m}^2}\right)(8\text{ m})(12\text{ m}) + \left(9810 \frac{\text{N}}{\text{m}^3}\right)(8\text{ m})(144\text{ m}^2)$$

$$= 20.9 \times 10^6 \text{ N}.$$

(a) An element of area in the form of a horizontal strip.

The moment about the y axis due to the pressure on the plate is

$$M = \int_A xp\, dA = \int_6^{18} x(p_{atm} + \gamma x)(8\text{ m})\, dx$$

$$= p_{atm}(8\text{ m}) \int_6^{18} x\, dx + \gamma(8\text{ m}) \int_6^{18} x^2\, dx$$

$$= 262 \times 10^6 \text{ N-m}.$$

The force F acting at the center of pressure (Fig. b) exerts a moment about the y axis equal to M:

$$x_p F = M.$$

Therefore, the location of the center of pressure is

(b) The center of pressure.

$$x_p = \frac{M}{F} = \frac{262 \text{ MN-m}}{20.9 \text{ MN}} = 12.5 \text{ m}.$$

Critical Thinking

Notice that the center of pressure does not coincide with the centroid of the area. The center of pressure of a plane area generally coincides with the centroid of the area only when the pressure is uniformly distributed. In this example, the pressure increases with depth, and as a result, the center of pressure is below the centroid.

Example 10.12 Determination of a Pressure Force (▶ *Related Problem 10.91*)

The container is filled with a liquid with weight density γ. Determine the force exerted by the pressure of the liquid on the cylindrical wall AB.

Strategy
The pressure of the liquid on the cylindrical wall varies with depth (Fig. a). It is the force exerted by this pressure distribution we want to determine. We could determine it by integrating over the cylindrical surface, but we can avoid that by drawing a free-body diagram of the quarter-cylinder of liquid to the right of A.

Solution
We draw the free-body diagram of the quarter-cylinder of liquid in Fig. b. The pressure distribution on the cylindrical surface of the liquid is the same one that acts on the cylindrical wall. If we denote the force exerted on the liquid by this pressure distribution by $\mathbf{F}_p$, the force exerted by the liquid on the cylindrical wall is $-\mathbf{F}_p$.

The other forces parallel to the x–y plane that act on the quarter-cylinder of liquid are its weight, atmospheric pressure at the upper surface, and the pressure distribution of the liquid on the left side. The volume of liquid is $(\frac{1}{4}\pi R^2)b$, so the force exerted on the free-body diagram by the weight of the liquid is $\frac{1}{4}\gamma\pi R^2 b\mathbf{i}$. The force exerted on the upper surface by atmospheric pressure is $Rbp_{\text{atm}}\mathbf{i}$.

We can integrate to determine the force exerted by the pressure on the left side of the free-body diagram. Its magnitude is

$$\int_A p\, dA = \int_0^R (p_{\text{atm}} + \gamma x)b\, dx = Rb\left(p_{\text{atm}} + \frac{1}{2}\gamma R\right).$$

From the equilibrium equation

$$\Sigma\mathbf{F} = \frac{1}{4}\gamma\pi R^2 b\mathbf{i} + Rbp_{\text{atm}}\mathbf{i} + Rb\left(p_{\text{atm}} + \frac{1}{2}\gamma R\right)\mathbf{j} + \mathbf{F}_p = \mathbf{0},$$

we obtain the force exerted on the wall AB by the pressure of the liquid:

$$-\mathbf{F}_p = Rb\left(p_{\text{atm}} + \frac{\pi}{4}\gamma R\right)\mathbf{i} + Rb\left(p_{\text{atm}} + \frac{1}{2}\gamma R\right)\mathbf{j}.$$

Critical Thinking
The need to integrate over a curved surface to calculate a pressure force can often be avoided by choosing a suitable free-body diagram as we have done in this example.

(a) The pressure of the liquid on the wall AB.

(b) Free-body diagram of the liquid to the right of A.

Problems

10.73 An engineer planning a water system for a new community estimates that at maximum expected usage, the pressure drop between the central system and the farthest planned fire hydrant will be 25 psi. Firefighting personnel indicate that a gage pressure of 40 psi at the fire hydrant is required. The weight density of the water is $\gamma = 62.4$ lb/ft^3. How tall would a water tower at the central system have to be to provide the needed pressure?

10.74 A cube of material is suspended below the surface of a liquid of weight density γ. By calculating the forces exerted on the faces of the cube by pressure, show that their sum is an upward force of magnitude γb^3.

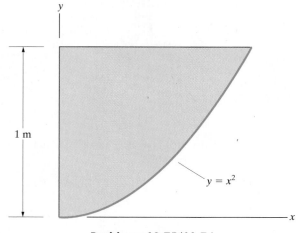

Problem 10.74

10.75 The area shown is subjected to a *uniform* pressure $p_{atm} = 1 \times 10^5$ Pa.
(a) What is the total force exerted on the area by the pressure?
(b) What is the moment about the y-axis due to the pressure on the area?

10.76 The area shown is subjected to a *uniform* pressure. Determine the coordinates of the center of pressure.

Problems 10.75/10.76

10.77 The area shown is subjected to a *uniform* pressure $p_{atm} = 14.7$ psi.
(a) What is the total force exerted on the area by the pressure?
(b) What is the moment about the y axis due to the pressure on the area?

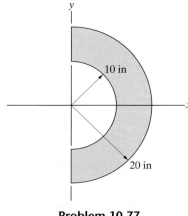

Problem 10.77

▶ **10.78** In Active Example 10.10, suppose that the water depth relative to point A is increased from 2 ft to 3 ft. Determine the reactions on the gate at the supports A and B.

▶ **10.79** The top of the rectangular plate is 2 m below the surface of a lake. Atmospheric pressure is $p_{atm} = 1 \times 10^5$ Pa and the mass density of the water is $\rho = 1000$ kg/m^3.
(a) What is the maximum pressure exerted on the plate by the water?
(b) Determine the force exerted on a face of the plate by the pressure of the water. (See Example 10.11.)

▶ **10.80** In Problem 10.79, how far below the top of the plate is the center of pressure located? (See Example 10.11.)

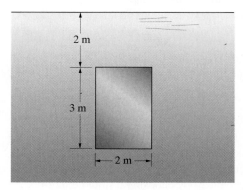

Problems 10.79/10.80

10.81 The width of the dam (the dimension into the page) is 100 m. The mass density of the water is $\rho = 1000$ kg/m³. Determine the force exerted on the dam by the gage pressure of the water

(a) by integration;

(b) by calculating the "volume" of the pressure distribution.

10.82 In Problem 10.81, how far down from the surface of the water is the center of pressure due to the gage pressure of the water on the dam?

Problems 10.81/10.82

10.83 The width of the gate (the dimension into the page) is 3 m. Atmospheric pressure is $p_{atm} = 1 \times 10^5$ Pa and the mass density of the water is $\rho = 1000$ kg/m³. Determine the horizontal force and couple exerted on the gate by its built-in support A.

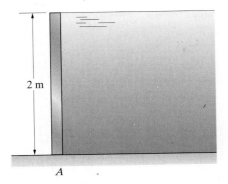

Problem 10.83

10.84 The homogeneous gate weighs 100 lb, and its width (the dimension into the page) is 3 ft. The weight density of the water is $\gamma = 62.4$ lb/ft³, and atmospheric pressure is $p_{atm} = 2120$ lb/ft². Determine the reactions at A and B.

Problem 10.84

10.85 The width of the gate (the dimension into the page) is 2 m and there is water of depth $d = 1$ m on the right side. Atmospheric pressure is $p_{atm} = 1 \times 10^5$ Pa and the mass density of the water is $\rho = 1000$ kg/m³. Determine the horizontal forces exerted on the gate at A and B.

10.86 The gate in Problem 10.85 is designed to rotate and release the water when the depth d exceeds a certain value. What is that depth?

Problems 10.85/10.86

10.87* The dam has water of depth 4 ft on one side. The width of the dam (the dimension into the page) is 8 ft. The weight density of the water is $\gamma = 62.4$ lb/ft^3, and atmospheric pressure $p_{atm} = 2120$ lb/ft^2. If you neglect the weight of the dam, what are the reactions at A and B?

Problem 10.87

10.88* The dam has water of depth 4 ft on one side. The width of the dam (the dimension into the page) is 8 ft. The weight density of the water is $\gamma = 62.4$ lb/ft^3, and atmospheric pressure is $p_{atm} = 2120$ lb/ft^2. If you neglect the weight of the dam, what are the reactions at A and B?

Problem 10.88

10.89 Consider a plane, vertical area A below the surface of a liquid. Let p_0 be the pressure at the surface.

(a) Show that the force exerted by pressure on the area is $F = \bar{p}A$, where $\bar{p} = p_0 + \gamma\bar{x}$ is the pressure of the liquid at the centroid of the area.

(b) Show that the x coordinate of the center of pressure is

$$x_p = \bar{x} + \frac{\gamma I_{y'}}{\bar{p}A},$$

where $I_{y'}$ is the moment of inertia of the area about the y' axis through its centroid.

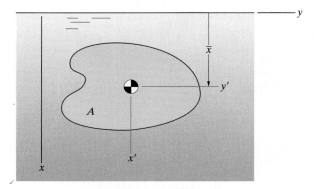

Problem 10.89

10.90 A circular plate of 1-m radius is below the surface of a stationary pool of water. Atmospheric pressure is $p_{atm} = 1 \times 10^5$ Pa, and the mass density of the water is $\rho = 1000$ kg/m^3. Determine (a) the force exerted on a face of the plate by the pressure of the water; (b) the x coordinate of the center of pressure. (See Problem 10.89.)

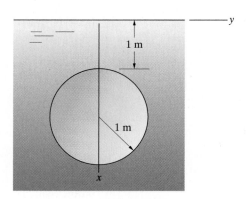

Problem 10.90

▶ **10.91*** The tank consists of a cylinder with hemispherical ends. It is filled with water ($\rho = 1000$ kg/m³). The pressure of the water at the top of the tank is 140 kPa. Determine the magnitude of the force exerted by the pressure of the water on each hemispherical end of the tank. (See Example 10.12.)

Problem 10.91

10.92 An object of volume V and weight W is suspended below the surface of a stationary liquid of weight density γ (Fig. a). Show that the tension in the cord is $W - V\gamma$. In other words, show that the pressure distribution on the surface of the object exerts an upward force equal to the product of the object's volume and the weight density of the water. This result is due to Archimedes (287–212 B.C.).

Strategy: Draw the free-body diagram of a volume of liquid that has the same shape and position as the object (Fig. b).

(a)   (b)

Problem 10.92

Review Problems

10.93 Determine the internal forces and moment at B (a) if $x = 250$ mm; (b) if $x = 750$ mm.

Problem 10.93

10.94 Determine the internal forces and moment (a) at B; (b) at C.

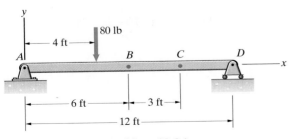

Problem 10.94

10.95 (a) Determine the maximum bending moment in the beam and the value of x where it occurs.

(b) Show that the equations for V and M as functions of x satisfy the equation $V = dM/dx$.

10.96 Draw the shear force and bending moment diagrams for the beam in Problem 10.95.

Problems 10.95/10.96

10.97 Determine the shear force and bending moment diagrams for the beam.

Problem 10.97

10.98 Determine V and M as functions of x for the beam ABC.

Problem 10.98

10.99 Draw the shear force and bending moment diagrams for beam ABC.

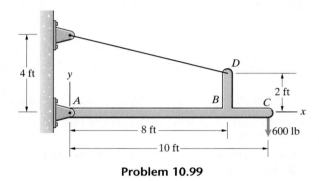

Problem 10.99

10.100 Determine the internal forces and moments at A.

10.101 Draw the shear force and bending moment diagrams of beam BC.

Problems 10.100/10.101

10.102 Determine the internal forces and moment at B
(a) if $x = 250$ mm;
(b) if $x = 750$ mm.

Problem 10.102

10.103 Draw the shear force and bending moment diagrams.

Problem 10.103

10.104 The homogeneous beam weighs 1000 lb. What are the internal forces and bending moment at its midpoint?

10.105 The homogeneous beam weighs 1000 lb. Draw the shear force and bending moment diagrams.

Problems 10.104/10.105

10.106 At A the main cable of the suspension bridge is horizontal and its tension is 1×10^8 lb.

(a) Determine the distributed load acting on the cable.

(b) What is the tension at B?

Problem 10.106

10.107 The power line has a mass of 1.4 kg/m. If the line will safely support a tension of 5 kN, determine whether it will safely support an ice accumulation of 0.4 kg/m.

Problem 10.107

10.108 The water depth at the center of the elliptical window is 20 ft. Determine the magnitude of the net force exerted on the window by the pressure of the seawater ($\gamma = 64$ lb/ft³) and the atmospheric pressure of the air on the opposite side. (See Problem 10.89.)

10.109 The water depth at the center of the elliptical window is 20 ft. Determine the magnitude of the net moment exerted on the window about the horizontal axis L by the pressure of the seawater ($\gamma = 64$ lb/ft³) and the atmospheric pressure of the air on the opposite side. (See Problem 10.89.)

Problems 10.108/10.109

10.110* The gate has water of 2-m depth on one side. The width of the gate (the dimension into the page) is 4 m, and its mass is 160 kg. The mass density of the water is $\rho = 1000$ kg/m³, and atmospheric pressure is $p_{atm} = 1 \times 10^5$ Pa. Determine the reactions on the gate at A and B. (The support at B exerts only a horizontal reaction on the gate.)

Problem 10.110

10.111 A spherical tank of 400-mm inner radius is full of water ($\rho = 1000$ kg/m³). The pressure of the water at the top of the tank is 4×10^5 Pa.

(a) What is the pressure of the water at the bottom of the tank?

(b) What is the total force exerted on the inner surface of the tank by the pressure of the water?

Strategy: For (b), draw a free-body diagram of the sphere of water in the tank.

Problem 10.111

Virtual Work and Potential Energy

When a spring is stretched, the work performed is stored in the spring as potential energy. Raising a load with a crane increases its gravitational potential energy. In this chapter we define work and potential energy and introduce a general and powerful result called the principle of virtual work.

◀ The torsional spring stores potential energy that drives the clock mechanism. In this chapter we use the concepts of virtual work and potential energy to analyze objects in equilibrium.

(a)

(b)

(c)

Figure 11.1
(a) A force $\mathbf{F}$ acting on an object.
(b) A displacement $d\mathbf{r}$ of P.
(c) The work $dU = (|\mathbf{F}| \cos \theta)|d\mathbf{r}|$.

11.1 Virtual Work

BACKGROUND

The principle of virtual work is a statement about work done by forces and couples when an object or structure is subjected to various hypothetical motions. Before we can introduce this principle, we must define work.

Work

Consider a force acting on an object at a point P (Fig. 11.1a). Suppose that the object undergoes an infinitesimal motion, so that P has a differential displacement $d\mathbf{r}$ (Fig. 11.1b). The *work* dU done by $\mathbf{F}$ as a result of the displacement $d\mathbf{r}$ is defined to be

$$dU = \mathbf{F} \cdot d\mathbf{r}. \tag{11.1}$$

From the definition of the dot product, $dU = (|\mathbf{F}| \cos \theta)|d\mathbf{r}|$, where θ is the angle between $\mathbf{F}$ and $d\mathbf{r}$ (Fig. 11.1c). The work is equal to the product of the component of $\mathbf{F}$ in the direction of $d\mathbf{r}$ and the magnitude of $d\mathbf{r}$. Notice that if the component of $\mathbf{F}$ parallel to $d\mathbf{r}$ points in the direction opposite to $d\mathbf{r}$, the work is negative. Also notice that if $\mathbf{F}$ is perpendicular to $d\mathbf{r}$, the work is zero. The dimensions of work are (force) $\times$ (length).

Now consider a couple acting on an object (Fig. 11.2a). The moment due to the couple is $M = Fh$ in the counterclockwise direction. If the object rotates through an infinitesimal counterclockwise angle $d\alpha$ (Fig. 11.2b), the points of application of the forces are displaced through differential distances $\frac{1}{2}h\,d\alpha$. Consequently, the total work done is $dU = F(\frac{1}{2}h\,d\alpha) + F(\frac{1}{2}h\,d\alpha) = M\,d\alpha$.

We see that when an object acted on by a couple M is rotated through an angle $d\alpha$ in the same direction as the couple (Fig. 11.2c), the resulting work is

$$dU = M\,d\alpha. \tag{11.2}$$

If the direction of the couple is opposite to the direction of $d\alpha$, the work is negative.

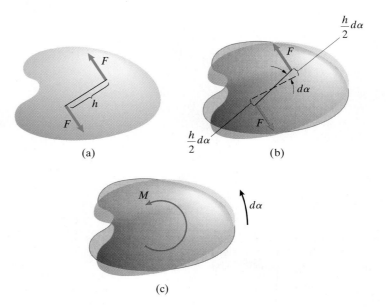

(a)

(b)

(c)

Figure 11.2
(a) A couple acting on an object.
(b) An infinitesimal rotation of the object.
(c) An object acted on by a couple M rotating through an angle $d\alpha$.

Principle of Virtual Work

Now that we have defined the work done by forces and couples, we can introduce the principle of virtual work. Before stating it, we first discuss an example to give you context for understanding the principle.

The homogeneous bar in Fig. 11.3a is supported by the wall and by the pin support at A and is loaded by a couple M. The free-body diagram of the bar is shown in Fig. 11.3b. The equilibrium equations are

$$\Sigma F_x = A_x - N = 0, \qquad (11.3)$$

$$\Sigma F_y = A_y - W = 0, \qquad (11.4)$$

$$\Sigma M_{\text{point } A} = NL \sin \alpha - W\frac{1}{2}L \cos \alpha - M = 0. \qquad (11.5)$$

We can solve these three equations for the reactions A_x, A_y, and N. However, we have a different objective.

Consider the following question: If the bar is acted on by the forces and couple in Fig. 11.3b and we subject it to a hypothetical infinitesimal translation in the x direction, as shown in Fig. 11.4, what work is done? The hypothetical displacement δx is called a *virtual displacement* of the bar, and the resulting work δU is called the *virtual work*. The pin support and the wall prevent the bar from actually moving in the x direction: the virtual displacement is a theoretical artifice. Our objective is to calculate the resulting virtual work:

$$\delta U = A_x \delta_x + (-N)\delta_x = (A_x - N)\delta_x. \qquad (11.6)$$

The forces A_y and W do no work because they are perpendicular to the displacements of their points of application. The couple M also does no work, because the bar does not rotate. Comparing this equation with Eq. (11.3), we find that the virtual work equals zero.

Next, we give the bar a virtual translation in the y direction (Fig. 11.5). The resulting virtual work is

$$\delta U = A_y \delta y + (-W)\delta y = (A_y - W)\delta y. \qquad (11.7)$$

From Eq. (11.4), the virtual work again equals zero.

Finally, we give the bar a virtual rotation while holding point A fixed (Fig. 11.6a). The forces A_x and A_y do no work because their point of application does not move. The work done by the couple M is $-M\,\delta\alpha$, because its direction is opposite to that of the rotation. The displacements of the points of application of the forces N and W are shown in Fig. 11.6b, and the components of the forces in the direction of the displacements are shown in Fig. 11.6c. The work done by N is $(N \sin \alpha)(L\,\delta\alpha)$, and the work done by W is $(-W \cos \alpha)(\frac{1}{2}L\,\delta\alpha)$. The total work is

$$\delta U = (N \sin \alpha)(L\delta\alpha) + (-W \cos \alpha)\left(\frac{1}{2}L\delta\alpha\right) - M\,\delta\alpha$$

$$= \left(NL \sin \alpha - W\frac{1}{2}L \cos \alpha - M\right)\delta\alpha. \qquad (11.8)$$

From Eq. (11.5), the virtual work resulting from the virtual rotation is also zero.

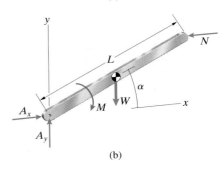

Figure 11.3
(a) A bar subjected to a couple M.
(b) Free-body diagram of the bar.

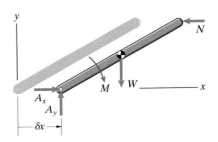

Figure 11.4
A virtual displacement δx.

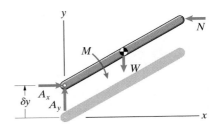

Figure 11.5
A virtual displacement δy.

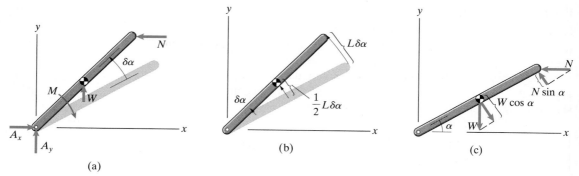

Figure 11.6
(a) A virtual rotation $\delta\alpha$.
(b) Displacements of the points of application of N and W.
(c) Components of N and W in the direction of the displacements.

We have shown that for three virtual motions of the bar, the virtual work is zero. These results are examples of a form of the principle of virtual work:

If an object is in equilibrium, the virtual work done by the external forces and couples acting on it is zero for any virtual translation or rotation:

$$\delta U = 0. \tag{11.9}$$

As our example illustrates, this principle can be used to derive the equilibrium equations for an object. Subjecting the bar to virtual translations δx and δy and a virtual rotation $\delta\alpha$ results in Eqs. (11.6)–(11.8). Because the virtual work must be zero in each case, we obtain Eqs. (11.3)–(11.5). But there is no advantage to this approach compared to simply drawing the free-body diagram of the object and writing the equations of equilibrium in the usual way. The advantages of the principle of virtual work become evident when we consider structures.

Application to Structures

The principle of virtual work stated in the preceding section applies to each member of a structure. By subjecting certain types of structures in equilibrium to virtual motions and calculating the total virtual work, we can determine unknown reactions at their supports as well as internal forces in their members. The procedure involves finding virtual motions that result in virtual work being done both by known loads and by unknown forces and couples.

Suppose that we want to determine the axial load in member BD of the truss in Fig. 11.7a. The other members of the truss are subjected to the 4-kN load and the forces exerted on them by member BD (Fig. 11.7b). If we give the structure a virtual rotation $\delta\alpha$ as shown in Fig. 11.7c, virtual work is done by the force T_{BD} acting at B and by the 4-kN load at C. Furthermore, the virtual work done by these two forces is the total virtual work done on the members of the structure, because the virtual work done by the internal forces they exert on each other cancels out. For example, consider joint C (Fig. 11.7d). The force T_{BC} is the axial load in member BC. The virtual work done at C on member BC is $T_{BC}(1.4 \text{ m})\,\delta\alpha$, and the work done at C on member CD is $(4 \text{ kN} - T_{BC})(1.4 \text{ m})\,\delta\alpha$. When we add up the virtual work done on the members to obtain the total virtual work on the structure, the virtual work due to the internal force T_{BC} cancels out. (If the members exerted

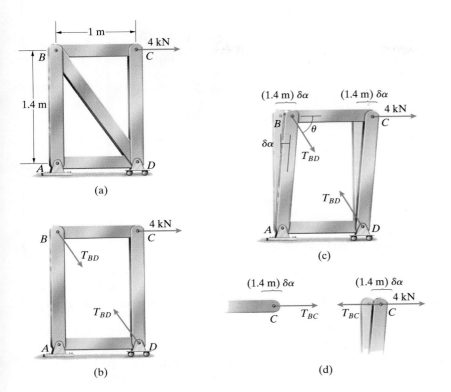

Figure 11.7
(a) A truss with a 4-kN load.
(b) Forces exerted by member BD.
(c) A virtual motion of the structure.
(d) Calculating the virtual work on members BC and CD at the joint C.

an internal *couple* on each other at C—for example, as a result of friction in the pin support—the virtual work would not cancel out.) Therefore, we can ignore internal forces in calculating the total virtual work on the structure:

$$\delta U = (T_{BD} \cos \theta)(1.4 \text{ m}) \, \delta\alpha + (4 \text{ kN})(1.4 \text{ m}) \, \delta\alpha = 0.$$

The angle $\theta = \arctan(1.4/1) = 54.5°$. Solving this equation, we obtain $T_{BD} = -6.88$ kN.

RESULTS

Work

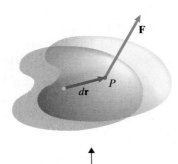

$dU = \mathbf{F} \cdot d\mathbf{r}.$ (11.1) ◀—— Work done by a force $\mathbf{F}$ as a result of a virtual displacement $d\mathbf{r}$ of its point of application.

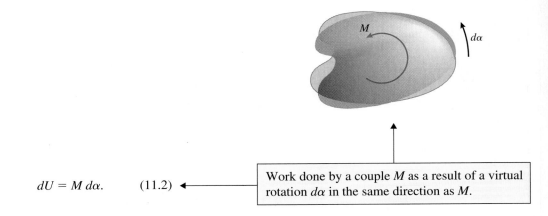

$$dU = M\,d\alpha. \qquad (11.2)$$

Work done by a couple M as a result of a virtual rotation $d\alpha$ in the same direction as M.

Principle of Virtual Work

If an object is in equilibrium, the virtual work done by the external forces and couples acting on it is zero for any virtual translation or rotation:

$$\delta U = 0. \qquad (11.9)$$

A virtual displacement or rotation is a *hypothetical* infinitesimal displacement or rotation.

The principle of virtual work can be applied to structures if no net work is done by internal forces and couples the members exert on each other. This involves two steps:

1. Choose a virtual motion—Identify a virtual motion of the structure that results in virtual work being done by known loads and by an unknown force or couple that is to be determined.
2. Determine the virtual work—Calculate the total virtual work resulting from the virtual motion to obtain an equation for the unknown force or couple.

Active Example 11.1 **Applying Virtual Work to a Structure** (▶ *Related Problems 11.12–11.16*)

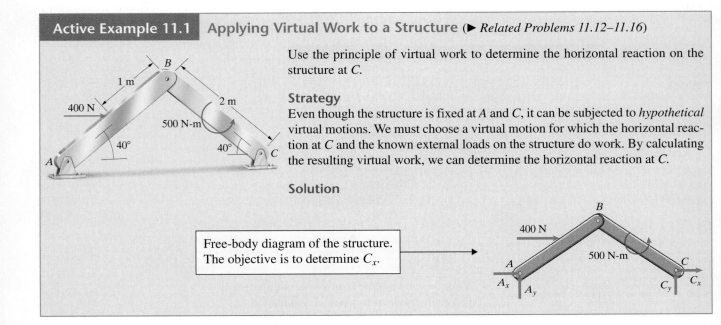

Use the principle of virtual work to determine the horizontal reaction on the structure at C.

Strategy
Even though the structure is fixed at A and C, it can be subjected to *hypothetical* virtual motions. We must choose a virtual motion for which the horizontal reaction at C and the known external loads on the structure do work. By calculating the resulting virtual work, we can determine the horizontal reaction at C.

Solution

Free-body diagram of the structure. The objective is to determine C_x.

Choose a virtual motion—Hold point A fixed and let point C undergo a horizontal virtual displacement δx. As a result, bar AB undergoes a clockwise virtual rotation $\delta\alpha$.

The work done by the 400-N force is $(400 \sin 40° \text{ N})(1 \text{ m})\delta\alpha$.

Bar BC undergoes a rotation $\delta\alpha$ in the counterclockwise direction, so the work done by the couple is $(500 \text{ N-m})\delta\alpha$. The work done by the reaction C_x is $C_x \delta x$. The total virtual work is

$$\delta U = (400 \sin 40° \text{ N})(1 \text{ m})\delta\alpha +$$
$$(500 \text{ N-m})\delta\alpha + C_x \delta x = 0.$$

To obtain C_x from this equation, the relationship between $\delta\alpha$ and δx must be determined.

> Determine the virtual work.

From the geometry of the structure,

$$x = 2 (2 \cos \alpha).$$

The derivative of this equation with respect to α is

$$\frac{dx}{d\alpha} = -4 \sin \alpha,$$

so an infinitesimal change in x is related to an infinitesimal change in α by

$$dx = -4 \sin \alpha \, d\alpha.$$

Because the clockwise virtual rotation $\delta\alpha$ is a *decrease* in α, δx is related to $\delta\alpha$ by

$$\delta x = 4 \sin 40° \, \delta\alpha.$$

Substituting this expression into the equation for the total virtual work gives

$$\delta U = \left[(400 \sin 40° \text{ N})(1 \text{ m}) + (500 \text{ N-m}) \right.$$
$$\left. + (4 \sin 40° \text{ m})C_x\right]\delta\alpha = 0.$$

Solving yields $C_x = -294$ N.

> Obtain the relationship between $\delta\alpha$ and δx.

Practice Problem Use the principle of virtual work to determine the vertical reaction at C. Do so by holding point A fixed and subjecting the *entire structure* to a rigid clockwise virtual rotation $\delta\alpha$.

Answer: $C_y = -79.3$ N.

Example 11.2   Applying Virtual Work to a Machine (▶ *Related Problem 11.21*)

The extensible platform is raised and lowered by the hydraulic cylinder *BC*. The total weight of the platform and men is *W*. The weights of the beams supporting the platform can be neglected. What axial force must the hydraulic cylinder exert to hold the platform in equilibrium in the position shown?

Strategy
We can use a virtual motion that coincides with the actual motion of the platform and beams when the length of the hydraulic cylinder changes. By calculating the virtual work done by the hydraulic cylinder and by the weight of the men and platform, we can determine the force exerted by the hydraulic cylinder.

Solution

Choose a Virtual Motion
We draw the free-body diagram of the platform and beams in Fig. a. Our objective is to determine the force *F* exerted by the hydraulic cylinder. If we hold point *A* fixed and subject point *C* to a horizontal virtual displacement δx, the only external forces that do virtual work are *F* and the weight *W*. (The reaction due to the roller support at *C* is perpendicular to the virtual displacement.)

(a) Free-body diagram of the platform and supporting beams.

Determine the Virtual Work The virtual work done by the force F as point C undergoes a virtual displacement δx to the right (Fig. b) is $-F\,\delta x$. To determine the virtual work done by the weight W, we must determine the vertical displacement of point D in Fig. b when point C moves to the right a distance δx. The dimensions b and h are related by

$$b^2 + h^2 = L^2,$$

where L is the length of the beam AD. Taking the derivative of this equation with respect to b, we obtain

$$2b + 2h\frac{dh}{db} = 0,$$

which we can solve for dh in terms of db:

$$dh = -\frac{b}{h}db.$$

Thus, when b increases an amount δx, the dimension h *decreases* an amount $(b/h)\,\delta x$. Because there are three pairs of beams, the platform moves downward a distance $(3b/h)\,\delta x$, and the virtual work done by the weight is $(3b/h)W\,\delta x$. The total virtual work is

$$\delta U = \left[-F + \left(\frac{3b}{h}\right)W\right]\delta x = 0,$$

and we obtain $F = (3b/h)W$.

Critical Thinking

We designed this example to demonstrate how advantageous the method of virtual work can be for certain types of problems. You can see that it would be very tedious to draw the free-body diagrams of the individual members of the frame supporting the platform and solve the equilibrium equations to determine the force exerted by the hydraulic cylinder. In contrast, it was relatively simple to determine the virtual work done by the external forces acting on the frame.

(b) A virtual displacement in which A remains fixed and C moves horizontally.

Problems

The following problems are to be solved using the principle of virtual work.

11.1 Determine the reactions at A.

Strategy: Subject the beam to three virtual motions: **1.** a horizontal displacement δx; **2.** a vertical displacement δy; and **3.** a rotation $\delta\theta$ about A.

Problem 11.1

11.2 (a) Determine the virtual work done by the 2-kN force and the 2.4 kN-m couple when the beam is rotated through a counterclockwise angle $\delta\theta$ about point A.

(b) Use the result of (a) to determine the reaction at B.

Problem 11.2

11.3 Determine the tension in the cable.

Problem 11.3

11.4 The L-shaped bar is in equilibrium. Determine F.

Problem 11.4

11.5 The dimension $L = 4$ ft and $w_0 = 300$ lb/ft. Determine the reactions at A and B.

Strategy: To determine the virtual work done by the distributed load, represent it by an equivalent force.

Problem 11.5

11.6 Determine the reactions at A and B.

Problem 11.6

11.7 The mechanism is in equilibrium. Determine the force R in terms of F.

Problem 11.7

11.8 Determine the reaction at the roller support.

Problem 11.8

11.9 Determine the couple M necessary for the mechanism to be in equilibrium.

Problem 11.9

11.10 The system is in equilibrium. The total mass of the suspended load and assembly A is 120 kg.

(a) By using equilibrium, determine the force F.

(b) Using the result of (a) and the principle of virtual work, determine the distance the suspended load rises if the cable is pulled upward 300 mm at B.

Problem 11.10

11.11 Determine the force P necessary for the mechanism to be in equilibrium.

Problem 11.11

▶ 11.12* Show that δx is related to $\delta \alpha$ by

$$\delta x = (L_1 \tan \beta)\, \delta \alpha.$$

(See Active Example 11.1.)

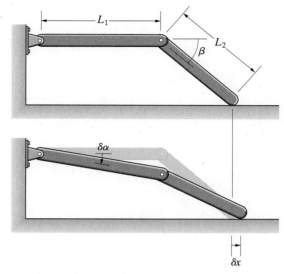

Problem 11.12

▶ 11.13 The horizontal surface is smooth. Determine the horizontal force F necessary for the system to be in equilibrium. (See Active Example 11.1.)

Problem 11.13

▶ **11.14*** Show that δx is related to $\delta\alpha$ by

$$\delta x = \frac{L_1 x \sin\alpha}{x - L_1 \cos\alpha}\delta\alpha.$$

Strategy: Write the law of cosines in terms of α and take the derivative of the resulting equation with respect to α. (See Active Example 11.1.)

Problem 11.14

▶ **11.15** The linkage is in equilibrium. What is the force F? (See Active Example 11.1.)

Problem 11.15

▶ **11.16** The linkage is in equilibrium. What is the force F? (See Active Example 11.1.)

Problem 11.16

11.17 Bar AC is connected to bar BD by a pin that fits in the smooth vertical slot. The masses of the bars are negligible. If $M_A = 30$ N-m, what couple M_B is necessary for the system to be in equilibrium?

Problem 11.17

11.18 The angle $\alpha = 20°$, and the force exerted on the stationary piston by pressure is 4 kN toward the left. What couple M is necessary to keep the system in equilibrium?

Problem 11.18

11.19 The structure is subjected to a 400-N load and is held in place by a horizontal cable. Determine the tension in the cable.

Problem 11.19

11.20 If the load on the car jack is $L = 6.5$ kN, what is the tension in the threaded shaft between A and B?

Problem 11.20

▶ **11.21** Determine the reactions at A and B. (Use the equilibrium equations to determine the horizontal components of the reactions, and use the procedure described in Example 11.2 to determine the vertical components.)

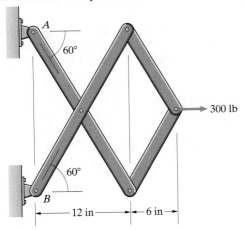

Problem 11.21

11.22 This device raises a load W by extending the hydraulic actuator DE. The bars AD and BC are each 2 m long, and the distances $b = 1.4$ m and $h = 0.8$ m. If $W = 4$ kN, what force must the actuator exert to hold the load in equilibrium?

Problem 11.22

11.23 Determine the force P necessary for the mechanism to be in equilibrium.

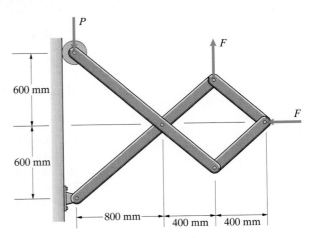

Problem 11.23

11.24 The collar A slides on the smooth vertical bar. The masses are $m_A = 20$ kg and $m_B = 10$ kg.

(a) If the collar A is given an upward virtual displacement δy, what is the resulting downward displacement of the mass B?

(b) Use virtual work to determine the tension in the spring.

Problem 11.24

11.2 Potential Energy

BACKGROUND

The work of a force $\mathbf{F}$ due to a differential displacement of its point of application is

$$dU = \mathbf{F} \cdot d\mathbf{r}.$$

If a function of position V exists such that for any $d\mathbf{r}$,

$$dU = \mathbf{F} \cdot d\mathbf{r} = -dV, \tag{11.10}$$

the function V is called the *potential energy* associated with the force $\mathbf{F}$, and $\mathbf{F}$ is said to be *conservative*. (The negative sign in this equation is in keeping with the interpretation of V as "potential" energy. Positive work results from a decrease in V.) If the forces that do work on a system are conservative, we will show that the total potential energy of the system can be used to determine its equilibrium positions.

Examples of Conservative Forces

Weights of objects and the forces exerted by linear springs are conservative. In the following sections, we derive the potential energies associated with these forces.

Weight In terms of a coordinate system with its y axis upward, the force exerted by the weight of an object is $\mathbf{F} = -W\mathbf{j}$ (Fig. 11.8a). If we give the object an arbitrary displacement $d\mathbf{r} = dx\mathbf{i} + dy\mathbf{j} + dz\mathbf{k}$ (Fig. 11.8b), the work done by its weight is

$$dU = \mathbf{F} \cdot d\mathbf{r} = (-W\mathbf{j}) \cdot (dx\mathbf{i} + dy\mathbf{j} + dz\mathbf{k}) = -W\,dy.$$

We seek a potential energy V such that

$$dU = -W\,dy = -dV \tag{11.11}$$

or

$$\frac{dV}{dy} = W.$$

If we neglect the variation in the weight with height and integrate, we obtain

$$V = Wy + C.$$

The constant C is arbitrary. Since this function satisfies Eq. (11.11) for any value of C, we will let $C = 0$. The position of the origin of the coordinate system can also be chosen arbitrarily. Thus, the potential energy associated with the weight of an object is

$$V = Wy, \tag{11.12}$$

where y is the height of the object above some chosen reference level, or *datum*.

Springs Consider a linear spring connecting an object to a fixed support (Fig. 11.9a). In terms of the stretch $S = r - r_0$, where r is the length of the spring and r_0 is its unstretched length, the force exerted on the object is kS (Fig. 11.9b). If the point at which the spring is attached to the object undergoes a differential displacement $d\mathbf{r}$ (Fig. 11.9c), the work done by the force on the object is

$$dU = -kS\,dS,$$

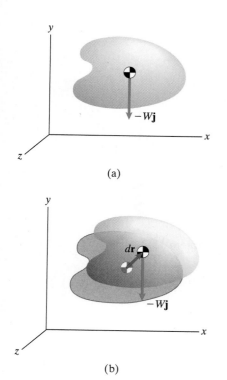

(a)

(b)

Figure 11.8
(a) Force exerted by the weight of an object.
(b) A differential displacement.

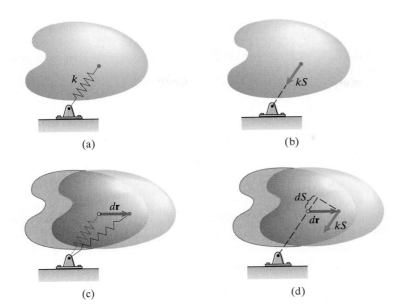

Figure 11.9
(a) A spring connected to an object.
(b) The force exerted on the object.
(c) A differential displacement of the object.
(d) The work done by the force is
$dU = -kS\,dS$.

where dS is the increase in the stretch of the spring resulting from the displacement (Fig. 11.9d). We seek a potential energy V such that

$$dU = -kS\,dS = -dV \tag{11.13}$$

or

$$\frac{dV}{dS} = kS.$$

Integrating this equation and letting the integration constant be zero, we obtain the potential energy associated with the force exerted by a linear spring:

$$V = \frac{1}{2}kS^2. \tag{11.14}$$

Notice that V is positive if the spring is either stretched (S is positive) or compressed (S is negative). Potential energy (the potential to do work) is stored in a spring by either stretching or compressing it.

Principle of Virtual Work for Conservative Forces

Because the work done by a conservative force is expressed in terms of its potential energy through Eq. (11.10), we can give an alternative statement of the principle of virtual work when an object is subjected to conservative forces:

Let an object be in equilibrium. If the forces and couples that do work on the object as the result of a virtual translation or rotation are conservative, the change in the total potential energy is zero:

$$\delta V = 0. \tag{11.15}$$

We emphasize that it is not necessary that all of the forces and couples acting on the object be conservative for this result to hold; it is necessary only that the forces and couples that do work be conservative. This principle also applies to a system of interconnected objects if the external forces that do work are conservative and the internal forces at the connections between objects either do no work or are conservative. Such a system is called a *conservative system*.

If the position of a system can be specified by a single coordinate q, the system is said to have one *degree of freedom*. The total potential energy of a conservative, one-degree-of-freedom system can be expressed in terms of q, and we can write Eq. (11.15) as

$$\delta V = \frac{dV}{dq} \delta q = 0.$$

Thus, when the object or system is in equilibrium, the derivative of its total potential energy with respect to q is zero:

$$\frac{dV}{dq} = 0. \tag{11.16}$$

We can use this equation to determine the values of q at which the system is in equilibrium.

Stability of Equilibrium

Suppose that a homogeneous bar of weight W and length L is suspended from a pin support at one end. In terms of the angle α shown in Fig. 11.10a, the height of the center of mass relative to the pinned end is $-\frac{1}{2}L \cos \alpha$. Choosing the level of the pin support as the datum, we can therefore express the potential energy associated with the weight of the bar as

$$V = -\frac{1}{2}WL \cos \alpha.$$

When the bar is in equilibrium,

$$\frac{dV}{d\alpha} = \frac{1}{2}WL \sin \alpha = 0.$$

This condition is satisfied when $\alpha = 0$ (Fig. 11.10b) and also when $\alpha = 180°$ (Fig. 11.10c).

There is a fundamental difference between the two equilibrium positions of the bar. In the position shown in Fig. 11.10b, if we displace the bar slightly from its equilibrium position and release it, the bar will remain near the equilibrium position. We say that this equilibrium position is *stable*. When the bar is in the position shown in Fig. 11.10c, if we displace it slightly and release it, the bar will move away from the equilibrium position. This equilibrium position is *unstable*.

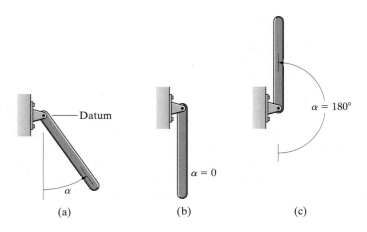

Figure 11.10
(a) A bar suspended from one end.
(b) The equilibrium position $\alpha = 0$.
(c) The equilibrium position $\alpha = 180°$.

The graph of the bar's potential energy V as a function of α is shown in Fig. 11.11a. The potential energy is a minimum at the stable equilibrium position $\alpha = 0$ and a maximum at the unstable equilibrium position $\alpha = 180°$. The derivative of V (Fig. 11.11b) equals zero at both equilibrium positions. The second derivative of V (Fig. 11.11c) is positive at the stable equilibrium position $\alpha = 0$ and negative at the unstable equilibrium position $\alpha = 180°$.

If a conservative, one-degree-of-freedom system is in equilibrium and the second derivative of V evaluated at the equilibrium position is positive, the equilibrium position is stable. If the second derivative of V is negative, it is unstable (Fig. 11.12).

$$\frac{dV}{dq} = 0, \quad \frac{d^2V}{dq^2} > 0: \qquad \text{Stable equilibrium}$$

$$\frac{dV}{dq} = 0, \quad \frac{d^2V}{dq^2} < 0: \qquad \text{Unstable equilibrium}$$

Proving these results requires analyzing the motion of the system near an equilibrium position.

Using potential energy to analyze the equilibrium of one-degree-of-freedom systems typically involves three steps:

1. **Determine the potential energy**—Express the total potential energy in terms of a single coordinate that specifies the position of the system.
2. **Find the equilibrium positions**—By calculating the first derivative of the potential energy, determine the equilibrium position or positions.
3. **Examine the stability**—Use the sign of the second derivative of the potential energy to determine whether the equilibrium positions are stable.

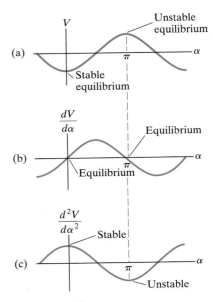

Figure 11.11
Graphs of V, $dV/d\alpha$, and $d^2V/d\alpha^2$.

Figure 11.12
Graphs of the potential energy V as a function of the coordinate q that exhibit stable and unstable equilibrium positions.

RESULTS

Potential Energy

If a function of position V exists such that, for any infinitesimal displacement $d\mathbf{r}$, the work done by a force $\mathbf{F}$ is

$$dU = \mathbf{F} \cdot d\mathbf{r} = -dV,$$

then V is called the *potential energy* associated with the force and $\mathbf{F}$ is said to be *conservative*.

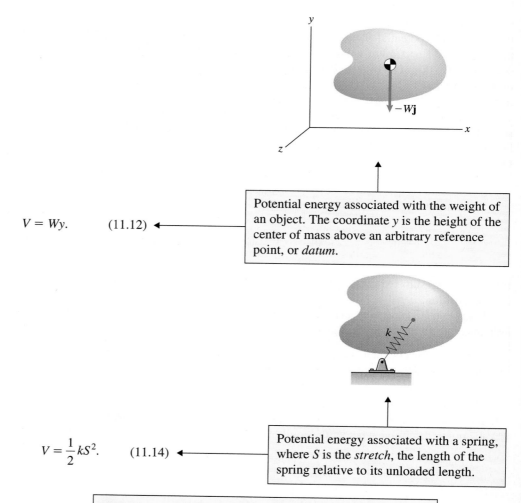

$$V = Wy. \qquad (11.12)$$

Potential energy associated with the weight of an object. The coordinate y is the height of the center of mass above an arbitrary reference point, or *datum*.

$$V = \frac{1}{2}kS^2. \qquad (11.14)$$

Potential energy associated with a spring, where S is the *stretch*, the length of the spring relative to its unloaded length.

Principle of Virtual Work for Conservative Forces

Let an object be in equilibrium. If the forces and couples that do work on the object as the result of a virtual translation or rotation are conservative, the change in the total potential energy is zero:

$$\delta V = 0. \qquad (11.15)$$

This principle also applies to a system of interconnected objects if the external forces that do work are conservative and the internal forces between objects either do no work or are conservative. Such a system is called a *conservative system*.

If the position of a system can be specified by a single coordinate q, the system is said to have one *degree of freedom*. When a conservative, one-degree-of-freedom system is in equilibrium,

$$\frac{dV}{dq} = 0. \qquad (11.16)$$

If the second derivative of V with respect to q is positive, the equilibrium position is stable, and if the second derivative is negative, the equilibrium position is unstable.

Active Example 11.3 **Stability of a Conservative System** (▶ *Related Problems 11.27–11.29*)

A crate of weight W is suspended from the ceiling by a spring. The coordinate x measures the vertical position of the center of mass of the crate relative to its position when the spring is unstretched. Determine the equilibrium position of the crate relative to its position when the spring is unstretched.

Strategy

The forces acting on the crate—its weight and the force exerted by the spring—are conservative. We can express the total potential energy in terms of the coordinate x and use Eq. (11.16) to determine the equilibrium position.

Solution

Let $x = 0$ be the datum. Because x is positive downward, the potential energy is $-Wx$.

> Potential energy associated with the weight.

The stretch of the spring equals x, so the potential energy is

$$\frac{1}{2}kx^2.$$

> Potential energy associated with the spring.

The total potential energy is

$$V = \frac{1}{2}kx^2 - Wx.$$

When the crate is equilibrium,

$$\frac{dV}{dx} = kx - W = 0.$$

> Apply Eq. (11.16).

The equilibrium position is

$$x = \frac{W}{k}.$$

Practice Problem Determine whether the equilibrium position of the crate is stable.

Answer: Yes.

Example 11.4 | **Stability of an Equilibrium Position** (▶ *Related Problems 11.31, 11.32*)

The homogeneous hemisphere is at rest on the plane surface. Show that it is in equilibrium in the position shown. Is the equilibrium position stable?

Strategy

To determine whether the hemisphere is in equilibrium and whether its equilibrium is stable, we must introduce a coordinate that specifies its orientation and express its potential energy in terms of that coordinate. We can use as the coordinate the angle of rotation of the hemisphere relative to the position shown.

Solution

Determine the Potential Energy Suppose that the hemisphere is rotated through an angle α relative to its original position (Fig. a). Then, from the datum shown, the potential energy associated with the weight W of the hemisphere is

$$V = -\frac{3}{8}RW \cos \alpha.$$

(a) The hemisphere rotated through an angle α.

Find the Equilibrium Positions When the hemisphere is in equilibrium,

$$\frac{dV}{d\alpha} = \frac{3}{8}RW \sin \alpha = 0,$$

which confirms that $\alpha = 0$ is an equilibrium position.

Examine the Stability The second derivative of the potential energy is

$$\frac{d^2V}{d\alpha^2} = \frac{3}{8}RW \cos \alpha.$$

This expression is positive at $\alpha = 0$, so the equilibrium position is stable.

Critical Thinking

Notice that we ignored the normal force exerted on the hemisphere by the plane surface. This force does no work and so does not affect the potential energy.

Example 11.5 Stability of an Equilibrium Position (▶ *Related Problems 11.41, 11.42*)

The pinned bars are held in place by the linear spring. Each bar has weight W and length L. The spring is unstretched when $\alpha = 0$, and the bars are in equilibrium when $\alpha = 60°$. Determine the spring constant k, and determine whether the equilibrium position is stable or unstable.

Strategy
The only forces that do work on the bars are their weights and the force exerted by the spring. By expressing the total potential energy in terms of α and using Eq. (11.16), we will obtain an equation we can solve for the spring constant k.

Solution
Determine the Potential Energy   If we use the datum shown in Fig. a, the potential energy associated with the weights of the two bars is

$$W\left(-\frac{1}{2}L \sin \alpha\right) + W\left(-\frac{1}{2}L \sin \alpha\right) = -WL \sin \alpha.$$

(a) Determining the total potential energy.

The spring is unstretched when $\alpha = 0$ and the distance between points A and B is $2L \cos \alpha$ (Fig. a), so the stretch of the spring is $2L - 2L \cos \alpha$. Therefore, the potential energy associated with the spring is $\frac{1}{2}k(2L - 2L \cos \alpha)^2$, and the total potential energy is

$$V = -WL \sin \alpha + 2kL^2(1 - \cos \alpha)^2.$$

When the system is in equilibrium,

$$\frac{dV}{d\alpha} = -WL \cos \alpha + 4kL^2 (\sin \alpha)(1 - \cos \alpha) = 0.$$

Because the system is in equilibrium when $\alpha = 60°$, we can solve this equation for the spring constant in terms of W and L:

$$k = \frac{W \cos \alpha}{4L (\sin \alpha)(1 - \cos \alpha)} = \frac{W \cos 60°}{4L (\sin 60°)(1 - \cos 60°)} = \frac{0.289W}{L}.$$

Examine the Stability   The second derivative of the potential energy is

$$\frac{d^2V}{d\alpha^2} = WL \sin \alpha + 4kL^2 (\cos \alpha - \cos^2 \alpha + \sin^2 \alpha)$$

$$= WL \sin 60° + 4kL^2 (\cos 60° - \cos^2 60° + \sin^2 60°)$$

$$= 0.866WL + 4kL^2.$$

This is a positive number, so the equilibrium position is stable.

Critical Thinking
How do you know when you can apply the principle of virtual work for conservative forces to a system? The system must be conservative, which means that the forces and couples that do work when the system undergoes a virtual motion are conservative. Conservative forces are forces for which a potential energy exists. In this example, work is done by the weights of the bars and the force exerted by the spring, which are conservative forces.

Problems

11.25 The potential energy of a conservative system is given by $V = 2x^3 + 3x^2 - 12x$.

(a) For what values of x is the system in equilibrium?

(b) Determine whether the equilibrium positions you found in (a) are stable or unstable.

11.26 The potential energy of a conservative system is given by $V = 2q^3 - 21q^2 + 72q$.

(a) For what values of q is the system in equilibrium?

(b) Determine whether the equilibrium positions you found in (a) are stable or unstable.

▶ **11.27** The mass $m = 2$ kg and the spring constant $k = 100$ N/m. The spring is unstretched when $x = 0$.

(a) Determine the value of x for which the mass is in equilibrium.

(b) Is the equilibrium position stable or unstable? (See Example 11.3.)

Problem 11.27

▶ **11.28** The *nonlinear* spring exerts a force $-kx + \varepsilon x^3$ on the mass, where k and ε are constants. Determine the potential energy V associated with the force exerted on the mass by the spring. (See Example 11.3.)

Problem 11.28

▶ **11.29** The 1-kg mass is suspended from the nonlinear spring described in Problem 11.28. The constants $k = 10$ and $\varepsilon = 1$, where x is in meters.

(a) Show that the mass is in equilibrium when $x = 1.12$ m and when $x = 2.45$ m.

(b) Determine whether the equilibrium positions are stable or unstable. (See Example 11.3.)

Problem 11.29

11.30 The two straight segments of the bar are each of weight W and length L. Determine whether the equilibrium position shown is stable if (a) $0 < \alpha_0 < 90°$; (b) $90° < \alpha_0 < 180°$.

Problem 11.30

▶ **11.31** The homogeneous composite object consists of a hemisphere and a cylinder. It is at rest on the plane surface. Show that this equilibrium position is stable only if $L < R/\sqrt{2}$. (See Example 11.4.)

Problem 11.31

▶ **11.32** The homogeneous composite object consists of a half-cylinder and a triangular prism. It is at rest on the plane surface. Show that this equilibrium position is stable only if $h < \sqrt{2}\,R$. (See Example 11.4.)

Problem 11.32

11.33 The homogeneous bar has weight W, and the spring is unstretched when the bar is vertical ($\alpha = 0$).

(a) Use potential energy to show that the bar is in equilibrium when $\alpha = 0$.

(b) Show that the equilibrium position $\alpha = 0$ is stable only if $2kL > W$.

11.34 Suppose that the bar in Problem 11.33 is in equilibrium when $\alpha = 20°$.

(a) Show that the spring constant $k = 0.490\ W/L$.

(b) Determine whether the equilibrium position is stable.

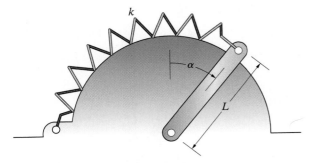

Problems 11.33/11.34

11.35 The bar AB has mass m and length L. The spring is unstretched when the bar is vertical ($\alpha = 0$). The light collar C slides on the smooth vertical bar so that the spring remains horizontal. Show that the equilibrium position $\alpha = 0$ is stable only if $2kL > mg$.

11.36 The bar AB in Problem 11.35 has mass $m = 4$ kg, length 2 m, and the spring constant is $k = 12$ N/m.

(a) Determine the value of α in the range $0 < \alpha < 90°$ for which the bar is in equilibrium.

(b) Is the equilibrium position determined in part (a) stable?

Problems 11.35/11.36

11.37 The bar AB has weight W and length L. The spring is unstretched when the bar is vertical ($\alpha = 0$). The light collar C slides on the smooth horizontal bar so that the spring remains vertical. Show that the equilibrium position $\alpha = 0$ is unstable.

11.38 The bar AB described in Problem 11.37 has a mass of 2 kg, and the spring constant is $k = 80$ N/m.

(a) Determine the value of α in the range $0 < \alpha < 90°$ for which the bar is in equilibrium.

(b) Is the equilibrium position determined in (a) stable?

Problems 11.37/11.38

11.39 Each homogeneous bar is of mass m and length L. The spring is unstretched when $\alpha = 0$. If $mg = kL$, determine the value of α in the range $0 < \alpha < 90°$ for which the system is in equilibrium.

11.40 Determine whether the equilibrium position found in Problem 11.39 is stable or unstable.

Problems 11.39/11.40

▶ **11.41** The pinned bars are held in place by the linear spring. Each bar has weight W and length L. The spring is unstretched when $\alpha = 90°$. Determine the value of α in the range $0 < \alpha < 90°$ for which the system is in equilibrium. (See Example 11.5.)

Problems 11.41/11.42

▶ **11.42** Determine whether the equilibrium position found in Problem 11.41 is stable or unstable. (See Example 11.5.)

11.43 The bar weighs 15 lb. The spring is unstretched when $\alpha = 0$. The bar is in equilibrium when $\alpha = 30°$. Determine the spring constant k.

11.44 Determine whether the equilibrium positions of the bar in Problem 11.43 are stable or unstable.

Problems 11.43/11.44

Review Problems

11.45 (a) Determine the couple exerted on the beam at A.
(b) Determine the vertical force exerted on the beam at A.

200 N-m   100 N

A 30°

2 m

Problem 11.45

11.46 The structure is subjected to a 20 kN-m couple. Determine the horizontal reaction at C.

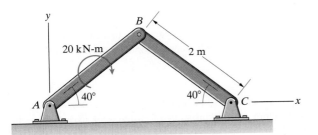

y

B

20 kN-m   2 m

A 40°   40° C x

Problem 11.46

11.47 The "rack and pinion" mechanism is used to exert a vertical force on a sample at A for a stamping operation. If a force $F = 30$ lb is exerted on the handle, use the principle of virtual work to determine the force exerted on the sample.

2 in

8 in

A F

Problem 11.47

11.48 If you were assigned to calculate the force exerted on the bolt by the pliers when the grips are subjected to forces F as shown in Fig. a, you could carefully measure the dimensions, draw free-body diagrams, and use the equilibrium equations. But another approach would be to measure the change in the distance between the jaws when the distance between the handles is changed by a small amount. If your measurements indicate that the distance d in Fig. b decreases by 1 mm when D is decreased 8 mm, what is the approximate value of the force exerted on the bolt by each jaw when the forces F are applied?

F

F

(a)

D d

(b)

Problem 11.48

11.49 The system is in equilibrium. The total weight of the suspended load and assembly A is 300 lb.

(a) By using equilibrium, determine the force F.

(b) Using the result of (a) and the principle of virtual work, determine the distance the suspended load rises if the cable is pulled downward 1 ft at B.

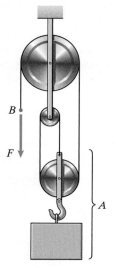

Problem 11.49

11.50 The system is in equilibrium.

(a) By drawing free-body diagrams and using equilibrium equations, determine the couple M.

(b) Using the result of (a) and the principle of virtual work, determine the angle through which pulley B rotates if pulley A rotates through an angle α.

Problem 11.50

11.51 The mechanism is in equilibrium. Neglect friction between the horizontal bar and the collar. Determine M in terms of F, α, and L.

Problem 11.51

11.52 In an injection casting machine, a couple M applied to arm AB exerts a force on the injection piston at C. Given that the horizontal component of the force exerted at C is 4 kN, use the principle of virtual work to determine M.

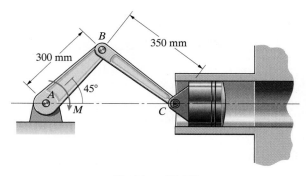

Problem 11.52

11.53 Show that if bar AB is subjected to a clockwise virtual rotation $\delta\alpha$, bar CD undergoes a counterclockwise virtual rotation $(b/a)\,\delta\alpha$.

11.54 The system is in equilibrium, $a = 800$ mm, and $b = 400$ mm. Use the principle of virtual work to determine the force F.

Problems 11.53/11.54

11.55 Show that if bar AB is subjected to a clockwise virtual rotation $\delta\alpha$, bar CD undergoes a clockwise virtual rotation $[ad/(ac + bc - bd)]\,\delta\alpha$.

11.56 The system is in equilibrium, $a = 300$ mm, $b = 350$ mm, $c = 350$ mm, and $d = 200$ mm. Use the principle of virtual work to determine the couple M.

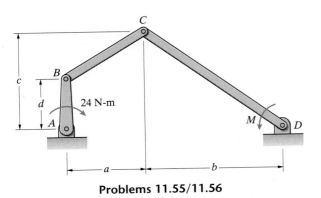

24 N-m

Problems 11.55/11.56

11.57 The mass of the bar is 10 kg, and it is 1 m in length. Neglect the masses of the two collars. The spring is unstretched when the bar is vertical ($\alpha = 0$), and the spring constant is $k = 100$ N/m. Determine the values of α at which the bar is in equilibrium.

11.58 Determine whether the equilibrium positions of the bar in Problem 11.57 are stable or unstable.

Problems 11.57/11.58

11.59 The spring is unstretched when $\alpha = 90°$. Determine the value of α in the range $0 < \alpha < 90°$ for which the system is in equilibrium.

11.60 Determine whether the equilibrium position found in Problem 11.59 is stable or unstable.

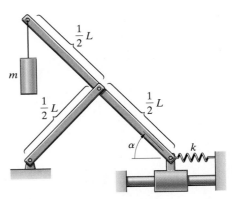

Problems 11.59/11.60

11.61 The hydraulic cylinder C exerts a horizontal force at A, raising the weight W. Determine the magnitude of the force the hydraulic cylinder must exert to support the weight in terms of W and α.

Problem 11.61

11.62 The homogeneous composite object consists of a hemisphere and a cone. It is at rest on the plane surface. Show that this equilibrium position is stable only if $h < \sqrt{3}R$.

Problem 11.62

APPENDIX

A

Review of Mathematics

A.1 Algebra

Quadratic Equations

The solutions of the quadratic equation

$$ax^2 + bx + c = 0$$

are

$$x = \frac{-b \pm \sqrt{b^2 - 4ac}}{2a}.$$

Natural Logarithms

The natural logarithm of a positive real number x is denoted by $\ln x$. It is defined to be the number such that

$$e^{\ln x} = x,$$

where $e = 2.7182\ldots$ is the base of natural logarithms.

Logarithms have the following properties:

$$\ln(xy) = \ln x + \ln y,$$
$$\ln(x/y) = \ln x - \ln y,$$
$$\ln y^x = x \ln y.$$

A.2 Trigonometry

The trigonometric functions for a right triangle are

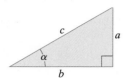

$$\sin \alpha = \frac{1}{\csc \alpha} = \frac{a}{c}, \quad \cos \alpha = \frac{1}{\sec \alpha} = \frac{b}{c}, \quad \tan \alpha = \frac{1}{\cot \alpha} = \frac{a}{b}.$$

The sine and cosine satisfy the relation

$$\sin^2 \alpha + \cos^2 \alpha = 1,$$

and the sine and cosine of the sum and difference of two angles satisfy

$$\sin(\alpha + \beta) = \sin \alpha \cos \beta + \cos \alpha \sin \beta,$$
$$\sin(\alpha - \beta) = \sin \alpha \cos \beta - \cos \alpha \sin \beta,$$
$$\cos(\alpha + \beta) = \cos \alpha \cos \beta - \sin \alpha \sin \beta,$$
$$\cos(\alpha - \beta) = \cos \alpha \cos \beta + \sin \alpha \sin \beta.$$

The **law of cosines** for an arbitrary triangle is

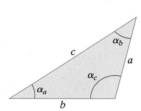

$$c^2 = a^2 + b^2 - 2ab \cos \alpha_c,$$

and the **law of sines** is

$$\frac{\sin \alpha_a}{a} = \frac{\sin \alpha_b}{b} = \frac{\sin \alpha_c}{c}.$$

A.3 Derivatives

$$\frac{d}{dx} x^n = nx^{n-1}$$

$$\frac{d}{dx} \sin x = \cos x$$

$$\frac{d}{dx} \sinh x = \cosh x$$

$$\frac{d}{dx} e^x = e^x$$

$$\frac{d}{dx} \cos x = -\sin x$$

$$\frac{d}{dx} \cosh x = \sinh x$$

$$\frac{d}{dx} \ln x = \frac{1}{x}$$

$$\frac{d}{dx} \tan x = \frac{1}{\cos^2 x}$$

$$\frac{d}{dx} \tanh x = \frac{1}{\cosh^2 x}$$

A.4 Integrals

$$\int x^n \, dx = \frac{x^{n+1}}{n+1} \quad (n \neq -1)$$

$$\int x^{-1} \, dx = \ln x$$

$$\int (a + bx)^{1/2} \, dx = \frac{2}{3b}(a + bx)^{3/2}$$

$$\int x(a + bx)^{1/2} \, dx = -\frac{2(2a - 3bx)(a + bx)^{3/2}}{15b^2}$$

$$\int (1 + a^2 x^2)^{1/2} \, dx = \frac{1}{2}\left\{ x(1 + a^2 x^2)^{1/2} \right.$$
$$\left. + \frac{1}{a} \ln \left[x + \left(\frac{1}{a^2} + x^2 \right)^{1/2} \right] \right\}$$

$$\int x(1 + a^2 x^2)^{1/2} \, dx = \frac{a}{3}\left(\frac{1}{a^2} + x^2 \right)^{3/2}$$

$$\int x^2(1 + a^2 x^2)^{1/2} \, dx = \frac{1}{4}ax\left(\frac{1}{a^2} + x^2 \right)^{3/2}$$
$$- \frac{1}{8a^2}x(1 + a^2 x^2)^{1/2} - \frac{1}{8a^3}\ln\left[x + \left(\frac{1}{a^2} + x^2 \right)^{1/2} \right]$$

$$\int (1 - a^2 x^2)^{1/2} \, dx = \frac{1}{2}\left[x(1 - a^2 x^2)^{1/2} + \frac{1}{a}\arcsin ax \right]$$

$$\int x(1 - a^2 x^2)^{1/2} \, dx = -\frac{a}{3}\left(\frac{1}{a^2} - x^2 \right)^{3/2}$$

$$\int x^2(a^2 - x^2)^{1/2} \, dx = -\frac{1}{4}x(a^2 - x^2)^{3/2}$$
$$+ \frac{1}{8}a^2\left[x(a^2 - x^2)^{1/2} + a^2 \arcsin\frac{x}{a} \right]$$

$$\int \frac{dx}{(1 + a^2 x^2)^{1/2}} = \frac{1}{a}\ln\left[x + \left(\frac{1}{a^2} + x^2 \right)^{1/2} \right]$$

$$\int \frac{dx}{(1 - a^2 x^2)^{1/2}} = \frac{1}{a}\arcsin ax \quad \text{or} \quad -\frac{1}{a}\arccos ax$$

$$\int \sin x \, dx = -\cos x$$

$$\int \cos x \, dx = \sin x$$

$$\int \sin^2 x \, dx = -\frac{1}{2}\sin x \cos x + \frac{1}{2}x$$

$$\int \cos^2 x \, dx = \frac{1}{2}\sin x \cos x + \frac{1}{2}x$$

$$\int \sin^3 x \, dx = -\frac{1}{3}\cos x(\sin^2 x + 2)$$

$$\int \cos^3 x \, dx = \frac{1}{3}\sin x(\cos^2 x + 2)$$

$$\int \cos^4 x \, dx = \frac{3}{8}x + \frac{1}{4}\sin 2x + \frac{1}{32}\sin 4x$$

$$\int \sin^n x \cos x \, dx = \frac{(\sin x)^{n+1}}{n+1} \quad (n \neq -1)$$

$$\int \sinh x \, dx = \cosh x$$

$$\int \cosh x \, dx = \sinh x$$

$$\int \tanh x \, dx = \ln \cosh x$$

$$\int e^{ax} \, dx = \frac{e^{ax}}{a}$$

$$\int xe^{ax} \, dx = \frac{e^{ax}}{a^2}(ax - 1)$$

A.5 Taylor Series

The Taylor series of a function $f(x)$ is

$$f(a + x) = f(a) + f'(a)x + \frac{1}{2!}f''(a)x^2 + \frac{1}{3!}f'''(a)x^3 + \cdots,$$

where the primes indicate derivatives.

Some useful Taylor series are

$$e^x = 1 + x + \frac{x^2}{2!} + \frac{x^3}{3!} + \cdots,$$

$$\sin(a + x) = \sin a + (\cos a)x - \frac{1}{2}(\sin a)x^2 - \frac{1}{6}(\cos a)x^3 + \cdots,$$

$$\cos(a + x) = \cos a - (\sin a)x - \frac{1}{2}(\cos a)x^2 + \frac{1}{6}(\sin a)x^3 + \cdots,$$

$$\tan(a + x) = \tan a + \left(\frac{1}{\cos^2 a}\right)x + \left(\frac{\sin a}{\cos^3 a}\right)x^2$$

$$+ \left(\frac{\sin^2 a}{\cos^4 a} + \frac{1}{3\cos^2 a}\right)x^3 + \cdots.$$

APPENDIX

B

Properties of Areas and Lines

B.1 Areas

The coordinates of the centroid of the area A are

$$\bar{x} = \frac{\int_A x \, dA}{\int_A dA}, \qquad \bar{y} = \frac{\int_A y \, dA}{\int_A dA}.$$

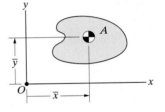

The moment of inertia about the x axis I_x, the moment of inertia about the y axis I_y, and the product of inertia I_{xy} are

$$I_x = \int_A y^2 \, dA, \qquad I_y = \int_A x^2 \, dA, \qquad I_{xy} = \int_A xy \, dA.$$

The polar moment of inertia about O is

$$J_O = \int_A r^2 \, dA = \int_A (x^2 + y^2) \, dA = I_x + I_y.$$

Area $= bh$

$$I_x = \frac{1}{3}bh^3, \qquad\qquad I_y = \frac{1}{3}hb^3, \qquad\qquad I_{xy} = \frac{1}{4}b^2h^2$$

$$I_{x'} = \frac{1}{12}bh^3, \qquad\qquad I_{y'} = \frac{1}{12}hb^3, \qquad\qquad I_{x'y'} = 0$$

Rectangular area

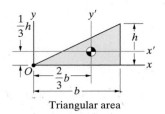

Triangular area

$$\text{Area} = \frac{1}{2}bh$$

$$I_x = \frac{1}{12}bh^3, \qquad I_y = \frac{1}{4}hb^3, \qquad I_{xy} = \frac{1}{8}b^2h^2$$

$$I_{x'} = \frac{1}{36}bh^3, \qquad I_{y'} = \frac{1}{36}hb^3, \qquad I_{x'y'} = \frac{1}{72}b^2h^2$$

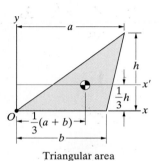

Triangular area

$$\text{Area} = \frac{1}{2}bh \qquad I_x = \frac{1}{12}bh^3, \qquad I_{x'} = \frac{1}{36}bh^3$$

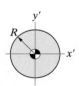

Circular area

$$\text{Area} = \pi R^2 \qquad I_{x'} = I_{y'} = \frac{1}{4}\pi R^4, \qquad I_{x'y'} = 0$$

Semicircular area

$$\text{Area} = \frac{1}{2}\pi R^2 \qquad I_x = I_y = \frac{1}{8}\pi R^4, \qquad I_{xy} = 0$$

$$I_{x'} = \frac{1}{8}\pi R^4, \qquad I_{y'} = \left(\frac{\pi}{8} - \frac{8}{9\pi}\right)R^4, \qquad I_{x'y'} = 0$$

Quarter-circular area

$$\text{Area} = \frac{1}{4}\pi R^2 \qquad I_x = I_y = \frac{1}{16}\pi R^4, \qquad I_{xy} = \frac{1}{8}R^4$$

$$I_{x'} = I_{y'} = \left(\frac{\pi}{16} - \frac{4}{9\pi}\right)R^4, \qquad I_{x'y'} = \left(\frac{1}{8} - \frac{4}{9\pi}\right)R^4$$

$$\text{Area} = \alpha R^2$$

$$I_x = \frac{1}{4}R^4\left(\alpha - \frac{1}{2}\sin 2\alpha\right), \qquad I_y = \frac{1}{4}R^4\left(\alpha + \frac{1}{2}\sin 2\alpha\right),$$

$$I_{xy} = 0$$

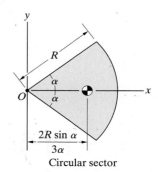

Circular sector

$$\text{Area} = \frac{1}{4}\pi ab$$

$$I_x = \frac{1}{16}\pi ab^3, \qquad I_y = \frac{1}{16}\pi a^3 b, \qquad I_{xy} = \frac{1}{8}a^2 b^2$$

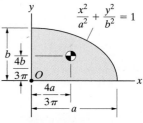

Quarter-elliptical area

$$\text{Area} = \frac{cb^{n+1}}{n+1}$$

$$I_x = \frac{c^3 b^{3n+1}}{9n+3}, \qquad I_y = \frac{cb^{n+3}}{n+3}, \qquad I_{xy} = \frac{c^2 b^{2n+2}}{4n+4}$$

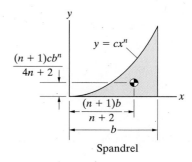

Spandrel

B.2 Lines

The coordinates of the centroid of the line L are

$$\bar{x} = \frac{\displaystyle\int_L x\, dL}{\displaystyle\int_L dL}, \qquad \bar{y} = \frac{\displaystyle\int_L y\, dL}{\displaystyle\int_L dL}, \qquad \bar{z} = \frac{\displaystyle\int_L z\, dL}{\displaystyle\int_L dL}.$$

Semicircular arc

Quarter-circular arc

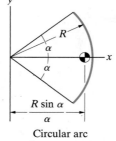

Circular arc

APPENDIX

C

Properties of Volumes and Homogeneous Objects

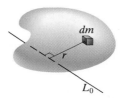

The coordinates of the centroid of the volume V are

$$\bar{x} = \frac{\int_V x \, dV}{\int_V dV}, \qquad \bar{y} = \frac{\int_V y \, dV}{\int_V dV}, \qquad \bar{z} = \frac{\int_V z \, dV}{\int_V dV}.$$

The center of mass of a homogeneous object coincides with the centroid of its volume.

The moment of inertia of the object about the axis L_0 is

$$I_0 = \int_m r^2 \, dm.$$

Slender bar

$$I_{x \text{ axis}} = 0, \qquad I_{y \text{ axis}} = I_{z \text{ axis}} = \frac{1}{3}ml^2$$

$$I_{x' \text{ axis}} = 0, \quad I_{y' \text{ axis}} = I_{z' \text{ axis}} = \frac{1}{12}ml^2$$

Thin circular plate

$$I_{x' \text{ axis}} = I_{y' \text{ axis}} = \frac{1}{4}mR^2, \qquad I_{z' \text{ axis}} = \frac{1}{2}mR^2$$

$$I_{x \text{ axis}} = \frac{1}{3}mh^2, \qquad I_{y \text{ axis}} = \frac{1}{3}mb^2, \qquad I_{z \text{ axis}} = \frac{1}{3}m(b^2 + h^2)$$

$$I_{x' \text{ axis}} = \frac{1}{12}mh^2, \qquad I_{y' \text{ axis}} = \frac{1}{12}mb^2, \qquad I_{z' \text{ axis}} = \frac{1}{12}m(b^2 + h^2)$$

Thin rectangular plate

$$I_{x \text{ axis}} = \frac{m}{A}I_x, \qquad I_{y \text{ axis}} = \frac{m}{A}I_y, \qquad I_{z \text{ axis}} = I_{x \text{ axis}} + I_{y \text{ axis}}$$

The terms I_x and I_y are the moments of inertia of the plate's cross-sectional area A about the x and y axes.

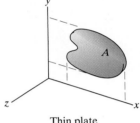

Thin plate

Volume $= abc$

$$I_{x' \text{ axis}} = \frac{1}{12}m(a^2 + b^2), \qquad I_{y' \text{ axis}} = \frac{1}{12}m(a^2 + c^2),$$

$$I_{z' \text{ axis}} = \frac{1}{12}m(b^2 + c^2)$$

Rectangular prism

Volume $= \pi R^2 l$

$$I_{x \text{ axis}} = I_{y \text{ axis}} = m\left(\frac{1}{3}l^2 + \frac{1}{4}R^2\right), \qquad I_{z \text{ axis}} = \frac{1}{2}mR^2$$

$$I_{x' \text{ axis}} = I_{y' \text{ axis}} = m\left(\frac{1}{12}l^2 + \frac{1}{4}R^2\right), \qquad I_{z' \text{ axis}} = \frac{1}{2}mR^2$$

Circular cylinder

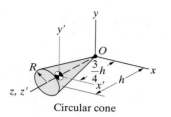

Circular cone

$$\text{Volume} = \frac{1}{3}\pi R^2 h$$

$$I_{x \text{ axis}} = I_{y \text{ axis}} = m\left(\frac{3}{5}h^2 + \frac{3}{20}R^2\right), \qquad I_{z \text{ axis}} = \frac{3}{10}mR^2$$

$$I_{x' \text{ axis}} = I_{y' \text{ axis}} = m\left(\frac{3}{80}h^2 + \frac{3}{20}R^2\right), \qquad I_{z' \text{ axis}} = \frac{3}{10}mR^2$$

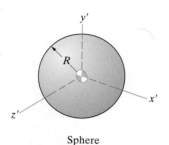

Sphere

$$\text{Volume} = \frac{4}{3}\pi R^3$$

$$I_{x' \text{ axis}} = I_{y' \text{ axis}} = I_{z' \text{ axis}} = \frac{2}{5}mR^2$$

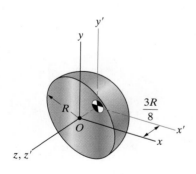

Hemisphere

$$\text{Volume} = \frac{2}{3}\pi R^3$$

$$I_{x \text{ axis}} = I_{y \text{ axis}} = I_{z \text{ axis}} = \frac{2}{5}mR^2$$

$$I_{x' \text{ axis}} = I_{y' \text{ axis}} = \frac{83}{320}mR^2, \quad I_{z' \text{ axis}} = \frac{2}{5}mR^2$$

Solutions to Practice Problems

Active Example 1.1

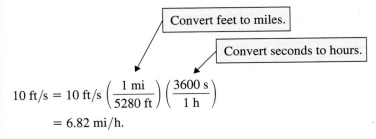

Convert feet to miles.

Convert seconds to hours.

$$10 \text{ ft/s} = 10 \text{ ft/s} \left(\frac{1 \text{ mi}}{5280 \text{ ft}} \right) \left(\frac{3600 \text{ s}}{1 \text{ h}} \right)$$

$$= 6.82 \text{ mi/h}.$$

Active Example 1.4

$$W = mg = (0.397 \text{ kg})(9.81 \text{ m/s}^2) = 3.89 \text{ N.}$$

Use Eq. (1.6) to calculate the weight in newtons.

Active Example 2.1

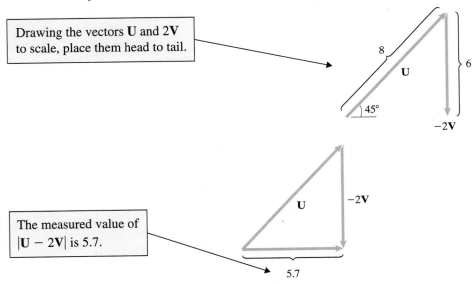

Drawing the vectors **U** and 2**V** to scale, place them head to tail.

8

U

6

45°

−2**V**

U

−2**V**

The measured value of $|\mathbf{U} - 2\mathbf{V}|$ is 5.7.

5.7

Active Example 2.3

$|F_y| = 3|F_x|$.

The magnitude of $\mathbf{F}$ is

$$900 \text{ N} = \sqrt{F_x^2 + F_y^2}$$

$$= \sqrt{F_x^2 + (3F_x)^2}.$$

Solving yields $F_x = 285$ N. The vector $\mathbf{F}$ in terms of its components is

$$\mathbf{F} = 285\mathbf{i} - 3(285)\mathbf{j} \text{ (N)}$$

$$= 285\mathbf{i} - 854\mathbf{j} \text{ (N)}.$$

> Required the magnitude of the y component of $\mathbf{F}$ to be three times the magnitude of the x component.

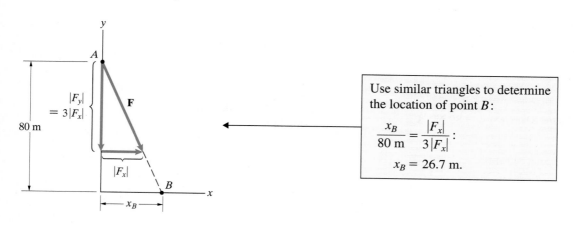

> Use similar triangles to determine the location of point B:
>
> $$\frac{x_B}{80 \text{ m}} = \frac{|F_x|}{3|F_x|} :$$
>
> $$x_B = 26.7 \text{ m}.$$

Active Example 2.6

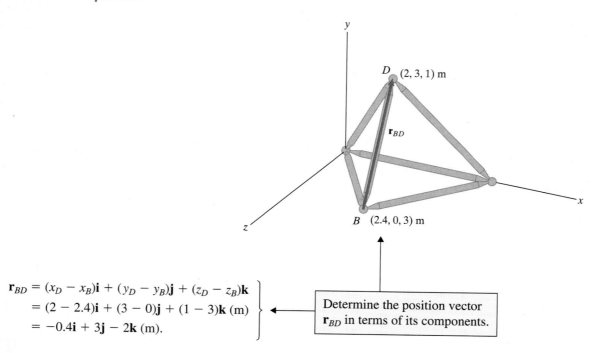

$\mathbf{r}_{BD} = (x_D - x_B)\mathbf{i} + (y_D - y_B)\mathbf{j} + (z_D - z_B)\mathbf{k}$

$= (2 - 2.4)\mathbf{i} + (3 - 0)\mathbf{j} + (1 - 3)\mathbf{k} \text{ (m)}$

$= -0.4\mathbf{i} + 3\mathbf{j} - 2\mathbf{k} \text{ (m)}.$

> Determine the position vector $\mathbf{r}_{BD}$ in terms of its components.

$$|\mathbf{r}_{BD}| = \sqrt{r^2_{BDx} + r^2_{BDy} + r^2_{BDz}}$$

$$= \sqrt{(-0.4 \text{ m})^2 + (3 \text{ m})^2 + (-2 \text{ m})^2}$$

$$= 3.63 \text{ m}.$$

Calculate the magnitude of $\mathbf{r}_{BD}$.

$$\mathbf{e}_{BD} = \frac{\mathbf{r}_{BD}}{|\mathbf{r}_{BD}|}$$

$$= \frac{-0.4\mathbf{i} + 3\mathbf{j} - 2\mathbf{k} \text{ (m)}}{3.63 \text{ (m)}}$$

$$= -0.110\mathbf{i} + 0.827\mathbf{j} - 0.551\mathbf{k}.$$

Divide $\mathbf{r}_{BD}$ by its magnitude to obtain $\mathbf{e}_{BD}$ in terms of its components.

Active Example 2.11

The vectors $\mathbf{U}$ and $\mathbf{V}$ are perpendicular if $\mathbf{U} \cdot \mathbf{V} = 0$. Use this condition to determine V_x.

$$\mathbf{U} \cdot \mathbf{V} = U_x V_x + U_y V_y + U_z V_z$$

$$= (6)V_x + (-5)(2) + (-3)(2)$$

$$= 6V_x - 16.$$

Calculate $\mathbf{U} \cdot \mathbf{V}$ in terms of the components of the vectors.

$$\mathbf{U} \cdot \mathbf{V} = 6V_x - 16 = 0,$$

$$V_x = 2.67.$$

Equate $\mathbf{U} \cdot \mathbf{V}$ to zero and solve for V_x.

Active Example 2.14

The cross product $\mathbf{U} \times \mathbf{V}$ is perpendicular to $\mathbf{U}$ and perpendicular to $\mathbf{V}$. By determining the vector $\mathbf{U} \times \mathbf{V}$ in terms of its components and dividing it by its magnitude $|\mathbf{U} \times \mathbf{V}|$, we can obtain the components of a unit vector that is perpendicular to $\mathbf{U}$ and perpendicular to $\mathbf{V}$.

$$\mathbf{U} \times \mathbf{V} = (U_y V_z - U_z V_y)\mathbf{i} - (U_x V_z - U_z V_x)\mathbf{j}$$

$$+ (U_x V_y - U_y V_x)\mathbf{k}$$

$$= [(2)(-4) - (-1)(-3)]\mathbf{i} - [(3)(-4) - (-1)(5)]\mathbf{j}$$

$$+ [(3)(-3) - (2)(5)]\mathbf{k}$$

$$= -11\mathbf{i} + 7\mathbf{j} - 19\mathbf{k}.$$

Calculate $\mathbf{U} \times \mathbf{V}$ in terms of the components of the vectors.

$$|\mathbf{U} \times \mathbf{V}| = \sqrt{(-11)^2 + (7)^2 + (-19)^2}$$

$$= 23.0.$$

$$\frac{\mathbf{U} \times \mathbf{V}}{|\mathbf{U} \times \mathbf{V}|} = \frac{-11\mathbf{i} + 7\mathbf{j} - 19\mathbf{k}}{23.0}$$

$$= -0.477\mathbf{i} + 0.304\mathbf{j} - 0.825\mathbf{k}.$$

Divide the vector $\mathbf{U} \times \mathbf{V}$ by its magnitude.

Active Example 3.1

Draw the Free-Body Diagram of the Car

Draw a sketch of the isolated car.

Complete the free-body diagram by
showing the forces exerted on the car
by its weight, the cable, and the ramp.

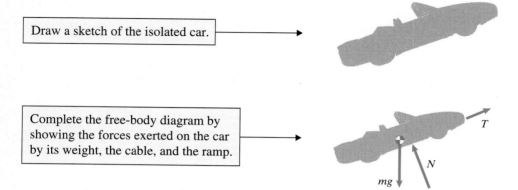

Apply the Equilibrium Equations

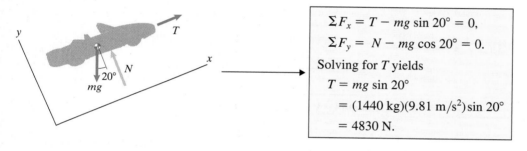

$$\Sigma F_x = T - mg \sin 20° = 0,$$
$$\Sigma F_y = N - mg \cos 20° = 0.$$

Solving for T yields

$$T = mg \sin 20°$$
$$= (1440 \text{ kg})(9.81 \text{ m/s}^2) \sin 20°$$
$$= 4830 \text{ N}.$$

Active Example 3.5

Draw the Free-Body Diagram and Apply Equilibrium

Isolate part of the cable system near
point A and show the forces exerted due to
the tensions in the cables. The sum of the
forces must equal zero:

$$\Sigma \mathbf{F} = \mathbf{T}_{AB} + \mathbf{T}_{AC} + \mathbf{T}_{AD} - (981 \text{ N})\mathbf{j} = \mathbf{0}.$$

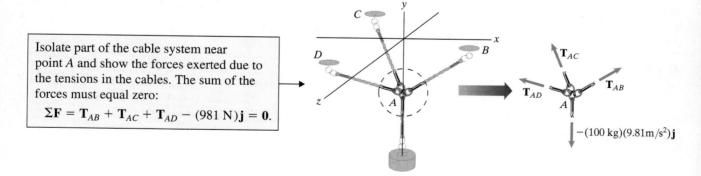

Write the Forces in Terms of Their Components

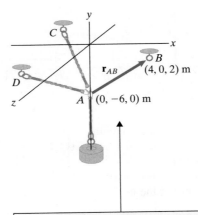

$$\mathbf{r}_{AB} = (x_B - x_A)\mathbf{i} + (y_B - y_A)\mathbf{j} + (z_B - z_A)\mathbf{k}$$
$$= 4\mathbf{i} + 6\mathbf{j} + 2\mathbf{k} \text{ (m)}.$$
$$\mathbf{e}_{AB} = \frac{\mathbf{r}_{AB}}{|\mathbf{r}_{AB}|} = 0.535\mathbf{i} + 0.802\mathbf{j} + 0.267\mathbf{k}.$$

> Obtain a unit vector that has the same direction as the force $\mathbf{T}_{AB}$ by dividing the position vector $\mathbf{r}_{AB}$ from point A to point B by its magnitude.

$$\mathbf{T}_{AB} = T_{AB}\mathbf{e}_{AB}$$
$$= T_{AB}(0.535\mathbf{i} + 0.802\mathbf{j} + 0.267\mathbf{k}),$$
$$\mathbf{T}_{AC} = T_{AC}(-0.302\mathbf{i} + 0.905\mathbf{j} - 0.302\mathbf{k}),$$
$$\mathbf{T}_{AD} = T_{AD}(-0.408\mathbf{i} + 0.817\mathbf{j} + 0.408\mathbf{k}).$$

> Express the force $\mathbf{T}_{AB}$ in terms of its components by writing it as the product of the tension T_{AB} in cable AB and the unit vector $\mathbf{e}_{AB}$. Express the forces $\mathbf{T}_{AC}$ and $\mathbf{T}_{AD}$ in terms of their components using the same procedure.

> Substitute these expressions into the equilibrium equation
> $$\mathbf{T}_{AB} + \mathbf{T}_{AC} + \mathbf{T}_{AD} - (981 \text{ N})\mathbf{j} = \mathbf{0}.$$
> Because the $\mathbf{i}$, $\mathbf{j}$, and $\mathbf{k}$ components must each equal zero, this results in three equations:

$$0.535T_{AB} - 0.302T_{AC} - 0.408T_{AD} = 0,$$
$$0.802T_{AB} + 0.905T_{AC} + 0.817T_{AD} - 981 \text{ N} = 0,$$
$$0.267T_{AB} - 0.302T_{AC} + 0.408T_{AD} = 0.$$

Solving these three equations yields $T_{AB} = 432$ N, $T_{AC} = 574$ N, and $T_{AD} = 141$ N.

Active Example 4.1

y

40 sin 30° kN   40 kN

A

30°

40 cos 30° kN

x

6 m

The magnitude of the moment of the horizontal
component about A is zero. The magnitude of
the moment of the vertical component is
$(6 \text{ m})(40 \sin 30° \text{ N}) = 120$ kN-m. Its direction
is counterclockwise, so the sum of the moments is
$M_A = 120$ kN-m.

> Calculate the sum of the moments
> of the components about A.

> Resolve the 40-kN force into
> horizontal and vertical components.

Active Example 4.4

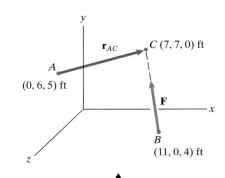

y

$\mathbf{r}_{AC}$ C (7, 7, 0) ft

A

(0, 6, 5) ft

F

x

B
(11, 0, 4) ft

z

$$\mathbf{r}_{AC} = (x_C - x_A)\mathbf{i} + (y_C - y_A)\mathbf{j} + (z_C - z_A)\mathbf{k}$$
$$= 7\mathbf{i} + \mathbf{j} - 5\mathbf{k} \text{ (ft)}.$$
$$\mathbf{M}_A = \mathbf{r}_{AC} \times \mathbf{F}$$

$$= \begin{vmatrix} \mathbf{i} & \mathbf{j} & \mathbf{k} \\ 7 & 1 & -5 \\ -40 & 70 & -40 \end{vmatrix}$$

$$= 310\mathbf{i} + 480\mathbf{j} + 530\mathbf{k} \text{ (ft-lb)}.$$

> (a) Apply Eq. (4.2) to determine
> the moment of $\mathbf{F}$ about point A.

$$D = \frac{|\mathbf{M}_A|}{|\mathbf{F}|}$$

$$= \frac{\sqrt{(310)^2 + (480)^2 + (530)^2} \text{ ft-lb}}{90 \text{ lb}}$$

$$= 8.66 \text{ ft}.$$

> (b) Use the relation $|\mathbf{M}_A| = D|\mathbf{F}|$, where
> D is the perpendicular distance from A to
> the line of action of $\mathbf{F}$.

Active Example 4.6

$\mathbf{r} = (x_A - x_C)\mathbf{i} + (y_A - y_C)\mathbf{j} + (z_A - z_C)\mathbf{k}$
$\quad = 4\mathbf{i} - 2\mathbf{j} + 2\mathbf{k} \text{ (m)}.$

> Determine the components of the vector from point C to the point of application of $\mathbf{F}$.

$\mathbf{M}_C = \mathbf{r} \times \mathbf{F}$

$= \begin{vmatrix} \mathbf{i} & \mathbf{j} & \mathbf{k} \\ 4 & -2 & 2 \\ -2 & 6 & 3 \end{vmatrix}$

$= -18\mathbf{i} - 16\mathbf{j} + 20\mathbf{k} \text{ (kN-m).}$

> Calculate the moment of $\mathbf{F}$ about point C.

$\mathbf{M}_{BC} = (\mathbf{e}_{BC} \cdot \mathbf{M}_C)\mathbf{e}_{BC}$
$\quad = [(0)(-18) + (0.8)(-16) + (-0.6)(20)]\mathbf{e}_{BC}$
$\quad = -24.8\mathbf{e}_{BC} \text{ (kN-m).}$

> Apply Eq. (4.4) to determine the moment of $\mathbf{F}$ about the axis BC. Although the moment of $\mathbf{F}$ about point C is not the same as the moment of $\mathbf{F}$ about point B, their components parallel to the axis BC are the same.

Active Example 4.9

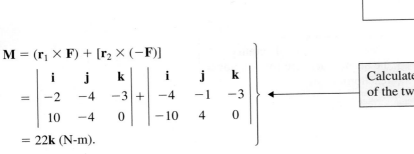

$\mathbf{M} = (\mathbf{r}_1 \times \mathbf{F}) + [\mathbf{r}_2 \times (-\mathbf{F})]$

$= \begin{vmatrix} \mathbf{i} & \mathbf{j} & \mathbf{k} \\ -2 & -4 & -3 \\ 10 & -4 & 0 \end{vmatrix} + \begin{vmatrix} \mathbf{i} & \mathbf{j} & \mathbf{k} \\ -4 & -1 & -3 \\ -10 & 4 & 0 \end{vmatrix}$

$= 22\mathbf{k} \text{ (N-m).}$

> Calculate the sum of the moments of the two forces about point P.

> The magnitude of the moment is 22 N-m. Pointing the thumb of the right hand in the direction of the unit vector $\mathbf{k}$, the direction of the moment in the x–y plane is counterclockwise.

Active Example 4.12

$$\left.\begin{aligned} \mathbf{F'} &= \mathbf{F} \\ &= 20\mathbf{i} + 15\mathbf{j} - 5\mathbf{k}\ (\text{kN}). \end{aligned}\right\}$$

The force $\mathbf{F'}$ must equal the sum of the forces in system 2.

$$\left.\begin{aligned} \mathbf{M'} &= \begin{vmatrix} \mathbf{i} & \mathbf{j} & \mathbf{k} \\ 4 & 3 & -2 \\ 20 & 15 & -5 \end{vmatrix} + (-105\mathbf{i} + 110\mathbf{j} + 90\mathbf{k}) \\ &= -90\mathbf{i} + 90\mathbf{j} + 90\mathbf{k}\ (\text{kN-m}). \end{aligned}\right\}$$

The couple $\mathbf{M'}$ must equal the sum of the moments about the origin due to the forces and moments in system 2.

Active Example 5.1

(a) Draw a diagram of the beam isolated from its pin and roller supports and show the reactions due to the support.

(b) Write the equilibrium equations,

$$\Sigma F_x = A_x = 0,$$
$$\Sigma F_y = A_y + B - 4\ \text{kN} = 0,$$
$$\Sigma M_{\text{left end}} = (3\ \text{m})B - (2\ \text{m})(4\ \text{kN}) = 0,$$

and solve them, obtaining

$$A_x = 0,\ A_y = 1.33\ \text{kN}, B = 2.67\ \text{kN}.$$

Active Example 5.5

Draw the free-body diagram of the beam. There are five unknown reactions.

$$\Sigma F_x = A_x + B_x = 0,$$
$$\Sigma F_y = A_y + B_y - 2\ \text{kN} = 0,$$
$$\Sigma M_{\text{point } A} = M_A + (5\ \text{m})B_y - (3\ \text{m})(2\ \text{kN}) = 0.$$

Write the equilibrium equations.

There are three independent equilibrium equations, so the beam is statically indeterminate and the degree of redundancy is $5 - 3 = 2$. We cannot determine any of the reactions from the equilibrium equations.

Active Example 5.7

Draw the Free-Body Diagram of the Bar

Isolate the bar and show the reactions exerted by the cables and the ball and socket support.

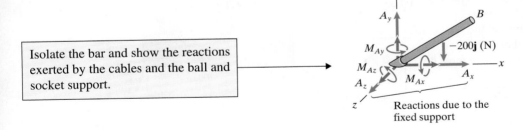

Reactions due to the fixed support

Apply the Equilibrium Equations

$$\Sigma F_x = A_x = 0,$$
$$\Sigma F_y = A_y - 200 \text{ N} = 0,$$
$$\Sigma F_z = A_z = 0.$$

The sums of the forces in each coordinate direction equal zero.

$$\Sigma \mathbf{M}_{\text{point } A} = M_{Ax}\mathbf{i} + M_{Ay}\mathbf{j} + M_{Az}\mathbf{k} + \left[\frac{1}{2} \, \mathbf{r}_{AB} \times (-200\mathbf{j}) \right]$$

$$= M_{Ax}\mathbf{i} + M_{Ay}\mathbf{j} + M_{Az}\mathbf{k} + \begin{vmatrix} \mathbf{i} & \mathbf{j} & \mathbf{k} \\ 0.5 & 0.3 & 0.2 \\ 0 & -200 & 0 \end{vmatrix}$$

$$= (M_{Ax} + 40)\mathbf{i} + M_{Ay}\mathbf{j} + (M_{Az} - 100)\mathbf{k}.$$

The sum of the moments about any point equals zero.

The components of this vector (the sums of the moments about the three coordinate axes) must each equal zero.

$$\Sigma M_x = M_{Ax} + 40 \text{ N-m} = 0,$$
$$\Sigma M_y = M_{Ay} = 0,$$
$$\Sigma M_z = M_{Az} - 100 \text{ N-m} = 0.$$

Solving the six scalar equilibrium equations yields $A_x = 0$, $A_y = 200$ N, $A_z = 0$, $M_{Ax} = -40$ N-m, $M_{Ay} = 0$, and $M_{Az} = 100$ N-m.

Active Example 5.10

The force exerted on the plate by the bar AB must be directed along the line between A and B, and the line of action of the weight of the plate is vertical, so the three forces on the plate are not parallel. Therefore they must be concurrent.

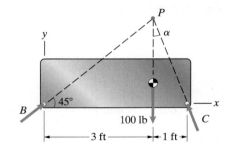

The angle $\alpha = \arctan(1/3) = 18.4°$.

$\Sigma F_x = B \sin 45° - C \sin \alpha = 0,$

$\Sigma F_y = B \cos 45° + C \cos \alpha - 100 \text{ lb} = 0.$

Solving yields the reactions $B = 35.4$ lb, $C = 79.1$ lb.

Apply the equilibrium equations.

Active Example 6.1

The angle $\alpha = \arctan(5/3) = 59.0°$.

$\Sigma F_x = T_{BC} \sin \alpha + T_{BD} + 3.33 \text{ kN} = 0,$

$\Sigma F_y = T_{BC} \cos \alpha = 0.$

Solving yields $T_{BC} = 0$ and $T_{BD} = -3.33$ kN. The axial force in member BC is zero and the axial force in member BD is 3.33 kN in compression, or

BC: zero, BD: 3.33 kN (C).

(Notice that joint C is one of the "special joints" we discussed. We could have determined by observation that $T_{BC} = 0$.)

Draw the free-body diagram of joint B and apply the equilibrium equations.

Active Example 6.3

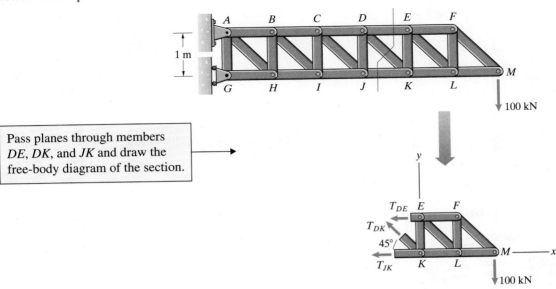

Pass planes through members
DE, DK, and JK and draw the
free-body diagram of the section.

$$\Sigma F_x = -T_{DE} - T_{DK} \cos 45° - T_{JK} = 0,$$
$$\Sigma F_y = T_{DK} \sin 45° - 100 \text{ kN} = 0,$$
$$\Sigma M_{\text{point } K} = (1 \text{ m})T_{DE} - (2 \text{ m})(100 \text{ kN}) = 0.$$

Solving yields $T_{DE} = 200$ kN, $T_{DK} = 141$ kN,
and $T_{JK} = -300$ kN. The axial loads are

  DE: 200 kN (T), DK: 141 kN (T),
  JK: 300 kN (C).

Apply the equilibrium equations.

Active Example 6.5

We can determine the axial forces in members AB and AC by analyzing joint A.

Draw the free-body diagram of joint A.

Divide the position vector from A to B by its magnitude to obtain a unit vector $\mathbf{e}_{AB}$ that points from A toward B. Express the axial force in member AB in terms of its components by writing it as $T_{AB}\mathbf{e}_{AB}$. Express the axial forces in members AC and AD in terms of their components in the same way.

$$\mathbf{r}_{AB} = -5\mathbf{i} - 3\mathbf{j} - 2\mathbf{k} \ (\text{ft}).$$

$$\mathbf{e}_{AB} = \frac{\mathbf{r}_{AB}}{|\mathbf{r}_{AB}|} = -0.811\mathbf{i} - 0.487\mathbf{j} - 0.324\mathbf{k}.$$

$$T_{AB}\mathbf{e}_{AB} = -T_{AB}(0.811\mathbf{i} + 0.487\mathbf{j} + 0.324\mathbf{k}),$$

$$T_{AC}\mathbf{e}_{AC} = T_{AC}(0.196\mathbf{i} - 0.588\mathbf{j} + 0.784\mathbf{k}),$$

$$T_{AD}\mathbf{e}_{AD} = T_{AD}(0.811\mathbf{i} - 0.487\mathbf{j} - 0.324\mathbf{k}).$$

$$T_{AB}\mathbf{e}_{AB} + T_{AC}\mathbf{e}_{AC} + T_{AD}\mathbf{e}_{AC} - (1200 \text{ lb})\mathbf{j} = 0.$$

The $\mathbf{i}$, $\mathbf{j}$, and $\mathbf{k}$ components of this equation must each equal zero, resulting in the three equations

$$-0.811\,T_{AB} + 0.196\,T_{AC} + 0.811\,T_{AD} = 0,$$

$$0.487\,T_{AB} + 0.588\,T_{AC} + 0.487\,T_{AD} + 1200 \text{ lb} = 0,$$

$$-0.324\,T_{AB} + 0.784\,T_{AC} - 0.324\,T_{AD} = 0.$$

Solving yields $T_{AB} = -904$ lb, $T_{AC} = -680$ lb, and $T_{AD} = -740$ lb. The axial forces are

AB: 904 lb (C), AC: 680 lb (C).

Apply equilibrium.

Active Example 6.6

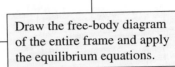

$$\Sigma F_x = A_x + C_x = 0,$$
$$\Sigma F_y = A_y + C_y = 0,$$
$$\Sigma M_{\text{point } A} = -200 \text{ N-m} + (0.4 \text{ m})C_x + (1 \text{ m})C_y = 0.$$

We can't determine any reactions from these equations. The free-body diagram of the entire frame is statically indeterminate.

> Draw the free-body diagram of the entire frame and apply the equilibrium equations.

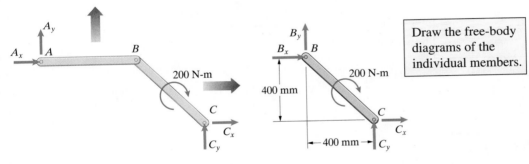

> Draw the free-body diagrams of the individual members.

$$\Sigma F_x = A_x - B_x = 0,$$
$$\Sigma F_y = A_y - B_y = 0,$$
$$\Sigma M_{\text{point } A} = -(0.6 \text{ m})B_y = 0.$$

Solving yields $A_y = 0$, $B_y = 0$, and $A_x = B_x$. (Notice that AB is a two-force member. We could have obtained these results by observation.)

> Apply equilibrium to member AB.

$$\Sigma F_x = B_x + C_x = 0,$$
$$\Sigma F_y = B_y + C_y = 0,$$
$$\Sigma M_{\text{point } B} = -200 \text{ N-m} + (0.4 \text{ m})C_x + (0.4 \text{ m})C_y = 0.$$

Because it has already been determined that $B_y = 0$, these equations can be solved for B_x, C_x, and C_y. The results are $B_x = -500$ N, $C_x = 500$ N, and $C_y = 0$, which completes the solution.

> Apply equilibrium to member BC.

Active Example 7.1

$$\bar{y} = \frac{\int_A y\,dA}{\int_A dA} = \frac{\int_0^b \frac{1}{2}\left(\frac{h}{b}x\right)\left(\frac{h}{b}x\,dx\right)}{\int_0^b \frac{h}{b}x\,dx} = \frac{\frac{1}{2}\left(\frac{h}{b}\right)^2\left[\frac{x^3}{3}\right]_0^b}{\frac{h}{b}\left[\frac{x^2}{2}\right]_0^b} = \frac{1}{3}h.$$

The area of the strip is $dA = (h/b)x\,dx$. The height of the midpoint of the strip is $y = (1/2)(h/b)x$. Use these expressions to evaluate Eq. (7.7).

Active Example 7.3

Choose the Parts
Divide the area into simple parts. The y coordinates of the centroids of the parts are shown.

Determine the Values for the Parts
Tabulate the terms needed to apply Eq. (7.9)$_2$.

	$\bar{y}_i$	A_i	$\bar{y}_i A_i$
Part 1 (triangle)	$\frac{1}{3}(2R)$	$\frac{1}{2}b(2R)$	$\left[\frac{1}{3}(2R)\right]\left[\frac{1}{2}b(2R)\right]$
Part 2 (rectangle)	R	$c(2R)$	$R[c(2R)]$
Part 3 (semicircle)	R	$\frac{1}{2}\pi R^2$	$R\left(\frac{1}{2}\pi R^2\right)$

$$\bar{y} = \frac{\bar{y}_1 A_1 + \bar{y}_2 A_2 + \bar{y}_3 A_3}{A_1 + A_2 + A_3}$$

$$= \frac{\left[\frac{1}{3}(2R)\right]\left[\frac{1}{2}b(2R)\right] + R[c(2R)] + R\left(\frac{1}{2}\pi R^2\right)}{\frac{1}{2}b(2R) + c(2R) + \frac{1}{2}\pi R^2}.$$

> **Calculate the Centroid**
> Use Eq. $(7.9)_2$ to determine the y component of the centroid.

Active Example 7.5

(a)

$$w = ax + b.$$

> Write w as an arbitrary *linear* function of x.

$$0 = a(0) + b,$$
$$100 \text{ N/m} = a(12 \text{ m}) + b.$$

Solving yields $a = (100/12) \text{ N/m}^2$ and $b = 0$. Therefore

$$w = \frac{100}{12} x \text{ N/m}.$$

> Use the known values of w at $x = 0$ and at $x = 12$ m to determine the constants a and b.

(b)

$$F = \int_L w \, dx$$

$$= \int_0^{12} \frac{100}{12} x \, dx$$

$$= 600 \text{ N}.$$

> Apply Eq. (7.10) to determine the downward force exerted by the distributed load.

$$M = \int_L xw \, dx$$

$$= \int_0^{12} \frac{100}{12} x^2 \, dx$$

$$= 4800 \text{ N-m}.$$

> Apply Eq. (7.11) to determine the clockwise moment about the origin exerted by the distributed load.

Active Example 7.8

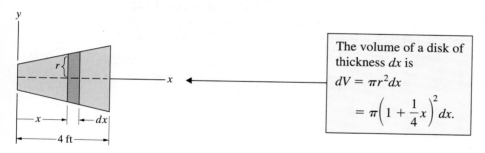

The volume of a disk of thickness dx is

$$dV = \pi r^2 dx$$

$$= \pi \left(1 + \frac{1}{4}x\right)^2 dx.$$

$$\bar{x} = \frac{\displaystyle\int_V x\, dV}{\displaystyle\int_V dV} = \frac{\displaystyle\int_0^4 x\pi \left(1 + \frac{1}{4}x\right)^2 dx}{\displaystyle\int_0^4 \pi \left(1 + \frac{1}{4}x\right)^2 dx} = 2.43 \text{ ft.}$$

Apply Eq. $(7.15)_1$.

Active Example 7.11

Choose the Parts
Divide the volume into simple parts.
The x coordinates of the centroids of
the parts are shown. See Appendix C.

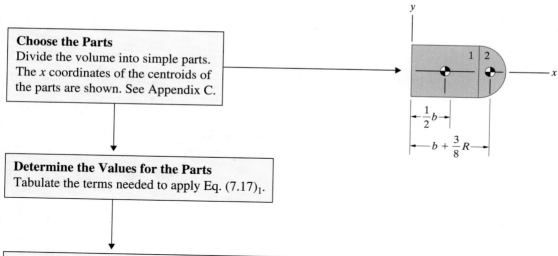

Determine the Values for the Parts
Tabulate the terms needed to apply Eq. $(7.17)_1$.

	$\bar{x}_i$	V_i	$\bar{x}_i V_i$
Part 1 (cylinder)	$\frac{1}{2}b$	$\pi R^2 b$	$\left(\frac{1}{2}b\right)(\pi R^2 b)$
Part 2 (hemisphere)	$b + \frac{3}{8}R$	$\frac{2}{3}\pi R^3$	$\left(b + \frac{3}{8}R\right)\left(\frac{2}{3}\pi R^3\right)$

$$\bar{x} = \frac{\bar{x}_1 V_1 + \bar{x}_2 V_2}{V_1 + V_2}$$

$$= \frac{\left(\frac{1}{2}b\right)\left(\pi R^2 b\right) + \left(b + \frac{3}{8}R\right)\left(\frac{2}{3}\pi R^3\right)}{\pi R^2 b + \frac{2}{3}\pi R^3}.$$

Calculate the Centroid
Use Eq. (7.17)₁ to determine the x component of the centroid.

Active Example 7.14

Revolving this triangle area about the x axis generates the volume of the cone. The y coordinate of the centroid of the area is shown. The area of the triangle is $A = \frac{1}{2}hR$. The volume of the cone is

$$V = 2\pi \bar{y}_T A = \frac{1}{3}\pi h R^2.$$

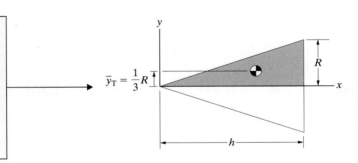

$$\bar{y}_T = \frac{1}{3}R$$

Active Example 7.16

$$\Sigma F_x = A_x - B = 0,$$

$$\Sigma F_y = A_y - (80)(9.81)\text{ N} = 0,$$

$$\Sigma M_{\text{point } A} = (1\text{m})B - \frac{2(1\text{ m})}{\pi}[(80)(9.81)\text{ N}] = 0.$$

Solving yields $A_x = 500$ N, $A_y = 785$ N, and $B = 500$ N.

Place the weight of the bar at its center of mass (the centroid of its axis; see Appendix B.2) and apply equilibrium.

Active Example 7.18

The center of mass coincides
with the centroid of the volume
of the bar, so

$$\bar{y}_1 = \frac{1}{2}(240 \text{ mm}) = 120 \text{ mm.}$$

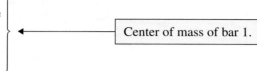 Center of mass of bar 1.

The y coordinate of the centroid of the
volume is

$$\bar{y}_2 = \frac{1}{2}(80 \text{ mm}) = 40 \text{ mm.}$$

Center of mass of bar 2.

$$
\begin{aligned}
\bar{y} &= \frac{\bar{y}_1 m_1 + \bar{y}_2 m_2}{m_1 + m_2} \\
&= \frac{(120 \text{ mm})(10.8 \text{ kg}) + (40 \text{ mm})(5.99 \text{ kg})}{10.8 \text{ kg} + 5.99 \text{ kg}} \\
&= 91.4 \text{ mm.}
\end{aligned}
$$

Apply Eq. (7.27)$_2$.

Active Example 8.1

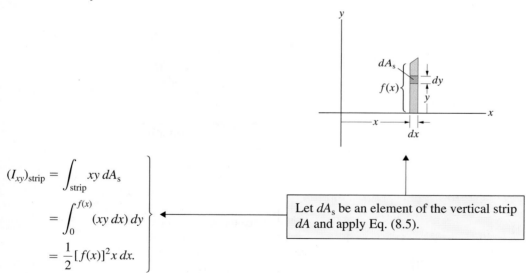

$$
\begin{aligned}
(I_{xy})_{\text{strip}} &= \int_{\text{strip}} xy \, dA_s \\
&= \int_0^{f(x)} (xy \, dx) \, dy \\
&= \frac{1}{2}[f(x)]^2 x \, dx.
\end{aligned}
$$

Let dA_s be an element of the vertical strip
dA and apply Eq. (8.5).

$$
\begin{aligned}
I_{xy} &= \int_0^b \frac{1}{2}[f(x)]^2 x \, dx \\
&= \int_0^b \frac{1}{2}\left(\frac{h}{b}x\right)^2 x \, dx \\
&= \frac{1}{8} b^2 h^2.
\end{aligned}
$$

Integrate the expression for $(I_{xy})_{\text{strip}}$ with
respect to x from $x = 0$ to $x = b$ to
determine I_x for the triangle.

Active Example 8.3

From Appendix B, the product of inertia of area 1 in terms of the $x'y'$ coordinate system is

$(I_{x'y'})_1 = 0.$

Therefore the product of inertia of area 1 about the xy coordinate system is

$(I_{xy})_1 = 0 + (0.5 \text{ m})(2 \text{ m})(1 \text{ m})(4 \text{ m}) = 4 \text{ m}^4.$

> Apply Eq. (8.12) to area 1.

The product of inertia of area 2 in terms of the $x'y'$ coordinate system is

$(I_{x'y'})_2 = 0.$

The product of inertia of area 2 in terms of the xy coordinate system is

$(I_{xy})_2 = 0 + (2 \text{ m})(0.5 \text{ m})(2 \text{ m})(1 \text{ m}) = 2 \text{ m}^4.$

> Apply Eq. (8.12) to area 2.

The product of inertia of the composite area in terms of the xy coordinate system is

$$I_{xy} = (I_{xy})_1 + (I_{xy})_2$$
$$= 4 \text{ m}^4 + 2 \text{ m}^4$$
$$= 6 \text{ m}^4.$$

> Sum the values for the parts.

Active Example 8.6

$$\tan 2\theta_p = \frac{2I_{xy}}{I_y - I_x} = \frac{2(6)}{16 - 9} = 1.71.$$

This yields $\theta_p = 29.9°$.

Determine θ_p from Eq. (8.26).

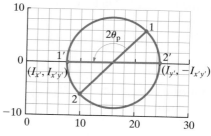

$$I_{x'} = \frac{I_x + I_y}{2} + \frac{I_x - I_y}{2} \cos 2\theta - I_{xy} \sin 2\theta$$

$$= \left(\frac{9 + 16}{2}\right) + \left(\frac{9 - 16}{2}\right) \cos[2(29.9°)] - (6)\sin[2(29.9°)]$$

$$= 5.55 \text{ m}^4,$$

$$I_{y'} = \frac{I_x + I_y}{2} - \frac{I_x - I_y}{2} \cos 2\theta + I_{xy} \sin 2\theta$$

$$= \left(\frac{9 + 16}{2}\right) - \left(\frac{9 - 16}{2}\right) \cos[2(29.9°)] + (6)\sin[2(29.9°)]$$

$$= 19.4 \text{ m}^4.$$

Calculate the principal moments of inertia from Eqs. (8.23) and (8.24).

Active Example 8.8

Place the point $1'$ at one of the points where the Mohr's circle intersects the horizontal axis. The principal moments of inertia are $I_{x'} = 7.5 \text{ ft}^4$, $I_{y'} = 24.5 \text{ ft}^4$. The angle measured counterclockwise from point 1 to point $1'$ is $2\theta_p = 135°$, so $\theta_p = 67.5°$.

Active Example 8.9

From Appendix B,

$$I_y = \frac{1}{4}hb^3.$$

Determine the moment of inertia of the area of the plate about the y axis.

The moment of inertia of the plate about the y axis is

$$I_{y\,axis} = \frac{m}{A}I_y$$

$$= \frac{m}{\frac{1}{2}bh}\left(\frac{1}{4}hb^3\right)$$

$$= \frac{1}{2}mb^2.$$

Apply Eq. (8.30).

Active Example 8.11

Treat the object as a composite object made up of the bars 1 and 2. The distance between the axis L_O and parallel axes through the centers of mass of the two bars are shown.

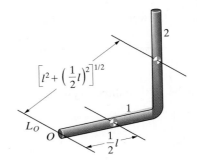

$$(I_O)_1 = I + d^2m$$

$$= \frac{1}{12}ml^2 + \left(\frac{1}{2}l\right)^2 m$$

$$= \frac{1}{3}ml^2.$$

Apply the parallel-axis theorem to bar 1.

$$(I_O)_2 = I + d^2m$$

$$= \frac{1}{12}ml^2 + \left[l^2 + \left(\frac{1}{2}l\right)^2\right]m$$

$$= \frac{4}{3}ml^2.$$

Apply the parallel-axis theorem to bar 2.

$$I_O = (I_O)_1 + (I_O)_2$$

$$= \frac{1}{3}ml^2 + \frac{4}{3}ml^2$$

$$= \frac{5}{3}ml^2.$$

Sum the results.

Active Example 9.1

Draw the free-body diagram of the crate. It is assumed that slip of the crate *up* the ramp is impending, so the direction of the friction force on the crate is *down* the ramp and its magnitude is $\mu_s N$.

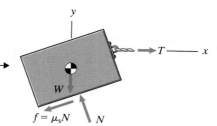

$$\Sigma F_x = T - N\sin 20° - \mu_s N\cos 20° = 0,$$
$$\Sigma F_y = N\cos 20° - \mu_s N\sin 20° - W = 0.$$

Solving these equations yields $T = 161$ lb.

Apply equilibrium.

Active Example 9.4

Draw the free-body diagram of the wedge assuming that $F = 0$ and that slip of the wedge out of the log is impending.

The sum of the forces in the vertical direction is

$$2N\sin\left(\frac{\alpha}{2}\right) - 2\mu_s N\cos\left(\frac{\alpha}{2}\right) = 0.$$

The wedge is in equilibrium if

$$\mu_s = \tan\left(\frac{\alpha}{2}\right) = \tan\left(\frac{10°}{2}\right) = 0.0875.$$

This is the *minimum* static coefficient of friction necessary for the wedge to remain in place in the log, so it will not slip out.

Apply equilibrium.

Active Example 9.5

The force $F = 200$ lb, the slope of the thread is $\alpha = 1.14°$, and the angle of friction is

$$\theta_k = \arctan\mu_k = \arctan(0.22) = 12.4°.$$

Substituting these values into Eq. (9.11),

$$M = rF\tan(\theta_k - \alpha)$$
$$= (1.6\text{ in})(200\text{ lb})\tan(12.4° - 1.14°)$$
$$= 63.8\text{ in-lb.}$$

Apply Eq. (9.11).

Active Example 9.6

$M = (6 \text{ in})(W - T).$ ◄————————

> The pulley moves in the counterclockwise direction. Express the counterclockwise couple on the pulley in terms of T and W.

The angle of kinetic friction is

$\theta_k = \arctan \mu_k = \arctan(0.2) = 11.3°.$

Equation (9.12) is

$M = rF \sin \theta_k:$

$(6 \text{ in})(W - T) = (0.5 \text{ in})\sqrt{(W + T \sin 45°)^2 + (T \cos 45°)^2} \sin 11.3°.$

Setting $W = 1000$ lb and solving yields

$T = 970$ lb.

> Apply Eq. (9.12).

Active Example 9.7

The radii $r_o = 1.75$ in and $r_i = 0.5$ in.

$\alpha = \arctan[b/(r_o - r_i)] = \arctan[5/(1.75 - 0.5)] = 76.0°.$

> Determine the angle α.

$M = \dfrac{2\mu_k F}{3 \cos \alpha} \dfrac{r_o^3 - r_i^3}{r_o^2 - r_i^2}$

$= \dfrac{2(0.18)(200 \text{ lb})}{3 \cos 76.0°} \left[\dfrac{(1.75 \text{ in})^3 - (0.5 \text{ in})^3}{(1.75 \text{ in})^2 - (0.5 \text{ in})^2} \right]$

$= 184$ in-lb.

> Apply Eq. (9.13).

Active Example 9.9

$T = We^{\mu_s \beta} = (100 \text{ lb})e^{(0.2)(\pi/2)} = 137$ lb. ◄————

> Apply Eq. (9.17) to the left cylinder. Assume that slip of the rope in the direction of the force T is impending.

$F = Te^{(0.4)(\pi/2)} = (137 \text{ lb})e^{(0.4)(\pi/2)} = 257$ lb. ◄————

> Apply Eq. (9.17) to the right cylinder. Assume that slip of the rope in the direction of the force F is impending.

Active Example 10.1

$$\Sigma F_x = -P_C = 0,$$

$$\Sigma F_y = V_C - F + \frac{3}{4}F = 0,$$

$$\Sigma M_{\text{point } C} = -M_C - \left(\frac{1}{2}L\right)F + \left(\frac{3}{4}L\right)\left(\frac{3}{4}F\right) = 0.$$

Solving yields $P_C = 0$, $V_C = \frac{1}{4}F$, and $M_C = \frac{1}{16}LF$.

Draw the free-body diagram of the part of the beam to the right of C. (Notice the defined positive directions of P_C, V_C, and M_C.) Apply equilibrium to determine the internal forces and moment.

Active Example 10.3

(a) Pass a plane through the beam at an arbitrary position x between B and C. The simplest free-body diagram is obtained by isolating the part of the beam to the right of the plane.

$$\Sigma F_x = -P = 0,$$
$$\Sigma F_y = V + 60 = 0,$$
$$\Sigma M_{\text{left end}} = -M + 60(4 - x) = 0.$$

Solving yields

$$\left.\begin{array}{l} V = -60 \text{ kN} \\ M = 60(4 - x) \text{ kN-m} \end{array}\right\} \; 2 < x < 4 \text{ m}.$$

Apply equilibrium to determine V and M.

Active Example 10.4

No couple is applied to the beam at A, so the bending moment at A is zero. The shear force between A and B is $V = 100 - 40x$ kN. With this expression, Eq. (10.6) can be integrated to determine the bending moment between A and B:

$$\int_0^M dM = \int_0^x (100 - 40x)\,dx:$$

$$M = 100x - 20x^2 \text{ kN-m.}$$

The value of M at B is

$$100(2) - 20(2)^2 = 120 \text{ kN-m.}$$

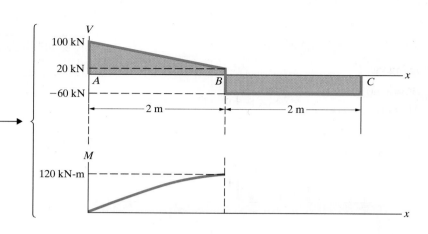

The shear force between B and C is $V = -60$ kN. Because V is constant, Eq. (10.6) indicates that the slope of the bending moment is constant—the diagram is a straight line. Because no couple is applied to the beam at C, the bending moment at C is zero. Therefore M decreases linearly from 120 kN-m at B to zero at C. This result can also be obtained by integrating Eq. (10.6):

$$\int_{120}^M dM = \int_2^x -60\,dx:$$

$$M = 240 - 60x \text{ kN-m.}$$

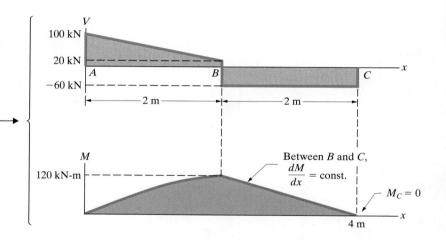

Active Example 10.6

The tension is given by Eq. (10.11) in terms of the tension at the lowest point and the horizontal coordinate relative to the lowest point. From Eq. (10.11), the maximum tension clearly occurs where the horizontal distance from the lowest point is the greatest, which in this example is the left attachment point. The maximum tension is

$$T = T_0\sqrt{1 + a^2 x_L^2}$$

$$= (686 \text{ lb})\sqrt{1 + (0.146 \text{ ft}^{-1})^2 (-23.4 \text{ ft})^2}$$

$$= 2440 \text{ lb.}$$

Apply Eq. (10.11) to determine the maximum tension.

Active Example 10.8

The maximum tension occurs where the
horizontal distance from the lowest point
is the greatest, at $x = 10$ m:

$T = T_0 \cosh ax$

$\quad = (50 \text{ N}) \cosh[(0.196 \text{ m}^{-1})(10 \text{ m})]$

$\quad = 181 \text{ N}.$

Apply Eq. (10.21).

Active Example 10.9

The angle α is

$\alpha = \arctan\left(\dfrac{h_2 - 1 \text{ m}}{1 \text{ m}}\right)$

$\quad = \arctan\left(\dfrac{1.25 \text{ m} - 1 \text{ m}}{1 \text{ m}}\right)$

$\quad = 14.0°.$

The sum of the horizontal forces is

$T_2 \cos\alpha - T_h = 0,$

yielding

$T_2 = \dfrac{T_h}{\cos\alpha}$

$\quad = \dfrac{131 \text{ N}}{\cos 14.0°}$

$\quad = 135 \text{ N}.$

Cut the cable at the left
attachment point and within
segment 2 and sum forces in
the horizontal direction.

Active Example 10.10

The gage pressure $p_g = \gamma x$ increases linearly from $p_g = 0$ at the surface of the water to $p_g = (2\ \text{ft})\gamma$ at the bottom of the gate. The centroid of the distribution is shown.

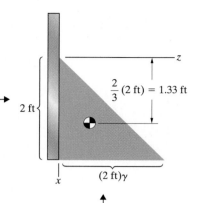

Determine the total force exerted by the gage pressure by calculating the "volume" of the pressure distribution. The "volume" is the product of the "area" of the triangle in the above figure and the dimension of the gate into the page.

$$F = \frac{1}{2}(2\ \text{ft})\,[(2\text{ft})(62.4\ \text{lb/ft}^2)]\,(3\ \text{ft})$$

$$= 374\ \text{lb}.$$

Draw the free-body diagram of the gate, placing the total force exerted by gage pressure at the center of pressure. Apply equilibrium to determine the reactions at A and B.

$$\Sigma F_x = A_x + 100\ \text{lb} = 0,$$

$$\Sigma F_z = A_z + B - 374\ \text{lb} = 0,$$

$$\Sigma M_{y\,\text{axis}} = (1\ \text{ft})B - (2\ \text{ft})A_z + (1.33\ \text{ft})(374\ \text{lb}) = 0.$$

Solving yields $A_x = -100\ \text{lb}$, $A_z = 291\ \text{lb}$, and $B = 83.2\ \text{lb}$.

Active Example 11.1

The work done by the 400-N force is $(400 \sin 40° \text{ N})(1 \text{ m})\delta\alpha$.
Bar BC undergoes a rotation $\delta\alpha$ in the clockwise direction, so
the work done by the couple is $-(500 \text{ N-m})\delta\alpha$. The work done
by the reaction C_y is $-C_y 2(2 \cos 40°)\delta\alpha$. The total virtual work is

$$\delta U = (400 \sin 40° \text{ N})(1 \text{ m})\delta\alpha - (500 \text{ N-m})\delta\alpha - C_y 2(2 \cos 40°)\delta\alpha = 0.$$

Solving yields $C_y = -79.3 \text{ N}$.

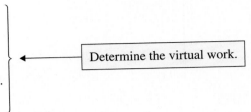

Determine the virtual work.

Active Example 11.3

The derivative of the potential energy
with respect to the coordinate x is

$$\frac{dV}{dx} = kx - W.$$

The second derivative is

$$\frac{d^2V}{dx^2} = k,$$

which is positive. The equilibrium
position is stable.

Determine whether the second derivative of
V is positive (stable) or negative (unstable).

Answers to Even-Numbered Problems

Chapter 1

1.2	(a) $e = 2.7183$; (b) $e^2 = 7.3891$; (c) $e^2 = 7.3892$.
1.4	17.8 m^2.
1.6	The 1-in wrench fits the 25-mm nut.
1.8	(a) 267 mi/h; (b) 392 ft/s.
1.10	310 N-m.
1.12	$g = 32.2 \text{ ft/s}^2$.
1.14	(a) 0.0208 m^2; (b) 32.2 in^2.
1.16	2.07×10^6 Pa.
1.18	27.4 lb/ft.
1.20	(a) kg-m/s; (b) 2.70 slug-ft/s.
1.22	(a) 0.397 kg; (b) 0.643 N.
1.24	(a) 4.60×10^{19} slugs; (b) 6.71×10^{20} kg.
1.26	163 lb.
1.28	32.1 km.
1.30	345,000 km.

Chapter 2

2.2	$|\mathbf{F}_{AB} + \mathbf{F}_{AC}| = 146$ kN, direction is 32° above the horizontal.
2.4	$|\mathbf{F}_A + \mathbf{F}_B + \mathbf{F}_C| = 83$ N.
2.6	$|\mathbf{r}_{AC}| = 181$ mm.
2.8	$|\mathbf{F}_B| = 86.6$ N, $|\mathbf{F}_C| = 50.0$ N.
2.10	$|\mathbf{L}| = 453$ lb, $|\mathbf{D}| = 211$ lb.
2.12	$|\mathbf{F}_{BA}| = 174$ lb.
2.14	$|\mathbf{r}_{BC}| = 390$ m, $\alpha = 21.2°$.
2.18	$F_y = -102$ MN.
2.20	$|\mathbf{F}| = 447$ kip.
2.22	$V_x = 16, \quad V_y = 12$ or $V_x = -16, \quad V_y = -12$.
2.24	(a) $\mathbf{F} = 56.4\mathbf{i} + 20.5\mathbf{j}$ (lb); (b) 97.4 lb.
2.26	$\mathbf{r}_{AD} = -1.8\mathbf{i} - 0.3\mathbf{j}$ (m), $|\mathbf{r}_{AD}| = 1.825$ m.
2.28	$\mathbf{r}_{AB} - \mathbf{r}_{BC} = \mathbf{i} - 1.73\mathbf{j}$ (m).
2.30	(a) $\mathbf{r}_{AB} = 48\mathbf{i} + 15\mathbf{j}$ (in); (b) $\mathbf{r}_{BC} = -53\mathbf{i} + 5\mathbf{j}$ (in); (c) $|\mathbf{r}_{AB} + \mathbf{r}_{BC}| = 20.6$ in.
2.32	(a) $\mathbf{r}_{AB} = 52.0\mathbf{i} + 30\mathbf{j}$ (mm); (b) $\mathbf{r}_{AB} = -42.4\mathbf{i} - 42.4\mathbf{j}$ (mm).
2.34	$x_B = 785$ m, $y_B = 907$ m or $x_B = 255$ m, $y_B = 1173$ m.
2.36	$\mathbf{e}_{CA} = 0.458\mathbf{i} - 0.889\mathbf{j}$.
2.38	$\mathbf{e} = 0.806\mathbf{i} + 0.593\mathbf{j}$.
2.40	$\mathbf{F} = -937\mathbf{i} + 750\mathbf{j}$ (N).
2.42	14,500 lb.
2.44	$|\mathbf{F}_{BA}| = 802$ N.
2.46	$|\mathbf{F}_A| = 1720$ lb, $\alpha = 33.3°$.
2.48	$57.9° \leq \alpha \leq 90°$.
2.50	$|\mathbf{F}_A| = 10$ kN, $|\mathbf{F}_D| = 8.66$ kN.
2.52	$|\mathbf{L}| = 214$ lb, $|\mathbf{D}| = 85.4$ lb.

2.54	$|\mathbf{F}_A| = 68.2$ kN.
2.56	$|\mathbf{F}_{AC}| = 2.11$ kN, $|\mathbf{F}_{AD}| = 2.76$ kN.
2.58	$x = 75 - 0.880s, \quad y = 12 + 0.476s$.
2.60	$\mathbf{r} = (0.814s - 6)\mathbf{i} + (0.581s + 1)\mathbf{j}$ (m).
2.62	$e_z = \dfrac{2}{3}$ or $e_z = -\dfrac{2}{3}$.
2.64	$U_x = 3.61, \quad U_y = -7.22, \quad U_z = -28.89$ or $U_x = -3.61, \quad U_y = 7.22, \quad U_z = 28.89$.
2.66	(a) $|\mathbf{U}| = 7, \quad |\mathbf{V}| = 13$; (b) $|3\mathbf{U} + 2\mathbf{V}| = 27.5$.
2.68	(a) $\cos\theta_x = 0.333, \quad \cos\theta_y = -0.667,$ $\cos\theta_z = -0.667$; (b) $\mathbf{e} = 0.333\mathbf{i} - 0.667\mathbf{j} - 0.667\mathbf{k}$.
2.70	$\mathbf{F} = -0.5\mathbf{i} + 0.2\mathbf{j} + 0.843\mathbf{k}$.
2.72	$\mathbf{r}_{BD} = -\mathbf{i} + 3\mathbf{j} - 2\mathbf{k}$ (m), $|\mathbf{r}_{BD}| = 3.74$ m.
2.74	$\mathbf{e}_{CD} = -0.535\mathbf{i} + 0.802\mathbf{j} + 0.267\mathbf{k}$.
2.76	$\mathbf{F} = 300\mathbf{i} + 477\mathbf{j} + 205\mathbf{k}$ (lb).
2.78	(a) $|\mathbf{r}_{AB}| = 16.2$ m; (b) $\cos\theta_x = 0.615, \quad \cos\theta_y = -0.492,$ $\cos\theta_z = -0.615$.
2.80	$\mathbf{r}_{AR}$: $\cos\theta_x = 0.667, \quad \cos\theta_y = 0.667,$ $\cos\theta_z = 0.333$. $\mathbf{r}_{BR}$: $\cos\theta_x = -0.242,$ $\cos\theta_y = 0.970, \quad \cos\theta_z = 0$.
2.82	$h = 8848$ m (29,030 ft).
2.84	$|\mathbf{F}_A + \mathbf{F}_B| = 217$ lb.
2.86	$\mathbf{F} = 474\mathbf{i} + 516\mathbf{j} + 565\mathbf{k}$ (N).
2.88	(a) $\mathbf{e}_{BC} = -0.286\mathbf{i} - 0.857\mathbf{j} + 0.429\mathbf{k}$; (b) $\mathbf{F} = -2.29\mathbf{i} - 6.86\mathbf{j} + 3.43\mathbf{k}$ (kN).
2.90	$|\mathbf{F}| = 424$ lb.
2.92	259 lb.
2.94	$|\mathbf{F}_{AC}| = 1116$ N, $|\mathbf{F}_{AD}| = 910$ N.
2.96	$\mathbf{T} = -15.4\mathbf{i} + 27.0\mathbf{j} + 7.7\mathbf{k}$ (lb).
2.98	$\mathbf{T} = -41.1\mathbf{i} + 28.8\mathbf{j} + 32.8\mathbf{k}$ (N).
2.100	32.4°.
2.102	Either $|\mathbf{V}| = 0$ or $\mathbf{V}$ is perpendicular to $\mathbf{U}$.
2.104	$U_x = 2.857, \quad V_y = 0.857, \quad W_z = -3.143$.
2.108	$\theta = 62.3°$.
2.110	$\theta = 53.5°$.
2.112	$14.0\mathbf{i} + 11.2\mathbf{j} - 8.40\mathbf{k}$ (N).
2.114	(a) 42.5°; (b) $-423\mathbf{j} + 604\mathbf{k}$ (lb).
2.116	$\mathbf{F}_p = 5.54\mathbf{j} + 3.69\mathbf{k}$ (N), $\mathbf{F}_n = 10\mathbf{i} + 6.46\mathbf{j} - 9.69\mathbf{k}$ (N).
2.118	$\mathbf{T}_n = -37.1\mathbf{i} + 31.6\mathbf{j} + 8.2\mathbf{k}$ (N).
2.120	$\mathbf{F}_p = -0.1231\mathbf{i} + 0.0304\mathbf{j} - 0.1216\mathbf{k}$ (lb).
2.122	$\mathbf{v}_p = -1.30\mathbf{i} - 1.68\mathbf{j} - 3.36\mathbf{k}$ (m/s).
2.124	(a) $\mathbf{U} \times \mathbf{V} = 44\mathbf{i} + 56\mathbf{j} - 16\mathbf{k}$.
2.126	$2180\mathbf{i} + 1530\mathbf{j} - 1750\mathbf{k}$ (ft-lb).
2.128	Either $|\mathbf{V}| = 0$ or $\mathbf{V}$ is parallel to $\mathbf{U}$.
2.130	(a), (c) $\mathbf{U} \times \mathbf{V} = -51.8\mathbf{k}$; (b), (d) $\mathbf{V} \times \mathbf{U} = 51.8\mathbf{k}$.

2.134 (a) $\mathbf{r}_{OA} \times \mathbf{r}_{OB} = -4\mathbf{i} + 36\mathbf{j} + 32\mathbf{k}$ (m^2);
(b) $-0.083\mathbf{i} + 0.745\mathbf{j} + 0.662\mathbf{k}$
or $0.083\mathbf{i} - 0.745\mathbf{j} - 0.662\mathbf{k}$.

2.136 $\mathbf{r}_{AB} \times \mathbf{F} = -2400\mathbf{i} + 9600\mathbf{j} + 7200\mathbf{k}$ (ft-lb).

2.138 $\mathbf{r}_{CA} \times \mathbf{T} = -4.72\mathbf{i} - 3.48\mathbf{j} - 7.96\mathbf{k}$ (N-m).

2.140 $x_B = 2.81$ m, $y_B = 6.75$ m, $z_B = 3.75$ m.

2.144 1.8×10^6 mm^2.

2.146 $U_y = -2$.

2.148 $|\mathbf{A}| = 1110$ lb, $\alpha = 29.7°$.

2.150 $|\mathbf{E}| = 313$ lb, $|\mathbf{F}| = 140$ lb.

2.152 $\mathbf{e}_{AB} = 0.625\mathbf{i} - 0.469\mathbf{j} - 0.625\mathbf{k}$.

2.154 $\mathbf{F}_p = 8.78\mathbf{i} - 6.59\mathbf{j} - 8.78\mathbf{k}$ (lb).

2.156 $\mathbf{r}_{BA} \times \mathbf{F} = -70\mathbf{i} + 40\mathbf{j} - 100\mathbf{k}$ (ft-lb).

2.158 (a), (b) $686\mathbf{i} - 486\mathbf{j} - 514\mathbf{k}$ (ft-lb).

2.160 (a) $\mathbf{F} = 139\mathbf{i} + 58.2\mathbf{j} + 80\mathbf{k}$ (lb); (b) $\theta_x = 35.5°$, $\theta_y = 70°$, $\theta_z = 62.0°$.

2.162 $\mathbf{F}_p = 1.29\mathbf{i} - 3.86\mathbf{j} + 2.57\mathbf{k}$ (kN), $\mathbf{F}_n = -1.29\mathbf{i} - 2.14\mathbf{j} - 2.57\mathbf{k}$ (kN).

2.164 $\mathbf{r}_{AG} \times \mathbf{W} = -16.4\mathbf{i} - 82.4\mathbf{k}$ (N-m).

2.166 $\mathbf{r}_{BC} \times \mathbf{T} = 33.3\mathbf{i} - 125\mathbf{j} - 183\mathbf{k}$ (N-m).

Chapter 3

3.2 $F_2 = 4.77$ lb, $\alpha = 35.2°$.

3.4 $T_{AB} = T_{AC} = 1.53$ kN.

3.6 $T = 785$ N, $P = 823$ N.

3.8 $k = 1960$ N/m, $m_A = 4$ kg, $m_B = 6$ kg.

3.10 (a) $|N_{\text{crane}}| = 197$ kN, $|f_{\text{crane}}| = 0.707$ kN;
(b) $|N_{\text{caisson}}| = 3.22$ kN, $|f_{\text{caisson}}| = 0.707$ kN.

3.12 (a) $|N| = 11.06$ kN, $|f| = 4.03$ kN;
(b) $\alpha = 31.0°$.

3.14 (a) 254 lb; (b) 41.8°.

3.16 5.91 kN.

3.18 (a) 128 N; (b) 98.1 N.

3.20 $T_{\text{left}} = 299$ lb, $T_{\text{right}} = 300$ lb.

3.22 188 lb.

3.24 (a) 66.1 lb; (b) 12.3 lb.

3.26 $T_{AB} = 2.75$ kN, $T_{BC} = 2.06$ kN.

3.28 Upper cable tension is $0.828W$, lower cable tension is $0.132W$.

3.30 $T_{AB} = 1.21$ N, $T_{AD} = 2.76$ N.

3.32 $m = 12.2$ kg.

3.34 $F_B = 3680$ lb, $F_C = 2330$ lb.

3.36 $h = b$.

3.38 $T_{AB} = 688$ lb.

3.40 $T_{AB} = 64.0$ kN, $T_{BC} = 61.0$ kN.

3.44 $\alpha = 79.7°$, $T_{AB} = 120$ N, $T_{BC} = 21.4$ N, $T_{CD} = 62.6$ N.

3.46 $W_1 = 133$ lb.

3.48 (b) Left surface: 36.6 lb; right surface: 25.9 lb.

3.50 $k = 1420$ N/m.

3.52 $T = mgL/(h + R)$.

3.56 $m_2 = 12.5$ kg.

3.58 (a) $T = W/2$; (b) $T = W/4$; (c) $T = W/8$.

3.60 $L = 131.1$ kN, $D = 36.0$ kN.

3.62 (a) $\gamma = -14.0°$; (b) 4 km.

3.64 $T_{AB} = 405$ lb, $T_{AC} = 395$ lb, $T_{AD} = 103$ lb.

3.66 $T_{AB} = 1.54$ lb, $T_{AC} = 1.85$ lb.

3.68 Two at B, three at C, and three at D.

3.70 $T_{AB} = 9390$ lb, $T_{AC} = 5390$ lb, $T_{AD} = 10,980$ lb.

3.72 $D = 1176$ N, $T_{OA} = 6774$ N.

3.74 $T_{BC} = 1.61$ kN, $T_{BD} = 1.01$ kN.

3.76 $T_{EF} = T_{EG} = 738$ kN.

3.78 (a) The tension $= 2.70$ kN;
(b) The force exerted by the bar $= 1.31\mathbf{i} - 1.31\mathbf{k}$ (kN).

3.80 $T_{AB} = 357$ N.

3.82 $F = 36.6$ N.

3.84 $W = 25.0$ lb.

3.86 (a) 83.9 lb; (b) 230.5 lb.

3.88 $T = mg/26$.

3.90 $F = 162.0$ N.

3.92 $T_{AB} = 420$ N, $T_{AC} = 533$ N, $|\mathbf{F}_S| = 969$ N.

3.94 $N = 2580$ lb, $f = 995$ lb.

3.96 $T_{AC} = 16.7$ lb, $T_{AD} = 17.2$ lb, $T_{AE} = 9.21$ lb.

3.98 Normal force $= 12.15$ kN, friction force $= 4.03$ kN.

Chapter 4

4.2 134 N-m.

4.4 $F = 36.2$ N.

4.6 25.0 kN-m clockwise.

4.8 $L = 2.4$ m.

4.10 $15.8° \leq \alpha \leq 37.3°$.

4.12 0.961 kN-m counterclockwise.

4.14 $M_S = 611$ in-lb.

4.16 $M_P = 298$ N-m.

4.18 410 N-m counterclockwise.

4.20 (a) $F_B = 37.5$ lb, $F_C = 22.5$ lb, $F_D = 26.0$ lb;
(b) Zero.

4.22 (a) $A = 56.6$ lb, $B = 24.4$ lb, $C = 12.2$ lb;
(b) Zero.

4.24 640 lb.

4.26 $M = 2.39$ kN-m.

4.28 (a) $A_x = 18.1$ kN, $A_y = -29.8$ kN, $B = -20.4$ kN;
(b) Zero.

4.30 (a) $A_x = 300$ lb, $A_y = 240$ lb, $B = 280$ lb;
(b) Zero.

4.32 60.4 ft-lb.

4.34 -22.3 ft-lb.

4.36 $M = -2340$ N-m.

4.38 $T_{AB} = T_{AC} = 223$ kN.

4.40 617 N-m.

4.42 $M_A = -3.00$ kN-m, $M_D = 7.50$ kN-m.

4.44 796 N.

4.46 (a), (b) $480\mathbf{k}$ (N-m).

4.48 (a) $800\mathbf{k}$ (kN-m);
(b) $-400\mathbf{k}$ (kN-m).

4.50 $\mathbf{F} = 20\mathbf{i} + 40\mathbf{j}$ (N).

4.52 $\mathbf{M}_O = -5600\mathbf{k}$ (ft-lb).

4.54 (a), (b) 1270 N-m.

4.56 128 ft-lb.

4.58 985 ft-lb.

4.60 58.0 kN.

4.62 (a) $|\mathbf{F}| = 1586$ N;
(b) $|\mathbf{F}| = 1584$ N.

4.64 $-16.4\mathbf{i} - 111.9\mathbf{k}$ (N-m).

4.66 $\mathbf{F} = 4\mathbf{i} - 4\mathbf{j} + 2\mathbf{k}$ (kN) or
$\mathbf{F} = 4\mathbf{i} - 3.38\mathbf{j} + 2.92\mathbf{k}$ (kN).

4.68 $\mathbf{M}_D = 1.25\mathbf{i} + 1.25\mathbf{j} - 6.25\mathbf{k}$ (kN-m).

4.70 $T_{AC} = 2.23$ kN, $T_{AD} = 2.43$ kN.

4.72 $T_{AB} = 1.60$ kN, $T_{AC} = 1.17$ kN.

4.74 $T_{BC} = 886$ N, $T_{BD} = 555$ N.

4.76 $\mathbf{M} = 482\mathbf{k}$ (kN-m).

4.78 (a) $\mathbf{M}_{x\,\text{axis}} = 80\mathbf{i}$ (N-m);
(b) $\mathbf{M}_{y\,\text{axis}} = -140\mathbf{j}$ (N-m);
(c) $\mathbf{M}_{z\,\text{axis}} = \mathbf{0}$.

4.80 (a) Zero; (b) $2.7\,\mathbf{k}$ (kN-m).

4.82 (a) $\mathbf{M}_{x\,\text{axis}} = -16\mathbf{i}$ (kN-m);
(b) $\mathbf{M}_{z\,\text{axis}} = 15\mathbf{k}$ (kN-m).

4.84 $\mathbf{F} = 80\mathbf{i} + 80\mathbf{j} + 40\mathbf{k}$ (lb).

4.86 $-16.4\mathbf{i}$ (N-m).

4.88 (a), (b) $\mathbf{M}_{AB} = -76.1\mathbf{i} - 95.1\mathbf{j}$ (N-m).

4.90 $\mathbf{M}_{AO} = 119.1\mathbf{j} + 79.4\mathbf{k}$ (N-m).

4.92 $\mathbf{M}_{AB} = 77.1\mathbf{j} - 211.9\mathbf{k}$ (ft-lb).

4.94 $\mathbf{M}_{y\,\text{axis}} = 215\mathbf{j}$ (N-m).

4.96 $\mathbf{M}_{x\,\text{axis}} = 44\mathbf{i}$ (N-m).

4.98 $-338\mathbf{j}$ (ft-lb).

4.100 $|\mathbf{F}| = 13$ lb.

4.102 $\mathbf{M}_{\text{axis}} = -478\mathbf{i} - 174\mathbf{k}$ (N-m).

4.104 1 N-m.

4.106 $124\mathbf{k}$ (ft-lb).

4.108 28 N-m clockwise.

4.110 $\alpha = 30.9°$ or $\alpha = 71.8°$.

4.112 (b) $FL\cos 30°$.

4.114 40 ft-lb clockwise, or $-40\mathbf{k}$ (ft-lb).

4.116 2200 ft-lb clockwise.

4.118 (a) $C = 26$ kN-m; (b) Zero.

4.120 (a) $\mathbf{M} = -14\mathbf{i} - 10\mathbf{j} - 8\mathbf{k}$ (kN-m); (b) $D = 6.32$ m.

4.122 356 ft-lb.

4.124 $|\mathbf{M}| = 6.13$ kN-m.

4.126 $M_{Cy} = 7$ kN-m, $M_{Cz} = -2$ kN-m.

4.128 Yes.

4.130 Systems 1, 2, and 4 are equivalent.

4.134 $F = 265$ N.

4.136 $F = 70$ lb, $M = 130$ in-lb.

4.138 (a) $\mathbf{F} = -10\mathbf{j}$ (lb), $M = -10$ ft-lb; (b) $D = 1$ ft.

4.140 $\mathbf{F} = 200\mathbf{i} + 180\mathbf{j}$ (N), $d = 0.317$ m.

4.142 (a) $A_x = 12$ kip, $A_y = 10$ kip, $B = -10$ kip;
(b) $\mathbf{F} = -12\mathbf{i}$ (kip), intersects at $y = 5$ ft;
(c) They are both zero.

4.144 $\mathbf{F} = 104\mathbf{j}$ (kN), $M = 13.2$ kN-m counterclockwise.

4.146 $\mathbf{F} = 100\mathbf{j}$ (lb), $\mathbf{M} = \mathbf{0}$.

4.148 (a) $\mathbf{F} = 920\mathbf{i} - 390\mathbf{j}$ (N), $M = -419$ N-m;
(b) intersects at $y = 456$ mm.

4.150 $\mathbf{F} = 800\mathbf{j}$ (lb), intersects at $x = 7.5$ in.

4.152 (a) $-360\mathbf{k}$ (in-lb);
(b) $-36\mathbf{j}$ (in-lb);
(c) $\mathbf{F} = 10\mathbf{i} - 30\mathbf{j} + 3\mathbf{k}$ (lb),
$\mathbf{M} = -36\mathbf{j} - 360\mathbf{k}$ (in-lb).

4.154 $x = 2.00$ ft, $z = -0.857$ ft.

4.156 $\mathbf{F} = 100\mathbf{j} + 80\mathbf{k}$ (N), $\mathbf{M} = 240\mathbf{j} - 300\mathbf{k}$ (N-m).

4.158 (a) $\mathbf{F} = \mathbf{0}$, $\mathbf{M} = rA\mathbf{i}$;
(b) $\mathbf{F}' = \mathbf{0}$, $\mathbf{M}' = rA\mathbf{i}$.

4.160 (a) $\mathbf{F} = \mathbf{0}$, $\mathbf{M} = 4.60\mathbf{i} + 1.86\mathbf{j} - 3.46\mathbf{k}$ (kN-m);
(b) 6.05 kN-m.

4.162 $\mathbf{F} = -20\mathbf{i} + 20\mathbf{j} + 10\mathbf{k}$ (lb),
$\mathbf{M} = 50\mathbf{i} + 250\mathbf{j} + 100\mathbf{k}$ (in-lb).

4.164 (a) $\mathbf{F} = 28\mathbf{k}$ (kip), $\mathbf{M} = 96\mathbf{i} - 192\mathbf{j}$ (ft-kip);
(b) $x = 6.86$ ft, $y = 3.43$ ft.

4.166 $\mathbf{F} = 100\mathbf{i} + 20\mathbf{j} - 20\mathbf{k}$ (N),
$\mathbf{M} = -143\mathbf{i} + 406\mathbf{j} - 280\mathbf{k}$ (N-m).

4.168 $\mathbf{M}_\text{p} = 0$, line of action intersects at $y = 0$, $z = 2$ ft.

4.170 $x = 2.41$ m, $y = 3.80$ m.

4.172 $\mathbf{F} = 40.8\mathbf{i} + 40.8\mathbf{j} + 81.6\mathbf{k}$ (N),
$\mathbf{M} = -179.6\mathbf{i} + 391.9\mathbf{j} - 32.7\mathbf{k}$ (N-m).

4.174 (a) $320\mathbf{i}$ (in-lb);
(b) $\mathbf{F} = -20\mathbf{k}$ (lb), $\mathbf{M} = 320\mathbf{i} + 660\mathbf{j}$ (in-lb);
(c) $\mathbf{M}_t = 0$, $x = 33$ in, $y = -16$ in.

4.176 $|\mathbf{M}_P| = 244$ N-m.

4.178 (a) -76.2 N-m;
(b) -66.3 N-m.

4.180 $|\mathbf{F}| = 224$ lb, $|\mathbf{M}| = 1600$ ft-lb.

4.182 501 lb.

4.184 $-228.1\mathbf{i} - 68.4\mathbf{k}$ (N-m).

4.186 $\mathbf{M}_{x\,\text{axis}} = -153\mathbf{i}$ (ft-lb).

4.188 $\mathbf{M}_{CD} = -173\mathbf{i} + 1038\mathbf{k}$ (ft-lb).

4.190 (a) $\mathbf{T}_{AB} = \mathbf{T}_{CD} = 173.2$ lb;
(b) $\mathbf{F} = 300\mathbf{j}$ (lb) at $x = 4$ ft.

4.192 $\mathbf{F} = -20\mathbf{i} + 70\mathbf{j}$ (N), $M = 22$ N-m.

4.194 $\mathbf{F}' = -100\mathbf{i} + 40\mathbf{j} + 30\mathbf{k}$ (lb),
$\mathbf{M} = -80\mathbf{i} + 200\mathbf{k}$ (in-lb).

4.196 $\mathbf{F} = 1166\mathbf{i} + 566\mathbf{j}$ (N), $y = 13.9$ m.

4.198 $\mathbf{F} = 190\mathbf{j}$ (N), $\mathbf{M} = -98\mathbf{i} + 184\mathbf{k}$ (N-m).

4.200 $\mathbf{F} = -0.364\mathbf{i} + 4.908\mathbf{j} + 1.090\mathbf{k}$ (kN),
$\mathbf{M} = -0.131\mathbf{i} - 0.044\mathbf{j} + 1.112\mathbf{k}$ (kN-m).

Chapter 5

5.2 $A_x = -1$ kN, $A_y = -5.73$ kN,
$M_A = -22.9$ kN-m.

5.4 Tension is 386 lb, $B_x = 493$ lb, $B_y = 186$ lb.

5.6 (b) $A_x = 0$, $A_y = -1.85$ kN, $B_y = 2.74$ kN.

5.8 (b) $A_x = 0$, $A_y = -5$ kN, $B_y = 15$ kN.

5.10 (b) $A = 100$ lb, $B = 200$ lb.

5.12 (b) $A_x = 502$ N, $A_y = 870$ N.

5.14 (b) $A_x = 4$ kN, $A_y = -2.8$ kN, $B_y = 2.8$ kN.

5.16 On each hand, 66.3 lb. On each foot, 23.7 lb.

5.18 $A_x = -100$ lb, $A_y = -225$ lb, $E = 625$ lb.

5.20 $k = 3380$ N/m, $B_x = -188.0$ N, $B_y = 98.7$ N.

5.22 5.93 kN.

5.24 $R = 12.5$ lb, $B_x = 11.3$ lb, $B_y = 15.3$ lb.

5.26 (a) 21.2 lb; (b) 30 lb.

5.28 $W_L = 1125$ lb.

5.30 6.23 lb.

5.32 $T = 3.68$ lb.

5.34 $T_{AE} = 28.6$ lb, $D_x = -26.9$ lb, $D_y = 32.2$ lb.

5.36 $A_x = -1.83$ kN, $A_y = 2.10$ kN, $B_y = 2.46$ kN.

5.38 $A_x = -200$ lb, $A_y = -100$ lb, $M_A = 1600$ ft-lb.

5.40 $k = 3.21$ lb/ft.

5.42 $A_x = 3.46$ kN, $A_y = -2$ kN,
$B_x = -3.46$ kN, $B_y = 2$ kN.

5.44 $\mathbf{F} = 28.3\mathbf{i} + 58.3\mathbf{j}$ (lb), $D = 7.03$ ft, $A_x = -28.3$ lb,
$A_y = -58.3$ lb, $M_A = -410$ ft-lb.

5.46 $A_x = -1.57$ kN, $A_y = 1.57$ kN, $E_x = 1.57$ kN.

5.48 $A_x = 0$, $A_y = 200$ lb, $M_A = 900$ ft-lb.

5.50 $A_x = 57.7$ lb, $A_y = -13.3$ lb, $B = 15.3$ lb.

5.52 $W = 15$ kN.

5.54   (b) $C_x = 500$ N, $C_y = -200$ N.

5.56 $T_{BC} = 5.45$ lb, $A_x = 5.03$ lb, $A_y = 7.90$ lb.

5.58   20.3 kN.

5.60 $W_2 = 2484$ lb, $A_x = -2034$ lb, $A_y = 2425$ lb.

5.62 $W = 46.2$ N, $A_x = 22.3$ N, $A_y = 61.7$ N.

5.64 $F = 44.5$ lb, $A_x = 25.3$ lb, $A_y = -1.9$ lb.

5.66 $W = 132$ lb.

5.68

5.76   (1) and (2) are improperly supported. For (3), reactions
are $A = F/2$, $B = F/2$, $C = F$.

5.78   (b) $A_x = -6.53$ kN, $A_y = -3.27$ kN,
$A_z = 3.27$ kN, $M_{Ax} = 0$, $M_{Ay} = -6.53$ kN-m,
$M_{Az} = -6.53$ kN-m.

5.80   374 lb.

5.82 $C_x = -349$ lb, $C_y = 698$ lb,
$C_z = 175$ lb, $M_{Cx} = -3490$ ft-lb,
$M_{Cy} = -2440$ ft-lb, $M_{Cz} = 2790$ ft-lb.

5.84   (a) $-17.8\mathbf{i} - 62.8\mathbf{k}$ (N-m);
(b) $A_x = 0$, $A_y = 360$ N, $A_z = 0$,
$M_{Ax} = 17.8$ N-m, $M_{Ay} = 0$, $M_{Az} = 62.8$ N-m.

5.86 $A_x = 166.7$ N, $A_y = 200$ N, $A_z = 66.7$ N,
$T_{BC} = 100$ N, $T_{BD} = 170$ N.

5.88 $|\mathbf{F}| = 10.9$ kN.

5.90 $T_{AB} = 553$ lb, $T_{AC} = 289$ lb,
$O_x = 632$ lb, $O_y = 574$ lb, $O_z = 0$.

5.92 $x = 0.1$ m, $z = 0.133$ m.

5.94 $T_{BD} = 50.2$ lb, $A_x = -34.4$ lb,
$A_y = 17.5$ lb, $A_z = -24.1$ lb,
$M_{Ax} = 0$, $M_{Ay} = 192.5$ in-lb.

5.96 $\mathbf{F} = 4\mathbf{j}$ (kN) at $x = 0$, $z = 0.15$ m.

5.98   (b) $A_x = -0.74$ kN, $A_y = 1$ kN, $A_z = -0.64$ kN,
$B_x = 0.74$ kN, $B_z = 0.64$ kN.

5.100 $F_y = 34.5$ lb.

5.102 $T_{BD} = 1.47$ kN, $T_{BE} = 1.87$ kN,
$A_x = 0$, $A_y = 4.24$ kN, $A_z = 0$.

5.104 $T = 139$ lb, $A_x = 46.4$ lb, $A_y = -26.8$ lb,
$A_z = 31.7$ lb, $M_{Ax} = -63.4$ ft-lb,
$M_{Ay} = -110$ ft-lb.

5.106   Tension is 60 N, $B_x = -10$ N, $B_y = 90$ N,
$B_z = 10$ N, $M_{By} = 1$ N-m, $M_{Bz} = -3$ N-m.

5.108   Tension is 60 N, $B_x = -10$ N, $B_y = 75$ N,
$B_z = 15$ N, $C_y = 15$ N, $C_z = -5$ N.

5.110 $A_x = -2.86$ kip, $A_y = 17.86$ kip, $A_z = -8.10$ kip,
$B_y = 3.57$ kip, $B_z = 12.38$ kip.

5.112 $A_x = 0$, $A_y = 400$ N, $B_x = 1000$ N,
$B_y = -400$ N, $B_z = 0$, $T = 1080$ N.

5.114 $|\mathbf{A}| = 8.54$ kN, $|\mathbf{B}| = 10.75$ kN.

5.116 $A_x = 3.62$ kN, $A_y = 5.89$ kN, $A_z = 5.43$ kN,
$C_x = 8.15$ kN, $C_y = 0$, $C_z = 0.453$ kN.

5.118 $T_{AB} = 488$ lb, $T_{CD} = 373$ lb, reaction is
$31\mathbf{i} + 823\mathbf{j} - 87\mathbf{k}$ (lb).

5.120 $A_x = -76.7$ N, $A_y = 97.0$ N, $A_z = -54.3$ N,
$M_{Ax} = -2.67$ N-m, $M_{Ay} = 6.39$ N-m,
$M_{Az} = 2.13$ N-m.

5.122   (a) 60 lb;
(b) $A_x = 38.1$ lb, $A_y = 46.3$ lb   or $A_x = -38.1$ lb,
$A_y = -46.3$ lb.

5.124   Tension is 33.3 lb; magnitude of reaction is 44.1 lb.

5.126 $\alpha = 10.9°$, $F_A = 1.96$ kN, $F_B = 2.27$ kN.

5.128   (a) No, because of the 3 kN-m couple; (b) magnitude at
A is 7.88 kN; magnitude at B is 6.66 kN; (c) no.

5.130   (b) $A_x = -8$ kN, $A_y = 2$ kN, $C_x = 8$ kN.

5.134   (b) $T_A = 7.79$ lb, $T_B = 10.28$ lb;   (c) 6.61 lb.

5.136   (a) There are four unknown reactions and three equilibrium equations;   (b) $A_x = -50$ lb, $B_x = 50$ lb.

5.138   (b) Force on nail = 55 lb, normal force = 50.77 lb,
friction force = 9.06 lb.

5.140 $k = 13,500$ N/m.

5.142 $A_y = 727$ lb, $H_x = 225$ lb, $H_y = 113$ lb.

5.144 $\alpha = 0$ and $\alpha = 59.4°$.

5.146   The force is 800 N upward; its line of action passes
through the midpoint of the plate.

5.148 $m = 67.2$ kg.

5.150 $\alpha = 90°$, $T_{BC} = W/2$, $A = W/2$.

Chapter 6

6.2 AB: 915 N (C); AC: 600 N (C); BC: 521 N (T).

6.4 BC: 800 lb (T); CD: 600 lb (C).

6.6   (a) Tension: 2.43 kN in AB and BD.
Compression: 2.88 kN in CD.
(b) Tension: 1.74 kN in BD.
Compression: 1.60 kN in CD.

6.8   Tension, 31.9 kip in AC, CE, EG, and GH. Compression, 42.5 kip in BD and DF.

6.10 BD: zero; CD: 10 kN (T); CE: 16 kN (C).

6.12   (a) Tension: 5540 lb in BD. Compression: 7910 lb in CE.
(b) Tension: 2770 lb in BD. Compression: 3760 lb in CE.

6.14 $F = 8.33$ kN.

6.16 DE: 3.66 kN (C); DF: 1.45 kN (C); DG: 3.36 kN (T).

6.18 AB: 10.56 kN (T); AC: 17.58 kN (C); BC: 6.76 kN (T);
BD: 1.81 kN (T); CD: 16.23 kN (C).

6.20 AB: 375 lb (C); AC: 625 lb (T); BC: 300 lb (T).

6.22 BC: 90.1 kN (T); CD: 90.1 kN (C); CE: 300 kN (T).

6.24 BC: 1200 kN (C); BI: 300 kN (T); BJ: 636 kN (T).

6.26 AB: 2520 lb (C); BC: 2160 lb (C); CD: 1680 lb (C).

6.32 BC: 400 kN (T), BI: 141 kN (T), HI: 500 kN (C).

6.34 (a), (b) 141 kN (C).

6.36 AB: $1.33F$ (C); BC: $1.33F$ (C); CE: $1.33F$ (T).

6.38 BD: 95.6 kip (C); BE: 41.1 kip (T); CE: 58.4 kip (T).

6.40 DF: 69.1 kip (C); DG: 29.4 kip (C); EG: 95.6 kip (T).

6.42 96.2 kN (T).

6.44 AC: 2000 lb (C); BC: 800 lb (T); BD: 1000 lb (T).

6.46 DF: 16 kN (T); DG: 6.67 kN (C); EG: 26.7 kN (C).

6.48 2.50 kN (C).

6.50 CE: 680 kN (T); CF: 374 kN (C); DF: 375 kN (C).

6.52 (a) 1160 lb (C).

6.54 IL: 16 kN (C); KM: 24 kN (T).

6.58 AD: 4.72 kN (C); BD: 4.16 kN CD (C); CD: 4.85 kN (C).

6.60 AB, AC, AD: $0.408F$ (C).

6.62 AB: 379 lb (C); AC: 665 lb (C); AD: 160 lb (C).

6.64 BC: 32.7 kN (T); BD: 45.2 kN (T); BE: 112.1 kN (C).

6.66 $P_3 = -315$ kN.

6.68 5.59 kN (C) in each member.

6.70 $A_x = 400$ N, $A_y = -900$ N, $B_x = -400$ N, $B_y = 900$ N, $M_A = -540$ N-m.

6.72 $C_x = 736$ N, $C_y = 2450$ N, $E_x = 245$ N, $E_y = -1720$ N.

6.74 $C_x = 66.7$ lb, $C_y = 24$ lb.

6.76 $A_x = 0$, $A_y = -400$ N, $C_x = -600$ N, $C_y = -300$ N, $D_x = 0$, $D_y = 1000$ N.

6.78 $D_x = -1475$ N, $D_y = -516$ N, $E_x = 0$, $E_y = -516$ N, $M_E = 619$ N-m.

6.80 $A_x = -2.35$ kN, $A_y = 2.35$ kN, $B_x = 0$, $B_y = -4.71$ kN, $C_x = 2.35$ kN, $C_y = 2.35$ kN.

6.82 Tension $= 62.5$ lb, $F_x = -75$ lb, $F_y = 25$ lb.

6.84 $B_x = -400$ lb, $B_y = -300$ lb, $C_x = 400$ lb, $C_y = 200$ lb, $D_x = 0$, $D_y = 100$ lb.

6.86 $A_x = -150$ lb, $A_y = 120$ lb, $B_x = 180$ lb, $B_y = -30$ lb, $D_x = -30$ lb, $D_y = -90$ lb.

6.88 $A_x = -310$ lb, $A_y = -35$ lb, $B_x = 80$ lb, $B_y = -80$ lb, $C_x = 310$ lb, $C_y = 195$ lb, $D_x = -80$ lb, $D_y = -80$ lb.

6.90 $A_x = 170$ lb, $A_y = 129$ lb, $B_x = -170$ lb, $B_y = -209$ lb.

6.94 $A_x = -22$ lb, $A_y = 15$ lb, $C_x = -14$ lb, $C_y = 3$ lb.

6.96 300 lb (C).

6.98 B: 73.5 N; C: 88.8 N.

6.100 $T_{BC} = 1410$ N, $T_{DF} = 625$ N.

6.102 $A_x = 2$ kN, $A_y = -1.52$ kN, $B_x = -2$ kN, $B_y = 1.52$ kN.

6.104 $E_x = 604$ lb, $E_y = 179$ lb, axial force is 616 lb.

6.106 100 N.

6.108 At B: 1750 N. DE: 1320 N (C).

6.110 742 lb.

6.112 1150 lb.

6.114 $K_x = 847$ N, $K_y = 363$ N.

6.116 $T_{AB} = 7.14$ kN (C), $T_{AC} = 5.71$ kN (T), $T_{BC} = 10$ kN (T).

6.118 BC: 120 kN (C); BG: 42.4 kN (T); FG: 90 kN (T).

6.120 AB: 125 lb (C); AC: zero; BC: 188 lb (T); BD: 225 lb (C); CD: 125 lb (C); CE: 225 lb (T).

6.122 $T_{BD} = 13.3$ kN (T), $T_{CD} = 11.7$ kN (T), $T_{CE} = 28.3$ kN (C).

6.124 AC: 480 N (T); CD: 240 N (C); CF: 300 N (T).

6.126 Tension: member AC, 480 lb (T); Compression: member BD, 633 lb (C).

6.128 CD: 11.42 kN (C); CJ: 4.17 kN (C); IJ: 12.00 kN (T).

6.130 AB: 7.20 kN (C); AC: 4.56 kN (C).

6.132 $A_x = -1.57$ kN, $A_y = 1.18$ kN, $B_x = 0$, $B_y = -2.35$ kN, $C_x = 1.57$ kN, $C_y = 1.18$ kN.

6.134 $B_x = 3820$ lb, $B_y = 6690$ lb, $C = 9020$ lb, $D_x = -1390$ lb, $D_y = -1930$ lb.

6.136 973 N.

6.138 $A_x = -52.33$ kN, $A_y = -43.09$ kN, $E_x = 0.81$ kN, $E_y = -14.86$ kN.

Chapter 7

7.2 $\bar{x} = 3/8$.

7.4 $\bar{x} = 1.25$, $\bar{y} = 0.825$.

7.8 $\bar{x} = 0.711$ ft, $y = 0.584$ ft.

7.10 $\bar{x} = 0$, $y = 1.6$ ft.

7.12 $\bar{x} = 8$, $\bar{y} = 3.6$.

7.14 $\bar{x} = 0.533$.

7.16 $\bar{x} = 1$.

7.18 $\bar{y} = -7.6$.

7.20 $\bar{y} = 2.53$.

7.22 $a = 0.656$, $b = 6.56 \times 10^{-5}$ m^{-2}.

7.24 $\bar{x} = \bar{y} = 4R/3\pi$.

7.26 $\bar{x} = 3.31$.

7.28 $\bar{x} = 116$ mm.

7.30 $\bar{x} = 9.90$ in, $\bar{y} = 0$.

7.32 $\bar{x} = 23.9$ in, $\bar{y} = 33.3$ in.

7.34 $\bar{x} = 2.88$ ft, $\bar{y} = 3.20$ ft.

7.36 $\bar{x} = 3.67$ mm, $\bar{y} = 21.52$ mm.

7.38 $b = 39.6$ mm, $h = 18.2$ mm.

7.40 $\bar{x} = 9.64$ m, $\bar{y} = 4.60$ m.

7.44 $\bar{x} = 6.47$ ft, $\bar{y} = 10.60$ ft.

7.46 $A_x = 0$, $A_y = 160$ N, $B = 200$ N.

7.48 $A_x = -1200$ N, $A_y = 800$ N, $B = 2200$ N.

7.50 $A_x = 0$, $A_y = 10$ kN, $M_A = -31.3$ kN-m.

7.52 $A_x = 0$, $A_y = 4.17$ kN, $B_y = 8.83$ kN.

7.54 $A_y = 3267$ lb, $B_x = -800$ lb, $B_y = -1267$ lb.

7.56 BD: 21.3 kN (C); CD: 3.77 kN (C); CE: 24 kN (T).

7.58 $A_x = -18$ kN, $A_y = 20$ kN, $B_x = 0$, $B_y = -4$ kN, $C_x = 18$ kN, $C_y = -16$ kN.

7.60 $V = 275$ m^3, height $= 2.33$ m.

7.62 $V = 4.16$ m^3, $\bar{x} = 1.41$ m.

7.64 $\bar{x} = 0.675R$, $\bar{y} = 0$, $\bar{z} = 0$.

7.66 $\bar{y} = 0.410$.

7.68 $\bar{x} = 3.24$.

7.70 $\bar{x} = R\sin\alpha/\alpha$, $\bar{y} = R(1 - \cos\alpha)/\alpha$.

7.72 $\bar{x} = 38.3$ mm.

7.74 $\bar{x} = -128$ mm, $\bar{y} = \bar{z} = 0$.

7.76 $\bar{x} = 0$, $\bar{y} = 43.7$ mm, $\bar{z} = 38.2$ mm.

7.78 $\bar{x} = 229.5$ mm, $\bar{y} = \bar{z} = 0$.

7.80 $\bar{x} = 23.65$ mm, $\bar{y} = 36.63$ mm, $\bar{z} = 3.52$ mm.

7.82 $\bar{x} = 6$ m, $\bar{y} = 1.83$ m.

7.84 $\bar{x} = 65.9$ mm, $\bar{y} = 21.7$ mm, $\bar{z} = 68.0$ mm.

7.86 $A = \frac{3}{4}\pi R\sqrt{h^2 + R^2}$.

7.88 $\bar{y}_s = 4R/3\pi$.

7.90 $\bar{y} = 0.410$.

7.92 $A = 138$ ft^2.

7.94 $V = 0.0377$ m^3.

7.96 $V = 2.48 \times 10^6$ mm^3.

7.98 Volume $= 0.0266$ m^3.

7.100 $A_x = 0$, $A_y = 294$ N, $B_y = 196$ N.

7.102 $A_x = 0$, $A_y = 316$ N, $B = 469$ N.

7.104 $\bar{x} = 6.59$ in, $\bar{y} = 2.17$ in, $\bar{z} = 6.80$ in.

7.106 $A_x = 0$, $A_y = 3.16$ kN, $M_A = 1.94$ kN-m.

7.108 $\bar{x} = 121$ mm, $\bar{y} = 0$, $\bar{z} = 0$.

7.110 $\bar{x}_3 = 82$ mm, $\bar{y}_3 = 122$ mm, $\bar{z}_3 = 16$ mm.

7.112 (a) $\bar{x} = 5.17$ m; (b) $A_x = -50$ kN, $A_y = -25.0$ kN, $G = 33.8$ kN.

7.114 Mass $= 408$ kg, $\bar{x} = 2.5$ m, $\bar{y} = -1.5$ m.

7.116 $\bar{x} = 20.10$ in, $\bar{y} = 8.03$ in, $\bar{z} = 15.35$ in.

7.118 $\bar{x} = 3/8$, $\bar{y} = 3/5$.

7.120 $\bar{x} = 87.3$ mm, $\bar{y} = 55.3$ mm.

7.122 917 N (T).

7.124 $A_x = 7$ kN, $A_y = -6$ kN, $D_x = 4$ kN, $D_y = 0$.

7.126 $\bar{x} = 1.87$ m.

7.128 $A = 682$ in^2.

7.130 $\bar{x} = 110$ mm.

7.132 $\bar{x} = 1.70$ m.

7.134 $\bar{x} = 25.24$ mm, $\bar{y} = 8.02$ mm, $\bar{z} = 27.99$ mm.

7.136 (a) $\bar{x} = 1.511$ m; (b) $\bar{x} = 1.611$ m.

7.138 $A = 80.7$ kN, $B = 171.6$ kN.

Chapter 8

8.2 $I_x = 0.0288$ m^4, $k_x = 0.346$ m.

8.4 (a) $I_y = 12.8 \times 10^5$ mm^4; (b) $I_{y'} = 3.2 \times 10^5$ mm^4.

8.6 $I_y = 0.175$ m^4, $k_y = 0.624$ m.

8.8 $I_{xy} = 0.0638$ m^4.

8.10 $I_x = 1.69$.

8.12 $I_{xy} = 0.583$.

8.14 $I_x = 1330$, $k_x = 4.30$.

8.16 $I_{xy} = 2070$.

8.18 $I_x = 953$, $k_x = 6.68$.

8.20 (a) $I_x = \frac{1}{8}\pi R^4$, $k_x = \frac{1}{2}R$.

8.22 $I_y = 49.09$ m^4, $k_y = 2.50$ m.

8.24 $I_y = 522$, $k_y = 2.07$.

8.28 $I_y = 10$ m^4, $k_y = 1.29$ m.

8.30 $I_x = 6.00 \times 10^6$ mm^4, $k_x = 23.5$ mm.

8.32 $I_y = 0.0125$ m^4, $k_y = 0.177$ m.

8.34 $I_y = 3.6 \times 10^5$ mm^4, $J_O = 1 \times 10^6$ mm^4.

8.36 $I_x = 2.65 \times 10^8$ mm^4, $k_x = 129$ mm.

8.38 $I_x = 7.79 \times 10^7$ mm^4, $k_x = 69.8$ mm.

8.40 $I_{xy} = 1.08 \times 10^7$ mm^4.

8.42 $J_O = 363$ ft^4, $k_O = 4.92$ ft.

8.44 $I_x = 10.7$ ft^4, $k_x = 0.843$ ft.

8.46 $I_{xy} = 7.1$ ft^4.

8.48 $J_O = 5.63 \times 10^7$ mm^4, $k_O = 82.1$ mm.

8.50 $I_x = 1.08 \times 10^7$ mm^4, $k_x = 36.0$ mm.

8.52 $J_O = 1.58 \times 10^7$ mm^4, $k_O = 43.5$ mm.

8.54 $J_O = 2.35 \times 10^5$ in^4, $k_O = 15.1$ in.

8.56 $I_x = 49.7$ m^4, $k_x = 2.29$ m.

8.58 $I_y = 2.55 \times 10^6$ in^4, $k_y = 27.8$ in.

8.60 $I_{xy} = 2.54 \times 10^6$ in^4.

8.62 $I_x = 1.26 \times 10^6$ in^4, $k_x = 19.5$ in.

8.64 $I_y = 4.34 \times 10^4$ in^4, $k_y = 10.5$ in.

8.66 $I_{xy} = 4.83 \times 10^4$ in^4.

8.68 $J_O = 4.01 \times 10^4$ in^4, $k_O = 14.6$ in.

8.70 $I_x = 8.89 \times 10^3$ in^4, $k_x = 7.18$ in.

8.72 $I_y = 3.52 \times 10^3$ in^4, $k_y = 4.52$ in.

8.74 $I_{xy} = 995$ in^4.

8.76 $J_O = 5.80 \times 10^6$ mm^4, $k_O = 37.5$ mm.

8.78 $I_x = 1470$ in^4, $I_y = 3120$ in^4.

8.80 $I_x = 4020$ in^4, $I_y = 6980$ in^4, or $I_x = 6820$ in^4, $I_y = 4180$ in^4.

8.82 $I_x = 4.01 \times 10^6$ mm^4.

8.86 $I_x = 59.8 \times 10^6$ mm^4, $I_y = 18.0 \times 10^6$ mm^4.

8.88 $I_{x'} = 7.80$ ft^4, $I_{y'} = 24.2$ ft^4, $I_{x'y'} = -2.20$ ft^4.

8.90 $I_{x'} = 1.20 \times 10^6$ in^4, $I_{y'} = 7.18 \times 10^5$ in^4, $I_{x'y'} = 2.11 \times 10^5$ in^4.

8.92 $\theta_p = -12.1°$, principal moments of inertia are 80.2×10^{-6} m^4 and 27.7×10^{-6} m^4.

8.94 $I_{x'} = 7.80$ ft^4 $I_{y'} = 24.2$ ft^4 $I_{x'y'} = -2.20$ ft^4.

8.96 $I_{x'} = 1.20 \times 10^6$ in^4, $I_{y'} = 7.18 \times 10^5$ in^4, $I_{x'y'} = 2.11 \times 10^5$ in^4.

8.98 $\theta_p = -12.1°$, principal moments of inertia are 80.2×10^{-6} m^4 and 27.7×10^{-6} m^4.

8.100 $I_O = 14$ kg-m^2.

8.102 $I_{z\,axis} = 15.1$ kg-m^2.

8.104 $I_{x\,axis} = 0.667$ kg-m^2, $I_{y\,axis} = 2.67$ kg-m^2.

8.106 $I_{y\,axis} = 1.99$ slug-ft^2.

8.108 20.8 kg-m^2.

8.110 $I_O = \frac{17}{12}ml^2$.

8.112 $I_{z\,axis} = 47.0$ kg-m^2.

8.114 $I_{z\,axis} = 0.0803$ slug-ft^2.

8.116 3810 slug-ft^2.

8.118 $I_{z\,\text{axis}} = 9.00$ kg-m^2.

8.120 $I_{y\,\text{axis}} = 0.0881$ slug-ft^2.

8.122 $I_{x\,\text{axis}} = m\left(\frac{1}{3}l^2 + \frac{1}{4}R^2\right)$.

8.124 $I_{x\,\text{axis}} = I_{y\,\text{axis}} = m\left(\frac{3}{20}R^2 + \frac{3}{5}h^2\right)$.

8.126 $I_{x\,\text{axis}} = \frac{1}{6}mh^2 + \frac{1}{3}ma^2$.

8.128 $I_{x\,\text{axis}} = 0.221$ kg-m^2.

8.130 $I_{x'} = 0.995$ kg-m^2, $\ I_{y'} = 20.1$ kg-m^2.

8.132 $I_{z\,\text{axis}} = 0.00911$ kg-m^2.

8.134 $I_O = 0.00367$ kg-m^2.

8.136 $I_{z\,\text{axis}} = 0.714$ slug-ft^2.

8.138 $I_y = \frac{1}{5}, \ k_y = \sqrt{\frac{3}{5}}$.

8.140 $J_O = \frac{26}{105}, \ k_O = \sqrt{\frac{26}{35}}$.

8.142 $I_y = 12.8, \ k_y = 2.19$.

8.144 $I_{xy} = 2.13$.

8.146 $I_{x'} = 0.183, \ k_{x'} = 0.262$.

8.148 $I_y = 2.75 \times 10^7$ mm^4, $\ k_y = 43.7$ mm.

8.150 $I_x = 5.03 \times 10^7$ mm^4, $\ k_x = 59.1$ mm.

8.152 $I_y = 94.2$ ft^4, $\ k_y = 2.24$ ft.

8.154 $I_x = 396$ ft^4, $\ k_x = 3.63$ ft.

8.156 $\theta_p = 19.5°, \ 20.3$ m^4, $\ 161$ m^4.

8.158 $I_{y\,\text{axis}} = 0.0702$ kg-m^2.

8.160 $I_{z\,\text{axis}} = \frac{1}{10}mw^2$.

8.162 $I_{x\,\text{axis}} = 3.83$ slug-ft^2.

8.164 0.537 kg-m^2.

Chapter 9

9.2 1.04 lb.

9.4 (a) $\alpha = 38.7°$; (b) $\alpha = 11.3°$.

9.6 (a) No; (b) 20.4 lb.

9.8 177 N.

9.10 20 lb.

9.12 $\alpha = 14.0°$.

9.14 (a) $T = 56.5$ N.

9.16 (a) Yes. The force is $\mu_s W$; (b) $3\mu_s W$.

9.18 $89.6 \le T \le 110.4$ lb.

9.20 $F = 267$ N.

9.22 $M = hrF\mu_k/[2(h + b\mu_k)]$.

9.24 9.40 ft-lb.

9.26 $\alpha = 33.4°$.

9.28 $\alpha = 28.3°$.

9.30 (a) $M = 162$ in-lb; (b) $M = 135$ in-lb.

9.32 $M = \mu_s RW[\sin\alpha + \mu_s(1 - \cos\alpha)]/[(1 + \mu_s^2)\sin\alpha]$.

9.34 $\alpha = 39.6°$.

9.36 (a) $T = 9.42$ lb; (b) $T = 33.3$ lb.

9.40 $y = 234$ mm.

9.42 $\alpha = 9.27°$.

9.44 $F = 44$ lb.

9.48 $\alpha = 1.54°, \ P = 202$ N.

9.50 (a) $F = \mu_s W$;
　　(b) $F = (W/2)(\mu_{sA} + \mu_{sB})/[1 + (h/b)(\mu_{sA} - \mu_{sB})]$.

9.52 $F/2$.

9.54 333 N.

9.56 $F = 74.3$ lb.

9.58 (a) $f = 24.5$ N; (b) $\mu_s = 0.503$.

9.60 (a) $f = 8$ kN; (b) $\mu_s = 0.533$.

9.62 $\mu_s = 0.432$.

9.64 $\mu_s = 0.901$.

9.66 $F = 139$ lb.

9.68 $F = 102$ lb.

9.70 $F = 1360$ lb.

9.72 $F = 156$ N.

9.74 343 kg.

9.76 No. The minimum value of μ_s required is 0.176.

9.78 $F = 1160$ N.

9.80 1.84 N-m.

9.82 (a) 0.967 in-lb; (b) 0.566 in-lb.

9.84 (a) 2.39 ft-lb; (b) 1.20 ft-lb.

9.86 11.8 ft-lb.

9.88 108 in-lb.

9.90 27.4 in-lb.

9.92 4.18 N-m.

9.94 4.88 N-m.

9.96 17.4 N-m.

9.98 $W = 1.55$ lb.

9.100 106 N.

9.102 51.9 lb.

9.104 $T = 40.9$ N.

9.106 $F_B = 207$ N.

9.108 $M = 1.92$ ft-lb.

9.110 $T = 346$ N.

9.112 $M = 160$ in-lb.

9.114 $M = 12.7$ N-m.

9.116 $M = 7.81$ N-m.

9.118 $M = 5.20$ N-m.

9.120 (a) $M = 93.5$ N-m; (b) 8.17 percent.

9.122 9.51 ft-lb.

9.124 80.1 lb.

9.126 $T_C = 107$ N.

9.128 $M = rW(e^{\pi\mu_k} - 1)$.

9.130 (a) 14.2 lb; (b) 128.3 lb.

9.132 13.1 lb.

9.134 $M_A = 65.2$ N-m, $\ M_B = 32.6$ N-m.

9.136 (a) $f = 10.3$ lb.

9.138 $F = 290$ lb.

9.140 $\alpha = 65.7°$.

9.142 $\alpha = 24.2°$.

9.144 $b = (h/\mu_s - t)/2$.

9.146 $h = 5.82$ in.

9.148 286 lb.

9.150 1130 kg, torque $= 2.67$ kN-m.

9.152 $f = 2.63$ N.

9.154 $\mu_s = 0.272$.

9.156 $M = 1.13$ N-m.

9.158 $P = 43.5$ N.

9.160 146 lb.

9.162 (a) $W = 106$ lb; (b) $W = 273$ lb.

Chapter 10

10.2 $P_A = 0$, $V_A = 100$ N, $M_A = 40$ N-m.

10.4 $P_A = 0$, $V_A = 400$ lb, $M_A = -1900$ ft-lb.

10.6 (a) $P_A = 0$, $V_A = 4$ kN, $M_A = 4$ kN-m;
(b) $P_A = 0$, $V_A = 2$ kN, $M_A = 3$ kN-m.

10.8 $P_B = 0$, $V_B = 40$ N, $M_B = 373$ N-m.

10.10 $P_A = 0$, $V_A = -400$ lb, $M_A = 267$ ft-lb.

10.12 (a) $P_B = 0$, $V_B = -31$ lb, $M_B = 572$ ft-lb;
(b) $P_B = 0$, $V_B = 24$ lb, $M_B = 600$ ft-lb.

10.14 $P_A = 0$, $V_A = -2$ kN, $M_A = 6$ kN-m.

10.16 $P_A = 300$ N, $V_A = -150$ N, $M_A = 330$ N-m.

10.18 $P_A = 4$ kN, $V_A = 6$ kN, $M_A = 4.8$ kN-m.

10.20 $P_A = 0$, $V_A = -6$ kN, $M_A = 6$ kN-m.

10.22 $V = 400$ lb, $M = 400x$ ft-lb.

10.24 (a) $V = (5/2)(12 - x)^2$ lb,
$M = -(5/6)(12 - x)^3$ ft-lb.

10.26 $V = -600$ N, $M = -600x$ N-m.

10.28 (a) $0 < x < 6$ ft, $P = 0$, $V = 50$ lb,
$M = 50x$ ft-lb; $6 < x < 12$ ft,
$P = 0$, $V = 50 - (25/3)(x - 6)^2$ lb,
$M = 50x - (25/9)(x - 6)^3$ ft-lb;
(b)

10.30 No. The maximum bending moment magnitude is 8 kN-m.

10.32 $M = 54.2$ N-m at $x = 233$ mm.

10.34

10.36 $V = -100x^2$ lb, $M = -33.3x^3 + 1800$ ft-lb.

10.38

10.40

10.42

10.44

10.46

10.48

10.50   759 kip.
10.52   (a) T_{max} = 86.2 kN;   (b) 36.14 m.
10.54 AC: 1061 N (T), BC: 1200 N (C).
10.56   Length = 108.3 m, h = 37.2 m.
10.58

10.60   22.8 m.
10.62   (a) h_1 = 4.95 m, h_2 = 2.19 m;
        (b) T_{AB} = 1.90 kN, T_{BC} = 1.84 kN.
10.64 T_1 = 185 N, T_3 = 209 N.
10.66   (a) h_2 = 4 ft;   (b) 90.1 lb.
10.68 h_1 = 1.739 m, h_3 = 0.957 m.
10.70 h_2 = 464 mm, h_3 = 385 mm.
10.72 h_2 = 8.38 ft, h_3 = 12.08 ft.
10.76 x_p = 3/8 m, y_p = 3/5 m.
10.78 A_x = −100 lb, A_z = 562 lb, B = 281 lb.
10.80   1.55 m.
10.82   6.67 m.
10.84 A: 257 lb to the right, 248 lb upward; B: 136 lb.
10.86 d = 1.5 m.
10.88 A_x = 2160 lb, A_y = 2000 lb, B_x = 1830 lb.
10.90   (a) 376 kN;   (b) x_p = 2.02 m.
10.94   (a) P_B = 0, V_B = −26.7 lb, M_B = 160 ft-lb;
        (b) P_C = 0, V_C = −26.7 lb, M_C = 80 ft-lb.

10.96

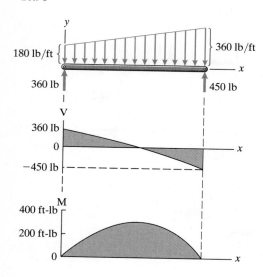

10.98 $0 < x < 2$ m, $P = 0$, $V = 1.33$ kN,
$M = 1.33x$ kN-m;
$2 < x < 6$ m, $P = 0$, $V = -2.67$ kN,
$M = 2.67(6 - x)$ kN-m.

10.100 $P_A = 0$, $V_A = 8$ kN, $M_A = -8$ kN-m.

10.102 (a) $P_B = 0$, $V_B = -40$ N, $M_B = 10$ N-m;
(b) $P_B = 0$, $V_B = -40$ N, $M_B = 10$ N-m.

10.104 $P = 0$, $V = -100$ lb, $M = -50$ ft-lb.

10.106 (a) $w = 74{,}100$ lb/ft; (b) 1.20×10^8 lb.

10.108 84.4 kip.

10.110 A: 44.2 kN to the left, 35.3 kN upward; B: 34.3 kN.

Chapter 11

11.2 (a) Work $= -3.20\,\delta\theta$ kN-m; (b) $B = 2.31$ kN.

11.4 $F = 217$ N.

11.6 $A_x = 0$, $A_y = -237$ lb, $B_y = 937$ lb.

11.8 $F = 450$ N.

11.10 (a) $F = 392$ N; (b) 100 mm.

11.16 $F = 360$ lb.

11.18 $M = 270$ N-m.

11.20 12 kN.

11.22 9.17 kN.

11.24 (a) $0.625\,\delta y$; (b) 216 N.

11.26 (a) $q = 3$, $q = 4$;
(b) $q = 3$ is unstable and $q = 4$ is stable.

11.28 $V = \frac{1}{2}kx^2 - \frac{1}{4}\varepsilon x^4$.

11.30 (a) Stable; (b) Unstable.

11.34 (b) It is stable.

11.36 (a) $\alpha = 35.2°$; (b) No.

11.38 (a) $\alpha = 28.7°$; (b) Yes.

11.40 Stable.

11.42 Unstable.

11.44 $\alpha = 0$ is unstable and $\alpha = 30°$ is stable.

11.46 $C_x = -7.78$ kN.

11.48 $8F$.

11.50 (a) $M = 800$ N-m; (b) $\alpha/4$.

11.52 $M = 1.50$ kN-m.

11.54 $F = 5$ kN.

11.56 $M = 63$ N-m.

11.58 $\alpha = 0$ is unstable and $\alpha = 59.4°$ is stable.

11.60 Unstable.

11.62 $\alpha = 30°$.

Index

Properties of Areas and Lines

Areas

The coordinates of the centroid of the area A are

$$\bar{x} = \frac{\int_A x \, dA}{\int_A dA}, \qquad \bar{y} = \frac{\int_A y \, dA}{\int_A dA}.$$

The moment of inertia about the x axis I_x, the moment of inertia about the y axis I_y, and the product of inertia I_{xy} are

$$I_x = \int_A y^2 \, dA, \qquad I_y = \int_A x^2 \, dA, \qquad I_{xy} = \int_A xy \, dA.$$

The polar moment of inertia about O is

$$J_O = \int_A r^2 \, dA = \int_A (x^2 + y^2) \, dA = I_x + I_y.$$

Triangular area

$$\text{Area} = \frac{1}{2}bh \qquad I_x = \frac{1}{12}bh^3, \qquad I_{x'} = \frac{1}{36}bh^3$$

Rectangular area

$$\text{Area} = bh$$

$$I_x = \frac{1}{3}bh^3, \qquad I_y = \frac{1}{3}hb^3, \qquad I_{xy} = \frac{1}{4}b^2h^2$$

$$I_{x'} = \frac{1}{12}bh^3, \qquad I_{y'} = \frac{1}{12}hb^3, \qquad I_{x'y'} = 0$$

Circular area

$$\text{Area} = \pi R^2 \qquad I_{x'} = I_{y'} = \frac{1}{4}\pi R^4, \qquad I_{x'y'} = 0$$

Triangular area

$$\text{Area} = \frac{1}{2}bh$$

$$I_x = \frac{1}{12}bh^3, \qquad I_y = \frac{1}{4}hb^3, \qquad I_{xy} = \frac{1}{8}b^2h^2$$

$$I_{x'} = \frac{1}{36}bh^3, \qquad I_{y'} = \frac{1}{36}hb^3, \qquad I_{x'y'} = \frac{1}{72}b^2h^2$$

Semicircular area

$$\text{Area} = \frac{1}{2}\pi R^2 \qquad I_x = I_y = \frac{1}{8}\pi R^4, \qquad I_{xy} = 0$$

$$I_{x'} = \frac{1}{8}\pi R^4, \qquad I_{y'} = \left(\frac{\pi}{8} - \frac{8}{9\pi}\right)R^4, \qquad I_{x'y'} = 0$$